Complete Solutions Manual

Introductory Chemistry:
A Foundation, Introductory Chemistry, Basic Chemistry

EIGHTH EDITION

Steven S. Zumdahl
University of Illinois, Urbana-Champaign

Donald J. DeCoste
University of Illinois, Urbana-Champaign

Prepared by

Gretchen M. Adams
University of Illinois, Urbana-Champaign

James F. Hall
University of Massachusetts Lowell

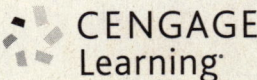

Australia • Brazil • Mexico • Singapore • United Kingdom • United States

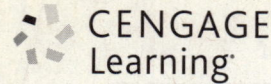

© 2015 Cengage Learning

ALL RIGHTS RESERVED. No part of this work covered by the copyright herein may be reproduced, transmitted, stored, or used in any form or by any means graphic, electronic, or mechanical, including but not limited to photocopying, recording, scanning, digitizing, taping, Web distribution, information networks, or information storage and retrieval systems, except as permitted under Section 107 or 108 of the 1976 United States Copyright Act, without the prior written permission of the publisher except as may be permitted by the license terms below.

ISBN-13: 978-1-285-85814-2
ISBN-10: 1-285-85814-X

Cengage Learning
200 First Stamford Place, 4th Floor
Stamford, CT 06902
USA

Cengage Learning is a leading provider of customized learning solutions with office locations around the globe, including Singapore, the United Kingdom, Australia, Mexico, Brazil, and Japan. Locate your local office at: **www.cengage.com/global**.

Cengage Learning products are represented in Canada by Nelson Education, Ltd.

To learn more about Cengage Learning Solutions, visit **www.cengage.com**.

To find online supplements and other instructional support, please visit **www.cengagebrain.com**.

National Geographic Learning/Cengage Learning is pleased to distribute our college-level materials to high schools for Advanced Placement*, honors, and electives courses. To contact your National Geographic Learning representative, please call us toll-free at **1-888-915-3276** or visit us at **http://ngl.cengage.com**.

For permission to use material from this text or product, submit all requests online at **www.cengage.com/permissions**
Further permissions questions can be emailed to **permissionrequest@cengage.com**.

NOTE: UNDER NO CIRCUMSTANCES MAY THIS MATERIAL OR ANY PORTION THEREOF BE SOLD, LICENSED, AUCTIONED, OR OTHERWISE REDISTRIBUTED EXCEPT AS MAY BE PERMITTED BY THE LICENSE TERMS HEREIN.

READ IMPORTANT LICENSE INFORMATION

Dear Professor or Other Supplement Recipient:

Cengage Learning has provided you with this product (the "Supplement") for your review and, to the extent that you adopt the associated textbook for use in connection with your course (the "Course"), you and your students who purchase the textbook may use the Supplement as described below. Cengage Learning has established these use limitations in response to concerns raised by authors, professors, and other users regarding the pedagogical problems stemming from unlimited distribution of Supplements.

Cengage Learning hereby grants you a nontransferable license to use the Supplement in connection with the Course, subject to the following conditions. The Supplement is for your personal, noncommercial use only and may not be reproduced, posted electronically or distributed, except that portions of the Supplement may be provided to your students IN PRINT FORM ONLY in connection with your instruction of the Course, so long as such students are advised that they may not copy or distribute any portion of the Supplement to any third party. You may not sell, license, auction, or otherwise redistribute the Supplement in any form. We ask that you take reasonable steps to protect the Supplement from unauthorized use, reproduction, or distribution. Your use of the Supplement indicates your acceptance of the conditions set forth in this Agreement. If you do not accept these conditions, you must return the Supplement unused within 30 days of receipt.

All rights (including without limitation, copyrights, patents, and trade secrets) in the Supplement are and will remain the sole and exclusive property of Cengage Learning and/or its licensors. The Supplement is furnished by Cengage Learning on an "as is" basis without any warranties, express or implied. This Agreement will be governed by and construed pursuant to the laws of the State of New York, without regard to such State's conflict of law rules.

Thank you for your assistance in helping to safeguard the integrity of the content contained in this Supplement. We trust you find the Supplement a useful teaching tool.

Printed in the United States of America
1 2 3 4 5 6 7 17 16 15 14 13

Contents

1. Chemistry: An Introduction..1
2. Measurements and Calculations..3
3. Matter...28
   Cumulative Review: Chapters 1, 2, and 3..34
4. Chemical Foundations: Elements, Atoms, and Ions...41
5. Nomenclature...54
   Cumulative Review: Chapters 4 and 5...73
6. Chemical Reactions: An Introduction..82
7. Reactions in Aqueous Solution..90
   Cumulative Review: Chapters 6 and 7...108
8. Chemical Composition...117
9. Chemical Quantities...163
   Cumulative Review: Chapters 8 and 9...204
10. Energy...213
11. Modern Atomic Theory..225
12. Chemical Bonding..238
    Cumulative Review: Chapters 10, 11, and 12..260
13. Gases..272
14. Liquids and Solids..304
15. Solutions..316
    Cumulative Review: Chapters 13, 14, and 15..351
16. Acids and Bases...363
17. Equilibrium..379
    Cumulative Review: Chapters 16 and 17...397
18. Oxidation–Reduction Reactions: Electrochemistry.......................................406
19. Radioactivity and Nuclear Energy...427
20. Organic Chemistry...437
21. Biochemistry..460

Preface

This guide contains complete solutions for the end-of-chapter problems in the eighth editions of *Introductory Chemistry*, *Introductory Chemistry: A Foundation*, and *Basic Chemistry* by Steven S. Zumdahl and Donald J. DeCoste. Several hundred new problems and questions have been prepared for the new editions of the text, which we hope will be of even greater help to your students in gaining an understanding of the fundamental principles of chemistry.

We have tried to give the most detailed solutions possible to all the problems, even though some problems give repeat, drill practice on the same subject. Our chief attempt at brevity is to give molar masses for compounds without showing the calculation (after the subject of molar mass itself has been discussed). We have also made a conscious effort in this guide to solve each problem in the manner discussed in the textbook. The instructor, of course, may wish to discuss alternative methods of solution with his or her students.

One topic that causes many students concern is the matter of significant figures, and the determination of the number of digits to which a solution to a problem should be reported. To avoid truncation errors in the solutions contained in this guide, the solutions typically report intermediate answers to one more digit that appropriate for the final answer. The final answer to each problem is then given to the correct number of significant figures based on the data provided in the problem.

Gretchen M. Adams
University of Illinois at Urbana-Champaign
gadams4@illinois.edu
and
James F. Hall
University of Massachusetts Lowell

CHAPTER 1

Chemistry: An Introduction

1. The specific answer will depend on student experiences. In general, students are intimidated by chemistry because they perceive it to be highly mathematical, requiring a great deal of memorization, and having a difficult technical vocabulary. Many students taking chemistry as a foundation science cannot see its relevance to their major.

2. The answer will depend on student examples.

3. There are obviously many such examples. Many new drugs and treatments have recently become available thanks to research in biochemistry and cell biology. New long-wearing, more comfortable contact lenses have been produced by research in polymer and plastics chemistry. Special plastics and metals were prepared for the production of compact discs to replace vinyl phonograph records. As for the "dark side," chemistry contributes increased global pollution if not conducted carefully.

4. Answer depends on student responses/examples.

5. This answer depends on your own experience.

6. This answer depends on your own experience, but consider the following examples: oven cleaner (the label says it contains sodium hydroxide; it converts the burned-on grease in the oven to a soapy material that washes away); drain cleaner (the label says it contains sodium hydroxide; it dissolves the clog of hair in the drain); stomach antacid (the label says it contains calcium carbonate; it makes me belch and makes my stomach feel better); hydrogen peroxide (the label says it is a 3% solution of hydrogen peroxide; when applied to a wound, it bubbles); depilatory cream (the label says it contains sodium hydroxide; it removes unwanted hair from skin).

7. David and Susan first recognized the problem (unexplained medical problems). A possible explanation was then proposed (the glaze on their china might be causing lead poisoning). The explanation was tested by experiment (it was determined that the china did contain lead). A full discussion of this scenario is given in the text.

8. The scientist must recognize the problem and state it clearly, propose possible solutions or explanations, and then decide through experimentation which solution or explanation is best.

9. law; theory

10. Answer depends on student response. A quantitative observation must include a number. For example "There are two windows in this room" represents a quantitative observation, but "The walls of this room are yellow" is a qualitative observation.

11. True. A theory is a set of tested hypotheses that provides an overall explanation of an observed phenomenon.

Chapter 1: Chemistry: An Introduction

12. False. Theories can be refined and changed because they are interpretations. They represent possible explanations of why nature behaves in a particular way. Theories are refined by performing experiments and making new observations, not by proving the existing observations as false (which is something that can be witnessed and recorded).

13. Answer depends on student responses/examples.

14. Scientists are human, too. When a scientist formulates a hypothesis, he or she wants it to be proven correct. In academic research, for example, scientists want to be able to publish papers on their work to gain renown and acceptance from their colleagues. In industrial situations, the financial success of the individual and of the company as a whole may be at stake. Politically, scientists may be under pressure from the government to "beat the other guy."

15. Chemistry is not just a set of facts that have to be memorized. To be successful in chemistry, you have to be able to apply what you have learned to new situations, new phenomena, and new experiments. Rather than just learning a list of facts or studying someone else's solution to a problem, your instructor hopes you will learn *how* to solve problems *yourself,* so that you will be able to apply what you have learned in future circumstances.

16. Chemistry is not merely a list of observations, definitions, and properties. Chemistry is the study of very real interactions among different samples of matter, whether within a living cell, or in a chemical factory. When we study chemistry, at least in the beginning, we try to be as general and as nonspecific as possible, so that the *basic principles* learned can be applied to many situations. In a beginning chemistry course, we learn to interpret and solve a basic set of very simple problems in the hope that the method of solving these simple problems can be extended to more complex real life situations later on. The actual solution to a problem, at this point, is not as important as learning how to recognize and interpret the problem, and how to propose reasonable, experimentally testable hypotheses.

17. In real life situations, the problems and applications likely to be encountered are not simple textbook examples. One must be able to observe an event, hypothesize a cause, and then test this hypothesis. One must be able to carry what has been learned in class forward to new, different situations.

18. A good student will: learn the background and fundamentals of the subject from their classes and textbook; will develop the ability to recognize and solve problems and to extend what was learned in the classroom to "real" situations; will learn to make careful observations; and will be able to communicate effectively. While some academic subjects may emphasize use of one or more of these skills, Chemistry makes extensive use of all of them.

CHAPTER 2

Measurements and Calculations

1. measurement

2. "Scientific notation" means we have to put the decimal point after the first significant figure, and then express the order of magnitude of the number as a power of ten. So we want to put the decimal point after the first 2:

   $2{,}421 \rightarrow 2.421 \times 10^{\text{to some power}}$

   To be able to move the decimal point three places to the left in going from 2,421 to 2.421, means I will need a power of 10^3 after the number, where the exponent 3 shows that I moved the decimal point 3 places to the left.

   $2{,}421 \rightarrow 2.421 \times 10^{\text{to some power}} = 2.421 \times 10^3$

3.
   a. 9.651
   b. 3.521
   c. 9.3241
   d. 1.002

4.
   a. 10^6
   b. 10^{-2}
   c. 10^{-4}
   d. 10^9

5.
   a. positive
   b. positive
   c. negative
   d. negative

6.
   a. negative
   b. zero
   c. negative
   d. positive

7.
   a. The decimal point must be moved one space to the right, so the exponent is negative; $0.5012 = 5.012 \times 10^{-1}$.
   b. The decimal point must be moved six spaces to the left, so the exponent is positive; $5{,}012{,}000 = 5.012 \times 10^6$.

Chapter 2: Measurements and Calculations

    c. The decimal point must be moved six spaces to the right, so the exponent is negative; $0.000005012 = 5.012 \times 10^{-6}$.

    d. The decimal point does not have to be moved, so the exponent is zero; $5.012 = 5.012 \times 10^{0}$.

    e. The decimal point must be moved three spaces to the left, so the exponent is positive; $5012 = 5.012 \times 10^{3}$.

    f. The decimal point must be moved three spaces to the right, so the exponent is negative; $0.005012 = 5.012 \times 10^{-3}$.

8.
    a. The decimal point must be moved three spaces to the right: 2,789

    b. The decimal point must be moved three spaces to the left: 0.002789

    c. The decimal point must be moved seven spaces to the right: 93,000,000

    d. The decimal point must be moved one space to the right: 42.89

    e. The decimal point must be moved 4 spaces to the right: 99,990

    f. The decimal point must be moved 5 spaces to the left: 0.00009999

9.
    a. six spaces to the right

    b. five spaces to the left

    c. one space to the right

    d. The decimal point does not have to be moved.

    e. 18 spaces to the right

    f. 16 spaces to the left

10.
    a. three spaces to the left

    b. one space to the left

    c. five spaces to the right

    d. one space to the left

    e. two spaces to the right

    f. two spaces to the left

11. To say that scientific notation is in *standard* form means that you have a number between 1 and 10, followed by an exponential term.

    a. The decimal point must be moved 4 spaces to the left, so the exponent will be 4: 9.782×10^{4}

    b. 42.14 must first be converted to 4.214×10^{1} and then the exponents combined: 4.214×10^{4}

    c. 0.08214 must first be converted to 8.214×10^{-2} and then the exponents combined: 8.214×10^{-5}

    d. The decimal point must be moved four spaces to the right, so the exponent will be –4: 3.914×10^{-4}

Chapter 2: Measurements and Calculations

- e. The decimal point must be moved two spaces to the left, so the exponent will be 2: 9.271×10^2
- f. The exponents must be combined: 4.781×10^{-1}

12.
- a. The decimal point must be moved 3 places to the right: 6,244
- b. The decimal point must be moved 2 spaces to the left: 0.09117
- c. The decimal point must be moved 1 space to the right: 82.99
- d. The decimal point must be moved 4 spaces to the left: 0.0001771
- e. The decimal point must be moved 2 spaces to the right: 545.1
- f. The decimal point must be moved 5 spaces to the left: 0.00002934

13.
- a. $1/1033 = 9.681 \times 10^{-4}$
- b. $1/10^5 = 1 \times 10^{-5}$
- c. $1/10^{-7} = 1 \times 10^7$
- d. $1/0.0002 = 5 \times 10^3$
- e. $1/3{,}093{,}000 = 3.233 \times 10^{-7}$
- f. $1/10^{-4} = 1 \times 10^4$
- g. $1/10^9 = 1 \times 10^{-9}$
- h. $1/0.000015 = 6.7 \times 10^4$

14.
- a. $1/0.00032 = 3.1 \times 10^3$
- b. $10^3/10^{-3} = 1 \times 10^6$
- c. $10^3/10^3 = 1\ (1 \times 10^0)$; any number divided by itself is unity.
- d. $1/55{,}000 = 1.8 \times 10^{-5}$
- e. $(10^5)(10^4)(10^{-4})/10^{-2} = 1 \times 10^7$
- f. $43.2/(4.32 \times 10^{-5}) = \dfrac{4.32 \times 10^1}{4.32 \times 10^{-5}} = 1.00 \times 10^6$
- g. $(4.32 \times 10^{-5})/432 = \dfrac{4.32 \times 10^{-5}}{4.32 \times 10^2} = 1.00 \times 10^{-7}$
- h. $1/(10^5)(10^{-6}) = 1/(10^{-1}) = 1 \times 10^1$

15. mass, kilogram; length, meter; temperature, kelvin

16.
- a. kilo
- b. milli
- c. nano
- d. mega

Chapter 2:   Measurements and Calculations

      e.   deci

      f.   micro

17. Since a meter is longer than a yard, the floor will require somewhat more than 25 square yards of linoleum. 25 m^2 = 5 m × 5 m = 5.47 yd × 5.47 yd = 30 yd^2

18. Since a pound is 453.6 grams, the 125-g can will be slightly more than ¼ pound.

19. Since a liter is slightly more than a quart, and since 4 quarts make 1 gallon, 48 liters will be approximately 12 gallons.

20. Since 1 inch = 2.54 cm, the nail is approximately an inch long.

21. $100 \text{ km} \times \dfrac{1 \text{ mi}}{1.6093 \text{ km}} = 62 \text{ km}$

22. $100. \text{ mi} \times \dfrac{1.6093 \text{ km}}{1 \text{ mi}} = 161 \text{ km}$

23. $2 \text{ m} \times \dfrac{100 \text{ cm}}{1 \text{ m}} = 200 \text{ cm}$; $2 \text{ m} \times \dfrac{100 \text{ cm}}{1 \text{ m}} \times \dfrac{1 \text{ in}}{2.54 \text{ cm}} = 79$ in (80 in to one significant figure)

24. 1.62 m is approximately 5 ft, 4 in. The woman is slightly taller.

25. a.   kilometers
    b.   meters
    c.   centimeters
    d.   micrometers

26. a.   inch
    b.   yard
    c.   mile

27. a.   about 4 liters
    b.   about half a liter (500 mL)
    c.   about 1/4 of a liter (250 mL)

28. b (the other units would give very small numbers for the length)

29. We estimate measurements between the smallest divisions on the scale; since this is our best estimate, the last significant digit recorded is uncertain.

30. When we use a measuring device with an analog scale, we estimate the reading to 0.1 of the smallest scale divisions on the measuring scale. Since this last reading is decided by the user, not by the divisions on the measuring scale, the final digit of the measurement is uncertain no matter how careful we may be in making the determination.

Chapter 2: Measurements and Calculations

31. The third figure in the length of the pin is uncertain because the measuring scale of the ruler has *tenths* as the smallest marked scale division. The length of the pin is given as 2.85 cm (rather than any other number) to indicate that the point of the pin appears to the observer to be *half way* between the smallest marked scale divisions.

32. The scale of the ruler shown is only marked to the nearest *tenth* of a centimeter; writing 2.850 would imply that the scale was marked to the nearest *hundredth* of a centimeter (and that the zero in the thousandths place had been estimated).

33.
   a. three
   b. two
   c. two
   d. four

34.
   a. probably only two
   b. infinite (definition)
   c. infinite (definition)
   d. probably one
   e. three (the race is defined to be 500. miles)

35. increase the preceding digit by 1

36. It is better to round off only the final answer, and to carry through extra digits in intermediate calculations. If there are enough steps to the calculation, rounding off in each step may lead to a cumulative error in the final answer.

37.
   a. 2.55×10^5
   b. 2.56×10^{-4}
   c. 4.79×10^4
   d. 8.21×10^3

38.
   a. 1.57×10^6
   b. 2.77×10^{-3}
   c. 7.76×10^{-2}
   d. 1.17×10^{-3}

39.
   a. 4.34×10^5
   b. 9.34×10^4
   c. 9.916×10^1
   d. 9.327×10^0

40.
   a. 3.42×10^{-4}
   b. 1.034×10^4

Chapter 2: Measurements and Calculations

      c. 1.7992×10^1

      d. 3.37×10^5

41. Since the only operations in the calculation are multiplication and division, the number of significant figures is limited by the factor of 0.15 that has only two significant figures.

42. 170. mL;

        18 mL
  + 128.7 mL
  + 23.45 mL
  = 170.15 mL

18 mL limits the precision to the ones place, thus the answer is rounded to 170. mL

43. three (based on 2.31 having 3 significant figures)

44. three; (b), (c), and (d); (a) contains two significant figures

45. two decimal places (based on 2.11 being known only to the second decimal place)

46. none (10,434 is only known to the nearest whole number)

47.
    a.   52.36 (the answer can only be given to the second decimal place because 0.81 is only known to the second decimal place)

    b.   10.90 (the answer can only be given to the second decimal place because 2.21 is only known to the second decimal place)

    c.   5.25 (the answer can only be given to the second decimal place because 4.14 is only known to the second decimal place)

    d.   6.5 (the answer can only be given to two significant figures because 3.1 is only known to two significant figures.

48.
    a.   2.3 (the answer can only be given to two significant figures because 3.1 is only known to two significant figures)

    b. 9.1×10^2: (the answer can only be given to the first decimal place because 4.1 is only given to the first decimal place; both numbers have the same power of ten)

    c. 1.323×10^3: (the numbers must be first expressed as the same power of ten; $1.091 \times 10^3 + 0.221 \times 10^3 + 0.0114 \times 10^3 = 1.323 \times 10^3$)

    d. 6.63×10^{-13} (the answer can only be given to three significant figures because 4.22×10^6 is only given to three significant figures)

49.
    a.   two (based on 1.1 having only two significant figures)

    b.   two (based on 0.22 having only 2 significant figures)

    c.   two (based on 0.00033 having only two significant figures)

    d.   three (assuming sum in numerator is considered to second decimal place)

Chapter 2: Measurements and Calculations

50. a. one (the factor of 2 has only one significant figure)
     b. four (the sum within the parentheses will contain four significant figures)
     c. two (based on the factor 4.7×10^{-6} only having two significant figures)
     d. three (based on the factor 63.9 having only three significant figures)

51. a. two (the factor of 2.1 has only two significant figures)
     b. two (the factor of 0.98 has only two significant figures)
     c. four (the factor of 3.014 has only four significant figures)
     d. three (the factor of 1.86×10^{-3} has only three significant figures)

52. a. $(2.0944 + 0.0003233 + 12.22)/7.001 = (14.31)/7.001 = 2.045$
     b. $(1.42 \times 10^2 + 1.021 \times 10^3)/(3.1 \times 10^{-1}) =$
        $(142 + 1021)/(3.1 \times 10^{-1}) = (1163)/(3.1 \times 10^{-1}) = 3752 = 3.8 \times 10^3$
     c. $(9.762 \times 10^{-3})/(1.43 \times 10^2 + 4.51 \times 10^1) =$
        $(9.762 \times 10^{-3})/(143 + 45.1) = (9.762 \times 10^{-3})/(188.1) = 5.19 \times 10^{-5}$
     d. $(6.1982 \times 10^{-4})^2 = (6.1982 \times 10^{-4})(6.1982 \times 10^{-4}) = 3.8418 \times 10^{-7}$

53. conversion factor

54. an infinite number (a definition)

55. $\dfrac{1 \text{ mi}}{1760 \text{ yd}}$ and $\dfrac{1760 \text{ yd}}{1 \text{ mi}}$

56. $\dfrac{5280 \text{ ft}}{1 \text{ mi}}$; $15.6 \text{ mi} \times \dfrac{5280 \text{ ft}}{1 \text{ mi}}$

     $\dfrac{1 \text{ mi}}{5280 \text{ ft}}$; $86.19 \text{ ft} \times \dfrac{1 \text{ mi}}{5280 \text{ ft}}$

57. $\dfrac{\$1.75}{\text{lb}}$; $5.3 \text{ lb} \times \dfrac{\$1.75}{\text{lb}}$

58. $\dfrac{\text{lb}}{\$1.75}$; $\$10.00 \times \dfrac{\text{lb}}{\$1.75}$

59. a. $12.5 \text{ in} \times \dfrac{2.54 \text{ cm}}{1 \text{ in}} = 31.8 \text{ cm}$

     b. $12.5 \text{ cm} \times \dfrac{1 \text{ in}}{2.54 \text{ cm}} = 4.92 \text{ in}$

     c. $2513 \text{ ft} \times \dfrac{1 \text{ mi}}{5280 \text{ ft}} = 0.4759 \text{ mi}$

Chapter 2: Measurements and Calculations

d. $4.53 \text{ ft} \times \dfrac{1 \text{ yd}}{3 \text{ ft}} \times \dfrac{1 \text{ m}}{1.0936 \text{ yd}} = 1.38 \text{ m}$

e. $6.52 \text{ min} \times \dfrac{60 \text{ sec}}{1 \text{ min}} = 391 \text{ sec}$

f. $52.3 \text{ cm} \times \dfrac{1 \text{ m}}{100 \text{ cm}} = 0.523 \text{ m}$

g. $4.21 \text{ m} \times \dfrac{1.0936 \text{ yd}}{1 \text{ m}} = 4.60 \text{ yd}$

h. $8.02 \text{ oz} \times \dfrac{1 \text{ lb}}{16 \text{ oz}} = 0.501 \text{ lb}$

60. a. $2.23 \text{ m} \times \dfrac{1.0936 \text{ yd}}{1 \text{ m}} = 2.44 \text{ yd}$

b. $46.2 \text{ yd} \times \dfrac{1 \text{ m}}{1.0936 \text{ yd}} = 42.2 \text{ m}$

c. $292 \text{ cm} \times \dfrac{1 \text{ in}}{2.54 \text{ cm}} = 115 \text{ in}$

d. $881.2 \text{ in} \times \dfrac{2.54 \text{ cm}}{1 \text{ in}} = 2238 \text{ cm}$

e. $1043 \text{ km} \times \dfrac{1 \text{ mi}}{1.6093 \text{ km}} = 648.1 \text{ mi}$

f. $445.5 \text{ mi} \times \dfrac{1.6093 \text{ km}}{1 \text{ mi}} = 716.9 \text{ km}$

g. $36.2 \text{ m} \times \dfrac{1 \text{ km}}{1000 \text{ m}} = 0.0362 \text{ km}$

h. $0.501 \text{ km} \times \dfrac{1000 \text{ m}}{1 \text{ km}} \times \dfrac{100 \text{ cm}}{1 \text{ m}} = 5.01 \times 10^4 \text{ cm}$

61. a. $1.75 \text{ mi} \times \dfrac{1.6093 \text{ km}}{1 \text{ mi}} = 2.82 \text{ km}$

b. $2.63 \text{ gal} \times \dfrac{4 \text{ qt}}{1 \text{ gal}} = 10.5 \text{ qt}$

c. $4.675 \text{ cal} \times \dfrac{4.184 \text{ J}}{1 \text{ cal}} = 19.56 \text{ J}$

d. $756.2 \text{ mm Hg} \times \dfrac{1 \text{ atm}}{760 \text{ mm Hg}} = 0.9950 \text{ atm}$

Chapter 2: Measurements and Calculations

e. $36.3 \text{ amu} \times \dfrac{1.66056 \times 10^{-27} \text{ kg}}{1 \text{ amu}} = 6.03 \times 10^{-26} \text{ kg}$

f. $46.2 \text{ in} \times \dfrac{2.54 \text{ cm}}{1 \text{ in}} = 117 \text{ cm}$

g. $2.75 \text{ qt} \times \dfrac{32 \text{ fl oz}}{1 \text{ qt}} = 88.0 \text{ fl oz}$

h. $3.51 \text{ yd} \times \dfrac{1 \text{ m}}{1.0936 \text{ yd}} = 3.21 \text{ m}$

62. a. $254.3 \text{ g} \times \dfrac{1 \text{ kg}}{1000 \text{ g}} = 0.2543 \text{ kg}$

b. $2.75 \text{ kg} \times \dfrac{1000 \text{ g}}{1 \text{ kg}} = 2750 \text{ g}$

c. $2.75 \text{ kg} \times \dfrac{2.2046 \text{ lb}}{1 \text{ kg}} = 6.06 \text{ lb}$

d. $2.75 \text{ kg} \times \dfrac{1000 \text{ g}}{1 \text{ kg}} \times \dfrac{16 \text{ oz}}{453.59 \text{ g}} = 97.0 \text{ oz}$

e. $534.1 \text{ g} \times \dfrac{1 \text{ kg}}{1000 \text{ g}} \times \dfrac{2.2046 \text{ lb}}{1 \text{ kg}} = 1.177 \text{ lb}$

f. $1.75 \text{ lb} \times \dfrac{1 \text{ kg}}{2.2046 \text{ lb}} \times \dfrac{1000 \text{ g}}{1 \text{ kg}} = 794 \text{ g}$

g. $8.7 \text{ oz} \times \dfrac{453.59 \text{ g}}{16 \text{ oz}} = 250 \text{ g}$

h. $45.9 \text{ g} \times \dfrac{16 \text{ oz}}{453.59 \text{ g}} = 1.62 \text{ oz}$

63. $1.89 \times 10^{25} \text{ C atoms} \times \dfrac{12.01 \text{ g}}{6.02 \times 10^{23} \text{ C atoms}} = 377 \text{ g}$

64. $2558 \text{ mi} \times \dfrac{1.6093 \text{ km}}{1 \text{ mi}} = 4117 \text{ km}$

65. To decide which train is faster, both speeds must be expressed in the *same unit* of distance (either miles or kilometers).

$\dfrac{225 \text{ km}}{1 \text{ hr}} \times \dfrac{1 \text{ mi}}{1.6093 \text{ km}} = 140. \text{ mi/hr}$

So the Boston-New York trains will be faster.

Chapter 2: Measurements and Calculations

66. $1 \times 10^{-10} \text{ m} \times \dfrac{100 \text{ cm}}{1 \text{ m}} = 1 \times 10^{-8} \text{ cm}$

    $1 \times 10^{-8} \text{ cm} \times \dfrac{1 \text{ in}}{2.54 \text{ cm}} = 4 \times 10^{-9} \text{ in.}$

    $1 \times 10^{-8} \text{ cm} \times \dfrac{1 \text{ m}}{100 \text{ cm}} \times \dfrac{10^9 \text{ nm}}{1 \text{ m}} = 0.1 \text{ nm}$

67. Celsius

68. freezing

69. 212°F; 100°C

70. 373

71. 100

72. Fahrenheit (F)

73. $T_K = T_C + 273 \qquad T_C = T_K - 273$

    a. 44.2°C + 273 = 317.2 K (317 K)

    b. 891 K − 273 = 618°C

    c. −20°C + 273 = 253 K

    d. 273.1 K − 273 = 0.1°C (0°C)

74. $T_C = (T_F - 32)/1.80 \qquad T_K = T_C + 273 \qquad T_C = T_K - 273$

    a. $T_c = (T_f - 32)/1.80 = (-201°F - 32)/1.80 = (-233)/1.80 = -129.4°C$

    −129.4°C + 273 = 144 K

    b. −201°C + 273 = 72 K

    c. $T_F = 1.80(T_c) + 32 = 1.80(351°C) + 32 = 664°F$

    d. $T_c = (T_f - 32)/1.80 = (-150°F - 32)/1.80 = -101°C$

75. $T_C = (T_F - 32)/1.80$

    a. (45 − 32)/1.80 = 13/1.80 = 7.2°C

    b. (115 − 32)/1.80 = 83/1.80 = 46°C

    c. (−10 − 32)/1.80 = −42/1.80 = −23°C

    d. Assuming 10,000°F to be known to two significant figures: (10,000 − 32)/1.80 = 5500°C

Chapter 2: Measurements and Calculations

76. $T_F = 1.80(T_C) + 32$

   a. $1.80(78.1) + 32 = 173°F$
   b. $1.80(40.) + 32 = 104°F$
   c. $1.80(-273) + 32 = -459°F$
   d. $1.80(32) + 32 = 90.°F$

77. a. Gallium is in the liquid state over the temperature range of this thermometer.
   b. $T_F = 1.80(T_C) + 32$
   $T_F = 1.80(50°C) + 32 = 122°F$
   $T_F = 1.80(500°C) + 32 = 932°F$

78. $T_F = 1.80(T_C) + 32$ $T_C = (T_F - 32)/1.80$ $T_K = T_C + 273$

   a. $275 - 273 = 2°C$
   b. $(82 - 32)/1.80 = 28°C$
   c. $1.80(-21) + 32 = -5.8°F\ (-6°F)$
   d. $(-40 - 32)/1.80 = -40\ °C$ (Celsius and Fahrenheit temperatures are the same at –40).

79. Density represents the mass per unit volume of a substance.

80. g/cm^3 (g/mL)

81. lead

82. 100 in.^{3}

83. smaller; gases are mostly empty space, so there is less mass in a given volume than for solids and liquids.

84. Density is a *characteristic* property, which is always the same for a pure substance.

85. Gold is the most dense; hydrogen is the least dense; 1 g of hydrogen would occupy the larger volume.

86. Silver is the most dense (10.5 g/cm^3).

87. density = $\dfrac{mass}{volume}$

   a. $d = \dfrac{452.1\ g}{292\ cm^3} = 1.55\ g/cm^3$

   b. $m = 0.14\ lb = 63.5\ g$ $v = 125\ mL = 125\ cm^3$

   $d = \dfrac{63.5\ g}{125\ cm^3} = 0.51\ g/cm^3$

Chapter 2:   Measurements and Calculations

    c. $m = 1.01$ kg $= 1010$ g

$$d = \frac{1010 \text{ g}}{1000 \text{ cm}^3} = 1.01 \text{ g/cm}^3$$

    d. $m = 225$ mg $= 0.225$ g $v = 2.51$ mL $= 2.51$ cm^3

$$d = \frac{0.225 \text{ g}}{2.51 \text{ cm}^3} = 0.0896 \text{ g/cm}^3$$

88. $\text{density} = \dfrac{\text{mass}}{\text{volume}}$

    a. $m = 4.53 \text{ kg} \times \dfrac{1000 \text{ g}}{1 \text{ kg}} = 4530$ g

$$d = \frac{4530 \text{ g}}{225 \text{ cm}^3} = 20.1 \text{ g/cm}^3$$

    b. $v = 25.0 \text{ mL} \times \dfrac{1 \text{ cm}^3}{1 \text{ mL}} = 25.0$ cm^3

$$d = \frac{26.3 \text{ g}}{25.0 \text{ cm}^3} = 1.05 \text{ g/cm}^3$$

    c. $m = 1.00 \text{ lb} \times \dfrac{1 \text{ kg}}{2.2046 \text{ lb}} \times \dfrac{1000 \text{ g}}{1 \text{ kg}} = 454$ g

$$d = \frac{454 \text{ g}}{500. \text{ cm}^3} = 0.907 \text{ g/cm}^3$$

    d. $m = 352 \text{ mg} \times \dfrac{1 \text{ g}}{1000 \text{ mg}} = 0.352$ g

$$d = \frac{0.352 \text{ g}}{0.271 \text{ cm}^3} = 1.30 \text{ g/cm}^3$$

89. $125 \text{ mL} \times \dfrac{3.12 \text{ g}}{1 \text{ mL}} = 390.$ g

$85.0 \text{ g} \times \dfrac{1 \text{ mL}}{3.12 \text{ g}} = 27.2$ mL

90. $4.50 \text{ L} \times \dfrac{1000 \text{ mL}}{1 \text{ L}} \times \dfrac{0.920 \text{ g}}{\text{mL}} = 4140$ g

$375 \text{ g} \times \dfrac{\text{mL}}{0.920 \text{ g}} \times \dfrac{1 \text{ L}}{1000 \text{ L}} = 0.408$ L

Chapter 2: Measurements and Calculations

91. $d = \dfrac{929 \text{ g}}{1000 \text{ mL}} = 0.929$ g/mL assuming 1000 mL is exact.

92. $m = 3.5 \text{ lb} \times \dfrac{453.59 \text{ g}}{1 \text{ lb}} = 1.59 \times 10^3$ g

    $v = 1.2 \times 10^4 \text{ in.}^3 \times \left(\dfrac{2.54 \text{ cm}}{1 \text{ in}}\right)^3 = 1.97 \times 10^5$ cm^3

    $d = \dfrac{1.59 \times 10^3 \text{ g}}{1.97 \times 10^5 \text{ cm}^3} = 8.1 \times 10^{-3}$ g/cm^3

    The material will float.

93. The volume of the iron can be calculated from its mass and density:

    $v = 52.4 \text{ g} \times \dfrac{1 \text{ cm}^3}{7.87 \text{ g}} = 6.66$ cm^3 = 6.66 mL.

    The liquid level in the graduated cylinder will rise by 6.66 mL when the piece of iron is added, giving a final volume of (75.0 + 6.66) = 81.7 mL

94. $25.75 \text{ g} \times \dfrac{\text{cm}^3}{19.32 \text{ g}} = 1.333$ cm^3 = 1.333 mL

    13.3 mL + 1.333 mL = 14.6 mL

95. a. $50.0 \text{ g} \times \dfrac{1 \text{ cm}^3}{2.16 \text{ g}} = 23.1$ cm^3

    b. $50.0 \text{ g} \times \dfrac{1 \text{ cm}^3}{13.6 \text{ g}} = 3.68$ cm^3

    c. $50.0 \text{ g} \times \dfrac{1 \text{ cm}^3}{0.880 \text{ g}} = 56.8$ cm^3

    d. $50.0 \text{ g} \times \dfrac{1 \text{ cm}^3}{10.5 \text{ g}} = 4.76$ cm^3

96. a. $50.0 \text{ cm}^3 \times \dfrac{19.32 \text{ g}}{1 \text{ cm}^3} = 966$ g

    b. $50.0 \text{ cm}^3 \times \dfrac{7.87 \text{ g}}{1 \text{ cm}^3} = 394$ g

    c. $50.0 \text{ cm}^3 \times \dfrac{11.34 \text{ g}}{1 \text{ cm}^3} = 567$ g

    d. $50.0 \text{ cm}^3 \times \dfrac{2.70 \text{ g}}{1 \text{ cm}^3} = 135$ g

Chapter 2: Measurements and Calculations

97. a. three
    b. three
    c. three

98. a. $3.011 \times 10^{23} = 301{,}100{,}000{,}000{,}000{,}000{,}000{,}000$
    b. $5.091 \times 10^{9} = 5{,}091{,}000{,}000$
    c. $7.2 \times 10^{2} = 720$
    d. $1.234 \times 10^{5} = 123{,}400$
    e. $4.32002 \times 10^{-4} = 0.000432002$
    f. $3.001 \times 10^{-2} = 0.03001$
    g. $2.9901 \times 10^{-7} = 0.00000029901$
    h. $4.2 \times 10^{-1} = 0.42$

99. a. 4.25×10^{2}
    b. 7.81×10^{-4}
    c. 2.68×10^{4}
    d. 6.54×10^{-4}
    e. 7.26×10^{1}

100. (e)

101. a. $1.25 \text{ in.} \times \dfrac{1 \text{ ft}}{12 \text{ in}} = 0.104 \text{ ft}$

   $1.25 \text{ in.} \times \dfrac{2.54 \text{ cm}}{1 \text{ in}} = 3.18 \text{ cm}$

   b. $2.12 \text{ qt} \times \dfrac{1 \text{ gal}}{4 \text{ qt}} = 0.530 \text{ gal}$

   $2.12 \text{ qt} \times \dfrac{1 \text{ L}}{1.0567 \text{ qt}} = 2.01 \text{ L}$

   c. $2640 \text{ ft} \times \dfrac{1 \text{ mi}}{5280 \text{ ft}} = 0.500 \text{ mi}$

   $2640 \text{ ft} \times \dfrac{1.6093 \text{ km}}{5280. \text{ ft}} = 0.805 \text{ km}$

   d. $1.254 \text{ kg} \times \dfrac{10^{3} \text{ g}}{1 \text{ kg}} \times \dfrac{1 \text{ cm}^{3}}{11.34 \text{ g}} = 110.6 \text{ cm}^{3}$

   e. $250. \text{ mL} \times 0.785 \text{ g/mL} = 196 \text{ g}$

Chapter 2: Measurements and Calculations

    f. $3.5 \text{ in.}^3 \times \left(\dfrac{2.54 \text{ cm}}{1 \text{ in}}\right)^3 = 57 \text{ cm}^3 = 57 \text{ mL}$

        $57 \text{ cm}^3 \times 13.6 \text{ g/cm}^3 = 7.8 \times 10^2 \text{ g} = 0.78 \text{ kg}$

102. a. $36.2 \text{ blim} \times \dfrac{1400 \text{ kryll}}{1 \text{ blim}} = 5.07 \times 10^4 \text{ kryll}$

    b. $170 \text{ kryll} \times \dfrac{1 \text{ blim}}{1400 \text{ kryll}} = 0.12 \text{ blim}$

    c. $72.5 \text{ kryll}^2 \times \left(\dfrac{1 \text{ blim}}{1400 \text{ kryll}}\right)^2 = 3.70 \times 10^{-5} \text{ blim}^2$

103. $110 \text{ km} \times \dfrac{1 \text{ hr}}{100 \text{ km}} = 1.1 \text{ hr}$

104. $\dfrac{45 \text{ mi}}{\text{hr}} \times \dfrac{1.61 \text{ km}}{1 \text{ mi}} \times \dfrac{1000 \text{ m}}{1 \text{ km}} \times \dfrac{1 \text{ hr}}{3600 \text{ s}} = 20. \text{ m/s}$

105. $45 \text{ mi} \times \dfrac{1.6093 \text{ km}}{1 \text{ mi}} = 72.4 \text{ km}$

      $38 \text{ mi} \times \dfrac{1.6093 \text{ km}}{1 \text{ mi}} = 61.2 \text{ km}$

      1 gal = 3.7854 L

      highway: 72.4 km/3.7854 L = 19 km/L

      city: 61.2 km/3.7854 L = 16 km/L

106. $1 \text{ lb} \times \dfrac{1 \text{ kg}}{2.205 \text{ lb}} \times \dfrac{2.76 \text{ euros}}{1 \text{ kg}} \times \dfrac{\$1.00}{1.44 \text{ euros}} = \$0.87$

107. $15.6 \text{ g} \times \dfrac{1 \text{ capsule}}{0.65 \text{ g}} = 24 \text{ capsules}$

108. °X = 1.26°C + 14

109. $v = \tfrac{4}{3}(\pi r^3) = \tfrac{4}{3}(3.1416)(0.5 \text{ cm})^3 = 0.52 \text{ cm}^3$

      $d = \dfrac{2.0 \text{ g}}{0.52 \text{ cm}^3} = 3.8 \text{ g/cm}^3$ (the ball will sink)

110. $d = \dfrac{36.8 \text{ g}}{10.5 \text{ L}} = 3.50 \text{ g/L}$ ($3.50 \times 10^{-3} \text{ g/cm}^3$)

Chapter 2: Measurements and Calculations

111. a. $25.0 \text{ g} \times \dfrac{1 \text{ cm}^3}{0.000084 \text{ g}} = 2.98 \times 10^5 \text{ cm}^3$

     b. $25.0 \text{ g} \times \dfrac{1 \text{ cm}^3}{13.6 \text{ g}} = 1.84 \text{ cm}^3$

     c. $25.0 \text{ g} \times \dfrac{1 \text{ cm}^3}{11.34 \text{ g}} = 2.20 \text{ cm}^3$

     d. $25.0 \text{ g} \times \dfrac{1 \text{ cm}^3}{1.00 \text{ g}} = 25.0 \text{ cm}^3$

112. For ethanol, $100. \text{ mL} \times \dfrac{0.785 \text{ g}}{1 \text{ mL}} = 78.5 \text{ g}$

     For benzene, $1000 \text{ mL} \times \dfrac{0.880 \text{ g}}{1 \text{ mL}} = 880. \text{ g}$

     total mass, $78.5 + 880. = 959 \text{ g}$

113. three

114. a. negative
     b. negative
     c. positive
     d. zero
     e. negative

115. a. positive
     b. negative
     c. negative
     d. zero

116. a. 2; positive
     b. 11; negative
     c. 3; positive
     d. 5; negative
     e. 5; positive
     f. 0; zero
     g. 1; negative
     h. 7; negative

Chapter 2: Measurements and Calculations

117.
- a. 4; positive
- b. 6; negative
- c. 0; zero
- d. 5; positive
- e. 2; negative

118.
- a. 1; positive
- b. 3; negative
- c. 0; zero
- d. 3; positive
- e. 9; negative

119.
- a. The decimal point must be moved two places to the left, so the exponent is positive 2; $529 = 5.29 \times 10^2$.
- b. The decimal point must be moved eight places to the left, so the exponent is positive 8; $240,000,000 = 2.4 \times 10^8$.
- c. The decimal point must be moved 17 places to the left, so the exponent is positive 17; $301,000,000,000,000,000 = 3.01 \times 10^{17}$.
- d. The decimal point must be moved four places to the left, so the exponent is positive 4; $78,444 = 7.8444 \times 10^4$.
- e. The decimal point must be moved four places to the right, so the exponent is negative 4; $0.0003442 = 3.442 \times 10^{-4}$.
- f. The decimal point must be moved 10 places to the right, so the exponent is negative 10; $0.000000000902 = 9.02 \times 10^{-10}$.
- g. The decimal point must be moved two places to the right, so the exponent is negative 2; $0.043 = 4.3 \times 10^{-2}$.
- h. The decimal point must be moved two places to the right, so the exponent is negative 2; $0.0821 = 8.21 \times 10^{-2}$.

120.
- a. The decimal point must be moved five places to the left; $2.98 \times 10^{-5} = 0.0000298$.
- b. The decimal point must be moved nine places to the right; $4.358 \times 10^9 = 4,358,000,000$.
- c. The decimal point must be moved six places to the left; $1.9928 \times 10^{-6} = 0.0000019928$.
- d. The decimal point must be moved 23 places to the right; $6.02 \times 10^{23} = 602,000,000,000,000,000,000,000$.
- e. The decimal point must be moved one place to the left; $1.01 \times 10^{-1} = 0.101$.
- f. The decimal point must be moved three places to the left; $7.87 \times 10^{-3} = 0.00787$.
- g. The decimal point must be moved seven places to the right; $9.87 \times 10^7 = 98,700,000$.
- h. The decimal point must be moved two places to the right; $3.7899 \times 10^2 = 378.99$.
- i. The decimal point must be moved one place to the left; $1.093 \times 10^{-1} = 0.1093$.

Chapter 2: Measurements and Calculations

    j.   The decimal point must be moved zero places; $2.9004 \times 10^0 = 2.9004$.

    k.   The decimal point must be moved four places to the left; $3.9 \times 10^{-4} = 0.00039$.

    l.   The decimal point must be moved eight places to the left; $1.904 \times 10^{-8} = 0.00000001904$.

121. To say that scientific notation is in *standard* form means that you have a number between 1 and 10, followed by an exponential term. The numbers given in this problem are *not* between 1 and 10 as written.

    a. $102.3 \times 10^{-5} = (1.023 \times 10^2) \times 10^{-5} = 1.023 \times 10^{-3}$

    b. $32.03 \times 10^{-3} = (3.203 \times 10^1) \times 10^{-3} = 3.203 \times 10^{-2}$

    c. $59933 \times 10^2 = (5.9933 \times 10^4) \times 10^2 = 5.9933 \times 10^6$

    d. $599.33 \times 10^4 = (5.9933 \times 10^2) \times 10^4 = 5.9933 \times 10^6$

    e. $5993.3 \times 10^3 = (5.9933 \times 10^3) \times 10^3 = 5.9933 \times 10^6$

    f. $2054 \times 10^{-1} = (2.054 \times 10^3) \times 10^{-1} = 2.054 \times 10^2$

    g. $32{,}000{,}000 \times 10^{-6} = (3.2 \times 10^7) \times 10^{-6} = 3.2 \times 10^1$

    h. $59.933 \times 10^5 = (5.9933 \times 10^1) \times 10^5 = 5.9933 \times 10^6$

122. a. $1/10^2 = 1 \times 10^{-2}$

    b. $1/10^{-2} = 1 \times 10^2$

    c. $55/10^3 = \dfrac{5.5 \times 10^1}{1 \times 10^3} = 5.5 \times 10^{-2}$

    d. $(3.1 \times 10^6)/10^{-3} = \dfrac{3.1 \times 10^6}{1 \times 10^{-3}} = 3.1 \times 10^9$

    e. $(10^6)^{1/2} = 1 \times 10^3$

    f. $(10^6)(10^4)/(10^2) = \dfrac{(1 \times 10^6)(1 \times 10^4)}{(1 \times 10^2)} = 1 \times 10^8$

    g. $1/0.0034 = \dfrac{1}{3.4 \times 10^{-3}} = 2.9 \times 10^2$

    h. $3.453/10^{-4} = \dfrac{3.453}{1 \times 10^{-4}} = 3.453 \times 10^4$

123. meter

124.

Chapter 2:   Measurements and Calculations

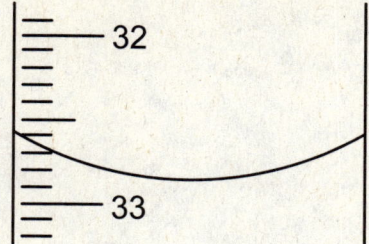

125.   100 km (See inside cover of textbook.)

126.   1 L = 1 dm^3 = 1000 cm^3 = 1000 mL

127.   250. mL

128.   0.105 m

129.   100 km/hr = 62.1 mi/hr; you would not violate the speed limit.

130.   They weigh the same.

$$1 \text{ mg} \times \frac{1 \text{ g}}{1000 \text{ mg}} = 0.001 \text{ g}$$

131.   4.25 g (425 mg = 0.425 g)

132.   5×10^{11} nm

$$500 \text{ m} \times \frac{10^9 \text{ nm}}{1 \text{ m}} = 5 \times 10^{11} \text{ nm}$$

133.   volume

134. $v = l \times h \times w$

0.310 m^3 = (0.7120 m)(0.52458 m) × w

w = 0.830 m (The answer is to three significant figures because the final volume of the box is reported to three significant figures. The other two measurements contain more significant figures and do not limit the precision of the volume.)

135.   a.   one
       b.   one
       c.   four
       d.   two
       e.   infinite (definition)
       f.   one

Chapter 2: Measurements and Calculations

136. a. 0.000426

 b. 4.02×10^{-5}

 c. 5.99×10^{6}

 d. 400.

 e. 0.00600

137. a. 0.7556

 b. 293

 c. 17.01

 d. 432.97

138. a. 2149.6 (the answer can only be given to the first decimal place, because 149.2 is only known to the first decimal place)

 b. 5.37×10^{3} (the answer can only be given to two decimal places because 4.34 is only known to two decimal places; moreover, since the power of ten is the same for each number, the calculation can be performed directly)

 c. Before performing the calculation, the numbers have to be converted so that they contain the same power of ten.
 $4.03 \times 10^{-2} - 2.044 \times 10^{-3} = 4.03 \times 10^{-2} - 0.2044 \times 10^{-2} = 3.83 \times 10^{-2}$ (the answer can only be given the second decimal place because 4.03×10^{-2} is only known to the second decimal place)

 d. Before performing the calculation, the numbers have to be converted so that they contain the same power of ten.
 $2.094 \times 10^{5} - 1.073 \times 10^{6} = 2.094 \times 10^{5} - 10.73 \times 10^{5} = -8.64 \times 10^{5}$

139. a. 5.57×10^{7} (the answer can only be given to three significant figures because 0.0432 and 4.43×10^{8} are only known to three significant figures)

 b. 2.38×10^{-1} (the answer can only be given to three significant figures because 0.00932 and 4.03×10^{2} are only known to three significant figures)

 c. 4.72 (the answer can only be given to three significant figures because 2.94 is only known to three significant figures)

 d. 8.08×10^{8} (the answer can only be given to three significant figures because 0.000934 is only known to three significant figures)

140. a. $(2.9932 \times 10^{4})(2.4443 \times 10^{2} + 1.0032 \times 10^{1}) =$
 $(2.9932 \times 10^{4})(24.443 \times 10^{1} + 1.0032 \times 10^{1}) =$
 $(2.9932 \times 10^{4})(25.446 \times 10^{1}) = 7.6166 \times 10^{6}$

 b. $(2.34 \times 10^{2} + 2.443 \times 10^{-1})/(0.0323) =$
 $(2.34 \times 10^{2} + 0.002443 \times 10^{2})/(0.0323) =$
 $(2.34 \times 10^{2})/(0.0323) = 7.24 \times 10^{3}$

Chapter 2: Measurements and Calculations

    c. $(4.38 \times 10^{-3})^2 = 1.92 \times 10^{-5}$

    d. $(5.9938 \times 10^{-6})^{1/2} = 2.4482 \times 10^{-3}$

141. $\dfrac{1 \text{ L}}{1000 \text{ cm}^3}$; $\dfrac{1000 \text{ cm}^3}{1 \text{ L}}$

142. $\dfrac{1 \text{ year}}{12 \text{ months}}$; $\dfrac{12 \text{ months}}{1 \text{ year}}$

143.   a. $8.43 \text{ cm} \times \dfrac{10 \text{ mm}}{1 \text{ cm}} = 84.3 \text{ mm}$

    b. $2.41 \times 10^2 \text{ cm} \times \dfrac{1 \text{ m}}{100 \text{ cm}} = 2.41 \text{ m}$

    c. $294.5 \text{ nm} \times \dfrac{1 \text{ m}}{10^9 \text{ nm}} \times \dfrac{100 \text{ cm}}{1 \text{ m}} = 2.945 \times 10^{-5} \text{ cm}$

    d. $404.5 \text{ m} \times \dfrac{1 \text{ km}}{1000 \text{ m}} = 0.4045 \text{ km}$

    e. $1.445 \times 10^4 \text{ m} \times \dfrac{1 \text{ km}}{1000 \text{ m}} = 14.45 \text{ km}$

    f. $42.2 \text{ mm} \times \dfrac{1 \text{ cm}}{10 \text{ mm}} = 4.22 \text{ cm}$

    g. $235.3 \text{ m} \times \dfrac{1000 \text{ mm}}{1 \text{ m}} = 2.353 \times 10^5 \text{ mm}$

    h. $903.3 \text{ nm} \times \dfrac{1 \text{ m}}{10^9 \text{ nm}} \times \dfrac{10^6 \text{ μm}}{1 \text{ m}} = 0.9033 \text{ μm}$

144.   a. $908 \text{ oz} \times \dfrac{1 \text{ lb}}{16 \text{ oz}} \times \dfrac{1 \text{ kg}}{2.2046 \text{ lb}} = 25.7 \text{ kg}$

    b. $12.8 \text{ L} \times \dfrac{1 \text{ qt}}{0.94633 \text{ L}} \times \dfrac{1 \text{ gal}}{4 \text{ qt}} = 3.38 \text{ gal}$

    c. $125 \text{ mL} \times \dfrac{1 \text{ L}}{1000 \text{ mL}} \times \dfrac{1 \text{ qt}}{0.94633 \text{ L}} = 0.132 \text{ qt}$

    d. $2.89 \text{ gal} \times \dfrac{4 \text{ qt}}{1 \text{ gal}} \times \dfrac{1 \text{ L}}{1.0567 \text{ qt}} \times \dfrac{1000 \text{ mL}}{1 \text{ L}} = 1.09 \times 10^4 \text{ mL}$

    e. $4.48 \text{ lb} \times \dfrac{453.59 \text{ g}}{1 \text{ lb}} = 2.03 \times 10^3 \text{ g}$

    f. $550 \text{ mL} \times \dfrac{1 \text{ L}}{1000 \text{ mL}} \times \dfrac{1.0567 \text{ qt}}{1 \text{ L}} = 0.58 \text{ qt}$

Chapter 2:   Measurements and Calculations

145. $9.3 \times 10^7 \text{ mi} \times \dfrac{1 \text{ km}}{0.62137 \text{ mi}} = 1.5 \times 10^8 \text{ km}$

$1.5 \times 10^8 \text{ km} \times \dfrac{1000 \text{ m}}{1 \text{ km}} \times \dfrac{100 \text{ cm}}{1 \text{ m}} = 1.5 \times 10^{13} \text{ cm}$

146. $5.3 \times 10^3 \text{ lbs} \times \dfrac{1 \text{ kg}}{2.2046 \text{ lbs}} \times \dfrac{1 \text{ metric ton}}{1000 \text{ kg}} = 2.4 \text{ metric tons}$

147. $T_K = T_C + 273$

    a. $0 + 273 = 273$ K

    b. $25 + 273 = 298$ K

    c. $37 + 273 = 310.$ K

    d. $100 + 273 = 373$ K

    e. $-175 + 273 = 98$ K

    f. $212 + 273 = 485$ K

148. a.   Celsius temperature = $(175 - 32)/1.80 = 79.4°C$

       Kelvin temperature = $79.4 + 273 = 352$ K

    b. $255 - 273 = -18$ °C

    c. $(-45 - 32)/1.80 = -43°C$

    d. $1.80(125) + 32 = 257°F$

149. density = $\dfrac{\text{mass}}{\text{volume}}$

    a. $d = \dfrac{234 \text{ g}}{2.2 \text{ cm}^3} = 110 \text{ g/cm}^3$

    b. $m = 2.34 \text{ kg} \times \dfrac{1000 \text{ g}}{1 \text{ kg}} = 2340 \text{ g}$

        $v = 2.2 \text{ m}^3 \times \left(\dfrac{100 \text{ cm}}{1 \text{ m}}\right)^3 = 2.2 \times 10^6 \text{ cm}^3$

        $d = \dfrac{2340 \text{ g}}{2.2 \times 10^6 \text{ cm}^3} = 1.1 \times 10^{-3} \text{ g/cm}^3$

    c. $m = 1.2 \text{ lb} \times \dfrac{453.59 \text{ g}}{1 \text{ lb}} = 544 \text{ g}$

        $v = 2.1 \text{ ft}^3 \times \left(\dfrac{12 \text{ in}}{1 \text{ ft}}\right)^3 \times \left(\dfrac{2.54 \text{ cm}}{1 \text{ in}}\right)^3 = 5.95 \times 10^4 \text{ cm}^3$

Chapter 2: Measurements and Calculations

$$d = \frac{544 \text{ g}}{5.95 \times 10^4 \text{ cm}^3} = 9.1 \times 10^{-3} \text{ g/cm}^3$$

d. $m = 4.3 \text{ ton} \times \dfrac{2000 \text{ lb}}{1 \text{ ton}} \times \dfrac{453.59 \text{ g}}{1 \text{ lb}} = 3.90 \times 10^6 \text{ g}$

$v = 54.2 \text{ yd}^3 \times \left(\dfrac{1 \text{ m}}{1.0936 \text{ yd}}\right)^3 \times \left(\dfrac{100 \text{ cm}}{1 \text{ m}}\right)^3 = 4.14 \times 10^7 \text{ cm}^3$

$d = \dfrac{3.90 \times 10^6 \text{ g}}{4.14 \times 10^7 \text{ cm}^3} = 9.4 \times 10^{-2} \text{ g/cm}^3$

150. $85.5 \text{ mL} \times \dfrac{0.915 \text{ g}}{1 \text{ mL}} = 78.2 \text{ g}$

151. $50.0 \text{ g} \times \dfrac{1 \text{ mL}}{1.31 \text{ g}} = 38.2 \text{ g}$

152. $m = 155 \text{ lb} \times \dfrac{453.59 \text{ g}}{1 \text{ lb}} = 7.031 \times 10^4 \text{ g}$

$v = 4.2 \text{ ft}^3 \times \left(\dfrac{12 \text{ in}}{1 \text{ ft}}\right)^3 \times \left(\dfrac{2.54 \text{ cm}}{1 \text{ in}}\right)^3 = 1.189 \times 10^5 \text{ cm}^3$

$d = \dfrac{7.031 \times 10^4 \text{ g}}{1.189 \times 10^5 \text{ cm}^3} = 0.59 \text{ g/cm}^3$

153. Volume = 21.6 mL − 12.7 mL = 8.9 mL

$d = \dfrac{33.42 \text{ g}}{8.9 \text{ mL}} = 3.8 \text{ g/mL}$

154. $T_F = 1.80(T_C) + 32$

  a. 23 °F
  b. 32 °F
  c. −321 °F
  d. −459 °F
  e. 187 °F
  f. −459 °F

155. a. 10^3
  b. 10^9
  c. 10^{-2}
  d. 10^{-3}

Chapter 2: Measurements and Calculations

156. a. The Mars Climate Orbiter dipped 100 km lower in the Mars atmosphere than was planned. Using the conversion factor between miles and kilometers found inside the cover of this text

$$100 \text{ km} \times \frac{1 \text{ mi}}{1.6093 \text{ km}} = 62 \text{ mi}$$

   b. The aircraft required 22,300 kg of fuel, but only 22,300 lb of fuel was loaded. Using the conversion factor between pounds and kilograms found inside the cover of this text, the amount of fuel required in pounds was

$$22,300 \text{ kg} \times \frac{2.2046 \text{ lb}}{1 \text{ kg}} = 49,163 \text{ lb}$$

   Therefore, $(49,163 - 22,300) = 26,863 = 2.69 \times 10^4$ lb additional fuel was needed.

157. a. The text mentions oxygen sensors in automobile exhaust systems; detection of nitrogen-containing compounds in airline baggage; use of sensory hair from crabs to detect low levels of hormones; use of pineapple extracts to detect hydrogen peroxide

   b. We can now detect the presence of impurities or contaminants to much lower levels than was possible in the past. Although that may seem helpful, we now have to determine whether these contaminants were always present and are not harmful or if they are something new that we should be concerned about.

158. $\dfrac{10^{-8} \text{ g}}{\text{L}} \times \dfrac{1 \text{ lb}}{453.59 \text{ g}} \times \dfrac{1 \text{ L}}{1.0567 \text{ qt}} \times \dfrac{4 \text{ qt}}{1 \text{ gal}} = 8 \times 10^{-11}$ lb/gal

159.

Scientific Notation	Number of Significant Figures
9.000×10^2	4
3.007×10^3	4
2.345×10^4	4
2.700×10^2	4
4.37×10^5	3

160.

Number of Significant Figures	Result
2	0.51
3	29.1
3	8.61
3	1.89
4	134.6

Chapter 2: Measurements and Calculations

      3   14.4

161. $4145 \text{ mi} \times \dfrac{5280 \text{ ft}}{1 \text{ mi}} \times \dfrac{1 \text{ fathom}}{6 \text{ ft}} \times \dfrac{1 \text{ cable length}}{100 \text{ fathoms}} = 3.648 \times 10^4$ cable lengths

   $4145 \text{ mi} \times \dfrac{1.6093 \text{ km}}{1 \text{ mi}} \times \dfrac{1000 \text{ m}}{1 \text{ km}} = 6.671 \times 10^6$ m

   3.648×10^4 cable lengths $\times \dfrac{1 \text{ nautical mile}}{10 \text{ cable lengths}} = 3648$ nautical miles

162. $1.25 \text{ mi} \times \dfrac{1.6093 \text{ km}}{1 \text{ mi}} \times \dfrac{1000 \text{ m}}{1 \text{ km}} = 2011.625$ m

   60 sec + 59.2 sec = 119.2 sec

   $\dfrac{2011.625 \text{ m}}{119.2 \text{ s}} = 16.9$ m/s

163. $T_C = (T_F - 32)/1.80$

   $T_C = (69.1 - 32)/1.80 = 20.6°C$

164. $T_C = (T_F - 32)/1.80$

   $T_C = (134 - 32)/1.80 = 56.7°C$

   Since the temperature is higher than the melting point (44°C), phosphorus would be a liquid.

165. $1.84 \text{ cm} \times 3.61 \text{ cm} \times 2.10 \text{ cm} = 13.9 \text{ cm}^3$

   $13.9 \text{ cm}^3 \times \dfrac{22.57 \text{ g}}{\text{cm}^3} = 315$ g

166. $69 \text{ pm} \times \dfrac{1 \text{ m}}{10^{12} \text{ pm}} \times \dfrac{100 \text{ cm}}{1 \text{ m}} = 6.9 \times 10^{-9}$ cm

   $V = \tfrac{4}{3}\pi(6.9 \times 10^{-9} \text{ cm})^3 = 1.4 \times 10^{-24} \text{ cm}^3$

   $d = \dfrac{mass}{volume} = \dfrac{3.35 \times 10^{-23} \text{ g}}{1.4 \times 10^{-24} \text{ cm}^3} = 24 \text{ g/cm}^3$

CHAPTER 3

Matter

1. has mass; occupies space
2. intermolecular forces
3. liquids; solids
4. solids
5. In solids, the particles are essentially fixed in position relative to one another and can only vibrate in place. In liquids, the particles are still in close proximity to one another, but are able to move in three dimensions relative to one another. In gases, the particles are *not* in close proximity to each other and move freely and independently of one another.
6. gaseous
7. Liquids and gases both have no rigid shape and take on the shape of their containers (because the molecules in them are free to move relative to each other). Liquids are essentially incompressible, whereas gases are readily compressible.
8. The *stronger* the inter-particle forces, the more rigid is the sample overall.
9. The gaseous 10-g sample of water has a much *larger* volume than either the solid or liquid samples. Although the 10-g sample of water vapor contains the same amount of water as the solid and liquid sample (the same number of water molecules), there is a great deal of empty space in the gaseous sample.
10. Gases are easily compressed into smaller volumes, whereas solids and liquids are not. Because a gaseous sample consists mostly of empty space, the gas particles are pushed closer together when pressure is applied to a gas.
11. Density and color are physical properties. The odor would often be described as a physical characteristic, but since odor results from a chemical interaction between the molecules of vapor and receptors in the nose, many scientists would classify the odor as a chemical property.
12. chemical change; New products are formed. The reactants are water molecules, which undergoes a chemical reaction to produce hydrogen and oxygen molecules. These are chemically different than water molecules.
13. Magnesium burns in air.
14. Magnesium is malleable and ductile.
15. The answer will depend on the students' examples.
16. (d); The identity of the molecules that make up (a), (b) and (c) does not change, just the rearrangement.
17. a. chemical; the grease/oil inside the oven is actually converted to soap, which can then be washed away
    b. physical; the molecules in the rubber band rearrange themselves when the rubber band is stretched, but the identity of the molecules does not change

Chapter 3:   Matter

    c.   chemical; "rust" represents the product of the reaction of iron with oxygen in the air

    d.   physical; the odor is characteristic of the acid itself and does not represent any change in the acid. However, because the perception of odor does involve a chemical process between the molecules of vapor and receptors in the nose, some scientists would classify odor as a chemical property.

    e.   chemical; the acid breaks bonds in the cellulose molecules that constitute the cotton fibers

    f.   physical; crystallizing a pure substance from a solution does not change the identity of the substance

    g.   chemical; the green patina is a compound of copper that results from the copper metal reacting with gases in the atmosphere

    h.   chemical; the "brown" color represents the chemical breakdown of carbohydrates (starches) in the bread

    i.   physical; evaporation is a physical change

    j.   chemical; the proteins, carbohydrates, and fats in the steak are converted to elemental carbon by overheating.

    k.   chemical; the "fizzing" represents the production of a new gaseous substance (oxygen gas) from the breakdown of the hydrogen peroxide.

18.   a.   physical; the iron is only being heated.

    b.   chemical; the sugars in the marshmallow are being reduced to carbon.

    c.   chemical; most strips contain a peroxide which decomposes.

    d.   chemical; the bleach oxidizes dyes in the fabric.

    e.   physical; evaporation is only a change of state.

    f   physical; the salt is only modifying the physical properties of the solution, not undergoing a chemical reaction.

    g.   chemical; the drain cleaner breaks bonds in the hair.

    h.   physical; students will most likely reply that this is a physical change since the perfume is evaporating; the sensation of smell, however, depends on chemical processes.

    i.   physical; the sublimation is only a change of state.

    j.   physical; the wood is only being physically divided into smaller pieces.

    k.   chemical; the cellulose in the wood is reacting with oxygen gas

19.   compounds

20.   Compounds consist of two or more elements combined together chemically in a fixed composition, no matter what their source may be. For example, water on earth consists of molecules containing one oxygen atom and two hydrogen atoms. Water on Mars (or any other planet) has the same composition.

21.   compounds

22.   compounds

23.   the same

Chapter 3: Matter

24. He, F_2, S_8; Elements cannot be broken down into other substances by chemical means. HCl is a compound because it can be broken down into hydrogen and chlorine.

25. This would represent a mixture: the iron and sulfur have only been placed together in the same container at this point, but no reaction has occurred. This is confirmed by the fact that the magnet can be used to remove the iron from the sulfur.

26. Given that the product of the process is no longer attracted by the magnet, this strongly suggests that the iron has been converted to an iron/sulfur compound—a pure substance.

27. The term *homogeneous* in this context means that there are no variations in composition in different areas of the mixture.

28. solutions: window cleaner, shampoo, rubbing alcohol

    mixtures: salad dressing, jelly beans, the change in my pocket

29. a. mixture
    b. mixture
    c. pure substance (hopefully)
    d. mixture (the hydrogen peroxide is dissolved in water)

30. a. mixture
    b. mixture
    c. mixture
    d. pure substance

31. a. heterogeneous
    b. homogeneous (macroscopically, assuming plain mayonnaise)
    c. heterogeneous
    d. heterogeneous
    e. heterogeneous

32. Concrete is a mixture: the various components of the particular concrete are still distinguishable within the concrete if examined closely.

33. Consider a salt solution (sodium chloride in water). Since water boils at a much lower temperature than sodium chloride, the water can be boiled off from the solution, collected, and subsequently condensed back into the liquid state. This separates the two chemical substances.

34. Consider a mixture of salt (sodium chloride) and sand. Salt is soluble in water, sand is not. The mixture is added to water and stirred to dissolve the salt and is then filtered. The salt solution passes through the filter; the sand remains on the filter. The water can then be evaporated from the salt.

35. If water is added to the sample, and the sample is then heated to boiling, this should dissolve the benzoic acid but not the charcoal. The hot sample could then be *filtered*, which would remove the charcoal. The solution that passed through the filter could then be cooled. This should cause some of the benzoic acid to crystallize, or the solution could be heated carefully to boil off the water leaving benzoic acid behind.

36. The chemical identities of the components of the mixture are not changed by filtration or distillation: the various components are separated by physical, not chemical, means.

37. compound

38. a. compound; pure substance
    b. element; pure substance
    c. homogeneous mixture

39. Chalk must be a compound, as it loses mass when heated, and appears to change into a substance with different physical properties (the hard chalk turns into a crumbly substance).

40. Because vaporized water is still the *same substance* as solid water, no chemical reaction has occurred. Sublimation is a physical change.

41. Liquids and gases both flow freely and take on the shape of their container. The molecules in liquids are relatively close together and interact with each other, whereas the molecules in gases are far apart from each other and do not interact with each other.

42. False. No reaction has taken place. The substances are merely separating, not changing into different substances. This is an example of a heterogeneous mixture.

43. physical

44. (b); P_4 is an element. Dissolving sugar in water is not a chemical change (both the sugar and water are still intact). NaCl is a compound.

45. chemical

46. physical

47. state

48. pure substance; compound; element

49. a. physical; milk contains protein, and when the vinegar is added, the acidity of the vinegar causes a change in the protein's shape, making it insoluble in water (see Chapter 21)

    b. chemical; exposure to the oxygen of the air allows bacteria to grow, which cause the chemical breakdown of components of the butter.

    c. physical; salad dressing is a physical mixture of water soluble and insoluble components, which only combine temporarily when the dressing is shaken.

    d. chemical; milk of magnesia is a *base*, which chemically reacts with and neutralizes the acid of the stomach.

    e. chemical; steel consists mostly of iron, which chemically reacts with the oxygen of the atmosphere.

    f. chemical; carbon monoxide combines chemically with the hemoglobin fraction of the blood, making it impossible for the hemoglobin to combine with oxygen.

    g. chemical; paper consists of the carbohydrate cellulose, which is broken down chemically by acids.

    h. physical; sweat consists mostly of water, which consumes heat from the body in evaporating.

    i. chemical; although the biochemical action of aspirin is not fully understood, the process is chemical in nature.

Chapter 3:   Matter

- j. physical; oil molecules are not water soluble, and are repelled by the moisture in skin.
- k. chemical; the fact that one substance is converted into two other substances demonstrates that this is a chemical process.

50.
- a. heterogeneous
- b. homogeneous
- c. heterogeneous
- d. heterogeneous
- e. homogeneous

51.
- a. heterogeneous
- b. homogeneous
- c. heterogeneous
- d. homogeneous (assuming there are no imperfections in the glass)
- e. heterogeneous

52. Answer depends on student response

53. Answer depends on student choices.

54. False. The substances in the mixture do not always combine to form a new product. Mixtures can be separated into pure substances, but this is not a chemical reaction.

55. Answer depends on student choices.

56. O_2 and P_4 are both still elements, even though the ordinary forms of these elements consist of molecules containing more than one atom (but all atoms in each respective molecule are the same). P_2O_5 is a compound, because it is made up of two or more different elements (not all the atoms in the P_2O_5 molecule are the same).

57. Answers depends on student response.

58. Assuming there is enough water present in the mixture to have dissolved all the salt, filter the mixture to separate out the sand from the mixture. Then distill the filtrate (consisting of salt and water), which will boil off the water, leaving the salt.

Chapter 3: Matter

59. See Figure 3.6 in the text.

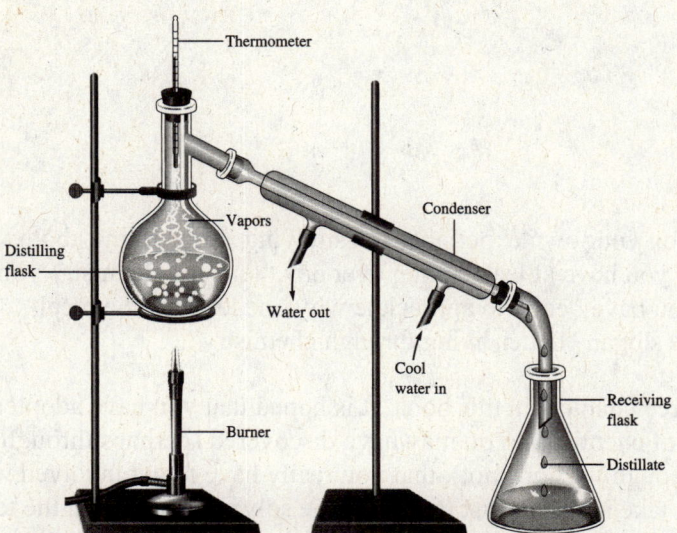

60. The most obvious difference is the physical states: water is a liquid under room conditions, hydrogen and oxygen are both gases. Hydrogen is flammable. Oxygen supports combustion. Water does neither.

61. (b); A compound consists of two or more different elements. Mixtures consist of two or more substances (elements and/or compounds).

62. a. False. A spoonful of sugar is a compound (sucrose, $C_{12}H_{22}O_{11}$).

　　b. False. Element and compounds are pure substances.

　　c. True.

　　d. False. Gasoline is a mixture.

　　e. True.

63. (c)

64. (a), (d)

CUMULATIVE REVIEW

Chapters 1, 2, and 3

1. Obviously, this answer depends on your own experiences in studying and learning about chemistry. We hope that by now you have at least gotten over any "fear" of chemistry you may have started out with. Perhaps you have begun to appreciate why one leading chemical manufacturer has as its corporate slogan "better living through chemistry".

2. By now, after having covered three chapters in this book, it is hoped that you have adopted an "active" approach to your study of chemistry. You may have discovered (perhaps through a disappointing grade on a quiz (though we hope not), that you really have to get involved with chemistry. You can't just sit and take notes, or just look over the solved examples in the textbook. You have to learn to solve problems. You have to learn how to interpret problems, and how to reduce them to the simple mathematical relationships you have studied. Whereas in some courses you might get by on just giving back on exams the facts or ideas presented in class, in chemistry you have to be able to extend and synthesize what has been discussed and to apply the material to new situations. Don't get discouraged if this is difficult at first: it's difficult for everyone at first.

3. The steps of the scientific method, in brief, are: (1) make observations of the system and state the problem clearly; (2) formulate an explanation or hypothesis to try to explain your observations; and (3) perform one or more experiments to test the validity of your hypothesis.

   For the case of the liquid material, we first have to state the problem: we have a sample of clear liquid that may be either a pure compound or a mixture, and we want to determine which of these is correct. The liquid is completely homogeneous, so we can't tell anything just by looking at it.

   If the unknown liquid is a mixture, we should be able to separate the components of the mixture by distillation (see Chapter 3). If the liquid is a mixture of two pure liquids, the liquids should boil at their characteristic temperatures during distillation. If the liquid is a solution of a solid material, then the liquid portion should boil off, leaving a solid residue behind. If the unknown liquid is a pure substance, rather than a mixture, it should boil away at a constant, uniform temperature. Let's hypothesize that the liquid is a mixture.

   Now we perform the experiment: We begin heating the liquid in a distillation apparatus, monitoring the temperature with a thermometer. At 65°C, the liquid begins to boil, and a clear, sweet-smelling liquid begins to collect in the receiving container. The temperature remains at 65°C until approximately half the liquid has distilled, whereupon the temperature rises suddenly to 100°C. At this point, we change receiving flasks. The temperature remains at 100°C until the remainder of the liquid boils. We notice that this second fraction of liquid collected has no odor.

   Based on the observations that two separate fractions were collected, which had different boiling points, and that only one of the fractions had a noticeable odor, we can conclude that our hypothesis that the unknown liquid was a mixture is correct.

4. It is difficult sometimes for students (especially beginning students) to understand why certain subjects are required for a given college major. The faculty of your major department, however, have collectively many years of experience in the subject in which you have chosen to specialize. They really do know what courses will be helpful to you in the future. They may have had trouble

with the same courses that now give you trouble, but they realize that all the work will be worth it in the end. Some courses you take, particularly in your major field itself, have obvious and immediate utility. Other courses, often times chemistry included, are provided to give you a general background knowledge, which may prove useful in understanding your own major or other subjects related to your major. In perhaps a burst of bravado, chemistry has been called "the central science" by one team of textbook authors. This moniker is very true however: in order to understand biology, physics, nutrition, farming, home economics, or whatever (it helps to have a general background in chemistry).

5. The table below shows the common units and prefixes used to show multiples/divisions:

Physical Quantity	Basic SI Unit
mass	kilogram
distance	meter
time	second
temperature	kelvin

Commonly Used Prefixes in the Metric System		
Prefix	Meaning	Power of Ten
mega-	million	10^6
kilo-	thousand	10^3
deci-	tenth	10^{-1}
centi-	hundredth	10^{-2}
milli-	thousandth	10^{-3}
nano-	billionth	10^{-9}

The metric system is in use in most of the world because its system of units and multiples is simple to remember, and the system permits easy conversion between units. The various multiples and subdivisions of the basic units are based on factors of *ten*, which is also the basis for our number system.

Some examples of using appropriate prefixes might include the following:

My English textbook measures 25 cm × 20 cm × 4 cm.

The aspirin tablet I just took contains 325 mg of aspirin.

The road signs in Europe give distances in kilometers, not miles.

6. Whenever a scientific measurement is made, we always employ the instrument or measuring device we are using to the limits of its precision. On a practical basis, this usually means that we *estimate* our reading of the last significant figure of the measurement. An example of the uncertainty in the last significant figure is given for measuring the length of a pin in the text in Figure 2.5. Scientists appreciate the limits of experimental techniques and instruments, and always assume that the last digit in a number representing a measurement has been estimated. Because the last significant figure in every measurement is assumed to be estimated, it is never possible to exclude uncertainty from measurements. The best we can do is to try to improve our techniques and instruments so that we get more significant figures for our measurements.

Cumulative Review: Chapters 1, 2, and 3

7. Scientists are careful about reporting their measurements to the appropriate number of significant figures so as to indicate to their colleagues the precision with which experiments were performed. That is, the number of significant figures reported indicates how "carefully" measurements were made. Suppose you were considering buying a new home, and the real estate agent told you that a prospective new house was "between 1000-2000 square feet of space" and had "five or six rooms, more or less" and stood on "maybe an acre or two of land". Would you buy the house or would you look for another real estate agent? When a scientist says that a sample of material "has a mass of 3.126 grams" he or she is narrowing down the limits as to the actual, true mass of the sample: the mass is clearly slightly more than half way between 3.12 and 3.13 grams.

   The rules for significant figures are covered in Section 2.5 of the text. In brief, these rules for experimentally measured numbers are as follows: (1) nonzero integers are *always* significant; (2) leading zeroes are *never* significant, captive zeroes are *always* significant, and trailing zeroes *may* be significant (if a decimal point is indicated); and (3) exact numbers (e.g., definitions) have an infinite number of significant figures.

   When we round off an answer to the correct number of significant figures (as limited by whatever measurement was least precise), we do this in a particular manner. If the digit to be removed is equal to or greater than 5, the preceding digit is increased by 1. If the digit to be removed is less than 5, the preceding digit is not changed. If you are going to perform a series of calculations involving a set of data, hold on to the digits in the intermediate calculations until you arrive at the final answer, and then round off the final answer to the appropriate number of significant figures.

   When doing arithmetic with experimentally determined numbers, the final answer is determined by the least precise measurement. In doing multiplication or division calculations, the number of significant figures in the result should be the same as the measurement with the fewest significant figures. In performing addition or subtraction, the number of significant figures in the result is limited by the measurement with the fewest decimal places.

8. Dimensional analysis is a method of problem solving that pays particular attention to the units of measurements and uses these units as if they were algebraic symbols that multiply, divide, and cancel. Consider the following example. A dozen eggs costs $1.25. Suppose we want to know how much one egg costs, and also how much three dozens of eggs will cost. To solve these problems, we need to make use of two equivalence statements:

   $$1 \text{ dozen eggs} = 12 \text{ eggs}$$
   $$1 \text{ dozen eggs} = \$1.25$$

   The first of these equivalence statements is obvious: everyone knows that 12 eggs is "equivalent" to one dozen. The second statement also expresses an equivalence: if you give the grocer $1.25, he or she will give you a dozen eggs. From these equivalence statements, we can construct the conversion factors we need to answer the two questions. For the first question (what does one egg cost) we can set up the calculation as follows

   $$\frac{\$1.25}{12 \text{ eggs}} = \$0.104 = \$0.10$$

   as the cost of one egg. Similarly, for the second question (the cost of 3 dozens eggs), we can set up the conversion as follows

   $$3 \text{ dozens} \times \frac{\$1.25}{1 \text{ dozen}} = \$3.75$$

as the cost of three dozens eggs. See Section 2.6 of the text for how we construct conversion factors from equivalence statements.

9. The Fahrenheit (°F) temperature scale is defined so that an ice/water bath has an equilibrium temperature of 32°F, and a boiling water bath a temperature of 212°F, under normal atmospheric pressure. There are 180 degree divisions between these two reference points. The Celsius (°C) temperature scale is defined so that an ice/water bath has a temperature of 0°C, whereas a boiling water bath has a temperature of 100°C, with 100 degree divisions between these reference points. Both the Fahrenheit and Celsius temperature scales are human inventions that choose convenient, stable, reproducible reference points as their definitions. The Kelvin (K) or Absolute temperature scale is based on a fundamental property of matter itself: the zero and lowest point (Absolute Zero) on the Kelvin temperature scale is the lowest possible temperature that can exist, and represents the temperature at which atoms and molecules are in their lowest possible energy states. All other temperatures on the Kelvin scale are positive relative to this. Since only this one reference point is used to define the Kelvin scale, the size of the Kelvin degree relative to this point was chosen to be the same size as the Celsius degree for convenience. The temperature of an ice/water bath is 273 K, and the temperature of a boiling water bath is 373 K.

10. Defining what scientists mean by "matter" often seems circular to students. Scientists say that matter is something that "has mass and occupies space", without ever really explaining what it means to "have mass" or to "occupy space"! The concept of matter is so basic and fundamental, that it becomes difficult to give a good textbook definition other than to say that matter is the "stuff" of which everything is made. Matter can be classified and subdivided in many ways, depending on what we are trying to demonstrate.

    On the most fundamental basis, all matter is composed of tiny particles (such as protons, electrons, neutrons, and the other subatomic particles). On one higher level, these tiny particles are combined in a systematic manner into units called atoms. Atoms, in turn, may be combined to constitute molecules. And finally, large groups of molecules may be placed together to form a bulk sample of substance that we can see.

    Matter can also be classified as to the physical state a particular substance happens to take. Some substances are solids, some are liquids, and some are gases. Matter can also be classified as to whether it is a pure substance (one type of molecule) or a mixture (more than one type of molecule), and furthermore whether a mixture is homogeneous or heterogeneous.

11. Your specific answer will depend on the chemical substance you chose.

    The physical properties of a substance are the inherent characteristics of the substance, which result in no change in the composition of the substance when we measure or study these properties. Such properties include color, odor, physical state, density, solubility, melting point, boiling point, etc. The chemical properties of a given substance indicate how that substance reacts with other substances. For example, when we say that sodium is a grayish-white, soft, low-density metal, we are describing some of sodium's physical properties. When we say that sodium metal reacts with chlorine gas to form sodium chloride, we are describing a chemical property of sodium.

12. Chemists tend to give a functional definition of what they mean by an "element": an element is a fundamental substance that cannot be broken down into any simpler substances by chemical methods. Compounds, on the other hand, can be broken down into simpler substances (the elements of which the compound is composed). For example, sulfur and oxygen are both elements (sulfur occurs as S_8 molecules and oxygen as O_2 molecules). When sulfur and oxygen

are placed together and heated, the compound sulfur dioxide (SO_2) forms. When we analyze the sulfur dioxide produced, we notice that each and every molecule consists of one sulfur atom and two oxygen atoms, and on a mass basis, consists of 50% each of sulfur and oxygen. We describe this by saying that sulfur dioxide has a constant composition. The fact that a given compound has constant composition is usually expressed in terms of the mass percentages of the elements present in the compound. The reason the mass percentages are constant is because of a constant number of atoms of each type present in the compound's molecules. If a scientist anywhere in the universe analyzed sulfur dioxide, he or she would find the same composition: if a scientist finds something that does not have the same composition, then the substance cannot be sulfur dioxide.

13. A mixture is a combination of two or more substances that may be varied in its composition. Most commonly in chemistry, a mixture is a combination of two or more pure substances (either elements or compounds). A solution is a particular type of mixture that appears completely homogeneous throughout. Although a solution is homogeneous in appearance, realize, however, that a solution is still a mixture of two or more pure substances: if it were possible to see the individual particles of a solution, we would notice that there were different types of molecules present.

In a sample of a pure substance, there is only one type of molecule present. There are two types of pure substances: elemental substances and compound substances. In an elemental substance, not only are all the molecules the same type, but all the atoms within those molecules are the same type. For example, the pure elemental substance oxygen consists of O_2 molecules (with no other type of molecule present). In addition, each O_2 molecule contains only one type of atom (O). In a sample of a compound substance, all the molecules are of the same type, but within each molecule are found atoms of different elements. For example, the pure compound substance water consists of H_2O molecules (with no other type of molecule present). Within each H_2O molecule, however, are found two different types of atoms (H and O). A compound is not a mixture, however, because the atoms of the different elements are chemically bonded (not just physically mixed), and the composition of the compound is not variable as to the relative amounts of each element present.

Two methods for separating mixtures are described in the text: filtration and distillation. Filtration can be used to separate a solid from a liquid. Distillation can be used to separate two liquids or a dissolved solid from a liquid. There are many other separation methods beyond the scope of this text.

14. a. The decimal point must be moved four places to the right: 8.917×10^{-4}

 b. The decimal point must be moved four places to the left: 0.0002795

 c. The decimal point must be moved three places to the right: 4913

 d. The decimal point must be moved seven places to the left: 8.51×10^7

 e. The arithmetic must be performed and then the exponents combined: 1.219×10^2

 f. The arithmetic must be performed and then the exponents combined: 3.396×10^{-9}

15. a. $493.2 \text{ g} \times \dfrac{1 \text{ kg}}{1000 \text{ g}} = 0.4932 \text{ kg}$

 b. $493.2 \text{ g} \times \dfrac{1 \text{ lb}}{453.59 \text{ g}} = 1.087 \text{ lb}$

c. $9.312 \text{ mi} \times \dfrac{1.6093 \text{ km}}{1 \text{ mi}} = 14.99 \text{ km}$

d. $9.312 \text{ mi} \times \dfrac{5280 \text{ ft}}{1 \text{ mi}} = 4.917 \times 10^4 \text{ ft}$

e. $4.219 \text{ m} \times \dfrac{1.0936 \text{ yd}}{1 \text{ m}} \times \dfrac{3 \text{ ft}}{1 \text{ yd}} = 13.84 \text{ ft}$

f. $4.219 \text{ m} \times \dfrac{100 \text{ cm}}{1 \text{ m}} = 421.9 \text{ cm}$

g. $429.2 \text{ mL} \times \dfrac{1 \text{ L}}{1000 \text{ mL}} = 0.4292 \text{ L}$

h. $2.934 \text{ L} \times \dfrac{1.0567 \text{ qt}}{1 \text{ L}} = 3.100 \text{ qt}$

16.
a. two (based on the factor of 2.1 in the denominator)
b. two (based on the factor of 5.2 in the numerator)
c. three (one before the decimal point, and two after the decimal point)
d. three (based on the sum of 5.338 and 2.11)
e. one (based on 9 only having one significant figure)
f. two (based on the sum of 4.2005 and 2.7)
g. two (based on the factor of 0.15)
h. three (two before the decimal point, and one after the decimal point)

17. $T_F = 1.80(T_C) + 32 \qquad T_C = (T_F - 32)/1.80 \qquad T_K = T_C + 273$

a. The sizes of the Kelvin and Celsius degrees are the same.

b. The Celsius degree is larger than the Fahrenheit degree by a factor of 1.80.

c. The normal freezing point of water is represented by 0°C, 273 K, and 32°F.

d. $T_K = T_C + 273 = 27.5°C + 273 = 300.5 \text{ (301) K}$

$T_F = 1.80(T_C) + 32 = 1.80(27.5°C) + 32 = 81.5°F$

e. $T_K = T_C + 273$ so $T_C = T_K - 273 = 298.1 \text{ K} - 273 = 25.1 \text{ (25)}°C$

$T_F = 1.80(T_C) + 32 = 1.80(25.1°C) + 32 = 77.1 \text{ (77)}°F$

f. $T_C = (T_F - 32)/1.80 = (98.6°F - 32)/1.80 = 37.0°C$

$T_K = T_C + 273 = 37°C + 273 = 310 \text{ K}$

Cumulative Review: Chapters 1, 2, and 3

18. density = mass/volume   mass = volume × density   volume = mass/density

   a. $\text{density} = \dfrac{78.5 \text{ g}}{100. \text{ mL}} = 0.785 \text{ g/mL}$

   b. $\text{volume} = \text{mass/density} = \dfrac{1.590 \text{ kg} \times \dfrac{1000 \text{ g}}{1 \text{ kg}}}{0.785 \text{ g/mL}} = 2025 \text{ mL} = 2.03 \text{ L}$

   c. $\text{mass} = \text{volume} \times \text{density} = 1.35 \text{ L} \times \dfrac{1000 \text{ mL}}{1 \text{ L}} \times \dfrac{0.785 \text{ g}}{1 \text{ mL}} = 1060 \text{ g} = 1.06 \text{ kg}$

   d. $\text{volume} = \text{mass/density} = \dfrac{25.2 \text{ g}}{2.70 \text{ g/cm}^3} = 9.33 \text{ cm}^3$

   e. volume = 12.0 cm × 2.5 cm × 2.5 cm = 75 cm^3
   mass = volume × density = 75 cm^3 × 2.70 g/cm^3 = 202.5 g = 2.0 × 10^2 g

19.
   a. physical
   b. chemical
   c. physical
   d. physical
   e. chemical
   f. physical
   g. chemical
   h. chemical

CHAPTER 4

Chemical Foundations: Elements, Atoms, and Ions

1. fire, earth, water, air

2. Robert Boyle

3. Boyle's most important contribution was his insistence that science should be firmly grounded in *experiment*. Boyle tried to limit the influence of any preconceptions about science and only accepted as fact what could be demonstrated.

4. oxygen, carbon, hydrogen

5. From Table 4.1: oxygen, silicon, aluminum, iron, calcium

6. a. Trace elements are those elements which are present in only tiny amounts in the body, but are critical for many bodily processes and functions.

    b. Answer depends on your choice of elements

7. B (boron); C (carbon); F (fluorine); H (hydrogen); I (iodine); K (potassium); N (nitrogen); O (oxygen); P (phosphorus); S (sulfur); U (uranium); V (vanadium); W (tungsten); Y (yttrium)

8. Sometimes the symbol for an element is based on its common name in another language. This is true for many of the more common metals since their existence was known to the ancients: some examples are iron, sodium, potassium, silver, and tin (the symbols come from their name in Latin); tungsten (the symbol comes from its name in German).

9. a. 4
    b. 11
    c. 12
    d. 3
    e. 8
    f. 9
    g. 13
    h. 7
    i. 6
    j. 5

10. a. copper
    b. cobalt
    c. calcium
    d. carbon
    e. chromium

Chapter 4: Chemical Foundations: Elements, Atoms, and Ions

	f.	cesium
	g.	chlorine
	h.	cadmium
11.	Co	cobalt
	Rb	rubidium
	Rn	radon
	Ra	radium
	U	uranium
12.	Si	silicon
	Ni	nickel
	Ag	silver
	K	potassium
	Ca	calcium
13.	a.	potassium
	b.	germanium
	c.	phosphorus
	d.	carbon
	e.	nitrogen
	f.	sodium
	g.	neon
	h.	iodine

14. B: barium, Ba; berkelium, Bk; beryllium, Be; bismuth, Bi; bohrium, Bh; boron, B; bromine, Br

N: neodymium, Nd; neon, Ne; neptunium, Np; nickel, Ni; niobium, Nb; nitrogen, N; nobelium, No

P: palladium, Pd; phosphorus, P; platinum, Pt; plutonium, Pu; polonium, Po; potassium, K; praseodymium, Pr; promethium, Pm; protactinium, Pa

S: samarium, Sm; scandium, Sc; seaborgium, Sg; selenium, Se; silicon, Si; silver, Ag; sodium, Na; strontium, Sr; sulfur, S

15. the law of constant composition

16.
   a. Elements are made of tiny particles called atoms.
   b. All the atoms of a given element are identical
   c. The atoms of a given element are different from those of any other element.
   d. A given compound always has the same numbers and types of atoms.
   e. Atoms are neither created nor destroyed in chemical processes. A chemical reaction simply changes the way the atoms are grouped together.

Chapter 4: Chemical Foundations: Elements, Atoms, and Ions

17. A compound is a distinct substance that is composed of two or more elements and always contains exactly the same relative masses of those elements.

18. According to Dalton, all atoms of the same element are *identical*; in particular, every atom of a given element has the same *mass* as every other atom of that element. If a given compound always contains the *same relative numbers* of atoms of each kind, and those atoms always have the *same masses*, then it follows that the compound made from those elements would always contain the same relative masses of its elements.

19.
 a. C_6H_6
 b. $AlCl_3$
 c. Na_2S
 d. N_2O_4
 e. $NaHCO_3$
 f. KI

20.
 a. CO_2
 b. CO
 c. $CaCO_3$
 d. H_2SO_4
 e. $BaCl_2$
 f. Al_2S_3

21.
 a. J. J. Thomson discovered the electron. Thomson postulated that, because negative particles had been detected in the atom, then there must also be positive particles to counterbalance the negative charge.

 b. William Thomson (Lord Kelvin) described the atom as a uniform pudding of positive charge, with electrons scattered throughout (like the raisins in a pudding) to balance the electrical charge.

22. False. Rutherford's bombardment experiments with metal foil suggested that the alpha particles were being deflected by coming near a *dense, positively charged* atomic nucleus.

23. Neutrons are found in the nucleus and carry no electrical charge.

24. protons

25. The proton and the neutron have similar (but not identical) masses. Both of these particles have a mass approximately 2000 times greater than that of an electron. The combination of the protons and the neutrons make up the bulk of the mass of an atom, but the electrons make the greatest contribution to the chemical properties of the atom.

26. neutron; electron

27. 10^{-13} cm = 10^{-15} m

28. electrons

29. Although all atoms of a given element contain the same number of protons in the nucleus, some atoms of a given element may have different numbers of neutrons. Isotopes are atoms of the same element with different mass numbers.

30. False. The mass number represents the total number of protons and neutrons in the nucleus.

31. An isolated atom has no charge, therefore the number of negatively charged electrons must equal the number of positively charged protons.

32. Neutrons are uncharged and contribute only to the mass.

Chapter 4: Chemical Foundations: Elements, Atoms, and Ions

33. Dalton's original assumption was reasonable for his time, but as mass determination techniques improved, it was discovered that a given element may be composed of several isotopes. Isotopes have the same number of protons and electrons, and so are chemical identical, but differ in the number of neutrons, which causes some physical differences.

34. Atoms of the same element (i.e., atoms with the same number of protons in the nucleus) may have different numbers of neutrons, and so will have different masses.

35.

Z	Symbol	Name
8	O	oxygen
29	Cu	copper
78	Pt	platinum
15	P	phosphorus
17	Cl	chlorine
50	Sn	tin
30	Zn	zinc

36.

Z	Symbol	Name
32	Ge	germanium
30	Zn	zinc
24	Cr	chromium
74	W	tungsten
38	Sr	strontium
27	Co	cobalt
4	Be	beryllium

37.
a. $^{13}_{6}C$

b. $^{12}_{6}C$

c. $^{14}_{6}C$

d. $^{11}_{5}B$

e. $^{10}_{5}B$

f. $^{10}_{5}B$

38.
a. $^{54}_{26}Fe$

b. $^{56}_{26}Fe$

c. $^{57}_{26}Fe$

d. $^{14}_{7}N$

e. $^{15}_{7}N$

f. $^{15}_{7}N$

39.
a. 56 protons, 74 neutrons, 56 electrons

b. 56 protons, 80 neutrons, 56 electrons

c. 22 protons, 24 neutrons, 22 electrons

d. 22 protons, 26 neutrons, 22 electrons

e. 3 protons, 3 neutrons, 3 electrons

f. 3 protons, 4 neutrons, 3 electrons

40. The relative amounts of ^{2}H and ^{18}O in a person's hair, compared to other isotopes of these elements, vary significantly from region to region in the United States and is related to the isotopic abundances in the drinking water in a region.

41. Isotopes are atoms that contain the same number of protons but a different number of neutrons. The text gives an example in which ivory from African elephants was identified as coming from a specific region based on the ratio of ^{13}C to ^{12}C in the ivory.

42.

Name	Symbol	Atomic Number	Mass Number	Number of neutrons
oxygen	$^{17}_{8}$O	8	17	9
oxygen	$^{17}_{8}$O	8	17	9
neon	$^{20}_{10}$Ne	10	20	10
iron	$^{56}_{26}$Fe	26	56	30
plutonium	$^{244}_{94}$Pu	94	244	150
mercury	$^{202}_{80}$Hg	80	202	122
cobalt	$^{59}_{27}$Co	27	59	32
nickel	$^{56}_{28}$Ni	28	56	28
fluorine	$^{19}_{9}$F	9	19	10
chromium	$^{50}_{24}$Cr	24	50	26

43. False. The elements are listed in the periodic table in order of increasing atomic number (number of protons in the nucleus; nuclear charge), so that elements with similar properties form vertical groups.

44. Elements with similar chemical properties are aligned *vertically* in families known as *groups*.

45. Metals are excellent conductors of heat and electricity, and are malleable, ductile, and generally shiny (lustrous) when a fresh surface is exposed.

46. True

47. Mercury is a liquid at room temperature.

48. The gaseous nonmetallic elements are hydrogen, nitrogen, oxygen, fluorine, chlorine, plus all the group 8 elements (noble gases). There are no gaseous metallic elements under room conditions.

49. The only metal that ordinarily occurs as a liquid is mercury. The only nonmetallic element that occurs as a liquid at room temperature is bromine (elements such as oxygen and nitrogen are frequently obtainable as liquids, but these result from compression of the gases into cylinders at very low temperatures).

50. metalloids or semimetals

51. a. Group 1; alkali metals

    b. Group 2; alkaline earth elements

Chapter 4: Chemical Foundations: Elements, Atoms, and Ions

- c. Group 8; noble gases
- d. Group 7; halogens
- e. Group 2; alkaline earth elements
- f. Group 8; noble gases
- g. Group 1; alkali metals

52.
- a. fluorine, chlorine, bromine, iodine, astatine
- b. lithium, sodium, potassium, rubidium, cesium, francium
- c. beryllium, magnesium, calcium, strontium, barium, radium
- d. helium, neon, argon, krypton, xenon, radon

53.
- a. Sr; $Z = 38$; Group 2; metal
- b. I; $Z = 53$; Group 7; nonmetal
- c. Si; $Z = 14$; Group 4; metalloid
- d. Cs; $Z = 55$; Group 1; metal
- e. S; $Z = 16$; Group 6; nonmetal

54. Arsenic, atomic number 33, is located on the dividing line between the metallic elements and the non-metallic elements, and is therefore classified as a metalloid. Arsenic is in Group 5 of the periodic table, whose other principal members are N, P, Sb, and Bi.

55. compounds (and mixtures of compounds)

56. Most of the elements are too reactive to be found in the uncombined form in nature and are found only in compounds.

57. argon

58. These elements are found *uncombined* in nature and do not readily react with other elements. For many years it was thought that these elements formed no compounds at all, although this has now been shown to be untrue.

59. diatomic

60. diatomic gases: H_2, N_2, O_2, Cl_2, and F_2

   monatomic gases: He, Ne, Kr, Xe, Rn, and Ar

61. electricity

62. chlorine

63. liquids: bromine, mercury, gallium

   gases: hydrogen, nitrogen, oxygen, fluorine, chlorine, and the noble gases (helium, neon, argon, krypton, xenon, radon)

64. carbon

65. zero

66. electrons

67. loses three

Chapter 4: Chemical Foundations: Elements, Atoms, and Ions

68. 2–
69. cations, anions
70. *-ide*
71. The answer will depend on the student's selection of elements; in general, the metallic elements are the ones that form positively charged ions.
72. False. N^{3-} contains 7 protons and 10 electrons. P^{3-} contains 15 protons and 18 electrons.
73.
    a. 54
    b. 18
    c. 23
    d. 10
    e. 54
    f. 80
74. number of protons = 26; number of electrons = 23; number of neutrons = 30
75.
    a. Ca: 20 protons, 20 electrons Ca^{2+}: 20 protons, 18 electrons
    b. P: 15 protons, 15 electrons P^{3-}: 15 protons, 18 electrons
    c. Br: 35 protons, 35 electrons Br^-: 35 protons, 36 electrons
    d. Fe: 26 protons, 26 electrons Fe^{3+}: 26 protons, 23 electrons
    e. Al: 13 protons, 13 electrons Al^{3+}: 13 protons, 10 electrons
    f. N: 7 protons, 7 electrons N^{3-}: 7 protons, 10 electrons
76.
    a. two electrons gained
    b. three electrons gained
    c. three electrons lost
    d. two electrons lost
    e. one electron lost
    f. two electrons lost.
77.
    a. I^-
    b. Sr^{2+}
    c. Cs^+
    d. Ra^{2+}
    e. F^-
    f. Al^{3+}
78.
    a. P^{3-}
    b. Ra^{2+}
    c. At^-
    d. no ion
    e. Cs^+
    f. Se^{2-}

Chapter 4: Chemical Foundations: Elements, Atoms, and Ions

79. A compound that has a high melting point (many hundreds of degrees) and which conducts an electrical current when melted or dissolved in water almost certainly consists of ions. Nonionic compounds have lower melting points than ionic compounds and do not conduct electricity when melted or in solution.

80. Sodium chloride is an ionic compound, consisting of Na^+ and Cl^- ions. When NaCl is dissolved in water, these ions are set free and can move independently to conduct the electric current.

81. In the solid state, although ions are present, they are rigidly held in fixed positions in the crystal of the substance. In order for ionic substances to be able to pass an electrical current, the ions must be able to *move*, which is possible when the solid is converted to the liquid state.

82. The total number of positive charges must equal the total number of negative charges so that there will be *no net charge* on the crystals of an ionic compound. A macroscopic sample of compound must ordinarily not have any net charge.

83.
   a. KCl, K_2S, K_3N
   b. $MgCl_2$, MgS, Mg_3N_2
   c. $AlCl_3$, Al_2S_3, AlN
   d. $CaCl_2$, CaS, Ca_3N_2
   e. $LiCl$, Li_2S, Li_3N

84.
   a. CsI, BaI_2, AlI_3
   b. Cs_2O, BaO, Al_2O_3
   c. Cs_3P, Ba_3P_2, AlP
   d. Cs_2Se, $BaSe$, Al_2Se_3
   e. CsH, BaH_2, AlH_3

85.
   a. At; $Z = 85$
   b. Xe; $Z = 54$
   c. Ra; $Z = 88$
   d. Sr; $Z = 38$
   e. Pb; $Z = 82$
   f. Se; $Z = 34$
   g. Ar; $Z = 18$
   h. Cs; $Z = 55$

86.
   a. 7; halogens
   b. 8; noble gases
   c. 2; alkaline earth elements
   d. 2; alkaline earth elements
   e. 4
   f. 6; (the members of group 6 are sometimes called the chalcogens)
   g. 8; noble gases
   h. 1; alkali metals

87.

	Element	Symbol	Atomic Number
Group 1	hydrogen	H	1
	lithium	Li	3
	sodium	Na	11
	potassium	K	19
Group 2	beryllium	Be	4
	magnesium	Mg	12
	calcium	Ca	20
	strontium	Sr	38
Group 6	oxygen	O	8
	sulfur	S	16
	selenium	Se	34
	tellurium	Te	52
Group 7	fluorine	F	9
	chlorine	Cl	17
	bromine	Br	35
	iodine	I	53

88. (d); Mass number is the sum of protons and neutrons. The number of protons identifies the element.

89. The atomic number represents the number of protons in the nucleus of an atom. The mass number represents the total number of protons and neutrons. No two different elements have the same atomic number. If the *total* number of protons and neutrons happens to be the same for two atoms, then the atoms will have the same mass number.

90. Most of the mass of an atom is concentrated in the nucleus: the *protons* and *neutrons* that constitute the nucleus have similar masses, and these particles are nearly two thousand times heavier than electrons. The chemical properties of an atom depend on the number and location of the *electrons* it possesses. Electrons are found in the outer regions of the atom and are the particles most likely to be involved in interactions between atoms.

91. Yes. For example, carbon and oxygen form carbon monoxide (CO) and carbon dioxide (CO_2). The existence of more than one compound between the same elements does not in any way contradict Dalton's theory. For example, the relative mass of carbon in different samples of CO is always the same, and the relative mass of carbon in different samples of CO_2 is also always the same. Dalton did not say, however, that two different compounds would have to have the same relative masses of the elements present. In fact, Dalton said that two different compounds of the same elements would have to have different relative masses of the elements.

92. $C_6H_{12}O_6$

93. FeO and Fe_2O_3

94. a. 29 protons; 34 neutrons; 29 electrons

   b. 35 protons; 45 neutrons; 35 electrons

   c. 12 protons; 12 neutrons; 12 electrons

Chapter 4: Chemical Foundations: Elements, Atoms, and Ions

95.

Mass Number	Symbol	Number of Neutrons
24	$^{24}_{13}Al$	11
25	$^{25}_{13}Al$	12
26	$^{26}_{13}Al$	13
28	$^{28}_{13}Al$	15
29	$^{29}_{13}Al$	16
30	$^{30}_{13}Al$	17

They are all considered aluminum atoms because the identity of the element is defined by the atomic number, which is the same for all of the isotopes listed.

96. The chief use of gold in ancient times was as *ornamentation*, whether in statuary or in jewelry. Gold possesses an especially beautiful luster, and because it is relatively soft and malleable, it could be worked finely by artisans. Among the metals, gold is particularly inert to attack by most substances in the environment.

97. Boyle defined a substance as an element if it could not be broken down into simpler substances by chemical means.

98. $^{81}_{35}Br^-$; Since the ion has a 1– charge (one extra electron versus protons), the number of protons is 35 (–36 e^- + 35 p^+ = –1 charge). This identifies the element as bromine. Mass number = 35 p^+ + 46 n = 81

99. a. Ba
    b. K
    c. Cs
    d. Pb
    e. Pt
    f. Au

100. (a), (b), (c), and (d); Cesium has 55 protons. Mass number = 133 = 55 p^+ + #n. #n = 78; Column 1A contains the alkali metals. –54 e^- + 55 p^+ = +1 charge

101. a. Ag
    b. Al
    c. Cd
    d. Sb
    e. Sn
    f. As

102. The metal ion is Cu^{2+}. Since the metal ion has 27 electrons and contains a 2+ charge, this means that it has two less electrons as compared to protons. Therefore the number of protons is 29. The number of protons is also the atomic number, identifying the metal ion as copper. Mass number = 29 p^+ + 34 n = 63

Chapter 4: Chemical Foundations: Elements, Atoms, and Ions

103.
   a. tellurium
   b. palladium
   c. zinc
   d. silicon
   e. cesium
   f. bismuth
   g. fluorine
   h. titanium

104.
   a. CO_2
   b. $AlCl_3$
   c. $HClO_4$
   d. SCl_6

105.
   a. nitrogen, N
   b. neon, Ne
   c. sodium, Na
   d. nickel, Ni
   e. titanium, Ti
   f. argon, Ar
   g. krypton, Kr
   h. xenon, Xe

106.
   a. $^{13}_{6}C$
   b. $^{13}_{6}C$
   c. $^{13}_{6}C$
   d. $^{44}_{19}K$
   e. $^{41}_{20}Ca$
   f. $^{35}_{19}K$

107.
   a. 22 protons, 19 neutrons, 22 electrons
   b. 30 protons, 34 neutrons, 30 electrons
   c. 32 protons, 44 neutrons, 32 electrons
   d. 36 protons, 50 neutrons, 36 electrons
   e. 33 protons, 42 neutrons, 33 electrons
   f. 19 protons, 22 neutrons, 19 electrons

Chapter 4:   Chemical Foundations: Elements, Atoms, and Ions

108.
Symbol	Protons	Neutrons	Mass Number
$^{41}_{20}$Ca	20	21	41
$^{55}_{25}$Mn	25	30	55
$^{109}_{47}$Ag	47	62	109
$^{45}_{21}$Sc	21	24	45

109.
a.   C; Z = 6; nonmetal

b.   Se; Z = 34; nonmetal

c.   Rn; Z = 86; nonmetal; noble gases

d.   Be; Z = 4; metal; alkaline earth elements

110.   **Cu-63**: 29 protons, 29 electrons, 34 neutrons, $^{63}_{29}$Cu

Cu-65: 29 protons, 29 electrons, 36 neutrons, $^{65}_{29}$Cu

111.   **Au**: gold
Kr: krypton
He: helium
C: carbon
Li: lithium
Si: silicon

112.   **tin**: Sn
beryllium: Be
hydrogen: H
chlorine: Cl
radium: Ra
xenon: Xe
zinc: Zn
oxygen: O

113.
# Protons	# Neutrons	Symbol
34	45	Se
19	20	K
53	74	I
4	5	Be
24	32	Cr

114.
Atom	G or L	Ion
O	G	O^{2-}
Mg	L	Mg^{2+}
Rb	L	Rb^+
Br	G	Br^-
Cl	G	Cl^-

Chapter 4: Chemical Foundations: Elements, Atoms, and Ions

115.

Atoms	# Protons	# Neutrons
$^{55}_{25}Mn$	25	30
$^{18}_{8}O$	8	10
$^{59}_{28}Ni$	28	31
$^{238}_{92}U$	92	146
$^{201}_{80}Hg$	80	121

116.

Atom/Ion	Protons	Neutrons	Electrons
$^{120}_{50}Sn$	50	70	50
$^{25}_{12}Mg^{2+}$	12	13	10
$^{56}_{26}Fe^{2+}$	26	30	24
$^{79}_{34}Se$	34	45	34
$^{35}_{17}Cl$	17	18	17
$^{63}_{29}Cu$	29	34	29

117. (a); Rutherford is the founder of the nuclear atom. A proton is heavier than an electron. The nucleus contains protons and neutrons.

CHAPTER 5

Nomenclature

1. a. It was very sweet-tasting.
   b. lead(II) acetate, plumbous acetate
   c. There are millions of known chemical compounds and a system of nomenclature is essential for communication among scientists.

2. A binary compound contains only two elements: the major types of binary compounds are *ionic* (compounds that contain a metal and a nonmetal) and *nonionic* (compounds containing two nonmetals).

3. positive; negative

4. cation

5. cation

6. Sodium chloride consists of Na^+ ions and Cl^- ions in an extended crystal lattice array. No discrete NaCl pairs are present.

7. –ous, –ic

8. Roman numeral

9. a. sodium bromide
   b. magnesium chloride
   c. aluminum phosphide
   d. strontium bromide
   e. silver iodide [or less frequently as silver(I) iodide]
   f. potassium sulfide

10. a. lithium iodide
    b. magnesium fluoride
    c. strontium oxide
    d. aluminum bromide
    e. calcium sulfide
    f. sodium oxide

11. a. correct
    b. incorrect; lead(II) chloride
    c. correct
    d. incorrect; sodium sulfide
    e. correct

12. a. correct
    b. incorrect; silver oxide or silver(I) oxide is acceptable
    c. incorrect; lithium oxide
    d. correct
    e. incorrect; cesium sulfide

13. a. As the chloride ion has a 1– charge, the tin ion must have a 4+ charge: the name is tin(IV) chloride.
    b. As the sulfide ion has a 2– charge, the iron ion must have a 3+ charge: the name is iron(III) sulfide.
    c. As the oxide ion has a 2– charge, the lead ion must have a 4+ charge: the name is lead(IV) oxide.
    d. As the sulfide ion has a 2– charge, each chromium ion must have a 3+ charge: the name is chromium(III) sulfide.
    e. As the oxide ion has a 2– charge, the copper ion must have a 2+ charge: the name is copper(II) oxide.
    f. As the oxide ion has a 2– charge, each copper ion must have a 1+ charge: the name is copper(I) oxide.

14. a. As the iodide ion has a 1– charge, the iron ion must have a 3+ charge: the name is iron(III) iodide.
    b. As the chloride ion has a 1– charge, the manganese ion must have a 2+ charge: the name is manganese(II) chloride.
    c. As the oxide ion has a 2– charge, the mercury ion must have a 2+ charge: the name is mercury(II) oxide.
    d. As the sulfide ion has a 2– charge, each copper ion must have a 1+ charge: the name is copper(I) sulfide.
    e. As the oxide ion has a 2– charge, the cobalt ion must have a 2+ charge: the name is cobalt(II) oxide.
    f. As the bromide ion has a 1– charge, the tin ion must have a 4+ charge: the name is tin(IV) bromide.

Chapter 5: **Nomenclature**

15. a. As each chloride ion has a 1– charge, the copper ion must have a 1+ charge: cupr*ous* chloride.

    b. As each oxide ion has a 2– charge, the iron ion must have a 3+ charge: the name is ferr*ic* oxide.

    c. As each chloride ion has a 1– charge, each mercury atom must have a 1+ charge: the name is mercur*ous* chloride.

    d. As each chloride ion has a 1– charge, the manganese ion must have a 2+ charge: the name is mangan*ous* chloride.

    e. As each oxide ion has a 2– charge, the titanium ion must have a 4+ charge: the name is titan*ic* oxide.

    f. As the oxide ion has a 2– charge, the lead ion must have a 2+ charge: the name is plumb*ous* oxide.

16. a. As each chloride ion has a 1– charge, the cobalt ion must have a 2+ charge: the name is cobalt*ous* chloride.

    b. As each bromide ion has a 1– charge, the chromium ion must have a 3+ charge: the name is chrom*ic* bromide.

    c. As each oxide ion has a 2– charge, the lead ion must have a 2+ charge: the name is plumb*ous* oxide.

    d. As each oxide ion has a 2– charge, the tin ion must have a 4+ charge: the name is stann*ic* oxide.

    e. As the oxide ion has a 2– charge, the cobalt ion must have a 3+ charge: the name is cobalt*ic* oxide.

    f. As the chloride ion has a 1– charge, the iron ion must have a 3+ charge: the name is ferr*ic* chloride.

17. Remember that for this type of compound of nonmetals, numerical prefixes are used to indicate how many of each type of atom are present. However, if only one atom of the first element mentioned in the compound is present in a molecule, the prefix *mono*– is not needed.

    a. krypton difluoride

    b. diselenium hexasulfide

    c. arsenic trihydride

    d. xenon tetr(*a*)oxide (the *a* is usually omitted for easier pronunciation)

    e. bromine trifluoride

    f. diphosphorus pentasulfide

18. Remember that for this type of compound of nonmetals, numerical prefixes are used to indicate how many of each type of atom are present. However, if only one atom of the first element mentioned in the compound is present in a molecule, the prefix *mono*– is not needed.

    a. germanium tetrahydride

    b. dinitrogen tetrabromide

     c.   diphosphorus pentoxide

     d.   carbon dioxide

     e.   ammonia

     f.   silicon dioxide

19.   a.   iron(II) phosphide, ferrous phosphide – ionic

     b.   calcium bromide – ionic

     c.   dinitrogen pentoxide – nonionic

     d.   lead(IV) chloride, plumbic chloride – ionic

     e.   disulfur decafluoride – nonionic

     f.   copper(I) oxide, cuprous oxide – ionic

20. Na_2O: sodium oxide; N_2O: dinitrogen monoxide; For Na_2O, the compound contains a metal and a nonmetal in which the charges must balance. When forming this compound, Na always forms a 1+ charge and oxygen always forms a 2– charge. Therefore, the prefixes are not needed. For N_2O, the compound contains only nonmetals and the charges do not have to balance. Therefore prefixes are needed to tell us how many of each atom are present.

21.   a.   magnesium sulfide – ionic

     b.   aluminum chloride – ionic

     c.   phosphorus trihydride (the common name *phosphine* is always used)

     d.   chlorine monobromide – nonionic

     e.   lithium oxide – ionic

     f.   tetraphosphorus decoxide – nonionic

22.   a.   radium chloride – ionic

     b.   selenium dichloride – nonionic

     c.   phosphorus trichloride – nonionic

     d.   sodium phosphide – ionic

     e.   manganese(II) fluoride (or manganous fluoride) – ionic

     f.   zinc oxide – ionic

23.   A polyatomic ion is a group of atoms bound together that, as a unit, carries an electrical charge. Examples will depend on student responses.

24.   An oxyanion is a polyatomic ion containing a given element and one or more oxygen atoms. The oxyanions of chlorine and bromine are given below:

Oxyanion	Name	Oxyanion	Name
ClO^-	hypochlorite	BrO^-	hypobromite
ClO_2^-	chlorite	BrO_2^-	bromite
ClO_3^-	chlorate	BrO_3^-	bromate
ClO_4^-	perchlorate	BrO_4^-	perbromate

Chapter 5: Nomenclature

25. one fewer oxygen atom

26. For a series of oxyanions, the prefix *hypo–* is used for the anion with the fewest oxygen atoms, and the prefix *per–* is used for the anion with the most oxygen atoms.

27. ClO_4^- perchlorate
    ClO^- hypochlorite
    ClO_3^- chlorate
    ClO_2^- chlorite

28. IO^- hypoiodite
    IO_2^- iodite
    IO_3^- iodate
    IO_4^- periodate

29. a. P^{3-}
    b. PO_4^{3-}
    c. PO_3^{3-}
    d. HPO_4^{2-}

30. a. Cl^-
    b. ClO^-
    c. ClO_3^-
    d. ClO_4^-

31. a. $MgCl_2$
    b. $Ca(ClO)_2$
    c. $KClO_3$
    d. $Ba(ClO_4)_2$

32. CN^- cyanide
    CO_3^{2-} carbonate
    HCO_3^- hydrogen carbonate (or bicarbonate)
    $C_2H_3O_2^-$ acetate

33. a. hydrogen carbonate, bicarbonate
    b. acetate
    c. cyanide
    d. hydroxide
    e. nitrite

Chapter 5:   Nomenclature

        f.   hydrogen phosphate

34.   a.   ammonium
       b.   dihydrogen phosphate
       c.   sulfate
       d.   hydrogen sulfite (also called *bi*sulfite)
       e.   perchlorate
       f.   iodate

35.   a.   ammonium nitrate
       b.   calcium hydrogen carbonate, calcium bicarbonate
       c.   magnesium sulfate
       d.   sodium hydrogen phosphate
       e.   potassium perchlorate
       f.   barium acetate

36.   a.   sodium permanganate
       b.   aluminum phosphate
       c.   chromium(II) carbonate, chromous carbonate
       d.   calcium hypochlorite
       e.   barium carbonate
       f.   calcium chromate

37.   An acid is a substance that produces hydrogen ions, H^+, when dissolved in water.

38.   oxygen (commonly referred to as *oxy*acids)

39.   a.   hydrochloric acid
       b.   sulfuric acid
       c.   nitric acid
       d.   hydroiodic acid
       e.   nitrous acid
       f.   chloric acid
       g.   hydrobromic acid
       h.   hydrofluoric acid
       i.   acetic acid

40.   a.   hypochlorous acid
       b.   sulfurous acid

Chapter 5: Nomenclature

    c.   bromic acid
    d.   hypoiodous acid
    e.   perbromic acid
    f.   hydrosulfuric acid
    g.   hydroselenic acid
    h.   phosphorous acid

41.
    a. $CoCl_2$
    b. $CoCl_3$
    c. Na_3P
    d. FeO
    e. CaH_2
    f. MnO_2
    g. MgI_2
    h. Cu_2S

42.
    a. MgF_2
    b. FeI_3
    c. HgS
    d. Ba_3N_2
    e. $PbCl_2$
    f. SnF_4
    g. Ag_2O
    h. K_2Se

43.
    a. CS_2
    b. H_2O
    c. N_2O_3
    d. Cl_2O_7
    e. CO_2
    f. NH_3
    g. XeF_4

44.
    a. N_2O
    b. NO_2
    c. N_2O_4
    d. SF_6

Chapter 5: Nomenclature

       e. PBr_3
       f. CI_4
       g. OCl_2

45.   a. NH_4NO_3
       b. $Mg(C_2H_3O_2)_2$
       c. CaO_2
       d. $KHSO_4$
       e. $FeSO_4$
       f. $KHCO_3$
       g. $CoSO_4$
       h. $LiClO_4$

46.   a. $NH_4C_2H_3O_2$
       b. $Fe(OH)_2$
       c. $Co_2(CO_3)_3$
       d. $BaCr_2O_7$
       e. $PbSO_4$
       f. KH_2PO_4
       g. Li_2O_2
       h. $Zn(ClO_3)_2$

47.   a. H_2S
       b. $HBrO_4$
       c. $HC_2H_3O_2$
       d. HBr
       e. $HClO_2$
       f. H_2Se
       g. H_2SO_3
       h. $HClO_4$

48.   a. HCN
       b. HNO_3
       c. H_2SO_4
       d. H_3PO_4
       e. $HClO$ or $HOCl$
       f. HBr

Chapter 5: Nomenclature

      g. $HBrO_2$

      h.   HF

49.   a. Na_2O_2

      b. $Ca(ClO_3)_2$

      c.   RbOH

      d. $Zn(NO_3)_2$

      e. $(NH_4)_2Cr_2O_7$

      f. $H_2S(aq)$

      g. $CaBr_2$

      h. $HOCl(aq)$

      i. K_2SO_4

      j. $HNO_3(aq)$

      k. $Ba(C_2H_3O_2)_2$

      l. Li_2SO_3

50.   a. $Ca(HSO_4)_2$

      b. $Zn_3(PO_4)_2$

      c. $Fe(ClO_4)_3$

      d. $Co(OH)_3$

      e. K_2CrO_4

      f. $Al(H_2PO_4)_3$

      g. $LiHCO_3$

      h. $Mn(C_2H_3O_2)_2$

      i. $MgHPO_4$

      j. $CsClO_2$

      k. BaO_2

      l. $NiCO_3$

51.

Formula	Roman Numeral Name	–ous/–ic Name
FeO	iron(II) oxide	ferrous oxide
Fe_2O_3	iron(III) oxide	ferric oxide
FeS	iron(II) sulfide	ferrous sulfide
Fe_2S_3	iron(III) sulfide	ferric sulfide
$FeCl_2$	iron(II) chloride	ferrous chloride
$FeCl_3$	iron(III) chloride	ferric chloride

52.   A moist paste of NaCl would contain Na^+ and Cl^- ions in solution, and would serve as a *conductor* of electrical impulses.

53. NO, nitrogen monoxide; NO$_2$, nitrogen dioxide; N$_2$O$_4$, dinitrogen tetr(a)oxide; N$_2$O$_5$, dinitrogen pent(a)oxide; N$_2$O, dinitrogen monoxide

54. H → H$^+$ (hydrogen ion: a cation) + e$^-$

    H + e$^-$ → H$^-$ (hydr*ide* ion: an anion)

55. (b); The correct name for (a) is iron(II) phosphate. The correct name for (c) is manganese(IV) oxide. The correct name for (d) is silicon dioxide.

56. missing oxyanions: IO$_3^-$; ClO$_2^-$

    missing oxyacids: HClO$_4$; HClO; HBrO$_2$

57.
    a. calcium acetate
    b. phosphorus trichloride
    c. copper(II) permanganate, cupric permanganate
    d. iron(III) carbonate, ferric carbonate
    e. lithium hydrogen carbonate, lithium bicarbonate
    f. chromium(III) sulfide, chromic sulfide
    g. calcium cyanide

58.
    a. gold(III) bromide, auric bromide
    b. cobalt(III) cyanide, cobaltic cyanide
    c. magnesium hydrogen phosphate
    d. diboron hexahydride (diborane is its common name)
    e. ammonia
    f. silver(I) sulfate (usually called silver sulfate)
    g. beryllium hydroxide

59.
    a. chloric acid
    b. cobalt(III) chloride; cobaltic chloride
    c. diboron trioxide
    d. water
    e. acetic acid
    f. iron(III) nitrate; ferric nitrate
    g. copper(II) sulfate; cupric sulfate

60. (b); iron(II) oxide has the formula FeO

61.
    a. K$_2$O
    b. MgO
    c. FeO

Chapter 5:   Nomenclature

      d. Fe_2O_3

      e. ZnO

      f. PbO

      g. Al_2O_3

62.   a. $M(C_2H_3O_2)_2$

      b. $M(MnO_4)_2$

      c. MO

      d. $MHPO_4$

      e. $M(OH)_2$

      f. $M(NO_2)_2$

63.   Answers are given, respectively, for the M^{1+}, M^{2+}, and M^{3+} ions:

      a. M_2CrO_4, $MCrO_4$, $M_2(CrO_4)_3$

      b. $M_2Cr_2O_7$, MCr_2O_7, $M_2(Cr_2O_7)_3$

      c. M_2S, MS, M_2S_3

      d. MBr, MBr_2, MBr_3

      e. $MHCO_3$, $M(HCO_3)_2$, $M(HCO_3)_3$

      f. M_2HPO_4, $MHPO_4$, $M_2(HPO_4)_3$

64.   a.   The metal ion is Mn^{2+}. Since the metal ion has 23 electrons and contains a 2+ charge, this means that it has two less electrons as compared to protons. Therefore the number of protons is 25. The number of protons is also the atomic number, identifying the metal ion as manganese.

      b.   The halogen ion is Cl^- with 18 electrons. The number of protons is 17, identifying the element as chlorine. Halogens form a 1– charge when bonding with a metal to form an ionic compound, thus the ion has one more electron as compared to protons.

      c.   Since the chloride ion has a 1– charge and the manganese ion has a 2+ charge, the formula is $MnCl_2$ and is named manganese(II) chloride (or manganous chloride). The charge on manganese must be specified using the Roman numeral.

65. Fe^{2+}:

$FeCO_3$	iron(II) carbonate; ferrous carbonate
$Fe(BrO_3)_2$	iron(II) bromate; ferrous bromate
$Fe(C_2H_3O_2)_2$	iron(II) acetate; ferrous acetate
$Fe(OH)_2$	iron(II) hydroxide; ferrous hydroxide
$Fe(HCO_3)_2$	iron(II) bicarbonate; ferrous bicarbonate
$Fe_3(PO_4)_2$	iron(II) phosphate; ferrous phosphate
$FeSO_3$	iron(II) sulfite; ferrous sulfite
$Fe(ClO_4)_2$	iron(II) perchlorate; ferrous perchlorate

$FeSO_4$	iron(II) sulfate; ferrous sulfate
FeO	iron(II) oxide; ferrous oxide
$FeCl_2$	iron(II) chloride; ferrous chloride

Al^{3+}:

$Al_2(CO_3)_3$	aluminum carbonate
$Al(BrO_3)_3$	aluminum bromate
$Al(C_2H_3O_2)_3$	aluminum acetate
$Al(OH)_3$	aluminum hydroxide
$Al(HCO_3)_3$	aluminum bicarbonate
$AlPO_4$	aluminum phosphate
$Al_2(SO_3)_3$	aluminum sulfite
$Al(ClO_4)_3$	aluminum perchlorate
$Al_2(SO_4)_3$	aluminum sulfate
Al_2O_3	aluminum oxide
$AlCl_3$	aluminum chloride

Na^+:

Na_2CO_3	sodium carbonate
$NaBrO_3$	sodium bromate
$NaC_2H_3O_2$	sodium acetate
$NaOH$	sodium hydroxide
$NaHCO_3$	sodium bicarbonate
Na_3PO_4	sodium phosphate
Na_2SO_3	sodium sulfite
$NaClO_4$	sodium perchlorate
Na_2SO_4	sodium sulfate
Na_2O	sodium oxide
$NaCl$	sodium chloride

Ca^{2+}:

$CaCO_3$	calcium carbonate
$Ca(BrO_3)_2$	calcium bromate
$Ca(C_2H_3O_2)_2$	calcium acetate
$Ca(OH)_2$	calcium hydroxide
$Ca(HCO_3)_2$	calcium bicarbonate
$Ca_3(PO_4)_2$	calcium phosphate
$CaSO_3$	calcium sulfite

Chapter 5: Nomenclature

$Ca(ClO_4)_2$	calcium perchlorate
$CaSO_4$	calcium sulfate
CaO	calcium oxide
$CaCl_2$	calcium chloride

NH_4^+:

$(NH_4)_2CO_3$	ammonium carbonate
NH_4BrO_3	ammonium bromate
$NH_4C_2H_3O_2$	ammonium acetate
NH_4OH	ammonium hydroxide
NH_4HCO_3	ammonium bicarbonate
$(NH_4)_3PO_4$	ammonium phosphate
$(NH_4)_2SO_3$	ammonium sulfite
NH_4ClO_4	ammonium perchlorate
$(NH_4)_2SO_4$	ammonium sulfate
$(NH_4)_2O$	ammonium oxide
NH_4Cl	ammonium chloride

Fe^{3+}:

$Fe_2(CO_3)_3$	iron(III) carbonate; ferric carbonate
$Fe(BrO_3)_3$	iron(III) bromate; ferric bromate
$Fe(C_2H_3O_2)_3$	iron(III) acetate; ferric acetate
$Fe(OH)_3$	iron(III) hydroxide; ferric hydroxide
$Fe(HCO_3)_3$	iron(III) bicarbonate; ferric bicarbonate
$FePO_4$	iron(III) phosphate; ferric phosphate
$Fe_2(SO_3)_3$	iron(III) sulfite; ferric sulfite
$Fe(ClO_4)_3$	iron(III) perchlorate; ferric perchlorate
$Fe_2(SO_4)_3$	iron(III) sulfate; ferric sulfate
Fe_2O_3	iron(III) oxide; ferric oxide
$FeCl_3$	iron(III) chloride; ferric chloride

Ni^{2+}:

$NiCO_3$	nickel(II) carbonate
$Ni(BrO_3)_2$	nickel(II) bromate
$Ni(C_2H_3O_2)_2$	nickel(II) acetate
$Ni(OH)_2$	nickel(II) hydroxide
$Ni(HCO_3)_2$	nickel(II) bicarbonate
$Ni_3(PO_4)_2$	nickel(II) phosphate

NiSO₃	nickel(II) sulfite
Ni(ClO₄)₂	nickel(II) perchlorate
NiSO₄	nickel(II) sulfate
NiO	nickel(II) oxide
NiCl₂	nickel(II) chloride

Hg_2^{2+}:

Hg₂CO₃	mercury(I) carbonate; mercurous carbonate
Hg₂(BrO₃)₂	mercury(I) bromate; mercurous bromate
Hg₂(C₂H₃O₂)₂	mercury(I) acetate; mercurous acetate
Hg₂(OH)₂	mercury(I) hydroxide; mercurous hydroxide
Hg₂(HCO₃)₂	mercury(I) bicarbonate; mercurous bicarbonate
(Hg₂)₃(PO₄)₂	mercury(I) phosphate; mercurous phosphate
Hg₂SO₃	mercury(I) sulfite; mercurous sulfite
Hg₂(ClO₄)₂	mercury(I) perchlorate; mercurous perchlorate
Hg₂SO₄	mercury(I) sulfate; mercurous sulfate
Hg₂O	mercury(I) oxide; mercurous oxide
Hg₂Cl₂	mercury(I) chloride; mercurous chloride

Hg^{2+}:

HgCO₃	mercury(II) carbonate; mercuric carbonate
Hg(BrO₃)₂	mercury(II) bromate; mercuric bromate
Hg(C₂H₃O₂)₂	mercury(II) acetate; mercuric acetate
Hg(OH)₂	mercury(II) hydroxide; mercuric hydroxide
Hg(HCO₃)₂	mercury(II) bicarbonate; mercuric bicarbonate
Hg₃(PO₄)₂	mercury(II) phosphate; mercuric phosphate
HgSO₃	mercury(II) sulfite; mercuric sulfite
Hg(ClO₄)₂	mercury(II) perchlorate; mercuric perchlorate
HgSO₄	mercury(II) sulfate; mercuric sulfate
HgO	mercury(II) oxide; mercuric oxide
HgCl₂	mercury(II) chloride; mercuric chloride

66.

Ca(NO₃)₂	CaSO₄	Ca(HSO₄)₂	Ca(H₂PO₄)₂	CaO	CaCl₂
Sr(NO₃)₂	SrSO₄	Sr(HSO₄)₂	Sr(H₂PO₄)₂	SrO	SrCl₂
NH₄NO₃	(NH₄)₂SO₄	NH₄HSO₄	NH₄H₂PO₄	(NH₄)₂O	NH₄Cl
Al(NO₃)₃	Al₂(SO₄)₃	Al(HSO₄)₃	Al(H₂PO₄)₃	Al₂O₃	AlCl₃
Fe(NO₃)₃	Fe₂(SO₄)₃	Fe(HSO₄)₃	Fe(H₂PO₄)₃	Fe₂O₃	FeCl₃
Ni(NO₃)₂	NiSO₄	Ni(HSO₄)₂	Ni(H₂PO₄)₂	NiO	NiCl₂
AgNO₃	Ag₂SO₄	AgHSO₄	AgH₂PO₄	Ag₂O	AgCl

Chapter 5: Nomenclature

$Au(NO_3)_3$	$Au_2(SO_4)_3$	$Au(HSO_4)_3$	$Au(H_2PO_4)_3$	Au_2O_3	$AuCl_3$
KNO_3	K_2SO_4	$KHSO_4$	KH_2PO_4	K_2O	KCl
$Hg(NO_3)_2$	$HgSO_4$	$Hg(HSO_4)_2$	$Hg(H_2PO_4)_2$	HgO	$HgCl_2$
$Ba(NO_3)_2$	$BaSO_4$	$Ba(HSO_4)_2$	$Ba(H_2PO_4)_2$	BaO	$BaCl_2$

67. unreactive

68. $(NH_4)_3PO_4$

69. two

70. iodine (solid), bromine (liquid), fluorine and chlorine (gases)

71. 2−

72. 1+

73. 3+

74. 2−

75.
- [1] e
- [2] a
- [3] a
- [4] g
- [5] g
- [6] f
- [7] g
- [8] a
- [9] e
- [10] j

76.
a. $Al(13e^-) \rightarrow Al^{3+}(10e^-) + 3e^-$
b. $S(16e^-) + 2e^- \rightarrow S^{2-}(18e^-)$
c. $Cu(29e^-) \rightarrow Cu^+(28e^-) + e^-$
d. $F(9e^-) + e^- \rightarrow F^-(10e^-)$
e. $Zn(30e^-) \rightarrow Zn^{2+}(28e^-) + 2e^-$
f. $P(15e^-) + 3e^- \rightarrow P^{3-}(18e^-)$

77.
a. none likely (element 36, Kr, is a noble gas)
b. Ga^{3+} (element 31, Ga, is in Group 3)
c. Te^{2-} (element 52, Te, is in Group 6)
d. Tl^{3+} (element 81, Tl, is in Group 3)

Chapter 5:   Nomenclature

  e. Br^- (element 35, Br, is in Group 7)

  f. Fr^+ (element 87, Fr, is in Group 1)

78. a. Two 1+ ions are needed to balance a 2– ion, so the formula must have two Na^+ ions for each S^{2-} ion: Na_2S.

  b. One 1+ ion exactly balances a 1– ion, so the formula should have an equal number of K^+ and Cl^- ions: KCl.

  c. One 2+ ion exactly balances a 2– ion, so the formula must have an equal number of Ba^{2+} and O^{2-} ions: BaO.

  d. One 2+ ion exactly balances a 2– ion, so the formula must have an equal number of Mg^{2+} and Se^{2-} ions: MgSe.

  e. One 2+ ion requires two 1– ions to balance charge, so the formula must have twice as many Br^- ions as Cu^{2+} ions: $CuBr_2$.

  f. One 3+ ion requires three 1– ions to balance charge, so the formula must have three times as many I^- ions as Al^{3+} ions: AlI_3.

  g. Two 3+ ions give a total of 6+, whereas three 2– ions will give a total of 6–. The formula then should contain two Al^{3+} ions and three O^{2-} ions: Al_2O_3.

  h. Three 2+ ions are required to balance two 3– ions, so the formula must contain three Ca^{2+} ions for every two N^{3-} ions: Ca_3N_2.

79. a. beryllium oxide

  b. magnesium iodide

  c. sodium sulfide

  d. aluminum oxide

  e. hydrogen chloride (gaseous); hydrochloric acid (aqueous)

  f. lithium fluoride

  g. silver(I) sulfide; usually called silver sulfide

  h. calcium hydride

80. a. silver(I) oxide or just silver oxide

  b. correct

  c. iron(III) oxide or ferric oxide

  d. lead(IV) oxide or plumbic oxide

  e. correct

81. a. As the bromide ion must have a 1– charge, the iron ion must be in the 2+ state: the name is iron(II) bromide.

  b. As sulfide ion always has a 2– charge, the cobalt ion must be in the 2+ state: the name is cobalt(II) sulfide.

  c. As sulfide ion always has a 2– charge, and as there are three sulfide ions present, each cobalt ion must be in the 3+ state: the name is cobalt(III) sulfide.

Chapter 5: Nomenclature

    d.   As oxide ion always has a 2– charge, the tin ion must be in the 4+ state: the name is tin(IV) oxide.

    e.   As chloride ion always has a 1– charge, each mercury ion must be in the 1+ state: the name is mercury(I) chloride.

    f.   As chloride ion always has a 1– charge, the mercury ion must be in the 2+ state: the name is mercury(II) chloride.

82.
- a. stannous chloride
- b. ferrous oxide
- c. stannic oxide
- d. plumbous sulfide
- e. cobaltic sulfide
- f. chromous chloride

83.
- a. xenon hexafluoride
- b. oxygen difluoride
- c. arsenic triiodide
- d. dinitrogen tetraoxide (tetroxide)
- e. dichlorine monoxide
- f. sulfur hexafluoride

84.
- a. iron(III) acetate, ferric acetate
- b. bromine monofluoride
- c. potassium peroxide
- d. silicon tetrabromide
- e. copper(II) permanganate, cupric permanganate
- f. calcium chromate

85.   nitr*ate* (the ending *–ate* always implies the larger number of oxygen atoms)

86.   (a); The correct name for (b) is chlorine dioxide. The correct name for (c) is lead(IV) iodide or plumbic iodide. The correct name for (d) is copper(II) sulfate.

87.
- a. Cr^{2+}
- b. CrO_4^{2-}
- c. Cr^{3+}
- d. $Cr_2O_7^{2-}$

88.
- a. carbonate
- b. chlorate
- c. sulfate

Chapter 5:   Nomenclature

      d.   phosphate

      e.   perchlorate

      f.   permanganate

89.   a.   lithium dihydrogen phosphate

      b.   copper(II) cyanide or cupric cyanide

      c.   lead(II) nitrate or plumbous nitrate

      d.   sodium hydrogen phosphate

      e.   sodium chlorite

      f.   cobalt(III) sulfate

90. $SrBr_2$; Alkaline earth family metals form a 2+ charge when bonding with a with a nonmetal to form an ionic compound. Since the ion contains 36 electrons and has a 2+ charge, this means it has two more protons as compared to electrons. The number of protons is therefore 38, identifying the ion as strontium.

91.   a. SO_2

      b. N_2O

      c. XeF_4

      d. P_4O_{10}

      e. PCl_5

      f. SF_6

      g. NO_2

92.   a. NaH_2PO_4

      b. $LiClO_4$

      c. $Cu(HCO_3)_2$

      d. $KC_2H_3O_2$

      e. BaO_2

      f. Cs_2SO_3

93.   a. $AgClO_4$

      b. $Co(OH)_3$

      c. $NaClO$ or $NaOCl$

      d. $K_2Cr_2O_7$

      e. NH_4NO_2

      f. $Fe(OH)_3$

      g. NH_4HCO_3

      h. $KBrO_4$

Chapter 5: Nomenclature

94.
Atom	G or L	Ion
K	L	K^+
Cs	L	Cs^+
Br	G	Br^-
S	G	S^{2-}
Se	G	Se^{2-}

95.
Compound Name	Formula
Carbon tetrabromide	CBr_4
Cobalt(II) phosphate	$Co_3(PO_4)_2$
Magnesium chloride	$MgCl_2$
Nickel(II) acetate	$Ni(C_2H_3O_2)_2$
Calcium nitrate	$Ca(NO_3)_2$

96.
Formula	Compound Name
$Co(NO_2)_2$	cobalt(II) nitrite or cobaltous nitrite
AsF_5	arsenic pentafluoride
LiCN	lithium cyanide
K_2SO_3	potassium sulfite
Li_3N	lithium nitride
$PbCrO_4$	lead(II) chromate or plumbous chromate

97.
Formula	Compound Name
H_2SO_3	sulfurous acid
$HC_2H_3O_2$	acetic acid
$HClO_4$	perchloric acid
HOCl	hypochlorous acid
HCN	hydrocyanic acid

98. (b) and (d); The symbols for the elements magnesium, aluminum, and xenon are *Mg*, *Al*, and *Xe*, respectively. Ga is expected to *lose* electrons to form ions in ionic compounds. The correct name for TiO_2 is *titanium(IV) oxide*.

CUMULATIVE REVIEW

Chapters 4 and 5

1. An element is a substance that cannot be broken down into simpler substances by chemical or physical means. The five most abundant elements on the earth are oxygen, silicon, aluminum, iron, and calcium. The five most abundant elements in the human body are oxygen, carbon, hydrogen, nitrogen, and calcium.

2. How many elements could you name? Although you certainly don't have to memorize all the elements, you should at least be able to give the symbol or name for the most common elements (listed in Table 4.3).

3. These symbols are derived from the name of the element in another language (Ag, argentum; Au, aurum; W, wolfram). Examples depend on student choices.

4. Dalton's atomic theory as presented in this text consists of five main postulates. Although Dalton's theory was exceptional scientific thinking for its time, some of the postulates have been modified as our scientific instruments and calculation methods have become increasingly more sophisticated. The main postulates of Dalton's theory are as follows: (1) Elements are made up of tiny particles called atoms; (2) all atoms of a given element are identical; (3) although all atoms of a given element are identical, these atoms are different from the atoms of all other elements; (4) atoms of one element can combine with atoms of another element to form a compound, and such a compound will always have the same relative numbers and types of atoms for its composition; and (5) atoms are merely rearranged into new groupings during an ordinary chemical reaction, and no atom is ever destroyed and no new atom is ever created during such a reaction.

5. A compound is a distinct, pure substance that is composed of two or more elements held together by chemical bonds. In addition, a given compound always contains exactly the same relative masses of its constituent elements. This latter statement is termed the law of constant composition. The law of constant composition is a result of the fact that a given compound is made up of molecules containing a particular type and number of each constituent atom. For example, water's composition by mass (88.8% oxygen, 11.2% hydrogen) is a result of the fact that each and every water molecule contains one oxygen atom (relative mass 16.0) and two hydrogen atoms (relative mass 1.008 each). The law of constant composition is important to our study of chemistry because it means that we can always assume that any sample of a given pure substance, from whatever source, will be identical to any other sample.

6. The expression *nuclear* atom indicates that we view the atom as having a dense center of positive charge (called the nucleus) around which the electrons move through primarily empty space. Rutherford's experiment involved shooting a beam of particles at a thin sheet of metal foil. According to the then current "plum pudding" model of the atom, most of these positively-charged particles should have passed right through the foil. However, Rutherford detected that a significant number of particles effectively bounced off something and were deflected backwards to the source of particles, and that other particles were deflected from the foil at large angles. Rutherford realized that his observations could be explained if the atoms of the metal foil had a

Cumulative Review: Chapters 4 and 5

small, dense, positively-charged nucleus, with a significant amount of empty space between nuclei. The empty space between nuclei would allow most of the particles to pass through the atom. However, if a particle hit a nucleus head-on, it would be deflected backwards at the source. If a positively-charged particle passed near a positively-charged nucleus (but did not hit the nucleus head-on), then the particle would be deflected by the repulsive forces between the positive charges. Rutherford's experiment conclusively disproved the "plum pudding" model for the atom, which envisioned the atom as a uniform sphere of positive charge, with enough negatively-charged electrons scattered through the atom to balance out the positive charge.

7. a. Protons and neutrons are found in the nucleus
    b. The neutron has the largest mass.
    c. The electron has the smallest mass.
    d. The electron is negatively charged.
    e. The neutron is electrically neutral

8. Isotopes represent atoms of the same element which have different atomic masses. Isotopes are a result of the fact that atoms of a given element may have different numbers of neutrons in their nuclei. Isotopes have the same atomic number (number of protons in the nucleus) but have different mass numbers (total number of protons and neutrons in the nucleus). The different isotopes of an atom are indicated by symbolism of the form $^A_Z X$ in which Z represents the atomic number, and A the mass number, of element X. For example, $^{13}_{6}C$ represents a nuclide of carbon with atomic number 6 (6 protons in the nucleus) and mass number 13 (reflecting 6 protons plus 7 neutrons in the nucleus). The various isotopes of an element have identical chemical properties because the chemical properties of an atom are a function of the electrons in the atom (*not* the nucleus). The physical properties of the isotopes of an element (and compounds containing those isotopes) may differ because of the difference in mass of the isotopes.

9.

Symbol	Name	Atomic Number	Group Number
Ca	calcium	20	2
I	iodine	53	7
Cs	cesium	55	1
S	sulfur	16	6
As	arsenic	33	5
Sr	strontium	38	2
Si	silicon	14	4
Rn	radon	86	8
Ra	radium	88	2
Se	selenium	34	6

10. Most elements are too reactive to be found in nature in other than the combined form. Aside from the noble metals gold, silver, and platinum, the only other elements commonly found in nature in the uncombined state are some of the gaseous elements (such as O2, N2, He, Ar, etc.), and the solid nonmetals carbon and sulfur.

11. Ions are electrically charged particles formed from atoms or molecules that have gained or lost one or more electrons. Isolated atoms typically do not form ions on their own, but are induced to gain or lose electrons by some other species that, in turn, loses or gains the electrons. Positively charged ions are called *cations*, whereas negative ions are termed *anions*. A positive ion forms when an atom or molecule *loses* one or more of its electrons (negative charges). For example, sodium atoms and magnesium atoms form ions as indicated below

$$Na(atom) \rightarrow Na^+(ion) + e^-$$

$$Mg(atom) \rightarrow Mg^{2+}(ion) + 2e^-$$

The resulting ions contain the same number of protons and neutrons in their nuclei as do the atoms from which they are formed, as the only change that has taken place involves the electrons (which are not in the nucleus). These ions obviously contain fewer electrons than the atoms from which they are formed, however. A negative ion forms when an atom or molecule *gains* one or more electrons from an outside source (another atom or molecule). For example, chlorine atoms and oxygen atoms form ions as indicated below:

$$Cl(atom) + e^- \rightarrow Cl^-(ion)$$

$$O(atom) + 2e^- \rightarrow O^{2-}(ion)$$

Because the periodic table is arranged in terms of the electronic structure of the elements, in particular with the elements in the same vertical column having *similar* electronic structures, the mere *location* of an element in the periodic table can be an indication of what simple ions the element forms. For example, the Group 1 elements all form 1+ ions (Li^+, Na^+, K^+, Rb^+, Cs^+), whereas the Group 7 elements all form 1– ions (F^-, Cl^-, Br^-, I^-). You will learn more about how the charge of an ion is related to an atom's electronic structure in a later chapter. For now, concentrate in learning the material shown in Figure 4.19.

12. Ionic compounds typically are hard, crystalline solids with high melting and boiling points. Ionic substances like sodium chloride, when dissolved in water or when melted, conduct electrical currents: chemists have taken this evidence to mean that ionic substances consist of positively– and negatively–charged particles (ions). Although an ionic substance is made up of positively– and negatively–charged particles, there is no net electrical charge on a sample of such a substance because the total number of positive charges is balanced by an equal number of negative charges. An ionic compound could not possibly exist of just cations or just anions: there must be a balance of charge or the compound would be very unstable (like charges repel each other).

13. The principle used when writing the formula of an ionic compound is sometimes called the "principle of electroneutrality". This is just a long word that means that a chemical compound must have an overall net electrical charge of *zero*. For ionic compounds, this means that the total number of positive charges on the positive ions present must *equal* the total number of negative charges on the negative ions present. For example, with sodium chloride, if we realize that an individual sodium ion has a 1+ charge, and that an individual chloride ion has a 1– charge, then if we combine one of each of these ions, the compound will have an overall net charge of zero: (1+) + (1–) = 0. On the other hand for magnesium iodide, when we realize that an individual magnesium ion has a 2+ charge, then clearly one iodide ion with its 1– charge will not lead to a compound with an overall charge of zero: we would need *two* iodide ions, each with its 1– charge, to balance the 2+ charge of the magnesium ion: (2+) + 2(1–) = 0. If we consider magnesium oxide, however, we would need only one oxide ion, with its 2– charge, to balance with one magnesium with its 2+ charge [(2+) + (2–) = 0], and so the formula of magnesium oxide is just MgO.

Cumulative Review: Chapters 4 and 5

14. When naming ionic compounds, we name the positive ion (cation) first. For simple binary Type I ionic compounds, the ending –*ide* is added to the root name of the element that is the negative ion (anion). For example, for the Type I ionic compound formed between potassium and sulfur, K_2S, the name would be potassium sulfide: potassium is the cation, sulfur is the anion (with the suffix –*ide* added). Type II compounds are named by either of two systems, the "*ous–ic*" system (which is falling out of use), and the "Roman numeral" system which is preferred by most chemists. Type II compounds involve elements that form more than one stable ion. It is therefore necessary to specify *which* ion is present in a given compound. For example, iron forms two types of stable ion: Fe^{2+} and Fe^{3+}. Iron can react with oxygen to form either of two stable oxides, FeO or Fe_2O_3, depending on which cation is involved. Under the Roman numeral naming system, FeO would be named iron(II) oxide to show that it contains Fe^{2+} ions; Fe_2O_3 would be named iron(III) oxide to indicate that it contains Fe^{3+} ions. The Roman numeral used in a name corresponds to the charge of the specific ion present in the compound. Under the less-favored "*ous–ic*" system, for an element that forms two stable ions, the ending –*ous* is used to indicate the lower-charged ion, whereas the ending –*ic* is used to indicate the higher-charged ion. FeO and Fe_2O_3 would thus be named ferr*ous* oxide and ferr*ic* oxide, respectively. The "*ous–ic*" system has fallen out of favor because it does not indicate the actual charge on the ion, but only that it is the lower or higher charged of the two. This can lead to confusion: for example Fe^{2+} is called ferrous ion in this system, but Cu^{2+} is called cupric ion (since there is also a Cu^+ stable ion).

15. Type III binary compounds represent compounds involving only nonmetallic elements. In writing the name for such compounds, the element listed first in the formula is named first (using the full name of the element), and then the second element in the formula is named as though it were an anion (with the –*ide* ending). This is similar, thus far, to the method used for naming ionic compounds (Type I). Since there often may be more than one compound possible involving the same two nonmetallic elements, the naming system for Type III compounds goes one step further than the system for ionic compounds, by explicitly stating (by means of a numerical prefix) the number of atoms of each of the nonmetallic elements present in the molecules of the compound. For example, carbon and oxygen (both nonmetals) form two common compounds, CO and CO_2. To indicate clearly which compound is being discussed, the names of these compounds indicate explicitly the number of oxygen atoms present by using a numerical prefix.

CO	carbon *mon*oxide (*mon* or *mono* is the prefix meaning "one")
CO_2	carbon *di*oxide (*di* is the prefix meaning "two")

    The prefix *mono* is not normally used for the first element named in a compound if there is only one atom of the element present, but numerical prefixes are used for the first element if there is more than one atom of that element present. For example, nitrogen and oxygen form many binary compounds. Study closely how the examples following are named:

NO	nitrogen *mon*oxide	
NO_2	nitrogen *di*oxide	
N_2O	*di*nitrogen *mon*oxide	
N_2O_4	*di*nitrogen *tetr*oxide	(*tetra* or *tetr* means "four")

16. A polyatomic ion is an ion containing more than one atom. Some common polyatomic ions you should be familiar with are listed in Table 5.4. Parentheses are used in writing formulas containing polyatomic ions to indicate unambiguously how many of the polyatomic ion are present in the formula, to make certain that there is no mistake as to what is meant by the formula. For example, consider the substance calcium phosphate. The correct formula for this substance is $Ca_3(PO_4)_2$, which indicates that three calcium ions are combined for every two

phosphate ions (check the total number of positive and negative charges to see why this is so). If we did not write the parenthesis around the formula for the phosphate ion, that is, if we had written Ca_3PO_{42}, people reading this formula might think that there were 42 oxygen atoms present!

17. Several families of polyatomic anions contain an atom of a given element, combined with differing numbers of oxygen atoms. Such anions are called "*oxyanions*". For example, sulfur forms two common oxyanions, SO_3^{2-} and SO_4^{2-}. When there are two oxyanions in such a series (as for sulfur), the name of the anion with fewer oxygen atoms ends in *–ite* and the name of the anion with more oxygen atoms ends in *–ate*. Under this method, SO_3^{2-} is named sulf*ite* and SO_4^{2-} is named sulf*ate*. When there are more than two members of such a series, the prefixes *hypo–* and *per–* are used to indicate the members of the series with the *fewest* and *largest* number of oxygen atoms. For example, bromine forms four common oxyanions. The formulas and names of these oxyanions are listed below.

Formula	Name	
BrO^-	*hypo*brom*ite*	(fewest number of oxygens)
BrO_2^-	brom*ite*	
BrO_3^-	brom*ate*	
BrO_4^-	*per*brom*ate*	(largest number of oxygens)

18. Acids, in general, are substances that produce protons (H^+ ions) when dissolved in water. For acids that do not contain oxygen, the prefix *hydro–* and the suffix *–ic* are used with the root name of the element present in the acid (for example: HCl, hydrochloric acid; H_2S, hydrosulfuric acid; HF, hydrofluoric acid). The nomenclature of acids whose anions contain oxygen is more complicated. A series of prefixes and suffixes is used with the name of the non-oxygen atom in the anion of the acid: these prefixes and suffixes indicate the relative (not actual) number of oxygen atoms present in the anion. Most of the elements that form oxyanions form two such anions: for example, sulfur forms sulfite ion (SO_3^{2-}) and sulfate ion (SO_4^{2-}), and nitrogen forms nitrite ion (NO_2^-) and nitrate ion (NO_3^-). For an element that forms two oxyanions, the acid containing the anions will have the ending *–ous* if the anion is the *–ite* anion and the ending *–ic* if the anion is the *–ate* anion. For example, HNO_2 is nitr*ous* acid and HNO_3 is nitr*ic* acid; H_2SO_3 is sulfur*ous* acid and H_2SO_4 is sulfur*ic* acid. The halogen elements (Group 7) each form four oxyanions, and consequently, four oxyacids. The prefix *hypo–* is used for the oxyacid that contains fewer oxygen atoms than the *–ite* anion, and the prefix *per–* is used for the oxyacid that contains more oxygen atoms than the *–ate* anion. For example,

Acid	Name	Anion	Anion name
HBrO	*hypo*brom*ous* acid	BrO^-	*hypo*brom*ite*
$HBrO_2$	brom*ous* acid	BrO_2^-	brom*ite*
$HBrO_3$	brom*ic* acid	BrO_3^-	brom*ate*
$HBrO_4$	*per*brom*ic* acid	BrO_4^-	*per*brom*ate*

Cumulative Review: Chapters 4 and 5

19.

Symbol	Name	Atomic Number	Group Number
Al	aluminum	13	3
Rn	radon	86	8
S	sulfur	16	6
Sr	strontium	38	2
Br	bromine	33	5
C	carbon	6	4
Ba	barium	56	2
Ra	radium	88	2
Na	sodium	11	1
K	potassium	19	1
Ge	germanium	32	4
Cl	chlorine	17	7

20. How many elements in each family could you name? Elements in the same family have the same type of electronic configuration, and tend to undergo similar chemical reactions with other groups. For example, Li, Na, K, Rb, Cs all react with elemental chlorine gas, Cl_2, to form an ionic compound of general formula M^+Cl^-.

21.
- a. magnesium, 12
- b. gallium, 31
- c. tin, 50
- d. antimony, 51
- e. strontium, 38
- f. silicon, 14
- g. cesium, 55
- h. calcium, 20
- i. chromium, 24
- j. cobalt, 27
- k. copper, 29
- l. silver, 47
- m. uranium, 92
- n. arsenic, 33
- o. astatine, 85
- p. argon, 18
- q. zinc, 30
- r. manganese, 25
- s. selenium, 34
- t. tungsten, 74
- u. radium, 88
- v. radon, 86
- w. cerium, 58
- x. zirconium, 40
- y. aluminum, 13
- z. palladium, 46

22.
- a. 8 electrons, 8 protons, 9 neutrons
- b. 92 electrons, 92 protons, 143 neutrons
- c. 17 electrons, 17 protons, 20 neutrons
- d. 1 electrons, 1 protons, 2 neutrons

e. 2 electrons, 2 protons, 2 neutrons
f. 50 electrons, 50 protons, 69 neutrons
g. 54 electrons, 54 protons, 70 neutrons
h. 30 electrons, 30 protons, 34 neutrons

23.
a. Mg Mg^{2+}
b. F F^-
c. Ag Ag^+
d. Al Al^{3+}
e. O O^{2-}
f. Ba Ba^{2+}
g. Na Na^+
h. Br Br^-
i. K K^+
j. Ca Ca^{2+}
k. S S^{2-}
l. Li Li^+
m. Cl Cl^-

24.
a. 12 protons, 10 electrons
b. 26 protons, 24 electrons
c. 26 protons, 23 electrons
d. 9 protons, 10 electrons
e. 28 protons, 26 electrons
f. 30 protons, 28 electrons
g. 27 protons, 24 electrons
h. 7 protons, 10 electrons
i. 16 protons, 18 electrons
j. 37 protons, 36 electrons
k. 34 protons, 36 electrons
l. 19 protons, 18 electrons

25. K_3N (potassium nitride); KBr (potassium bromide); KCl (potassium chloride); KH (potassium hydride); K_2O (potassium oxide); KI (potassium iodide)

Ca_3N_2 (calcium nitride); $CaBr_2$ (calcium bromide); $CaCl_2$ (calcium chloride); CaH_2 (calcium hydride); CaO (calcium oxide); CaI_2 (calcium iodide)

AlN (aluminum nitride); $AlBr_3$ (aluminum bromide); $AlCl_3$ (aluminum chloride); AlH_3 (aluminum hydride); Al_2O_3 (aluminum oxide); AlI_3 (aluminum iodide)

Cumulative Review: Chapters 4 and 5

Ag₃N (silver nitride); AgBr (silver bromide); AgCl (silver chloride); AgH (silver hydride); Ag₂O (silver oxide); AgI (silver iodide)

Realize that most of the hydrogen compounds of the nonmetallic elements that are given as ions in this question are, in fact, covalently bonded compounds (not ionic compounds): H_3N (NH_3, ammonia); HBr (hydrogen bromide); HCl (hydrogen chloride); H_2 (elemental hydrogen); H_2O (water); HI (hydrogen iodide)

Na_3N (sodium nitride); NaBr (sodium bromide); NaCl (sodium chloride); NaH (sodium hydride); Na_2O (sodium oxide); NaI (sodium iodide).

26.
- a. CuI
- b. $CoCl_2$
- c. Ag_2S
- d. Hg_2Br_2
- e. HgO
- f. Cr_2S_3
- g. PbO_2
- h. K_3N
- i. SnF_2
- j. Fe_2O_3

27.
- a. incorrect: Ag forms 1+ ions
- b. incorrect: iron forms only 2+ and 3+ ions
- c. incorrect: the $H_2PO_4^-$ ion is named *di*hydrogen phosphate
- d. incorrect: the ammonium ion has 1+ charge; the sulfide ion has 2– charge
- e. correct
- f. correct
- g. correct
- h. incorrect: Ba forms 2+ ions; the hydroxide ion has 1– charge
- i. correct: the peroxide ion, O_2^{2-}, has a 2– charge
- j. incorrect: Ca forms 2+ ions, the carbonate ion has 2– charge

28.
- a. NH_4^+, ammonium ion
- b. SO_3^{2-}, sulfite ion
- c. NO_3^-, nitrate ion
- d. SO_4^{2-}, sulfate ion
- e. NO_2^-, nitrite ion
- f. CN^-, cyanide ion
- g. OH^-, hydroxide ion

h. ClO_4^-, perchlorate ion

i. ClO^-, hypochlorite ion

j. PO_4^{3-}, phosphate ion

29. Na^+: $NaNO_2$, $NaNO_3$, Na_2SO_3, Na_2SO_4, $NaHSO_4$, $NaOH$, $NaCN$, Na_3PO_4, Na_2HPO_4, NaH_2PO_4, Na_2CO_3, $NaHCO_3$, $NaClO$, $NaClO_2$, $NaClO_3$, $NaClO_4$, $NaC_2H_3O_2$, $NaMnO_4$, $Na_2Cr_2O_7$, Na_2CrO_4, Na_2O_2

Ca^{2+}: $Ca(NO_2)_2$, $Ca(NO_3)_2$, $CaSO_3$, $CaSO_4$, $Ca(HSO_4)_2$, $Ca(OH)_2$, $Ca(CN)_2$, $Ca_3(PO_4)_2$, $CaHPO_4$, $Ca(H_2PO_4)_2$, $CaCO_3$, $Ca(HCO_3)_2$, $Ca(ClO)_2$, $Ca(ClO_2)_2$, $Ca(ClO_3)_2$, $Ca(ClO_4)_2$, $Ca(C_2H_3O_2)_2$, $Ca(MnO_4)_2$, $CaCr_2O_7$, $CaCrO_4$, CaO

30.
a. xenon dioxide
b. iodine pentachloride
c. phosphorus trichloride
d. carbon monoxide
e. oxygen difluoride
f. diphosphorus pentoxide
g. arsenic triiodide
h. sulfur trioxide

31.
a. $HgCl_2$
b. Fe_2O_3
c. $H_2SO_3(aq)$
d. CaH_2
e. KNO_3
f. AlF_3
g. N_2O
h. $H_2SO_4(aq)$
i. K_3N
j. NO_2
k. $AgC_2H_3O_2$
l. $HC_2H_3O_2(aq)$
m. $PtCl_4$
n. $(NH_4)_2S$
o. $CoBr_3$
p. $HF(aq)$

CHAPTER 6

Chemical Reactions: An Introduction

1. The types of evidence for a chemical reaction mentioned in the text are: a change in color, formation of a solid, evolution of a gas, and absorption or evolution of heat. Other bits of evidence that might also be observed include appearance or disappearance of a characteristic odor, or separation of the reaction mixture into layers of visibly different composition.

2. Most of these products contain a peroxide, which decomposes releasing oxygen gas.

3. The oven cleaner is a more or less clear liquid (sodium hydroxide) when applied, but turns into a thick, opaque, soapy layer after reacting with oils and greases on the oven walls. Chapter 21 discusses the formation of a soap by reaction of a fat or oil with sodium hydroxide.

4. Bubbling takes place as the hydrogen peroxide chemically decomposes into water and oxygen gas.

5. The container of a flashlight battery usually consists of zinc, which is one of the substances involved in the chemical reaction in the battery that generates the electricity. The fact that the zinc decays until the battery leaks is a sign that a chemical reaction has taken place.

6. The appearance of the black color actually signals the breakdown of starches and sugars in the bread to elemental carbon. You may also see steam coming from the bread (water produced by the breakdown of the carbohydrates).

7. The substances to the left of the arrow are called the "reactants," whereas those to the right of the arrow are termed the "products." The arrow indicates a chemical reaction has occurred.

8. a. N_2, H_2; The reactants are on the left side of the arrow.
   b. NH_3; The products are on the right side of the arrow.

9. the same as

10. Balancing an equation ensures that no atoms are created or destroyed during the reaction. The total mass after the reaction must be the same as the total mass before the reaction.

11. In many reactions, the physical state of the reactants or products may influence whether or not the reaction takes place. For example, some metallic elements do not react with cold water, but will react vigorously with steam.

12. gaseous

13. $Zn(s) + CuSO_4(aq) \rightarrow ZnSO_4(aq) + Cu(s)$

14. $CaCO_3(s) \rightarrow CaO(s) + CO_2(g)$

Chapter 6:   Chemical Reactions: An Introduction

15. $H_2(g) + O_2(g) \rightarrow H_2O(g)$

16. $N_2H_4(l) \rightarrow N_2(g) + H_2(g)$

17. $KI(aq) + H_2O(l) \rightarrow KOH(aq) + I_2(s) + H_2(g)$

18. $Ag_2O(s) \rightarrow Ag(s) + O_2(g)$

19. $B_2O_3(s) + Mg(s) \rightarrow B(g) + MgO(s)$

20. $CaCO_3(s) + HCl(aq) \rightarrow CaCl_2(aq) + H_2O(l) + CO_2(g)$

21. $P_4(s) + Cl_2(g) \rightarrow PCl_3(s)$

22. $SiO_2(s) + C(s) \rightarrow Si(s) + CO(g)$

23. $NH_4NO_3(s) \rightarrow N_2O(g) + H_2O(g)$

24. $Zn(s) + HCl(aq) \rightarrow H_2(g) + ZnCl_2(aq)$

25. $C_2H_2(g) + O_2(g) \rightarrow CO_2(g) + H_2O(g)$

26. $SO_2(g) + H_2O(l) \rightarrow H_2SO_3(aq)$

   $SO_3(g) + H_2O(l) \rightarrow H_2SO_4(aq)$

27. $BaO(s) + Al(s) \rightarrow Ba(s) + Al_2O_3(s)$

   $CaO(s) + Al(s) \rightarrow Ca(s) + Al_2O_3(s)$

   $SrO(s) + Al(s) \rightarrow Sr(s) + Al_2O_3(s)$

28. $NO(g) + O_3(g) \rightarrow NO_2(g) + O_2(g)$

29. $CH_4(g) + Cl_2(g) \rightarrow CCl_4(l) + HCl(g)$

30. $P_4(s) + O_2(g) \rightarrow P_2O_5(s)$

31. $CaO(s) + H_2O(g) \rightarrow Ca(OH)_2(s)$

32. $Xe(g) + F_2(g) \rightarrow XeF_4(s)$

33. $SnO_2(s) + C(s) \rightarrow Sn(l) + CO(g)$

34. $NH_3(g) + O_2(g) \rightarrow HNO_3(aq) + H_2O(l)$

35. The subscripts in a formula really define what compound is present, since the subscripts represent in what proportions the elements combine to form the compound. Changing the subscripts would be changing the identity of the compound.

Chapter 6: Chemical Reactions: An Introduction

36. We cannot change the identities or formulas of the reactants or products in a chemical equation when balancing the equation. The proposed equation has incorrectly changed one of the products from water to hydrogen gas. (Use molecular-level drawings to support your answer.)

37. a. $FeCl_3 + KOH \rightarrow Fe(OH)_3 + KCl$

    Balance chlorine: $FeCl_3 + KOH \rightarrow Fe(OH)_3 + \mathbf{3}KCl$

    Balance potassium: $FeCl_3 + \mathbf{3}KOH \rightarrow Fe(OH)_3 + \mathbf{3}KCl$

    Balanced equation: $FeCl_3(aq) + 3KOH(aq) \rightarrow Fe(OH)_3(s) + 3KCl(aq)$

    b. $Pb(C_2H_3O_2)_2 + KI \rightarrow PbI_2 + KC_2H_3O_2$

    Balance iodine: $Pb(C_2H_3O_2)_2 + \mathbf{2}KI \rightarrow PbI_2 + KC_2H_3O_2$

    Balance potassium: $Pb(C_2H_3O_2)_2 + \mathbf{2}KI \rightarrow PbI_2 + \mathbf{2}KC_2H_3O_2$

    Balanced equation: $Pb(C_2H_3O_2)_2(aq) + 2KI(aq) \rightarrow PbI_2(s) + KC_2H_3O_2(aq)$

    c. $P_4O_{10} + H_2O \rightarrow H_3PO_4$

    Balance phosphorus: $P_4O_{10} + H_2O \rightarrow \mathbf{4}H_3PO_4$

    Balance hydrogen: $P_4O_{10} + \mathbf{6}H_2O \rightarrow 4H_3PO_4$

    Balanced equation: $P_4O_{10}(s) + 6H_2O(l) \rightarrow 4H_3PO_4(aq)$

    d. $Li_2O + H_2O \rightarrow LiOH$

    Balance lithium: $Li_2O + H_2O \rightarrow \mathbf{2}LiOH$

    Balanced equation: $Li_2O(s) + H_2O(l) \rightarrow 2LiOH(aq)$

    e. $MnO_2 + C \rightarrow Mn + CO_2$

    The equation is already balanced.

    f. $Sb + Cl_2 \rightarrow SbCl_3$

    This equation is more difficult to balance than it may appear. The problem arises in the fact that there are two Cl atoms on the left side of the equation, whereas there are three Cl atoms on the right side of the equation. To balance the chlorine atoms, we need to know the smallest whole number into which both 2 and 3 divide. This number is 6: we need to adjust the coefficients of Cl_2 and $SbCl_3$ so that there will be 6 chlorine atoms on each side of the equation.

    Balance chlorine: $Sb + \mathbf{3}Cl_2 \rightarrow \mathbf{2}SbCl_3$

    Balance antimony: $\mathbf{2}Sb + 3Cl_2 \rightarrow 2SbCl_3$

    Balanced equation: $2Sb(s) + 3Cl_2(g) \rightarrow 2SbCl_3(s)$

    g. $CH_4 + H_2O \rightarrow CO + H_2$

    Balance hydrogen: $CH_4 + H_2O \rightarrow CO + \mathbf{3}H_2$

    Balanced equation: $CH_4(g) + H_2O(g) \rightarrow CO(g) + 3H_2(g)$

    h. $FeS + HCl \rightarrow FeCl_2 + H_2S$

    Balance chlorine: $FeS + \mathbf{2}HCl \rightarrow FeCl_2 + H_2S$

    Balanced equation: $FeS(s) + 2HCl(aq) \rightarrow FeCl_2(aq) + H_2S(g)$

Chapter 6: Chemical Reactions: An Introduction

38. $2K(s) + 2H_2O(l) \rightarrow H_2(g) + 2KOH(aq)$

39.
   a. $K_2SO_4(aq) + BaCl_2(aq) \rightarrow BaSO_4(s) + KCl(aq)$

   Balance chlorine: $K_2SO_4(aq) + BaCl_2(aq) \rightarrow BaSO_4(s) + \mathbf{2}KCl(aq)$

   Balanced equation: $K_2SO_4(aq) + BaCl_2(aq) \rightarrow BaSO_4(s) + 2KCl(aq)$

   b. $Fe(s) + H_2O(g) \rightarrow FeO(s) + H_2(g)$

   The equation is already balanced.

   c. $NaOH(aq) + HClO_4(aq) \rightarrow NaClO_4(aq) + H_2O(l)$

   The equation is already balanced.

   d. $Mg(s) + Mn_2O_3(s) \rightarrow MgO(s) + Mn(s)$

   Balance oxygen: $Mg(s) + Mn_2O_3(s) \rightarrow \mathbf{3}MgO(s) + Mn(s)$

   Balance magnesium: $\mathbf{3}Mg(s) + Mn_2O_3(s) \rightarrow 3MgO(s) + Mn(s)$

   Balance manganese: $3Mg(s) + Mn_2O_3(s) \rightarrow 3MgO(s) + \mathbf{2}Mn(s)$

   Balanced equation: $3Mg(s) + Mn_2O_3(s) \rightarrow 3MgO(s) + 2Mn(s)$

   e. $KOH(s) + KH_2PO_4(aq) \rightarrow K_3PO_4(aq) + H_2O(l)$

   Balance potassium: $\mathbf{2}KOH(s) + KH_2PO_4(aq) \rightarrow K_3PO_4(aq) + H_2O(l)$

   Balance hydrogen: $2KOH(s) + KH_2PO_4(aq) \rightarrow K_3PO_4(aq) + \mathbf{2}H_2O(l)$

   Balanced equation: $2KOH(s) + KH_2PO_4(aq) \rightarrow K_3PO_4(aq) + 2H_2O(l)$

   f. $NO_2(g) + H_2O(l) + O_2(g) \rightarrow HNO_3(aq)$

   Balance hydrogen: $NO_2(g) + H_2O(l) + O_2(g) \rightarrow \mathbf{2}HNO_3(aq)$

   Balance nitrogen: $\mathbf{2}NO_2(g) + H_2O(l) + O_2(g) \rightarrow 2HNO_3(aq)$

   Balance oxygen: $2NO_2(g) + H_2O(l) + \tfrac{1}{2}O_2(g) \rightarrow 2HNO_3(aq)$

   Balanced equation: $4NO_2(g) + 2H_2O(l) + O_2(g) \rightarrow 4HNO_3(aq)$

   g. $BaO_2(s) + H_2O(l) \rightarrow Ba(OH)_2(aq) + O_2(g)$

   Balance oxygen: $BaO_2(s) + H_2O(l) \rightarrow Ba(OH)_2(aq) + \tfrac{1}{2}O_2(g)$

   Balanced equation: $2BaO_2(s) + 2H_2O(l) \rightarrow 2Ba(OH)_2(aq) + O_2(g)$

   h. $NH_3(g) + O_2(g) \rightarrow NO(g) + H_2O(l)$

   Balance hydrogen: $\mathbf{2}NH_3(g) + O_2(g) \rightarrow NO(g) + \mathbf{3}H_2O(l)$

   Balance nitrogen: $2NH_3(g) + O_2(g) \rightarrow \mathbf{2}NO(g) + 3H_2O(l)$

   Balance oxygen: $2NH_3(g) + \tfrac{5}{2}O_2(g) \rightarrow 2NO(g) + 3H_2O(l)$

   Balanced equation: $4NH_3(g) + 5O_2(g) \rightarrow 4NO(g) + 6H_2O(l)$

Chapter 6: Chemical Reactions: An Introduction

40.
a. $Na_2SO_4(aq) + CaCl_2(aq) \rightarrow CaSO_4(s) + 2NaCl(aq)$
b. $3Fe(s) + 4H_2O(g) \rightarrow Fe_3O_4(s) + 4H_2(g)$
c. $Ca(OH)_2(aq) + 2HCl(aq) \rightarrow CaCl_2(aq) + 2H_2O(l)$
d. $Br_2(g) + 2H_2O(l) + SO_2(g) \rightarrow 2HBr(aq) + H_2SO_4(aq)$
e. $3NaOH(s) + H_3PO_4(aq) \rightarrow Na_3PO_4(aq) + 3H_2O(l)$
f. $2NaNO_3(s) \rightarrow 2NaNO_2(s) + O_2(g)$
g. $2Na_2O_2(s) + 2H_2O(l) \rightarrow 4NaOH(aq) + O_2(g)$
h. $4Si(s) + S_8(s) \rightarrow 2Si_2S_4(s)$

41.
a. $Fe_3O_4(s) + 4H_2(g) \rightarrow 3Fe(l) + 4H_2O(g)$
b. $K_2SO_4(aq) + BaCl_2(aq) \rightarrow BaSO_4(s) + 2KCl(aq)$
c. $2HCl(aq) + FeS(s) \rightarrow FeCl_2(aq) + H_2S(g)$
d. $Br_2(g) + 2H_2O(l) + SO_2(g) \rightarrow 2HBr(aq) + H_2SO_4(aq)$
e. $CS_2(l) + 3Cl_2(g) \rightarrow CCl_4(l) + S_2Cl_2(g)$
f. $Cl_2O_7(g) + Ca(OH)_2(aq) \rightarrow Ca(ClO_4)_2(aq) + H_2O(l)$
g. $PBr_3(l) + 3H_2O(l) \rightarrow H_3PO_3(aq) + 3HBr(g)$
h. $Ba(ClO_3)_2(s) \rightarrow BaCl_2(s) + 3O_2(s)$

42.
a. $4NaCl(s) + 2SO_2(g) + 2H_2O(g) + O_2(g) \rightarrow 2Na_2SO_4(s) + 4HCl(g)$
b. $3Br_2(l) + I_2(s) \rightarrow 2IBr_3(s)$
c. $Ca(s) + 2H_2O(g) \rightarrow Ca(OH)_2(aq) + H_2(g)$
d. $2BF_3(g) + 3H_2O(g) \rightarrow B_2O_3(s) + 6HF(g)$
e. $SO_2(g) + 2Cl_2(g) \rightarrow SOCl_2(l) + Cl_2O(g)$
f. $Li_2O(s) + H_2O(l) \rightarrow 2LiOH(aq)$
g. $Mg(s) + CuO(s) \rightarrow MgO(s) + Cu(l)$
h. $Fe_3O_4(s) + 4H_2(g) \rightarrow 3Fe(l) + 4H_2O(g)$

43.
a. $4KO_2(s) + 6H_2O(l) \rightarrow 4KOH(aq) + O_2(g) + 4H_2O_2(aq)$
b. $Fe_2O_3(s) + 6HNO_3(aq) \rightarrow 2Fe(NO_3)_3(aq) + 3H_2O(l)$
c. $4NH_3(g) + 5O_2(g) \rightarrow 4NO(g) + 6H_2O(g)$
d. $PCl_5(l) + 4H_2O(l) \rightarrow H_3PO_4(aq) + 5HCl(g)$
e. $C_2H_5OH(l) + 3O_2(g) \rightarrow 2CO_2(g) + 3H_2O(l)$
f. $2CaO(s) + 5C(s) \rightarrow 2CaC_2(s) + CO_2(g)$
g. $2MoS_2(s) + 7O_2(g) \rightarrow 2MoO_3(s) + 4SO_2(g)$
h. $FeCO_3(s) + H_2CO_3(aq) \rightarrow Fe(HCO_3)_2(aq)$

Chapter 6:   Chemical Reactions: An Introduction

44. a. $Ba(NO_3)_2(aq) + Na_2CrO_4(aq) \rightarrow BaCrO_4(s) + 2NaNO_3(aq)$

   b. $PbCl_2(aq) + K_2SO_4(aq) \rightarrow PbSO_4(s) + 2KCl(aq)$

   c. $C_2H_5OH(l) + 3O_2(g) \rightarrow 2CO_2(g) + 3H_2O(l)$

   d. $CaC_2(s) + 2H_2O(l) \rightarrow Ca(OH)_2(s) + C_2H_2(g)$

   e. $Sr(s) + 2HNO_3(aq) \rightarrow Sr(NO_3)_2(aq) + H_2(g)$

   f. $BaO_2(s) + H_2SO_4(aq) \rightarrow BaSO_4(s) + H_2O_2(aq)$

   g. $2AsI_3(s) \rightarrow 2As(s) + 3I_2(s)$

   h. $2CuSO_4(aq) + 4KI(s) \rightarrow 2CuI(s) + I_2(s) + 2K_2SO_4(aq)$

45. $C_2H_2(g) + O_2(g) \rightarrow CO_2(g) + H_2O(g)$

46. (a)

47. $KNO_3(s) + C(s) \rightarrow K_2CO_3(s) + CO(g) + N_2(g)$

48. whole numbers

49. $2H_2(g) + CO(g) \rightarrow CH_3OH(l)$

50. $2Al_2O_3(s) + 3C(s) \rightarrow 4Al(s) + 3CO_2(g)$

51. $Fe_3O_4(s) + 4H_2(g) \rightarrow 3Fe(s) + 4H_2O(g)$

   $Fe_3O_4(s) + 4CO(g) \rightarrow 3Fe(s) + 4CO_2(g)$

52. True. Coefficients can be fractions when balancing a chemical equation because the coefficients represent a ratio of the moles needed for the reaction to occur. As a result, moles can be fractions because it represents an amount. The key is to make sure the atoms are conserved from reactants to products. Take note that the accepted convention is that the "best" balanced equation is the one with the smallest integers (although not required).

53. $Fe(s) + O_2(g) \rightarrow FeO(s)$

   $Fe(s) + O_2(g) \rightarrow Fe_2O_3(s)$

54. $BaO_2(s) + H_2O(l) \rightarrow BaO(s) + H_2O_2(aq)$

55. $4B(s) + 3O_2(g) \rightarrow 2B_2O_3(s)$

   $B_2O_3(s) + 3H_2O(l) \rightarrow 2B(OH)_3(s)$

56. $2KClO_3(s) \rightarrow 2KCl(s) + 3O_2(g)$

57. $2H_2O_2(aq) \rightarrow 2H_2O(g) + O_2(g)$

58. $4FeO(s) + O_2(g) \rightarrow 2Fe_2O_3(s)$

59. $CaSiO_3(s) + 6HF(g) \rightarrow CaF_2(aq) + SiF_4(g) + 3H_2O(l)$

Chapter 6: Chemical Reactions: An Introduction

60. $3LiAlH_4(s) + AlCl_3(s) \rightarrow 4AlH_3(s) + 3LiCl(s)$

61. Many over-the-counter antacids contain either carbonate ion (CO_3^{2-}) or hydrogen carbonate ion (HCO_3^-). When either of these encounter stomach acid (primarily HCl), carbon dioxide gas is released.

62. $Fe(s) + S(s) \rightarrow FeS(s)$

63. $Na(s) + Cl_2(g) \rightarrow NaCl(s)$

64. $K_2CrO_4(aq) + BaCl_2(aq) \rightarrow BaCrO_4(s) + 2KCl(aq)$

65. $H_2S(g) + Pb(NO_3)_2(aq) \rightarrow PbS(s) + HNO_3(aq)$

66. $2NaCl(aq) + 2H_2O(l) \rightarrow 2NaOH(aq) + H_2(g) + Cl_2(g)$

    $2NaBr(aq) + 2H_2O(l) \rightarrow 2NaOH(aq) + H_2(g) + Br_2(g)$

    $2NaI(aq) + 2H_2O(l) \rightarrow 2NaOH(aq) + H_2(g) + I_2(g)$

67. $Mg(s) + O_2(g) \rightarrow MgO(s)$

68. (e); Subscripts cannot be changed to balance an equation. If the subscripts are changed, then the identity of at least one of the compounds will change.

69. $P_4(s) + O_2(g) \rightarrow P_4O_{10}(g)$

70. $CuO(s) + H_2SO_4(aq) \rightarrow CuSO_4(aq) + H_2O(l)$

71. $PbS(s) + O_2(g) \rightarrow PbO(s) + SO_2(g)$

72. (a)

73.
    a. $Cl_2(g) + 2KBr(aq) \rightarrow Br_2(l) + 2KCl(aq)$
    b. $4Cr(s) + 3O_2(g) \rightarrow 2Cr_2O_3(s)$
    c. $P_4(s) + 6H_2(g) \rightarrow 4PH_3(g)$
    d. $2Al(s) + 3H_2SO_4(aq) \rightarrow Al_2(SO_4)_3(aq) + 3H_2(g)$
    e. $PCl_3(l) + 3H_2O(l) \rightarrow H_3PO_3(aq) + 3HCl(aq)$
    f. $2SO_2(g) + O_2(g) \rightarrow 2SO_3(g)$
    g. $C_7H_{16}(l) + 11O_2(g) \rightarrow 7CO_2(g) + 8H_2O(g)$
    h. $2C_2H_6(g) + 7O_2(g) \rightarrow 4CO_2(g) + 6H_2O(g)$

74. $2Na_2S_2O_3(aq) + I_2(aq) \rightarrow Na_2S_4O_6(aq) + 2NaI(aq)$

75.
    a. $SiCl_4(l) + 2Mg(s) \rightarrow Si(s) + 2MgCl_2(s)$
    b. $2NO(g) + Cl_2(g) \rightarrow 2NOCl(g)$
    c. $3MnO_2(s) + 4Al(s) \rightarrow 3Mn(s) + 2Al_2O_3(s)$

Chapter 6: Chemical Reactions: An Introduction

    d. $16Cr(s) + 3S_8(s) \rightarrow 8Cr_2S_3(s)$

    e. $4NH_3(g) + 3F_2(g) \rightarrow 3NH_4F(s) + NF_3(g)$

    f. $Ag_2S(s) + H_2(g) \rightarrow 2Ag(s) + H_2S(g)$

    g. $3O_2(g) \rightarrow 2O_3(g)$

    h. $8Na_2SO_3(aq) + S_8(s) \rightarrow 8Na_2S_2O_3(aq)$

76.   Answers will vary but the following balanced equation should be reported:

$$4NH_3(g) + 3O_2(g) \rightarrow 2N_2(g) + 6H_2O(g)$$

77.   (d) and (e); When balancing a chemical equation, the coefficients can be changed as needed (but not the subscripts). The coefficients refer to the number of atoms/molecules or moles in the balanced equation in order to follow the law of conservation of matter. In a chemical equation, the reactants are on the left and the products are on the right.

78. $4Fe(s) + 3O_2(g) \rightarrow 2Fe_2O_3(s)$

       $2PbO_2(s) \rightarrow 2PbO(s) + O_2(g)$

       $2H_2O_2(l) \rightarrow O_2(g) + 2H_2O(l)$

79. $2MnO_2(s) + CO(g) \rightarrow Mn_2O_3(aq) + CO_2(g)$

       $2Al(s) + 3H_2SO_4(aq) \rightarrow Al_2(SO_4)_3(aq) + 3H_2(g)$

       $2C_4H_{10}(g) + 13O_2(g) \rightarrow 8CO_2(g) + 10H_2O(l)$

       $2NH_4I(aq) + Cl_2(g) \rightarrow 2NH_4Cl(aq) + I_2(g)$

       $2KOH(aq) + H_2SO_4(aq) \rightarrow K_2SO_4(aq) + 2H_2O(l)$

CHAPTER 7

Reactions in Aqueous Solution

1. Water is the most universal of all liquids. Water has a relatively large heat capacity and a relatively large liquid range, which means it can absorb the heat liberated by many reactions while still remaining in the liquid state. Water is very polar and dissolves well both ionic solutes and solutes with which it can hydrogen bond (this is especially important to the biochemical reactions of the living cell).

2. Driving forces are types of *changes* in a system that pull a reaction in the *direction of product formation*; driving forces discussed in Chapter 7 include: formation of a *solid*, formation of *water*, formation of a *gas*, and transfer of electrons.

3. precipitation

4. A reactant in aqueous solution is indicated with (*aq*). Formation of a solid is indicated with (*s*)

5. When an ionic solute such as NaCl (sodium chloride) is dissolved in water, the resulting solution consists of separate, individual, discrete hydrated sodium ions (Na^+) and separate, individual, discrete hydrated chloride ions (Cl^-). There are no identifiable NaCl units in such a solution and the positive and negative ions behave independently of one another.

6. Because each formula unit of $MgCl_2$ contains two chloride ions for each magnesium ion, that ratio will be preserved in the solution when $MgCl_2$ is dissolved in water.

7. A substance is said to be a strong electrolyte if *each* unit of the substance produces separated, distinct ions when the substance is dissolved in water. NaCl and KNO_3 are both strong electrolytes.

8. Chemists know that a solution contains independent ions because such a solution will readily allow an electrical current to pass through it. The simplest experiment that demonstrates this uses the sort of light–bulb conductivity apparatus described in the text: if the light bulb glows strongly, then the solution must contain a lot of ions to be conducting the electricity well.

9. The solubility rules are general rules describing the solubility of common ionic substances in water. They are based on countless observations of chemical compounds and reactions. For example, Solubility Rule 1 says that "most nitrate salts are soluble". So when we write an equation for the reaction of a nitrate salt with some other reagent, we know that any precipitate that forms will *not* involve the nitrate ion.

10. (a); The precipitate $BaSO_4$ will form.

11. a. insoluble (Rule 6: most sulfide salts are insoluble.)
    b. insoluble (Rule 5: most hydroxide compounds are insoluble)
    c. soluble (Rule 2: most salts of Na^+ are soluble; Rule 4: most sulfate salts are soluble.)

Chapter 7: Reactions in Aqueous Solution

      d.   soluble (Rule 2: most salts of NH_4^+ are soluble.)

      e.   insoluble (Rule 6: most carbonate salts are insoluble.)

      f.   insoluble (Rule 6: most phosphate salts are insoluble.)

      g.   insoluble (Exception to Rule 3)

      h.   insoluble (Exception to Rule 4)

12. a. soluble (Rule 1: most nitrate salts are soluble.)

      b.   soluble (Rule 2: most salts of K^+ are soluble.)

      c.   insoluble (Rule 4: most sulfate salts are soluble with $PbSO_4$ as an exception.)

      d.   insoluble (Rule 5: most hydroxide compounds are insoluble.)

      e.   soluble (Rule 2: most salts of K^+ are soluble.)

      f.   insoluble (Rule 3: most chloride salts are soluble with Hg_2Cl_2 as an exception.)

      g.   soluble (Rule 2: most salts of NH_4^+ are soluble.)

      h.   insoluble (Rule 6: most sulfide salts are insoluble.)

13. a. Rule 2: Most salts of K^+ are soluble.

      b.   Rule 1: Most nitrate salts are soluble.

      c.   Rule 2: Most salts of NH_4^+ are soluble.

      d.   Rule 4: Most sulfate salts are soluble.

      e.   Rule 3: Most chloride salts are soluble.

14. a. Rule 5: Most hydroxide compounds are insoluble.

      b.   Rule 6: Most carbonate salts are insoluble.

      c.   Rule 6: Most phosphate salts are insoluble.

      d.   Rule 3: Exception to the rule for chloride salts.

      e.   Rule 4: Exception to the rule for sulfate salts.

15. a. CuS: Rule 6 (most sulfide salts are insoluble).

      b. $Ba_3(PO_4)_2$: Rule 6 (most phosphate salts are insoluble).

      c.   AgCl: Rule 3 (exception to rule for chloride salts).

      d. $CoCO_3$: Rule 6 (most carbonate salts are insoluble).

      e. $CaSO_4$: Rule 4 (exception to rule for sulfate salts).

      f. Hg_2Cl_2: Rule 3 (exception to rule for chloride salts).

16. a. $MnCO_3$: Rule 6 (most carbonates are only slightly soluble).

      b. $CaSO_4$: Rule 4 (exception for sulfates).

      c. Hg_2Cl_2: Rule 3: (exception for chlorides).

      d.   soluble

Chapter 7: Reactions in Aqueous Solution

      e.   Ni(OH)$_2$: Rule 5 (most hydroxides are only slightly soluble).

      f.   BaSO$_4$: Rule 4 (exception for sulfates).

17.   The precipitates are marked in boldface type.

      a.   No precipitate: both (NH$_4$)$_2$SO$_4$ and HCl are soluble.

           NH$_4$Cl(*aq*) + H$_2$SO$_4$(*aq*) → no precipitate

      b.   Rule 6: Most carbonate salts are only slightly soluble.

           2K$_2$CO$_3$(*aq*) + SnCl$_4$(*aq*) → **Sn(CO$_3$)$_2$(*s*)** + 4KCl(*aq*)

      c.   Rule 3: exception to rule for chlorides

           2NH$_4$Cl(*aq*) + Pb(NO$_3$)$_2$(*aq*) → **PbCl$_2$(*s*)** + 2NH$_4$NO$_3$(*aq*)

      d.   Rule 5: Most hydroxide compounds are only slightly soluble.

           CuSO$_4$(*aq*) + 2KOH(*aq*) → **Cu(OH)$_2$(*s*)** + K$_2$SO$_4$(*aq*)

      e.   Rule 6: Most phosphate salts are only slightly soluble.

           Na$_3$PO$_4$(*aq*) + CrCl$_3$(*aq*) → **CrPO$_4$(*s*)** + 3NaCl(*aq*)

      f.   Rule 6: Most sulfide salts are only slightly soluble.

           3(NH$_4$)$_2$S(*aq*) + 2FeCl$_3$(*aq*) → **Fe$_2$S$_3$(*s*)** + 6NH$_4$Cl(*aq*)

18.   The formulas of the precipitates are in boldface type.

      a.   Rule 6: Most carbonate salts are insoluble.

           Na$_2$CO$_3$(*aq*) + CuSO$_4$(*aq*) → Na$_2$SO$_4$(*aq*) + **CuCO$_3$(*s*)**

      b.   Rule 3: Exception for chloride salts.

           HCl(*aq*) + AgC$_2$H$_3$O$_2$(*aq*) → HC$_2$H$_3$O$_2$(*aq*) + **AgCl(*s*)**

      c.   No precipitate

      d.   Rule 6: Most sulfide salts are insoluble.

           3(NH$_4$)$_2$S(*aq*) + 2FeCl$_3$(*aq*) → 6NH$_4$Cl(*aq*) + **Fe$_2$S$_3$(*s*)**

      e.   Rule 4: Exception for sulfate salts

           H$_2$SO$_4$(*aq*) + Pb(NO$_3$)$_2$(*aq*) → 2HNO$_3$(*aq*) + **PbSO$_4$(*s*)**

      f.   Rule 6: Most phosphate salts are insoluble.

           2K$_3$PO$_4$(*aq*) + 3CaCl$_2$(*aq*) → 6KCl(*aq*) + **Ca$_3$(PO$_4$)$_2$(*s*)**

19.   Hint: when balancing equations involving polyatomic ions, especially in precipitation reactions, balance the polyatomic ions as a *unit*, not in terms of the atoms the polyatomic ions contain (e.g., treat nitrate ion, NO$_3^-$ as a single entity, not as one nitrogen and three oxygen atoms). When finished balancing, however, be sure to count the individual number of atoms of each type on each side of the equation.

      a.   Na$_2$SO$_4$(*aq*) + CaCl$_2$(*aq*) → CaSO$_4$(*s*) + NaCl(*aq*)

           Balance sodium: Na$_2$SO$_4$(*aq*) + CaCl$_2$(*aq*) → CaSO$_4$(*s*) + **2**NaCl(*aq*)

Chapter 7:   Reactions in Aqueous Solution

Balanced equation: $Na_2SO_4(aq) + CaCl_2(aq) \rightarrow CaSO_4(s) + 2NaCl(aq)$

b. $Co(C_2H_3O_2)_2(aq) + Na_2S(aq) \rightarrow CoS(s) + NaC_2H_3O_2(aq)$

Balance acetate: $Co(C_2H_3O_2)_2(aq) + Na_2S(aq) \rightarrow CoS(s) + \mathbf{2}NaC_2H_3O_2(aq)$

Balanced equation: $Co(C_2H_3O_2)_2(aq) + Na_2S(aq) \rightarrow CoS(s) + 2NaC_2H_3O_2(aq)$

c. $KOH(aq) + NiCl_2(aq) \rightarrow Ni(OH)_2(s) + KCl(aq)$

Balance hydroxide: $\mathbf{2}KOH(aq) + NiCl_2(aq) \rightarrow Ni(OH)_2(s) + KCl(aq)$

Balance potassium: $2KOH(aq) + NiCl_2(aq) \rightarrow Ni(OH)_2(s) + \mathbf{2}KCl(aq)$

Balanced equation: $2KOH(aq) + NiCl_2(aq) \rightarrow Ni(OH)_2(s) + 2KCl(aq)$

20. Hint: when balancing equations involving polyatomic ions, especially in precipitation reactions, balance the polyatomic ions as a *unit*, not in terms of the atoms the polyatomic ions contain (e.g., treat nitrate ion, NO_3^- as a single entity, not as one nitrogen and three oxygen atoms). When finished balancing, however, be sure to count the individual number of atoms of each type on each side of the equation.

   a. $CaCl_2(aq) + AgNO_3(aq) \rightarrow Ca(NO_3)_2(aq) + AgCl(s)$

   balance chlorine: $CaCl_2(aq) + AgNO_3(aq) \rightarrow Ca(NO_3)_2(aq) + \mathbf{2}AgCl(s)$

   balance silver: $CaCl_2(aq) + \mathbf{2}AgNO_3(aq) \rightarrow Ca(NO_3)_2(aq) + 2AgCl(s)$

   balanced equation: $CaCl_2(aq) + 2AgNO_3(aq) \rightarrow Ca(NO_3)_2(aq) + 2AgCl(s)$

   b. $AgNO_3(aq) + K_2CrO_4(aq) \rightarrow Ag_2CrO_4(s) + KNO_3(aq)$

   balance silver: $\mathbf{2}AgNO_3(aq) + K_2CrO_4(aq) \rightarrow Ag_2CrO_4(s) + KNO_3(aq)$

   balance nitrate ion: $2AgNO_3(aq) + K_2CrO_4(aq) \rightarrow Ag_2CrO_4(s) + \mathbf{2}KNO_3(aq)$

   balanced equation: $2AgNO_3(aq) + K_2CrO_4(aq) \rightarrow Ag_2CrO_4(s) + 2KNO_3(aq)$

   c. $BaCl_2(aq) + K_2SO_4(aq) \rightarrow BaSO_4(s) + KCl(aq)$

   balance potassium: $BaCl_2(aq) + K_2SO_4(aq) \rightarrow BaSO_4(s) + \mathbf{2}KCl(aq)$

   balanced equation: $BaCl_2(aq) + K_2SO_4(aq) \rightarrow BaSO_4(s) + 2KCl(aq)$

21. The products are determined by having the ions "switch partners." For example, for a general reaction AB + CD →, the possible products are AD and CB if the ions switch partners. If either AD or CB is insoluble, then a precipitation reaction has occurred. In the following reaction, the formula of the precipitate is given in boldface type.

   a. $(NH_4)_2SO_4(aq) + Ba(NO_3)_2(aq) \rightarrow 2NH_4NO_3(aq) + \mathbf{BaSO_4(s)}$

   Rule 4: $BaSO_4$ is a listed exception.

   b. $H_2S(aq) + NiSO_4(aq) \rightarrow H_2SO_4(aq) + \mathbf{NiS(s)}$

   Rule 6: Most sulfide salts are only slightly soluble.

   c. $FeCl_3(aq) + 3NaOH(aq) \rightarrow 3NaCl(aq) + \mathbf{Fe(OH)_3(s)}$

   Rule 5: Most hydroxide compounds are only slightly soluble.

Chapter 7: Reactions in Aqueous Solution

22. The precipitate is lead(II) phosphate. The balanced equation is:
$$2Na_3PO_4(aq) + 3Pb(NO_3)_2(aq) \rightarrow Pb_3(PO_4)_2(s) + 6NaNO_3(aq)$$

23. The *net ionic equation* for a reaction in solution indicates only those components that are directly involved in the reaction. Other ions that may be present to balance charge, but which do not actively participate in the reaction are called *spectator ions* and are not indicated when writing the chemical equation for the reaction.

24. (e)

25. The products are determined by having the ions in the two aqueous ionic reagents "switch partners." For example, for a general reaction AB + CD →, the possible products are AD and CB if the ions switch partners. If either AD or CB is insoluble according to the solubility rules in Table 7.1, then a precipitation reaction has occurred. Answers will vary for each student.

26. Molecular: $K_2SO_4(aq) + Pb(NO_3)_2(aq) \rightarrow 2KNO_3(aq) + PbSO_4(s)$

    Complete Ionic: $2K^+(aq) + SO_4^{2-}(aq) + Pb^{2+}(aq) + 2NO_3^-(aq) \rightarrow 2K^+(aq) + 2NO_3^-(aq) + PbSO_4(s)$

    Net Ionic: $SO_4^{2-}(aq) + Pb^{2+}(aq) \rightarrow PbSO_4(s)$

27. $Cu^{2+}(aq) + CrO_4^{2-}(aq) \rightarrow CuCrO_4(s)$

    $Co^{3+}(aq) + CrO_4^{2-}(aq) \rightarrow Co_2(CrO_4)_3(s)$

    $Ba^{2+}(aq) + CrO_4^{2-}(aq) \rightarrow BaCrO_4(s)$

    $Fe^{3+}(aq) + CrO_4^{2-}(aq) \rightarrow Fe_2(CrO_4)_3(s)$

28. $Ag^+(aq) + Cl^-(aq) \rightarrow AgCl(s)$

    $Pb^{2+}(aq) + 2Cl^-(aq) \rightarrow PbCl_2(s)$

    $Hg_2^{2+}(aq) + 2Cl^-(aq) \rightarrow Hg_2Cl_2(s)$

29. $Ca^{2+}(aq) + C_2O_4^{2-}(aq) \rightarrow CaC_2O_4(s)$

30. $Co^{2+}(aq) + S^{2-}(aq) \rightarrow CoS(s)$

    $2Co^{3+}(aq) + 3S^{2-}(aq) \rightarrow Co_2S_3(s)$

    $Fe^{2+}(aq) + S^{2-}(aq) \rightarrow FeS(s)$

    $2Fe^{3+}(aq) + 3S^{2-}(aq) \rightarrow Fe_2S_3(s)$

31. Strong acids ionize completely in water. The strong acids are also strong electrolytes. Strong electrolytes dissociate completely in water.

32. Strong bases fully produce hydroxide ions when dissolved in water. The strong bases are also strong electrolytes. Strong electrolytes dissociate completely in water.

33. $H^+(aq) + OH^-(aq) \rightarrow H_2O$; formation of a water molecule

34. acids: HCl, H_2SO_4, HNO_3, $HClO_4$, HBr

    bases: NaOH, KOH, RbOH, CsOH

35. 1000; 1000

36. A salt is the ionic product remaining in solution when an acid neutralizes a base. For example, in the reaction HCl(aq) + NaOH(aq) → NaCl(aq) + H_2O(l) sodium chloride is the salt produced by the neutralization reaction.

37. Your textbook mentions four strong acids. You only had to give three of the following equations.

    $HCl(aq) \rightarrow H^+(aq) + Cl^-(aq)$

    $HNO_3(aq) \rightarrow H^+(aq) + NO_3^-(aq)$

    $H_2SO_4(aq) \rightarrow H^+(aq) + HSO_4^-(aq)$

    $HClO_4(aq) \rightarrow H^+(aq) + ClO_4^-(aq)$

38. $HBr(aq) \rightarrow H^+(aq) + Br^-(aq)$

    $HClO_4(aq) \rightarrow H^+(aq) + ClO_4^-(aq)$

39. The formulas of the salts are marked in boldface type. Remember that in an acid/base reaction in aqueous solution, *water* is always one of the products: keeping this in mind makes predicting the formula of the *salt* produced easy to do.

    a. HCl(aq) + KOH(aq) → H_2O(l) + **KCl**(aq)

    b. RbOH(aq) + HNO_3(aq) → H_2O(l) + **RbNO$_3$**(aq)

    c. $HClO_4$(aq) + NaOH(aq) → H_2O(l) + **NaClO$_4$**(aq)

    d. HBr(aq) + CsOH(aq) → H_2O(l) + **CsBr**(aq)

40. In general, the salt formed in an aqueous acid–base reaction consists of the *positive ion of the base* involved in the reaction, combined with the *negative ion of the acid*. The hydrogen ion of the strong acid combines with the hydroxide ion of the strong base to produce water, which is the other product of the acid–base reactions.

    a. $H_2SO_4(aq) + 2KOH(aq) \rightarrow K_2SO_4(aq) + 2H_2O(l)$

    b. $HNO_3(aq) + NaOH(aq) \rightarrow NaNO_3(aq) + H_2O(l)$

    c. $2HCl(aq) + Ca(OH)_2(aq) \rightarrow CaCl_2(aq) + 2H_2O(l)$

    d. $2HClO_4(aq) + Ba(OH)_2(aq) \rightarrow Ba(ClO_4)_2(aq) + 2H_2O(l)$

41. An oxidation–reduction reaction is one in which one species loses electrons (oxidation) and another species gains electrons (reduction). Electrons are transferred from the species being oxidized to the species being reduced.

42. Answer depends on student choice of example: $Na(s) + Cl_2(g) \rightarrow 2NaCl(s)$ is an example.

43. A driving force, in general, is an event that tends to help to convert the reactants of a process into the products. Some elements (metals) tend to lose electrons, whereas other elements (nonmetals)

Chapter 7: Reactions in Aqueous Solution

tend to gain electrons. A *transfer* of electrons from atoms of a metal to atoms of a nonmetal would be favorable and would result in a chemical reaction. A simple example of such a process is the reaction of sodium with chlorine: sodium atoms tend to each lose one electron (to form Na^+), whereas chlorine atoms tend to each gain one electron (to form Cl^-). The reaction of sodium metal with chlorine gas represents a transfer of electrons from sodium atoms to chlorine atoms to form sodium chloride.

44. The aluminum atoms lose 3 electrons to become Al^{3+} ions. Fe^{3+} ions gain 3 electrons to become Fe atoms.

45. Each calcium atom would lose two electrons. Each fluorine atom would gain one electron (so the F_2 molecule would gain two electrons). One calcium atom would be required to react with one fluorine, F_2, molecule. Calcium ions are charged 2+, fluoride ions are charged 1–.

46. Each magnesium atom would lose two electrons. Each oxygen atom would gain two electrons (so the O_2 molecule would gain four electrons). Two magnesium atoms would be required to react with each oxygen, O_2, molecule. Magnesium ions are charged 2+, oxide ions are charged 2–.

47. $MgCl_2$ is made up of Mg^{2+} ions and Cl^- ions. Magnesium atoms each lose two electrons to become Mg^{2+} ions. Chlorine atoms each gain one electron to become Cl^- ions (so each Cl_2 molecule gains two electrons to become two Cl^- ions).

48. $AlBr_3$ is made up of Al^{3+} ions and Br^- ions. Aluminum atoms each lose three electrons and bromine atoms each gain one electron (Br_2 gains two electrons).

49. a. $Co(s) + Br_2(l) \rightarrow CoBr_3(s)$

   Balance bromine: $Co(s) + 3Br_2(l) \rightarrow 2CoBr_3(s)$

   Balance cobalt: $2Co(s) + 3Br_2(l) \rightarrow 2CoBr_3(s)$

   Balanced equation: $2Co(s) + 3Br_2(l) \rightarrow 2CoBr_3(s)$

   cobalt is oxidized, bromine is reduced

   b. $Al(s) + H_2SO_4(aq) \rightarrow Al_2(SO_4)_3(aq) + H_2(g)$

   Balance sulfate ions: $Al(s) + 3H_2SO_4(aq) \rightarrow Al_2(SO_4)_3(aq) + H_2(g)$

   Balance hydrogen: $Al(s) + 3H_2SO_4(aq) \rightarrow Al_2(SO_4)_3(aq) + 3H_2(g)$

   Balance aluminum: $2Al(s) + 3H_2SO_4(aq) \rightarrow Al_2(SO_4)_3(aq) + 3H_2(g)$

   Balanced equation: $2Al(s) + 3H_2SO_4(aq) \rightarrow Al_2(SO_4)_3(aq) + 3H_2(g)$

   aluminum is oxidized, hydrogen is reduced

   c. $Na(s) + H_2O(l) \rightarrow NaOH(aq) + H_2(g)$

   Balance hydrogen: $Na(s) + 2H_2O(l) \rightarrow 2NaOH(aq) + H_2(g)$

   Balance sodium: $2Na(s) + 2H_2O(l) \rightarrow 2NaOH(aq) + H_2(g)$

   Balanced equation: $2Na(s) + 2H_2O(l) \rightarrow 2NaOH(aq) + H_2(g)$

   sodium is oxidized, hydrogen is reduced

   d. $Cu(s) + O_2(g) \rightarrow Cu_2O(s)$

Chapter 7: Reactions in Aqueous Solution

Balance copper: $2Cu(s) + O_2(g) \rightarrow Cu_2O(s)$

Balance oxygen: $2Cu(s) + \frac{1}{2}O_2(g) \rightarrow Cu_2O(s)$

Balanced equation: $\mathbf{4}Cu(s) + O_2(g) \rightarrow \mathbf{2}Cu_2O(s)$

copper is oxidized, oxygen is reduced

50. a. $P_4(s) + O_2(g) \rightarrow P_4O_{10}(s)$

balance oxygen: $P_4(s) + \mathbf{5}O_2(g) \rightarrow P_4O_{10}(s)$

balanced equation: $P_4(s) + 5O_2(g) \rightarrow P_4O_{10}(s)$

b. $MgO(s) + C(s) \rightarrow Mg(s) + CO(g)$

This equation is already balanced.

c. $Sr(s) + H_2O(l) \rightarrow Sr(OH)_2(aq) + H_2(g)$

balance oxygen: $Sr(s) + \mathbf{2}H_2O(l) \rightarrow Sr(OH)_2(aq) + H_2(g)$

balanced equation: $Sr(s) + 2H_2O(l) \rightarrow Sr(OH)_2(aq) + H_2(g)$

d. $Co(s) + HCl(aq) \rightarrow CoCl_2(aq) + H_2(g)$

balance hydrogen: $Co(s) + \mathbf{2}HCl(aq) \rightarrow CoCl_2(aq) + H_2(g)$

balanced equation: $Co(s) + 2HCl(aq) \rightarrow CoCl_2(aq) + H_2(g)$

51. a. In a double displacement reaction, two ionic solutes "switch partners" with the positive ion from one combining with the negative ion from the other to form the precipitate: for example, in the reaction $AgNO_3(aq) + HCl(aq) \rightarrow AgCl(s) + HNO_3(aq)$, silver ion from one solute combines with chloride ion from the other solute to form the precipitate. In a single displacement reaction, one element replaces another from its compound: in other words, a single displacement reaction is typically an oxidation–reduction reaction also: for example in the reaction $Zn(s) + CuSO_4(aq) \rightarrow Cu(s) + ZnSO_4(aq)$, zinc in the elemental form replaces copper in the copper compound, producing copper in the elemental form and a zinc compound. Many other examples are possible.

b. examples of formation of water:

$HCl(aq) + NaOH(aq) \rightarrow H_2O(l) + NaCl(aq)$
$H_2SO_4(aq) + 2KOH(aq) \rightarrow 2H_2O(l) + K_2SO_4(aq)$

examples of formation of a gaseous product:

$Mg(s) + 2HCl(aq) \rightarrow MgCl_2(aq) + H_2(g)$
$2KClO_3(s) \rightarrow 2KCl(s) + 3O_2(g)$

52. A reaction must be an oxidation–reduction reaction if any of the oxidation numbers of the atoms in the equation change. Aluminum changes oxidation state from 0 in Al to +3 (oxidation) in Al_2O_3 and $AlCl_3$; nitrogen changes oxidation state from –3 in NH_4^+ to +2 in NO (oxidation); chlorine changes oxidation state from +7 in ClO_4^- to –1 in $AlCl_3$ (reduction)

Chapter 7:   Reactions in Aqueous Solution

53. For each reaction, the type of reaction is first identified, followed by some of the reasoning that leads to this choice (there may be more than one way in which you can recognize a particular type of reaction).

   a. precipitation (from Table 7.1, $BaSO_4$ is insoluble).

   b. oxidation–reduction (Zn changes from the elemental to the combined state; hydrogen changes from the combined to the elemental state).

   c. precipitation (From Table 7.1, AgCl is insoluble.)

   d. acid–base (HCl is an acid; KOH is a base; water and a salt are produced.)

   e. oxidation–reduction (Cu changes from the combined to the elemental state; Zn changes from the elemental to the combined state.)

   f. acid–base (The $H_2PO_4^-$ ion behaves as an acid; NaOH behaves as a base; a salt and water are produced.)

   g. precipitation (From Table 7.1, $CaSO_4$ is insoluble); acid–base [$Ca(OH)_2$ is a base; H_2SO_4 is an acid; a salt and water are produced.]

   h. oxidation–reduction (Mg changes from the elemental to the combined state; Zn changes from the combined to the elemental state.)

   i. precipitation (From Table 7.1, $BaSO_4$ is insoluble.)

54. For each reaction, the type of reaction is first identified, followed by some of the reasoning that leads to this choice (there may be more than one way in which you can recognize a particular type of reaction).

   a. oxidation–reduction (Oxygen changes from the combined state to the elemental state.)

   b. oxidation–reduction (Zinc changes from the elemental to the combined state; hydrogen changes from the combined to the elemental state.)

   c. acid–base (H_2SO_4 is a strong acid and NaOH is a strong base; water and a salt are formed.)

   d. acid–base, precipitation (H_2SO_4 is a strong acid, and $Ba(OH)_2$ is a base; water and a salt are formed; an insoluble product forms.)

   e. precipitation (From the Solubility Rules of Table 7.1, AgCl is only slightly soluble.)

   f. precipitation (From the Solubility Rules of Table 7.1, $Cu(OH)_2$ is only slightly soluble.)

   g. oxidation–reduction (Chlorine and fluorine change from the elemental to the combined state.)

   h. oxidation–reduction (Oxygen changes from the elemental to the combined state.)

   i. acid–base (HNO_3 is a strong acid and $Ca(OH)_2$ is a strong base; a salt and water are formed.)

55. A combustion reaction is typically a reaction in which an element or compound reacts with oxygen so quickly and with so much release of energy that a flame results. In addition to the carbon dioxide and water chemical products, combustion reactions are a major source of heat energy.

56. oxidation–reduction

57. A synthesis reaction represents the production of a given compound from simpler substances (either elements or simpler compounds). For example,

$$O_2(g) + 2F_2(g) \rightarrow 2OF_2(g)$$

represents a simple synthesis reaction. Synthesis reactions may often (but not necessarily always) also be classified in other ways. For example, the reaction

$$C(s) + O_2(g) \rightarrow CO_2(g)$$

could also be classified as an oxidation–reduction reaction or as a combustion reaction (a special sub–classification of oxidation–reduction reaction that produces a flame). As another example, the reaction

$$2Fe(s) + 3Cl_2(g) \rightarrow 2FeCl_3(s)$$

is a synthesis reaction that also is an oxidation–reduction reaction.

58. A decomposition reaction is one in which a given compound is broken down into simpler compounds or constituent elements. The reactions

$$CaCO_3(s) \rightarrow CaO(s) + CO_2(g)$$

$$2HgO(s) \rightarrow 2Hg(l) + O_2(g)$$

both represent decomposition reactions. Such reactions often (but not necessarily always) may be classified in other ways. For example, the reaction of $HgO(s)$ is also an oxidation–reduction reaction.

59. Compounds like those in parts (a) and (b) of this problem, containing only carbon and hydrogen, are called *hydrocarbons*. When a hydrocarbon is reacted with oxygen (O_2), the hydrocarbon is almost always converted to carbon dioxide and water vapor. Because water molecules contain an odd number of oxygen atoms, and O_2 contains an even number of oxygen atoms, it is often difficult to balance such equations. For this reason, it is simpler to balance the equation using fractional coefficients if necessary, and then to multiply by a factor that will give whole number coefficients for the final balanced equation.

    a. $C_6H_6 + O_2 \rightarrow CO_2 + H_2O$

    Balance carbon: $C_6H_6 + O_2 \rightarrow \mathbf{6}CO_2 + H_2O$

    Balance hydrogen: $C_6H_6 + O_2 \rightarrow 6CO_2 + \mathbf{3}H_2O$

    Balance oxygen with fractional coefficient: $C_6H_6 + \frac{15}{2}O_2 \rightarrow 6CO_2 + 3H_2O$

    Balanced equation: $\mathbf{2}C_6H_6 + \mathbf{15}O_2 \rightarrow \mathbf{12}CO_2 + \mathbf{6}H_2O$

    b. $C_5H_{12} + O_2 \rightarrow CO_2 + H_2O$

    Balance carbon: $C_5H_{12} + O_2 \rightarrow \mathbf{5}CO_2 + H_2O$

    Balance hydrogen: $C_5H_{12} + O_2 \rightarrow 5CO_2 + \mathbf{6}H_2O$

    Balance oxygen: $C_5H_{12} + \mathbf{8}O_2 \rightarrow 5CO_2 + 6H_2O$

    Balanced equation: $C_5H_{12} + \mathbf{8}O_2 \rightarrow 5CO_2 + 6H_2O$

Chapter 7: Reactions in Aqueous Solution

    c. $C_2H_6O(l) + O_2(g) \rightarrow CO_2 + H_2O$

         Balance carbon: $C_2H_6O(l) + O_2(g) \rightarrow \mathbf{2}CO_2 + H_2O$

         Balance hydrogen: $C_2H_6O(l) + O_2(g) \rightarrow 2CO_2 + \mathbf{3}H_2O$

         Balance oxygen: $C_2H_6O(l) + \mathbf{3}O_2(g) \rightarrow 2CO_2 + 3H_2O$

         Balanced equation: $C_2H_6O(l) + 3O_2(g) \rightarrow 2CO_2 + 3H_2O$

60. Compounds like those in this problem, containing only carbon and hydrogen, are called *hydrocarbons*. When a hydrocarbon is reacted with oxygen (O_2), the hydrocarbon is almost always converted to carbon dioxide and water vapor. Because water molecules contain an odd number of oxygen atoms, and O_2 contains an even number of oxygen atoms, it is often difficult to balance such equations. For this reason, it is simpler to balance the equation using fractional coefficients if necessary, and then to multiply by a factor that will give whole number coefficients for the final balanced equation.

    a. $C_3H_8(g) + O_2(g) \rightarrow CO_2(g) + H_2O(g)$

         balance carbon: $C_3H_8(g) + O_2(g) \rightarrow \mathbf{3}CO_2(g) + H_2O(g)$

         balance hydrogen: $C_3H_8(g) + O_2(g) \rightarrow 3CO_2(g) + \mathbf{4}H_2O(g)$

         balance oxygen: $C_3H_8(g) + \mathbf{5}O_2(g) \rightarrow 3CO_2(g) + 4H_2O(g)$

         balanced equation: $C_3H_8(g) + 5O_2(g) \rightarrow 3CO_2(g) + 4H_2O(g)$

    b. $C_2H_4(g) + O_2(g) \rightarrow CO_2(g) + H_2O(g)$

         balance carbon: $C_2H_4(g) + O_2(g) \rightarrow \mathbf{2}CO_2(g) + H_2O(g)$

         balance hydrogen: $C_2H_4(g) + O_2(g) \rightarrow 2CO_2(g) + \mathbf{2}H_2O(g)$

         balance oxygen: $C_2H_4(g) + \mathbf{3}O_2(g) \rightarrow 2CO_2(g) + 2H_2O(g)$

         balanced equation: $C_2H_4(g) + 3O_2(g) \rightarrow 2CO_2(g) + 2H_2O(g)$

    c. $C_4H_{10}(g) + O_2(g) \rightarrow CO_2(g) + H_2O(g)$

         balance carbon: $C_4H_{10}(g) + O_2(g) \rightarrow \mathbf{4}CO_2(g) + H_2O(g)$

         balance hydrogen: $C_4H_{10}(g) + O_2(g) \rightarrow 4CO_2(g) + \mathbf{5}H_2O(g)$

         balance oxygen: $C_4H_{10}(g) + \frac{13}{2}O_2(g) \rightarrow 4CO_2(g) + 5H_2O(g)$

         balanced equation: $2C_4H_{10}(g) + 13O_2(g) \rightarrow 8CO_2(g) + 10H_2O(g)$

61. Specific examples will depend on the students' input. A typical combustion reaction is represented by the reaction of methane (CH_4) with oxygen gas

         $CH_4(g) + 2O_2(g) \rightarrow CO_2(g) + 2H_2O(g)$.

62. A reaction in which small molecules or atoms combine to make a larger molecule is called a *synthesis* reaction. An example would be the synthesis of sodium chloride from the elements

         $2Na(s) + Cl_2(g) \rightarrow 2NaCl(s)$.

A reaction in which a molecule is broken down into simpler molecules or atoms is called a *decomposition* reaction. An example would be the decomposition of sodium hydrogen carbonate when heated.

$2NaHCO_3(s) \rightarrow Na_2CO_3(s) + CO_2(g) + H_2O(g)$.

Specific examples will depend on the students' input.

63.
a. $CaO(s) + H_2O(l) \rightarrow Ca(OH)_2(s)$
b. $4Fe(s) + 3O_2(g) \rightarrow 2Fe_2O_3(s)$
c. $P_2O_5(s) + 3H_2O(l) \rightarrow 2H_3PO_4(aq)$

64.
a. $8Fe(s) + S_8(s) \rightarrow 8FeS(s)$
b. $4Co(s) + 3O_2(g) \rightarrow 2Co_2O_3(s)$
c. $Cl_2O_7(g) + H_2O(l) \rightarrow 2HClO_4(aq)$

65.
a. $CaSO_4(s) \rightarrow CaO(s) + SO_3(g)$
b. $Li_2CO_3(s) \rightarrow Li_2O(s) + CO_2(g)$
c. $2LiHCO_3(s) \rightarrow Li_2CO_3(s) + H_2O(g) + CO_2(g)$
d. $C_6H_6(l) \rightarrow 6C(s) + 3H_2(g)$
e. $4PBr_3(l) \rightarrow P_4(s) + 6Br_2(l)$

66.
a. $2Al(s) + 3Br_2(l) \rightarrow 2AlBr_3(s)$
b. $Zn(s) + 2HClO_4(aq) \rightarrow Zn(ClO_4)_2(aq) + H_2(g)$
c. $3Na(s) + P(s) \rightarrow Na_3P(s)$
d. $CH_4(g) + 4Cl_2(g) \rightarrow CCl_4(l) + 4HCl(g)$
e. $Cu(s) + 2AgNO_3(aq) \rightarrow Cu(NO_3)_2(aq) + 2Ag(s)$

67. A *molecular equation* uses the normal, uncharged formulas for the compounds involved. The *complete ionic equation* shows the compounds involved broken up into their respective ions (*all* ions present are shown). The *net ionic equation* shows only those ions that combine to form a precipitate, a gas, or a nonionic product such as water. The net ionic equation shows most clearly the species that are combining with each other.

68. (c); (a), (b), and (d) will form insoluble compounds with Pb^{2+}. For (e), a compound will not form between Na^+ and Pb^{2+}.

69.
a. $2Fe^{3+}(aq) + 3CO_3^{2-}(aq) \rightarrow Fe_2(CO_3)_3(s)$
b. $Hg_2^{2+}(aq) + 2\,Cl^-(aq) \rightarrow Hg_2Cl_2(s)$
c. no precipitate
d. $Cu^{2+}(aq) + S^{2-}(aq) \rightarrow CuS(s)$
e. $Pb^{2+}(aq) + 2Cl^-(aq) \rightarrow PbCl_2(s)$
f. $Ca^{2+}(aq) + CO_3^{2-}(aq) \rightarrow CaCO_3(s)$
g. $Au^{3+}(aq) + 3OH^-(aq) \rightarrow Au(OH)_3(s)$

Chapter 7: Reactions in Aqueous Solution

70. The formulas of the salts are indicated in boldface type.

    a. $HNO_3(aq) + KOH(aq) \rightarrow H_2O(l) + \mathbf{KNO_3}(aq)$

    b. $H_2SO_4(aq) + Ba(OH)_2(aq) \rightarrow 2H_2O(l) + \mathbf{BaSO_4}(s)$

    c. $HClO_4(aq) + NaOH(aq) \rightarrow H_2O(l) + \mathbf{NaClO_4}(aq)$

    d. $2HCl(aq) + Ca(OH)_2(aq) \rightarrow 2H_2O(l) + \mathbf{CaCl_2}(aq)$

71. For each cation, the precipitates that form with the anions listed in the right–hand column are given below. If no formula is listed, it should be assumed that the anion does *not* form a precipitate with the particular cation. See Table 7.1 for the Solubility Rules.

    Ag^+ ion: $AgCl, Ag_2CO_3, AgOH, Ag_3PO_4, Ag_2S, Ag_2SO_4$

    Ba^{2+} ion: $BaCO_3, Ba(OH)_2, Ba_3(PO_4)_2, BaS, BaSO_4$

    Ca^{2+} ion: $CaCO_3, Ca(OH)_2, Ca_3(PO_4)_2, CaS, CaSO_4$

    Fe^{3+} ion: $Fe_2(CO_3)_3, Fe(OH)_3, FePO_4, Fe_2S_3$

    Hg_2^{2+} ion: $Hg_2Cl_2, Hg_2CO_3, Hg_2(OH)_2, (Hg_2)_3(PO_4)_2, Hg_2S$

    Na^+ ion: all common salts are soluble

    Ni^{2+} ion: $NiCO_3, Ni(OH)_2, Ni_3(PO_4)_2, NiS$

    Pb^{2+} ion: $PbCl_2, PbCO_3, Pb(OH)_2, Pb_3(PO_4)_2, PbS, PbSO_4$

72. a. $2AgNO_3(aq) + H_2SO_4(aq) \rightarrow Ag_2SO_4(s) + 2HNO_3(aq)$

    b. $Ca(NO_3)_2(aq) + H_2SO_4(aq) \rightarrow CaSO_4(s) + 2HNO_3(aq)$

    c. $Pb(NO_3)_2(aq) + H_2SO_4(aq) \rightarrow PbSO_4(s) + 2HNO_3(aq)$

73. a. iron(III) hydroxide, $Fe(OH)_3$. Rule 5: Most hydroxide salts are only slightly soluble.

    b. nickel(II) sulfide, NiS. Rule 6: Most sulfide salts are only slightly soluble.

    c. silver chloride, AgCl. Rule 3: Although Most chloride salts are soluble, AgCl is a listed exception

    d. barium carbonate, $BaCO_3$. Rule 6: Most carbonate salts are only slightly soluble.

    e. mercury(I) chloride or mercurous chloride, Hg_2Cl_2. Rule 3: Although Most chloride salts are soluble, Hg_2Cl_2 is a listed exception

    f. barium sulfate, $BaSO_4$. Rule 4: Although Most sulfate salts are soluble, $BaSO_4$ is a listed exception

74. a. $H_2SO_4(aq) + 2NaOH(aq) \rightarrow 2H_2O(l) + Na_2SO_4(aq)$

    b. $HNO_3(aq) + RbOH(aq) \rightarrow H_2O(l) + RbNO_3(aq)$

    c. $HClO_4(aq) + KOH(aq) \rightarrow 2H_2O(l) + KClO_4(aq)$

    d. $HCl(aq) + KOH(aq) \rightarrow H_2O(l) + KCl(aq)$

Chapter 7: Reactions in Aqueous Solution

75. a. Rule 3: $Ag^+(aq) + Cl^-(aq) \rightarrow AgCl(s)$

b. Rule 6: $3Ca^{2+}(aq) + 2PO_4^{3-}(aq) \rightarrow Ca_3(PO_4)_2(s)$

c. Rule 3: $Pb^{2+}(aq) + 2Cl^-(aq) \rightarrow PbCl_2(s)$

d. Rule 6: $Fe^{3+}(aq) + 3OH^-(aq) \rightarrow Fe(OH)_3(s)$

76. Molecular: $Na_2SO_4(aq) + CaCl_2(aq) \rightarrow 2NaCl(aq) + CaSO_4(s)$

Complete Ionic: $2Na^+(aq) + SO_4^{2-}(aq) + Ca^{2+}(aq) + 2Cl^-(aq) \rightarrow 2Na^+(aq) + 2Cl^-(aq) + CaSO_4(s)$

Net Ionic: $SO_4^{2-}(aq) + Ca^{2+}(aq) \rightarrow CaSO_4(s)$

77. a. potassium hydroxide and perchloric acid

b. cesium hydroxide and nitric acid

c. potassium hydroxide and hydrochloric acid

d. sodium hydroxide and sulfuric acid

78. Aluminum atoms lose 3 electrons to become Al^{3+} ions. Iodine atoms gain 1 electron each to become I^- ions.

79. Fe_2S_3 is made up of Fe^{3+} and S^{2-} ions. Iron atoms each lose three electrons to become Fe^{3+} ions. Sulfur atoms each gain two electrons to become S^{2-} ions.

80. a. $Na + O_2 \rightarrow Na_2O_2$

Balance sodium: **2**$Na + O_2 \rightarrow Na_2O_2$

Balanced equation: $2Na(s) + O_2(g) \rightarrow Na_2O_2(s)$

b. $Fe(s) + H_2SO_4(aq) \rightarrow FeSO_4(aq) + H_2(g)$

Equation is already balanced!

c. $Al_2O_3 \rightarrow Al + O_2$

Balance oxygen: **2**$Al_2O_3 \rightarrow Al + 3O_2$

Balance aluminum: $2Al_2O_3 \rightarrow $ **4**$Al + 3O_2$

Balanced equation: $2Al_2O_3(s) \rightarrow 4Al(s) + 3O_2(g)$

d. $Fe + Br_2 \rightarrow FeBr_3$

Balance bromine: $Fe + $ **3**$Br_2 \rightarrow $ **2**$FeBr_3$

Balance iron: **2**$Fe + 3Br_2 \rightarrow 2FeBr_3$

Balanced equation: $2Fe(s) + 3Br_2(l) \rightarrow 2FeBr_3(s)$

e. $Zn + HNO_3 \rightarrow Zn(NO_3)_2 + H_2$

Balance nitrate ions: $Zn + $ **2**$HNO_3 \rightarrow Zn(NO_3)_2 + H_2$

Balanced equation: $Zn(s) + 2HNO_3(aq) \rightarrow Zn(NO_3)_2(aq) + H_2(g)$

Chapter 7: Reactions in Aqueous Solution

81. For each reaction, the type of reaction is first identified, followed by some of the reasoning that leads to this choice (there may be more than one way in which you can recognize a particular type of reaction).

   a. oxidation–reduction (Mg changes from the elemental state to the combined state in $MgSO_4$; hydrogen changes from the combined to the elemental state.)

   b. acid–base ($HClO_4$ is a strong acid and $RbOH$ is a strong base; water and a salt are produced.)

   c. oxidation–reduction (Both Ca and O_2 change from the elemental to the combined state.)

   d. acid–base (H_2SO_4 is a strong acid and $NaOH$ is a strong base; water and a salt are produced.)

   e. precipitation (From the Solubility Rules of Table 7.1, $PbCO_3$ is insoluble.)

   f. precipitation (From the Solubility Rules of Table 7.1, $CaSO_4$ is insoluble.)

   g. acid–base (HNO_3 is a strong acid and KOH is a strong base; water and a salt are produced.)

   h. precipitation (From the Solubility Rules of Table 7.1, NiS is insoluble.)

   i. oxidation–reduction (both Ni and Cl_2 change from the elemental to the combined state).

82. (a), (b), and (c); All three statements are true.

83.
   a. $4FeO(s) + O_2(g) \rightarrow 2Fe_2O_3(s)$
   b. $2CO(g) + O_2(g) \rightarrow 2CO_2(g)$
   c. $H_2(g) + Cl_2(g) \rightarrow 2HCl(g)$
   d. $16K(s) + S_8(s) \rightarrow 8K_2S(s)$
   e. $6Na(s) + N_2(g) \rightarrow 2Na_3N(s)$

84.
   a. $2NaHCO_3(s) \rightarrow Na_2CO_3(s) + H_2O(g) + CO_2(g)$
   b. $2NaClO_3(s) \rightarrow 2NaCl(s) + 3O_2(g)$
   c. $2HgO(s) \rightarrow 2Hg(l) + O_2(g)$
   d. $C_{12}H_{22}O_{11}(s) \rightarrow 12C(s) + 11H_2O(g)$
   e. $2H_2O_2(l) \rightarrow 2H_2O(l) + O_2(g)$

85. For simplicity, the physical states of the substances are omitted.

   $2Ba + O_2 \rightarrow 2BaO$

   $Ba + S \rightarrow BaS$

   $Ba + Cl_2 \rightarrow BaCl_2$

   $3Ba + N_2 \rightarrow Ba_3N_2$

   $Ba + Br_2 \rightarrow BaBr_2$

   $4K + O_2 \rightarrow 2K_2O$

   $2K + S \rightarrow K_2S$

Chapter 7: Reactions in Aqueous Solution

$2K + Cl_2 \rightarrow 2KCl$

$6K + N_2 \rightarrow 2K_3N$

$2K + Br_2 \rightarrow 2KBr$

$2Mg + O_2 \rightarrow 2MgO$

$Mg + S \rightarrow MgS$

$Mg + Cl_2 \rightarrow MgCl_2$

$3Mg + N_2 \rightarrow Mg_3N_2$

$Mg + Br_2 \rightarrow MgBr_2$

$4Rb + O_2 \rightarrow 2Rb_2O$

$2Rb + S \rightarrow Rb_2S$

$2Rb + Cl_2 \rightarrow 2RbCl$

$6Rb + N_2 \rightarrow 2Rb_3N$

$2Rb + Br_2 \rightarrow 2RbBr$

$2Ca + O_2 \rightarrow 2CaO$

$Ca + S \rightarrow CaS$

$Ca + Cl_2 \rightarrow CaCl_2$

$3Ca + N_2 \rightarrow Ca_3N_2$

$Ca + Br_2 \rightarrow CaBr_2$

$4Li + O_2 \rightarrow 2Li_2O$

$2Li + S \rightarrow Li_2S$

$2Li + Cl_2 \rightarrow 2LiCl$

$6Li + N_2 \rightarrow 2Li_3N$

$2Li + Br_2 \rightarrow 2LiBr$

86. $6Na + N_2 \rightarrow 2Na_3N$

87. For simplicity, the physical states of the substances are omitted.

    $Mg + Cl_2 \rightarrow MgCl_2$

    $Ca + Cl_2 \rightarrow CaCl_2$

    $Sr + Cl_2 \rightarrow SrCl_2$

    $Ba + Cl_2 \rightarrow BaCl_2$

    $Mg + Br_2 \rightarrow MgBr_2$

    $Ca + Br_2 \rightarrow CaBr_2$

    $Sr + Br_2 \rightarrow SrBr_2$

    $Ba + Br_2 \rightarrow BaBr_2$

Chapter 7: Reactions in Aqueous Solution

$$2Mg + O_2 \rightarrow 2MgO$$

$$2Ca + O_2 \rightarrow 2CaO$$

$$2Sr + O_2 \rightarrow 2SrO$$

$$2Ba + O_2 \rightarrow 2BaO$$

88. a. one
     b. one
     c. two
     d. two
     e. three

89. a. two; $O + 2e^- \rightarrow O^{2-}$
     b. one; $F + e^- \rightarrow F^-$
     c. three; $N + 3e^- \rightarrow N^{3-}$
     d. one; $Cl + e^- \rightarrow Cl^-$
     e. two; $S + 2e^- \rightarrow S^{2-}$

90. False. The balanced molecular equation is: $Ba(OH)_2(aq) + H_2SO_4(aq) \rightarrow BaSO_4(s) + 2H_2O(l)$. The complete ionic equation is: $Ba^{2+}(aq) + 2OH^-(aq) + 2H^+(aq) + SO_4^{2-}(aq) \rightarrow BaSO_4(s) + 2H_2O(l)$. The net ionic equation includes all species that take part in the chemical reaction. The OH^- and H^+ ions form water so they are also included in the net ionic equation. Thus the complete ionic equation and net ionic equation are the same.

91. a. $2I_4O_9(s) \rightarrow 2I_2O_6(s) + 2I_2(s) + 3O_2(g)$

     oxidation–reduction, decomposition

     b. $Mg(s) + 2AgNO_3(aq) \rightarrow Mg(NO_3)_2(aq) + 2Ag(s)$

     oxidation–reduction, single–displacement

     c. $SiCl_4(l) + 2Mg(s) \rightarrow 2MgCl_2(s) + Si(s)$

     oxidation–reduction, single–displacement

     d. $CuCl_2(aq) + 2AgNO_3(aq) \rightarrow Cu(NO_3)_2(aq) + 2AgCl(s)$

     precipitation, double–displacement

     e. $2Al(s) + 3Br_2(l) \rightarrow 2AlBr_3(s)$

     oxidation–reduction, synthesis

92. $3Na_2CrO_4(aq) + 2AlBr_3(aq) \rightarrow Al_2(CrO_4)_3(s) + 6NaBr(aq)$

93. $2Zn(s) + O_2(g) \rightarrow 2ZnO(s)$

     $4Al(s) + 3O_2(g) \rightarrow 2Al_2O_3(s)$

     $2Fe(s) + O_2(g) \rightarrow 2FeO(s)$; $4Fe(s) + 3O_2(g) \rightarrow 2Fe_2O_3(s)$

Chapter 7: Reactions in Aqueous Solution

$2Cr(s) + O_2(g) \rightarrow 2CrO(s); 4Cr(s) + 3O_2(g) \rightarrow 2Cr_2O_3(s)$

$2Ni(s) + O_2(g) \rightarrow 2NiO(s)$

94. $PbCl_2$; $PbSO_4$; $Pb_3(PO_4)_2$; $AgCl$; Ag_3PO_4

95. You have your choice of reactions here as long as they illustrate the correct reaction type. Those listed below are only examples:

   a. $C(s) + O_2(g) \rightarrow CO_2(g)$

   b. $AgNO_3(aq) + NaCl(aq) \rightarrow AgCl(s) + NaNO_3(aq)$

   c. $AgNO_3(aq) + NaCl(aq) \rightarrow AgCl(s) + NaNO_3(aq)$

   d. $H_2SO_4(aq) + 2NaOH(aq) \rightarrow Na_2SO_4(aq) + 2H_2O(l)$

   e. $C(s) + O_2(g) \rightarrow CO_2(g)$

   f. $C(s) + O_2(g) \rightarrow CO_2(g)$

   Note that some examples are repeated: a given reaction may sometimes be classified as more than one type of reaction. The reaction between carbon and oxygen gas, for example, is at the same time a combustion reaction (burning in oxygen), an oxidation–reduction reaction (the oxidation numbers of carbon and oxygen both change) and a synthesis reaction (two smaller entities unite to form a larger, more complex molecule)

96. $PbSO_4$; $AgCl$; none

97. $Sr_3(PO_4)_2$; Ag_2CO_3; none; $AgCl$; $PbCl_2$

CUMULATIVE REVIEW

Chapters 6 and 7

1. There are numerous ways we can recognize that a chemical reaction has taken place.

   In some reactions there may be a *color change*. For example, the ions of many of the transition metals are brightly colored in aqueous solution. If one of these ions undergoes a reaction in which the oxidation state changes, however, the characteristic color of the ion may be changed. For example, when a piece of zinc is added to an aqueous copper(II) ion solution (which is bright blue), the Cu^{2+} ions are reduced to copper metal, and the blue color of the solution fades as the reaction takes place.

   $$Zn(s) + Cu^{2+}(aq, \text{ blue solution}) \rightarrow Zn^{2+}(aq) + Cu(s, \text{ reddish-black solid})$$

   In many reactions of ionic solutes, a solid *precipitate* may form when the ions are combined. For example, when a clear, colorless aqueous solution of sodium chloride is added to a clear, colorless solution of silver nitrate, a white solid of silver chloride forms and settles out of the mixture.

   $$AgNO_3(aq) + NaCl(aq) \rightarrow NaNO_3(aq) + AgCl(s)$$

   In some reactions, *bubbles of a gaseous product* may form. For example, if a piece of magnesium metal is added to a solution of hydrochloric acid, bubbles of hydrogen gas form at the surface of the magnesium.

   $$Mg(s) + 2HCl(aq) \rightarrow MgCl_2(aq) + H_2(g)$$

   In some reactions, particularly in the combustion of organic chemical substances with oxygen gas, heat, light, and a flame may be produced. For example, when methane (natural gas) is burned in oxygen, a luminous flame is produced and heat energy is released:

   $$CH_4(g) + 2O_2 \rightarrow CO_2(g) + 2H_2O(g) + \text{energy}.$$

   All chemical reactions do produce some evidence that the reaction has occurred, but sometimes this evidence may *not* be visual, and may not be very obvious. For example, when very dilute aqueous solutions of acids and bases are mixed, the neutralization reaction

   $$H^+(aq) + OH^-(aq) \rightarrow H_2O(l)$$

   takes place. However, the only evidence for this reaction is the release of heat energy, which should be evident as a temperature change for the mixture. Because water has a relatively high specific heat capacity, however, if the acid and base solutions are very dilute, the temperature may only change by a fraction of a degree and may not be noticed.

2. A chemical equation indicates the substances necessary for a chemical reaction to take place, as well as what is produced by that chemical reaction. The substances to the left of the arrow in a chemical equation are called the reactants; those to the right of the arrow are referred to as the products. In addition, if a chemical equation has been balanced, then the equation indicates the relative proportions in which the reactant molecules combine to form the product molecules.

3. When we "balance" a chemical equation, we adjust the *coefficients* of the reactants and products in the equation so that the same total numbers of atoms of each element are present both before and after the reaction has taken place. Balancing chemical equations is so important because a balanced chemical equation shows us not only the identities of the reactants and products, but also the relative numbers of each involved in the process: this information is necessary if we are to do any sort of calculation involving the amounts of reactants required for a process or are to calculate the yield expected from a process. When we say that atoms must be *conserved* when writing a balanced chemical equations, we mean that the number of atoms of each element must be the same after the reaction is complete as before the reaction was attempted: no atoms are created or destroyed during a chemical reaction, they are just arranged into new combinations. We often indicate the physical states of the reactants and products for a chemical reaction, because sometimes the physical state is important for the reaction to be successful. The physical states are indicated by using letters in parentheses after the formula: (*s*), (*l*), (*g*), or (*aq*). For example, magnesium metal does not react to any appreciable extent with solid water (ice) or with liquid water at room temperature, but does react readily with gaseous water (steam) above 100°C.

$$Mg(s) + H_2O(s \text{ or } l) \rightarrow \text{ no reaction}$$

$$Mg(s) + 2H_2O(g) \rightarrow Mg(OH)_2(s) + H_2(g)$$

The amount of energy consumed or released by a chemical reaction is also strongly dependent on the physical states of the reactants and products.

4. It is *never* permissible to change the subscripts of a formula when balancing a chemical equation: changing the subscripts changes the *identity* of a substance from one chemical to another. For example, consider the unbalanced chemical equation

$$H_2(g) + O_2(g) \rightarrow H_2O(l).$$

If you changed the *formula* of the product from $H_2O(l)$ to $H_2O_2(l)$, the equation would appear to be "balanced". However, H_2O is water, whereas H_2O_2 is hydrogen peroxide–a completely different chemical substance (which is not prepared by reaction of the elements hydrogen and oxygen).

When we balance a chemical equation, it is permitted only to adjust the *coefficients* of a formula, so changing a coefficient merely changes the number of molecules of a substance being used in the reaction, without changing the identity of the substance. For the example above, we can balance the equation by putting coefficients of 2 in front of the formulas of H_2 and H_2O: these coefficients do not change the nature of what is reacting and what product is formed.

$$2H_2(g) + O_2(g) \rightarrow 2H_2O(l)$$

5. The concept of a "driving force" for chemical reactions, at this point, is a rather nebulous idea. Clearly there must be some reason why certain substances react when combined, and why other substances can be combined without anything happening. If you continue with further studies in chemistry, you will learn that there is a mathematical thermodynamic function that can be used to predict exactly what will happen when a given set of reactants is combined. Even at this point, however, we can use some generalizations about what sorts of events tend to make a reaction take place. A reaction is likely to occur if any of the following things occur as a result of the reaction: formation of a solid, formation of water (or another nonionized molecule), formation of a gas, or the transfer of electrons from one species to another. Here are examples of reactions illustrating each of these:

formation of a solid:

$BaCl_2(aq) + K_2CrO_4(aq) \rightarrow 2KCl(aq) + BaCrO_4(s)$

$Pb(NO_3)_2(aq) + 2NaCl(aq) \rightarrow PbCl_2(s) + 2NaNO_3(aq)$

formation of water:

$HCl(aq) + NaOH(aq) \rightarrow NaCl(aq) + H_2O(l)$

$CH_3COOH(aq) + KOH(aq) \rightarrow KCH_3COO(aq) + H_2O(l)$

formation of a gas:

$2NI_3(s) \rightarrow N_2(g) + 3I_2(s)$

$CaCO_3(s) \rightarrow CaO(s) + CO_2(g)$

transfer of electrons:

$Zn(s) + 2Ag^+(aq) \rightarrow Zn^{2+}(aq) + 2Ag(s)$

$Mg(s) + Cu^{2+}(aq) \rightarrow Mg^{2+}(aq) + Cu(s)$

6. A precipitation reaction is one in which a *solid* forms when the reactants are combined: the solid is called a precipitate. The driving force in such a reaction is the formation of the solid, thus *removing ions* from the solution. There are many examples of such precipitation reactions: consult the solubility rules in Table 7.1 if you need help. One example would be to combine barium nitrate and sodium carbonate solutions: a precipitate of barium carbonate would form.

The molecular equation for this reaction is:

$Ba(NO_3)_2(aq) + Na_2CO_3(aq) \rightarrow BaCO_3(s) + 2NaNO_3(aq)$

The net ionic equation for this reaction is:

$Ba^{2+}(aq) + CO_3^{2-}(aq) \rightarrow BaCO_3(s)$

7. A strong electrolyte is one which completely dissociates into ions when dissolved in water: that is, each unit of the substance that dissolves in water produces separated, free ions. Ionic compounds are strong electrolytes if they are soluble in water, because they already consist of ions in the solid state. Certain acids and bases also behave as strong electrolytes. A solution of a strong electrolyte actually consists of free, separated ions moving through the solvent independently of one another (there are no molecules or clusters of combined positive and negative ions). An apparatus for experimentally determining whether or not a substance is an electrolyte is shown in Figure 7.2 in the text.

8. In summary, nearly all compounds containing the nitrate, sodium, potassium, and ammonium ions are soluble in water. Most salts containing the chloride and sulfate ions are soluble in water, with specific exceptions (see Table 7.1 for these exceptions). Most compounds containing the hydroxide, sulfide, carbonate, and phosphate ions are not soluble in water, unless the compound also contains one of the cations mentioned above (Na^+, K^+, NH_4^+).

The solubility rules are phrased as if you had a sample of a given solute and wanted to see if you could dissolve it in water. These rules can also be applied, however, to predict the identity of the solid produced in a precipitation reaction: a given combination of ions will not be soluble in water whether you take a pure compound out of a reagent bottle or if you generate the insoluble combination of ions during a chemical reaction. For example, the solubility rules say that $BaSO_4$ is not soluble in water. This means not only that a pure sample of $BaSO_4$ taken from a reagent bottle will not dissolve in water, but also that if Ba^{2+} ion and SO_4^{2-} ion end up together in the same solution, they will precipitate as $BaSO_4$. If we were to combine barium chloride and sulfuric acid solutions

$$BaCl_2(aq) + H_2SO_4(aq) \rightarrow BaSO_4(s) + 2HCl(aq)$$

then, because barium sulfate is not soluble in water, a precipitate of $BaSO_4(s)$ would form. Because a precipitate of $BaSO_4(s)$ would form no matter what barium compound or what sulfate compound were mixed, we can write the net ionic equation for the reaction as

$$Ba^{2+}(aq) + SO_4^{2-}(aq) \rightarrow BaSO_4(s).$$

Thus if, for example, barium nitrate solution were combined with sodium sulfate solution, a precipitate of $BaSO_4$ would form. Barium sulfate is insoluble in water regardless of its source.

9. The spectator ions in a precipitation reaction are, basically, the ions in the solution that do *not* precipitate. Since we take the actual chemical reaction in a precipitation process to be the formation of the solid, and the spectator ions are not found in and do not participate in the formation of the solid, we leave them out of the net ionic equation for the reaction. Not including the spectator ions in the net ionic equation for a precipitation reaction also has another important implication: if we write the net ionic equation for the formation of silver chloride

$$Ag^+(aq) + Cl^-(aq) \rightarrow AgCl(s)$$

we are indicating that it is these specific ions that will combine to form silver chloride, regardless of where they come from. If any soluble silver salt is combined with any soluble chloride, we should get a precipitate of silver chloride. For example, silver nitrate (a common soluble silver salt) reacts in exactly the same manner with NaCl, KCl, NH_4Cl, and HCl, which are all soluble compounds containing the chloride ion.

Just because we leave the spectator ions out when writing a net ionic equation for a reaction does not mean that the spectator ions do not have to be present: the spectator ions are needed to provide a balance of charge in the reactant compounds for the ions that combine to form the precipitate. For the reaction above in which silver chloride is formed, it would not be possible to have a reagent bottle containing just silver ion (there would have to be some negative ion present) or just chloride ion (there would have to be some positive ion present).

10. Acids (such as the citric acid found in citrus fruits and the acetic acid found in vinegar) were first noted primarily because of their sour taste. The first bases noted were characterized by their bitter taste and slippery feel on the skin. Acids and bases chemically react with (neutralize) each other forming water: the net ionic equation is

$$H^+(aq) + OH^-(aq) \rightarrow H_2O(l).$$

The *strong* acids and bases fully ionize when they dissolve in water: because these substances fully ionize, they are strong electrolytes. The common strong acids are HCl(hydrochloric), HNO_3(nitric), H_2SO_4(sulfuric), and $HClO_4$(perchloric). The most common strong bases are the alkali metal hydroxides, particularly NaOH(sodium hydroxide) and KOH(potassium hydroxide).

11. A salt is basically any ionic compound that contains any ions other than H^+ and OH^- (compounds containing these ions are called acids and bases, respectively). In particular, a salt is formed in the neutralization reaction between an acid and a base. Your textbook describes acid–base neutralization reactions as reactions that result in the formation of water: the water results from the combination of the $H^+(aq)$ ion from the acid with the $OH^-(aq)$ ion from the base

$$H^+(aq) + OH^-(aq) \rightarrow H_2O(l)$$

Cumulative Review: Chapters 6 and 7

However, the H^+ ion must have been paired with some negative ion in the original acid solution, and the OH^- ion must have been paired with some positive ion in the original base solution. These counter-ions to the acid–base ions are what constitute the salt that is formed during the neutralization. Below are three acid–base neutralization reactions, with the salts that are formed indicated:

$$HCl(aq) + NaOH(aq) \rightarrow H_2O(l) + NaCl(aq, salt)$$

$$HNO_3(aq) + KOH(aq) \rightarrow H_2O(l) + KNO_3(aq, salt)$$

$$HC_2H_3O_2(aq) + NaOH(aq) \rightarrow H_2O(l) + NaC_2H_3O_2(aq, salt)$$

12. Oxidation–reduction reactions are electron-transfer reactions. Oxidation represents a loss of electrons by an atom, molecule, or ion, whereas reduction is the gain of electrons by such a species. Because an oxidation–reduction process represents the transfer of electrons between species, you can't have one without the other also taking place: the electrons lost by one species must be gained by some other species. An example of a simple oxidation–reduction reaction between a metal and a nonmetal could be the following

$$Mg(s) + F_2(g) \rightarrow MgF_2(s).$$

In this process, Mg atoms lose two electrons each to become Mg^{2+} ions in MgF_2: Mg is oxidized. Each F atom of F_2 gains one electron to become a F^- ion, for a total of two electrons gained for each F_2 molecule: F_2 is reduced.

$$Mg \rightarrow Mg^{2+} + 2e^-$$

$$2(F + e^- \rightarrow F^-)$$

13. Combustion reactions represent processes involving oxygen gas that release energy rapidly enough that a flame is produced. Combustion reactions are a special sub-class of oxidation–reduction reactions (the fact that elemental oxygen gas is a reactant but combined oxygen is a product shows this). The most common combustion reactions are those we make use of through the burning of petroleum products as sources of heat, light, or other forms of energy. For example, the burning of methane (natural gas) is shown below.

$$CH_4(g) + 2O_2(g) \rightarrow CO_2(g) + 2H_2O(g) + energy$$

Combustion reactions of other types of substances are also possible, however. For example, magnesium metal burns in oxygen gas (with a very bright, intense flame) to produce magnesium oxide.

$$2Mg(s) + O_2(g) \rightarrow 2MgO(s) + energy$$

14. In general, a synthesis reaction represents the reaction of elements or simple compounds to produce more complex substances. There are many examples of synthesis reactions, for example

$$N_2(g) + 3H_2(g) \rightarrow 2NH_3(g)$$

$$NaOH(aq) + CO_2(g) \rightarrow NaHCO_3(s).$$

Decomposition reactions represent the breakdown of a more complex substance into simpler substances. There are many examples of decomposition reactions, for example

$$2H_2O_2(aq) \rightarrow 2H_2O(l) + O_2(g).$$

Synthesis and decomposition reactions are very often also oxidation–reduction reactions, especially if an elemental substance reacts or is generated. It is not necessary, however, for

Cumulative Review: Chapters 6 and 7

synthesis and decomposition reactions to always involve oxidation–reduction. The reaction between NaOH and CO_2 given as an example of a synthesis reaction does *not* represent oxidation–reduction.

15. The different ways of classifying chemical reactions that have been discussed in the text are listed below, along with an example of each type of reaction:

formation of a solid (precipitation):

$FeCl_3(aq) + 3NaOH(aq) \rightarrow Fe(OH)_3(s) + 3NaCl(aq)$

formation of water (acid–base):

$H_2SO_4(aq) + 2NaOH(aq) \rightarrow Na_2SO_4(aq) + 2H_2O(l)$

transfer of electrons (oxidation–reduction):

$2Na(s) + Cl_2(g) \rightarrow 2NaCl(s)$

combustion:

$2C_2H_6(g) + 7O_2(g) \rightarrow 4CO_2(g) + 6H_2O(g) + energy$

synthesis (combination):

$Ca(s) + Cl_2(g) \rightarrow CaCl_2(s)$

decomposition:

$2HgO(s) \rightarrow 2Hg(l) + O_2(g)$

single displacement:

$Mg(s) + 2AgNO_3(aq) \rightarrow Mg(NO_3)_2(aq) + 2Ag(s)$

double displacement:

$Na_2SO_4(aq) + BaCl_2(aq) \rightarrow 2NaCl(aq) + BaSO_4(s)$

16.
   a. $C(s) + O_2(g) \rightarrow CO_2(g)$
   b. $2C(s) + O_2(g) \rightarrow 2CO(g)$
   c. $2Li(l) + 2C(s) \rightarrow Li_2C_2(s)$
   d. $FeO(s) + C(s) \rightarrow Fe(l) + CO(g)$
   e. $C(s) + 2F_2(g) \rightarrow CF_4(g)$

17.
   a. $Na_2SO_4(aq) + BaCl_2(aq) \rightarrow BaSO_4(s) + 2NaCl(aq)$
   b. $Zn(s) + H_2O(g) \rightarrow ZnO(s) + H_2(g)$
   c. $3NaOH(aq) + H_3PO_4(aq) \rightarrow Na_3PO_4(aq) + 3H_2O(l)$
   d. $2Al(s) + Mn_2O_3(s) \rightarrow Al_2O_3(s) + 2Mn(s)$
   e. $2C_7H_6O_2(s) + 15O_2(g) \rightarrow 14CO_2(g) + 6H_2O(g)$
   f. $2C_6H_{14}(l) + 19O_2(g) \rightarrow 12CO_2(g) + 14H_2O(g)$
   g. $2C_3H_8O(l) + 9O_2(g) \rightarrow 6CO_2(g) + 8H_2O(g)$
   h. $Mg(s) + 2HClO_4(aq) \rightarrow Mg(ClO_4)_2(aq) + H_2(g)$

Cumulative Review: Chapters 6 and 7

18. a. $Ba(NO_3)_2(aq) + K_2CrO_4(aq) \rightarrow BaCrO_4(s) + 2KNO_3(aq)$

    b. $NaOH(aq) + CH_3COOH(aq) \rightarrow H_2O(l) + NaCH_3COO(aq)$ (then evaporate the water from the solution)

    c. $AgNO_3(aq) + NaCl(aq) \rightarrow AgCl(s) + NaNO_3(aq)$

    d. $Pb(NO_3)_2(aq) + H_2SO_4(aq) \rightarrow PbSO_4(s) + 2HNO_3(aq)$

    e. $2NaOH(aq) + H_2SO_4(aq) \rightarrow Na_2SO_4(aq) + 2H_2O(l)$ (then evaporate the water from the solution)

    f. $Ba(NO_3)_2(aq) + Na_2CO_3(aq) \rightarrow BaCO_3(s) + 2NaNO_3(aq)$

19. $HCl(aq) + NaOH(aq) \rightarrow NaCl(aq) + H_2O(l)$

    $HNO_3(aq) + NaOH(aq) \rightarrow NaNO_3(aq) + H_2O(l)$

    $H_2SO_4(aq) + 2NaOH(aq) \rightarrow Na_2SO_4(aq) + 2H_2O(l)$

    $HCl(aq) + KOH(aq) \rightarrow KCl(aq) + H_2O(l)$

    $HNO_3(aq) + KOH(aq) \rightarrow KNO_3(aq) + H_2O(l)$

    $H_2SO_4(aq) + 2KOH(aq) \rightarrow K_2SO_4(aq) + 2H_2O(l)$

20. a. $FeO(s) + 2HNO_3(aq) \rightarrow Fe(NO_3)_2(aq) + H_2O(l)$

    acid–base, double-displacement

    b. $2Mg(s) + 2CO_2(g) + O_2(g) \rightarrow 2MgCO_3(s)$

    synthesis; oxidation–reduction

    c. $2NaOH(s) + CuSO_4(aq) \rightarrow Cu(OH)_2(s) + Na_2SO_4(aq)$

    precipitation, double-displacement

    d. $HI(aq) + KOH(aq) \rightarrow KI(aq) + H_2O(l)$

    acid–base, double–displacement

    e. $C_3H_8(g) + 5O_2(g) \rightarrow 3CO_2(g) + 4H_2O(g)$

    combustion; oxidation–reduction

    f. $Co(NH_3)_6Cl_2(s) \rightarrow CoCl_2(s) + 6NH_3(g)$

    decomposition

    g. $2HCl(aq) + Pb(C_2H_3O_2)_2(aq) \rightarrow 2HC_2H_3O_2(aq) + PbCl_2(s)$

    precipitation, double-displacement

    h. $C_{12}H_{22}O_{11}(s) \rightarrow 12C(s) + 11H_2O(g)$

    decomposition; oxidation–reduction

    i. $2Al(s) + 6HNO_3(aq) \rightarrow 2Al(NO_3)_3(aq) + 3H_2(g)$

    oxidation–reduction; single-displacement

    j. $4B(s) + 3O_2(g) \rightarrow 2B_2O_3(s)$

    synthesis; oxidation–reduction

21. For simplicity, the physical states of the substances are omitted.

$2Na + F_2 \rightarrow 2NaF$ $\qquad$ $4Na + O_2 \rightarrow 2Na_2O$

$2Na + S \rightarrow Ns_2S$ $\qquad$ $2Na + Cl_2 \rightarrow 2NaCl$

$Ca + F_2 \rightarrow CaF_2$ $\qquad$ $2Ca + O_2 \rightarrow 2CaO$

$Ca + S \rightarrow CaS$ $\qquad$ $Ca + Cl_2 \rightarrow CaCl_2$

$2Al + 3F_2 \rightarrow 2AlF_3$ $\qquad$ $4Al + 3O_2 \rightarrow 2Al_2O_3$

$2Al + 3S \rightarrow Al_2S_3$ $\qquad$ $2Al + 3Cl_2 \rightarrow 2AlCl_3$

$Mg + F_2 \rightarrow MgF_2$ $\qquad$ $2Mg + O_2 \rightarrow 2MgO$

$Mg + S \rightarrow MgS$ $\qquad$ $Mg + Cl_2 \rightarrow MgCl_2$

22. Specific examples will depend on students' responses. The following are general equations that illustrate each type of reaction:

    precipitation: typical when solutions of two ionic solutes are mixed, and one of the new combinations of ions is insoluble.

    $A^+B^-(aq) + C^+D^-(aq) \rightarrow AD(s) + C^+D^-(aq)$

    single displacement: one element replaces a less reactive element from a compound.

    $A(s) + B^+C^-(aq) \rightarrow A^+C^-(aq) + B(s)$

    combustion: a rapid oxidation reaction, most commonly involving $O_2(g)$. Most examples in the text involve the combustion of hydrocarbons or hydrocarbon derivatives.

    (hydrocarbon or derivative) $+ O_2(g) \rightarrow CO_2(g) + H_2O(g)$

    synthesis: elements or simple compounds combine to make more complicated molecules.

    $A(s) + B(s) \rightarrow AB(s)$

    oxidation–reduction: reactions in which electrons are transferred from one species to another. Examples of oxidation–reduction reactions include single-displacement, combustion, synthesis, and decomposition reactions.

    decomposition: a compound breaks down into elements and/or simpler compounds.

    $AB(s) \rightarrow A(s) + B(s)$

    acid–base neutralization: a neutralization takes place when a proton from an acid combines with a hydroxide ion from a base to make a water molecule.

23. Specific answers depend on student choice of examples.

24. a. no reaction (all combinations are soluble)
    b. $Ca^{2+}(aq) + SO_4^{2-}(aq) \rightarrow CaSO_4(s)$
    c. $Pb^{2+}(aq) + S^{2-}(aq) \rightarrow PbS(s)$
    d. $2Fe^{3+}(aq) + 3CO_3^{2-}(aq) \rightarrow Fe_2(CO_3)_3(s)$
    e. $Hg_2^{2+}(aq) + 2Cl^-(aq) \rightarrow Hg_2Cl_2(s)$
    f. $Ag^+(aq) + Cl^-(aq) \rightarrow AgCl(s)$

Cumulative Review: Chapters 6 and 7

    g. $3Ca^{2+}(aq) + 2PO_4^{3-}(aq) \rightarrow Ca_3(PO_4)_2(s)$

        Since phosphoric acid is not a very strong acid, a more realistic equation might be

        $3Ca^{2+}(aq) + 2H_3PO_4(aq) \rightarrow Ca_3(PO_4)_2(s) + 6H^+(aq)$

    h.   no reaction (all combinations are soluble)

25.   a. $Pb(NO_3)_2(aq) + Na_2S(aq) \rightarrow PbS(s) + 2NaNO_3(aq)$

    b. $AgNO_3(aq) + HCl(aq) \rightarrow AgCl(s) + HNO_3(aq)$

    c. $2Mg(s) + O_2(g) \rightarrow 2MgO(s)$

    d. $H_2SO_4(aq) + 2KOH(aq) \rightarrow K_2SO_4(aq) + 2H_2O(l)$

    e. $BaCl_2(aq) + H_2SO_4(aq) \rightarrow BaSO_4(s) + 2HCl(aq)$

    f. $Mg(s) + H_2SO_4(aq) \rightarrow MgSO_4(aq) + H_2(g)$

    g. $2Na_3PO_3(aq) + 3CaCl_2(aq) \rightarrow Ca_3(PO_3)_2(s) + 6NaCl$

    h. $2C_4H_{10}(l) + 13O_2(g) \rightarrow 8CO_2(g) + 10H_2O(g)$

CHAPTER 8

Chemical Composition

1. $100 \text{ washers} \times \dfrac{0.110 \text{ g}}{1 \text{ washer}} = 11.0 \text{ g}$ (assuming 100 washers is exact).

    $100. \text{ g} \times \dfrac{1 \text{ washer}}{0.110 \text{ g}} = 909 \text{ washers}$

2. The empirical formula is CFH from the structure given. The empirical formula represents the smallest whole number ratio of the number and types of atoms present.

3. The *atomic mass unit* (amu) is defined by scientists to more simply describe relative masses on an atomic or molecular scale. One amu is equivalent to 1.66×10^{-24} g.

4. The average atomic mass takes into account the various isotopes of an element and the relative abundances in which those isotopes are found.

5. a. $125 \text{ C atoms} \times \dfrac{12.01 \text{ amu}}{1 \text{ C atom}} = 1.50 \times 10^3 \text{ amu}$

    b. $5 \times 10^6 \text{ K atoms} \times \dfrac{39.10 \text{ amu}}{1 \text{ K atom}} = 1.955 \times 10^8 \text{ amu} = 2 \times 10^8 \text{ amu}$

    c. $1.04 \times 10^{22} \text{ Li atoms} \times \dfrac{6.941 \text{ amu}}{1 \text{ Li atom}} = 7.22 \times 10^{22} \text{ amu}$

    d. $1 \text{ Mg atom} \times \dfrac{24.31 \text{ amu}}{1 \text{ Mg atom}} = 24.31 \text{ amu}$

    e. $3.011 \times 10^{23} \text{ I atoms} \times \dfrac{126.9 \text{ amu}}{1 \text{ I atom}} = 3.821 \times 10^{25} \text{ amu}$

6. a. $40.08 \text{ amu Ca} \times \dfrac{1 \text{ Ca atom}}{40.08 \text{ amu}} = 1 \text{ Ca atom}$

    b. $919.5 \text{ amu W} \times \dfrac{1 \text{ W atom}}{183.9 \text{ amu}} = 5 \text{ W atoms}$

    c. $549.4 \text{ amu Mn} \times \dfrac{1 \text{ Mn atom}}{54.94 \text{ amu}} = 10 \text{ Mn atoms}$

    d. $6345 \text{ amu I} \times \dfrac{1 \text{ I atom}}{126.9 \text{ amu}} = 50 \text{ I atoms}$

    e. $2072 \text{ amu} \times \dfrac{1 \text{ Pb atom}}{207.2 \text{ amu}} = 10 \text{ Pb atoms}$

Chapter 8: Chemical Composition

7. One "average" iron atom has a mass of 55.85 amu

    299 iron atoms would weigh: $299 \text{ atoms} \times \dfrac{55.85 \text{ amu}}{1 \text{ atom}} = 1.67 \times 10^4 \text{ amu}$

    5529.2 amu represents: $5529.2 \text{ amu} \times \dfrac{1 \text{ atom}}{55.85 \text{ amu}} = 99 \text{ atoms}.$

8. One bromine atom has a mass of 79.90 amu.

    A sample containing 54 bromine atoms would weigh: $54 \text{ atoms} \times \dfrac{79.9 \text{ amu}}{1 \text{ atom}} = 4315 \text{ amu};$

    5672.9 amu of bromine would represent: $5672.9 \text{ amu} \times \dfrac{1 \text{ atom}}{79.9 \text{ amu}} = 71 \text{ atoms}.$

9. Avogadro's number (6.022×10^{23} atoms; 1.00 mol)

10. 118.7 g (1.00 mol)

11. molar masses: Na, 22.99 g; K, 39.10 g

    $11.50 \text{ g Na} \times \dfrac{1 \text{ mol Na}}{22.99 \text{ g}} = 0.5002 \text{ mol Na}$

    $0.5002 \text{ mol Na} \times \dfrac{6.033 \times 10^{23}}{1 \text{ mol Na}} = 3.012 \times 10^{23} \text{ atoms}$

    $0.5002 \text{ mol K} \times \dfrac{39.10 \text{ g}}{1 \text{ mol K}} = 19.56 \text{ g K}$

12. molar masses: Ag, 107.9 g; Cu, 63.55 g

    $300.0 \text{ g Ag} \times \dfrac{1 \text{ mol Ag}}{107.9 \text{ g Ag}} = 2.780 \text{ mol Ag}$

    $2.780 \text{ mol Ag} \times \dfrac{6.022 \times 10^{23}}{1 \text{ mol Ag}} = 1.674 \times 10^{24} \text{ atoms}$

    $2.780 \text{ mol Cu} \times \dfrac{63.55 \text{ g Cu}}{1 \text{ mol Cu}} = 176.7 \text{ g Cu}$

13. The ratio of the atomic mass of H to the atomic mass of N is (1.008 amu/14.01 amu), and the mass of hydrogen is given by

    $7.00 \text{ g N} \times \dfrac{1.008 \text{ amu}}{14.01 \text{ amu}} = 0.504 \text{ g H}.$

Chapter 8: Chemical Composition

14. The ratio of the atomic mass of Co to the atomic mass of F is (58.93 amu Co/19.00 amu F), and the mass of cobalt is given by

$$57.0 \text{ g F} \times \frac{58.93 \text{ amu Co}}{19.00 \text{ amu F}} = 177 \text{ g Co}.$$

15. mass of a magnesium atom = 3.82×10^{-23} g $\times \dfrac{24.31 \text{ amu Mg}}{22.99 \text{ amu Na}} = 4.04 \times 10^{-23}$ g

16. mass of a chlorine atom = 3.16×10^{-23} g $\times \dfrac{35.45 \text{ amu Cl}}{19.00 \text{ amu F}} = 5.90 \times 10^{-23}$ g

17. 1 mol of He atoms = 4.003 g

    4 mol of H atoms $\times \dfrac{1.008 \text{ g H}}{1 \text{ mol}} = 4.032$ g

    4 mol of H atoms has a mass slightly larger than 1 mol of He atoms.

18. 0.25 mol Xe atoms $\times \dfrac{131.3 \text{ g Xe}}{1 \text{ mol Xe}} = 32.83$ g Xe = 33 g Xe (two significant figures)

    2.0 mol C atoms $\times \dfrac{12.01 \text{ g C}}{1 \text{ mol C}} = 24.02$ g C = 24 g C (two significant figures)

    The carbon sample weighs less.

19. a. 4.95 g Ne $\times \dfrac{1 \text{ mol}}{20.18 \text{ g}} = 0.245$ mol Ne

    b. 72.5 g Ni $\times \dfrac{1 \text{ mol}}{58.69 \text{ g}} = 1.24$ mol Ni

    c. 115 mg Ag $\times \dfrac{1 \text{ g}}{1000 \text{ mg}} \times \dfrac{1 \text{ mol}}{107.9 \text{ g}} = 1.07 \times 10^{-3}$ mol Ag

    d. 6.22μg U $\times \dfrac{1 \text{ g}}{10^6 \mu\text{g}} \times \dfrac{1 \text{ mol}}{238.0 \text{ g}} = 2.61 \times 10^{-8}$ mol U

    e. 135 g I $\times \dfrac{1 \text{ mol}}{126.9 \text{ g}} = 1.06$ mol I

20. a. 49.2 g S $\times \dfrac{1 \text{ mol}}{32.07 \text{ g}} = 1.53$ mol of S

    b. 7.44×10^4 kg Pb $\times \dfrac{1000 \text{ g}}{1 \text{ kg}} \times \dfrac{1 \text{ mol}}{207.2 \text{ g}} = 3.59 \times 10^5$ mol Pb

    c. 3.27 mg Cl $\times \dfrac{1 \text{ g}}{1000 \text{ mg}} \times \dfrac{1 \text{ mol}}{35.45 \text{ g}} = 9.22 \times 10^{-5}$ mol Cl

Chapter 8:   **Chemical Composition**

    d. $4.01 \text{ g Li} \times \dfrac{1 \text{ mol}}{6.9419 \text{ g}} = 0.578 \text{ mol Li}$

    e. $100.0 \text{ g Cu} \times \dfrac{1 \text{ mol}}{63.55 \text{ g}} = 1.574 \text{ mol Cu}$

    f. $82.6 \text{ mg Sr} \times \dfrac{1 \text{ g}}{1000 \text{ mg}} \times \dfrac{1 \text{ mol}}{87.62 \text{ g}} = 9.43 \times 10^{-4} \text{ mol Sr}$

21.   a. $0.251 \text{ mol Li} \times \dfrac{6.941 \text{ g Li}}{1 \text{ mol Li}} = 1.74 \text{ g Li}$

    b. $1.51 \text{ mol Al} \times \dfrac{26.98 \text{ g Al}}{1 \text{ mol Al}} = 40.7 \text{ g Al}$

    c. $8.75 \times 10^{-2} \text{ mol Pb} \times \dfrac{207.2 \text{ g Pb}}{1 \text{ mol Pb}} = 18.1 \text{ g Pb}$

    d. $125 \text{ mol Cr} \times \dfrac{52.00 \text{ g Cr}}{1 \text{ mol Cr}} = 6.50 \times 10^{3} \text{ g Cr}$

    e. $4.25 \times 10^{3} \text{ mol Fe} \times \dfrac{55.85 \text{ g Fe}}{1 \text{ mol}} = 2.37 \times 10^{5} \text{ g Fe}$

    f. $0.000105 \text{ mol Mg} \times \dfrac{24.31 \text{ g Mg}}{1 \text{ mol Mg}} = 2.55 \times 10^{-3} \text{ g Mg}$

22.   a. $0.00552 \text{ mol Ca} \times \dfrac{40.08 \text{ g}}{1 \text{ mol}} = 0.221 \text{ g Ca}$

    b. $6.25 \text{ millimol B} \times \dfrac{1 \text{ mol}}{10^{3} \text{ millmol}} \times \dfrac{10.81 \text{ g}}{1 \text{ mol}} = 0.0676 \text{ g B}$

    c. $135 \text{ mol Al} \times \dfrac{26.98 \text{ g}}{1 \text{ mol}} = 3.64 \times 10^{3} \text{ g Al}$

    d. $1.34 \times 10^{-7} \text{ mol Ba} \times \dfrac{137.3 \text{ g}}{1 \text{ mol}} = 1.84 \times 10^{-5} \text{ g Ba}$

    e. $2.79 \text{ mol P} \times \dfrac{30.97 \text{ g}}{1 \text{ mol}} = 86.4 \text{ g P}$

    f. $0.0000997 \text{ mol As} \times \dfrac{74.92 \text{ g}}{1 \text{ mol}} = 7.47 \times 10^{-3} \text{ g As}$

23.   a. $1.50 \text{ g Ag} \times \dfrac{6.022 \times 10^{23} \text{ Ag atoms}}{107.9 \text{ g Ag}} = 8.37 \times 10^{21} \text{ Ag atoms}$

    b. $0.0015 \text{ mol Cu} \times \dfrac{6.022 \times 10^{23} \text{ Cu atoms}}{1 \text{ mol}} = 9.0 \times 10^{20} \text{ Cu atoms}$

Chapter 8:   Chemical Composition

c. $0.0015 \text{ g Cu} \times \dfrac{6.022 \times 10^{23} \text{ Cu atoms}}{63.55 \text{ g Cu}} = 1.4 \times 10^{19}$ Cu atoms

d. $2.00 \text{ kg} = 2.00 \times 10^3 \text{ g}$

$2.00 \times 10^3 \text{ g Mg} \times \dfrac{6.022 \times 10^{23} \text{ Mg atoms}}{24.31 \text{ g Mg}} = 4.95 \times 10^{25}$ Mg atoms

e. $1.000 \text{ oz} = 28.35 \text{ g}$

$2.34 \text{ oz} \times \dfrac{28.35 \text{ g}}{1.000 \text{ oz}} \times \dfrac{6.022 \times 10^{23} \text{ Ca atoms}}{40.08 \text{ g Ca}} = 9.97 \times 10^{23}$ Ca atoms

f. $2.34 \text{ g Ca} \times \dfrac{6.022 \times 10^{23} \text{ Ca atoms}}{40.08 \text{ g Ca}} = 3.52 \times 10^{22}$ Ca atoms

g. $2.34 \text{ mol Ca} \times \dfrac{6.022 \times 10^{23} \text{ Ca atoms}}{1 \text{ mol Ca}} = 1.41 \times 10^{24}$ Ca atoms

24. a. $125 \text{ Fe atoms} \times \dfrac{55.85 \text{ g Fe}}{6.022 \times 10^{23} \text{ Fe atoms}} = 1.16 \times 10^{-20} \text{ g}$

b. $125 \text{ Fe atoms} \times \dfrac{55.85 \text{ amu}}{1 \text{ Fe atom}} = 6.98 \times 10^3$ amu

c. $125 \text{ g Fe} \times \dfrac{1 \text{ mol Fe}}{55.85 \text{ g Fe}} = 2.24$ mol Fe

d. $125 \text{ mol Fe} \times \dfrac{55.85 \text{ g Fe}}{1 \text{ mol Fe}} = 6.98 \times 10^3$ g Fe

e. $125 \text{ g Fe} \times \dfrac{6.022 \times 10^{23} \text{ Fe atoms}}{55.85 \text{ g Fe}} = 1.35 \times 10^{24}$ Fe atoms

f. $125 \text{ mol Fe} \times \dfrac{6.022 \times 10^{23} \text{ Fe atoms}}{1 \text{ mol Fe}} = 7.53 \times 10^{25}$ Fe atoms

25. molar mass

26. (Answers will vary.) The molar mass is calculated by summing the individual atomic masses of the atoms in the formula. In the compound CH_4, the atomic mass of carbon and the atomic mass of four hydrogens are summed (giving a molar mass of 16.042 g/mol).

27. a. H_3PO_4 phosphoric acid

mass of 3 mol H = 3(1.008 g) = 3.024 g

mass of 1 mol P = 30.97 g

mass of 4 mol O = 4(16.00 g) = 64.00 g

molar mass of H_3PO_4 = (3.024 g + 30.97 g + 64.00 g) = 97.99 g

Chapter 8: Chemical Composition

    b. Fe_2O_3 ferric oxide, iron(III) oxide

          mass of 2 mol Fe = 2(55.85 g) = 111.7 g

          mass of 3 mol O = 3(16.00 g) = 48.00 g

          molar mass of Fe_2O_3 = (111.7 g + 48.00 g) = 159.7 g

    c. $NaClO_4$ sodium perchlorate

          mass of 1 mol Na = 22.99 g

          mass of 1 mol Cl = 35.45 g

          mass of 4 mol O = 4(16.00 g) = 64.00 g

          molar mass of $NaClO_4$ = (22.99 g + 35.45 g + 64.00 g) = 122.44 g

    d. $PbCl_2$ plumbous chloride, lead(II) chloride

          mass of 1 mol Pb = 207.2 g

          mass of 2 mol Cl = 2(35.45 g) = 70.90 g

          molar mass of $PbCl_2$ = (207.2 g + 70.90 g) = 278.1 g

    e.   HBr hydrogen bromide

          mass of 1 mol H = 1.008 g

          mass of 1 mol Br = 79.90 g

          molar mass of HBr = (1.008 g + 79.90 g) = 80.91 g

    f. $Al(OH)_3$ aluminum hydroxide

          mass of 1 mol Al = 26.98 g

          mass of 3 mol H = 3(1.008) = 3.024 g

          mass of 3 mol O = 3(16.00 g) = 48.00 g

          molar mass of $Al(OH)_3$ = (26.98 g + 3.024 g + 48.00 g) = 78.00 g

28.   a. $KHCO_3$ potassium hydrogen carbonate, potassium bicarbonate

          mass of 1 mol K = 39.10 g

          mass of 1 mol H = 1.008 g

          mass of 1 mol C = 12.01 g

          mass of 3 mol O = 3(16.00 g) = 48.00 g

          molar mass of $KHCO_3$ = (39.10 + 1.008 + 12.01 + 48.00) = 100.12 g

    b. Hg_2Cl_2 mercurous chloride, mercury(I) chloride

          mass of 2 mol Hg = 2(200.6 g) = 401.2 g

          mass of 2 mol Cl = 2(35.45 g) = 70.90 g

          molar mass of Hg_2Cl_2 = (401.2 g + 70.90 g) = 472.1 g

Chapter 8:   Chemical Composition

    c. H_2O_2 hydrogen peroxide

        mass of 2 mol H = 2(1.008 g) = 2.016 g

        mass of 2 mol O = 2(16.00 g) = 32.00 g

        molar mass of H_2O_2 = (2.016 g + 32.00 g) = 34.02 g

    d. $BeCl_2$ beryllium chloride

        mass of 1 mol Be = 9.012 g

        mass of 2 mol Cl = 2(35.45 g) = 70.90 g

        molar mass of $BeCl_2$ = (9.012 g + 70.90 g) = 79.91 g

    e. $Al_2(SO_4)_3$ aluminum sulfate

        mass of 2 mol Al = 2(26.98 g) = 53.96 g

        mass of 3 mol S = 3(32.07 g) = 96.21 g

        mass of 12 mol O = 12(16.00 g) = 192.0 g

        molar mass of $Al_2(SO_4)_3$ = (53.96 g + 96.21 g + 192.0 g) = 342.2 g

    f. $KClO_3$ potassium chlorate

        mass of 1 mol K = 39.10 g

        mass of 1 mol Cl = 35.45 g

        mass of 3 mol O = 3(16.00 g) = 48.00 g

        molar mass of $KClO_3$ = 122.55 g

29.   a. $BaCl_2$

        mass of 1 mol Ba = 137.3 g

        mass of 2 mol Cl = 2(35.45 g) = 70.90 g

        molar mass of $BaCl_2$ = (137.3 g + 70.90 g) = 208.2 g

    b. $Al(NO_3)_3$

        mass of 1 mol Al = 26.98 g

        mass of 3 mol N = 3(14.01 g) = 42.03 g

        mass of 9 mol O = 9(16.00 g) = 144.00 g

        molar mass of $Al(NO_3)_3$ = (26.98 g + 42.03 g + 144.00 g) = 213.0 g

    c. $FeCl_2$

        mass of 1 mol Fe = 55.85 g

        mass of 2 mol Cl = 2(35.45 g) = 70.90 g

        molar mass of $FeCl_2$ = (55.85 g + 70.90 g) = 126.75 g

Chapter 8: Chemical Composition

    d. SO_2

        mass of 1 mol S = 32.07 g

        mass of 2 mol O = 2(16.00 g) = 32.00 g

        molar mass of SO_2 = (32.07 g + 32.00 g) = 64.07 g

    e. $Ca(C_2H_3O_2)_2$

        mass of 1 mol Ca = 40.08 g

        mass of 4 mol C = 4(12.01 g) = 48.04 g

        mass of 6 mol H = 6(1.008 g) = 6.048 g

        mass of 4 mol O = 4(16.00 g) = 64.00 g

        molar mass of $Ca(C_2H_3O_2)_2$ = (40.08 g + 48.04 g + 6.048 g + 64.00 g) = 158.17 g

30.   a. $Ba(ClO_4)_2$

        mass of 1 mol Ba = 137.3 g

        mass of 2 mol Cl = 2(35.45 g) = 70.90 g

        mass of 8 mol O = 8(16.00 g) = 128.00 g

        molar mass of $Ba(ClO_4)_2$ = (137.3 g + 70.90 g + 128.00 g) = 336.2 g

    b. $MgSO_4$

        mass of 1 mol Mg = 24.31 g

        mass of 1 mol S = 32.07 g

        mass of 4 mol O = 4(16.00 g) = 64.00 g

        molar mass of $MgSO_4$ = (24.31 g + 32.07 g + 64.00 g) = 120.38 g

    c. $PbCl_2$

        mass of 1 mol Pb = 207.2 g

        mass of 2 mol Cl = 2(35.45 g) = 70.90 g

        molar mass of $PbCl_2$ = (207.2 g + 70.90 g) = 278.1 g

    d. $Cu(NO_3)_2$

        mass of 1 mol Cu = 63.55 g

        mass of 2 mol N = 2(14.01 g) = 28.02 g

        mass of 6 mol O = 6(16.00 g) = 96.00 g

        molar mass of $Cu(NO_3)_2$ = (63.55 g + 28.02 g + 96.00 g) = 187.57 g

    e. $SnCl_4$

        mass of 1 mol Sn = 118.7 g

        mass of 4 mol Cl = 4(35.45 g) = 141.8 g

        molar mass of $SnCl_4$ = (118.7 g + 141.8 g) = 260.5 g

Chapter 8: Chemical Composition

31. a. molar mass of NO_2 = 46.01 g;   21.4 mg = 0.0214 g

$$0.0214 \text{ g } NO_2 \times \frac{1 \text{ mol}}{46.01 \text{ g}} = 4.65 \times 10^{-4} \text{ mol } NO_2$$

  b. molar mass of $Cu(NO_3)_2$ = 187.6 g

$$1.56 \text{ g } Cu(NO_3)_2 \times \frac{1 \text{ mol}}{187.6 \text{ g}} = 8.32 \times 10^{-3} \text{ mol } Cu(NO_3)_2$$

  c. molar mass of CS_2 = 76.15 g

$$2.47 \text{ g } CS_2 \times \frac{1 \text{ mol}}{76.15 \text{ g}} = 0.0324 \text{ mol}$$

  d. molar mass of $Al_2(SO_4)_3$ = 342.2 g

$$5.04 \text{ g } Al_2(SO_4)_3 \times \frac{1 \text{ mol}}{342.2 \text{ g}} = 0.0147 \text{ mol } Al_2(SO_4)_3$$

  e. molar mass of $PbCl_2$ = 278.1 g

$$2.99 \text{ g} \times \frac{1 \text{ mol}}{278.1 \text{ g}} = 0.0108 \text{ mol } PbCl_2$$

  f. molar mass of $CaCO_3$ = 100.09 g

$$62.4 \text{ g } CaCO_3 \times \frac{1 \text{ mol}}{100.09 \text{ g}} = 0.623 \text{ mol } CaCO_3$$

32. a. molar mass Al_2O_3 = 101.96 g

$$47.2 \text{ g} \times \frac{1 \text{ mol}}{101.96 \text{ g}} = 0.463 \text{ mol}$$

  b. molar mass KBr = 119.00 g

$$1.34 \text{ kg} \times \frac{1000 \text{ g}}{1 \text{ kg}} \times \frac{1 \text{ mol}}{119.00 \text{ g}} = 11.3 \text{ mol}$$

  c. molar mass Ge = 72.59 g

$$521 \text{ mg} \times \frac{1 \text{ g}}{1000 \text{ mg}} \times \frac{1 \text{ mol}}{72.59 \text{ g}} = 7.18 \times 10^{-3} \text{ mol}$$

  d. molar mass of U = 238.0 g

$$56.2 \text{ μg} \times \frac{1 \text{ g}}{10^6 \text{ μg}} \times \frac{1 \text{ mol}}{238.0 \text{ g}} = 2.36 \times 10^{-7} \text{ mol}$$

  e. molar mass of $NaC_2H_3O_2$ = 82.03 g

$$29.7 \text{ g} \times \frac{1 \text{ mol}}{82.03 \text{ g}} = 1.69 \text{ mol} = 0.362 \text{ mol}$$

Chapter 8: Chemical Composition

    f.   molar mass of SO_3 = 80.07 g

$$1.03 \text{ g} \times \frac{1 \text{ mol}}{80.07 \text{ g}} = 0.0129 \text{ mol}$$

33.   a.   molar mass $MgCl_2$ = 95.21 g

$$41.5 \text{ g} \times \frac{1 \text{ mol}}{95.21 \text{ g}} = 0.436 \text{ mol}$$

    b.   molar mass Li_2O = 29.88 g; 135 mg = 0.135 g

$$0.135 \text{ g} \times \frac{1 \text{ mol}}{29.88 \text{ g}} = 4.52 \times 10^{-3} \text{ mol} = 4.52 \text{ mmol}$$

    c.   molar mass Cr = 52.00 g; 1.21 kg = 1210 g

$$1210 \text{ g} \times \frac{1 \text{ mol}}{52.00 \text{ g}} = 23.3 \text{ mol}$$

    d.   molar mass H_2SO_4 = 98.09 g

$$62.5 \text{ g} \times \frac{1 \text{ mol}}{98.09 \text{ g}} = 0.637 \text{ mol}$$

    e.   molar mass C_6H_6 = 78.11 g

$$42.7 \text{ g} \times \frac{1 \text{ mol}}{78.11 \text{ g}} = 0.547 \text{ mol}$$

    f.   molar mass H_2O_2 = 34.02 g

$$135 \text{ g} \times \frac{1 \text{ mol}}{34.02 \text{ g}} = 3.97 \text{ mol}$$

34.   a.   molar mass of Li_2CO_3 = 73.89 g

$$1.95 \times 10^{-3} \text{ g} \times \frac{1 \text{ mol}}{73.89 \text{ g}} = 2.64 \times 10^{-5} \text{ mol}$$

    b.   molar mass of $CaCl_2$ = 110.98 g

$$4.23 \text{ kg} \times \frac{1000 \text{ g}}{1 \text{ kg}} \times \frac{1 \text{ mol}}{110.98 \text{ g}} = 38.1 \text{ mol}$$

    c.   molar mass of $SrCl_2$ = 158.52 g

$$1.23 \text{ mg} \times \frac{1 \text{ g}}{1000 \text{ mg}} \times \frac{1 \text{ mol}}{158.52 \text{ g}} = 7.76 \times 10^{-6} \text{ mol}$$

    d.   molar mass of $CaSO_4$ = 136.15 g

$$4.75 \text{ g} \times \frac{1 \text{ mol}}{136.15 \text{ g}} = 3.49 \times 10^{-2} \text{ mol}$$

Chapter 8:   Chemical Composition

  e. molar mass of NO_2 = 46.01 g

$$96.2 \text{ mg} \times \frac{1 \text{ g}}{1000 \text{ mg}} \times \frac{1 \text{ mol}}{46.01 \text{ g}} = 2.09 \times 10^{-3} \text{ mol}$$

  f. molar mass of Hg_2Cl_2 = 472.1 g

$$12.7 \text{ g} \times \frac{1 \text{ mol}}{472.1 \text{ g}} = 0.0269 \text{ mol}$$

35. a. molar mass $AlCl_3$ = 133.3 g

$$1.25 \text{ mol} \times \frac{133.3 \text{ g}}{1 \text{ mol}} = 167 \text{ g}$$

  b. molar mass $NaHCO_3$ = 84.01 g

$$3.35 \text{ mol} \times \frac{84.01 \text{ g}}{1 \text{ mol}} = 281 \text{ g}$$

  c. molar mass HBr = 80.91 g

$$4.25 \text{ mmol} \times \frac{80.91 \text{ mg}}{1 \text{ mmol}} = 344 \text{ mg} = 0.344 \text{ g}$$

  d. molar mass U = 238.0 g

$$1.31 \times 10^{-3} \text{ mol} \times \frac{238.0 \text{ g}}{1 \text{ mol}} = 0.312 \text{ g}$$

  e. molar mass CO_2 = 44.01 g

$$0.00104 \text{ mol} \times \frac{44.01 \text{ g}}{1 \text{ mol}} = 0.0458 \text{ g}$$

  f. molar mass Fe = 55.85 g

$$1.49 \times 10^2 \text{ mol Fe} \times \frac{55.85 \text{ g}}{1 \text{ mol}} = 8.32 \times 10^3 \text{ g}$$

36. a. molar mass of SO_3 = 80.07 g

$$6.14 \times 10^{-4} \text{ mol} \times \frac{80.07 \text{ g}}{1 \text{ mol}} = 0.0492 \text{ g}$$

  b. molar mass of PbO_2 = 239.2 g

$$3.11 \times 10^5 \text{ mol} \times \frac{239.2 \text{ g}}{1 \text{ mol}} = 7.44 \times 10^7 \text{ g}$$

  c. molar mass of $CHCl_3$ = 119.368 g

$$0.495 \text{ mol} \times \frac{119.368 \text{ g}}{1 \text{ mol}} = 59.1 \text{ g}$$

  d. molar mass of $C_2H_3Cl_3$ = 133.394 g

Chapter 8: Chemical Composition

$$2.45 \times 10^{-8} \text{ mol} \times \frac{133.394 \text{ g}}{1 \text{ mol}} = 3.27 \times 10^{-6} \text{ g}$$

e. molar mass of LiOH = 23.949 g

$$0.167 \text{ mol} \times \frac{23.949 \text{ g}}{1 \text{ mol}} = 4.00 \text{ g}$$

f. molar mass of CuCl = 99.00 g

$$5.26 \text{ mol} \times \frac{99.00 \text{ g}}{1 \text{ mol}} = 521 \text{ g}$$

37. a. molar mass of C_2H_6O = 46.07 g

$$0.251 \text{ mol} \times \frac{46.07 \text{ g}}{1 \text{ mol}} = 11.6 \text{ g } C_2H_6O$$

b. molar mass of CO_2 = 44.01 g

$$1.26 \text{ mol} \times \frac{44.01 \text{ g}}{1 \text{ mol}} = 55.5 \text{ g } CO_2$$

c. molar mass of $AuCl_3$ = 303.4 g

$$9.31 \times 10^{-4} \text{ mol} \times \frac{303.4 \text{ g}}{1 \text{ mol}} = 0.282 \text{ g } AuCl_3$$

d. molar mass of $NaNO_3$ = 85.00 g

$$7.74 \text{ mol} \times \frac{85.00 \text{ g}}{1 \text{ mol}} = 658 \text{ g } NaNO_3$$

e. molar mass of Fe = 55.85 g

$$0.000357 \text{ mol} \times \frac{55.85 \text{ g}}{1 \text{ mol}} = 0.0199 \text{ g Fe}$$

38. a. molar mass C_6H_6 = 78.11 g

$$0.994 \text{ mol} \times \frac{78.11 \text{ g}}{1 \text{ mol}} = 77.6 \text{ g}$$

b. molar mass CaH_2 = 42.10 g

$$4.21 \text{ mol} \times \frac{42.10 \text{ g}}{1 \text{ mol}} = 177 \text{ g}$$

c. molar mass H_2O_2 = 34.02 g

$$1.79 \times 10^{-4} \text{ mol} \times \frac{34.02 \text{ g}}{1 \text{ mol}} = 6.09 \times 10^{-3} \text{ g}$$

d. molar mass $C_6H_{12}O_6$ = 180.16 g

$$1.22 \text{ mmol} \times \frac{1 \text{ mol}}{10^3 \text{ mmol}} \times \frac{105.99 \text{ g}}{1 \text{ mol}} = 0.220 \text{ g}$$

e. molar mass Sn = 118.7 g

$$10.6 \text{ mol} \times \frac{118.7 \text{ g}}{1 \text{ mol}} = 1.26 \times 10^3 \text{ g}$$

f. molar mass SrF_2 = 125.62 g

$$0.000301 \text{ mol} \times \frac{125.62 \text{ g}}{1 \text{ mol}} = 0.0378 \text{ g}$$

39. a. 4.75 millimol = 0.00475 mol

$$0.00475 \text{ mol} \times \frac{6.022 \times 10^{23} \text{ molecules}}{1 \text{ mol}} = 2.86 \times 10^{21} \text{ molecules}$$

b. molar mass of PH_3 = 33.99 g

$$4.75 \text{ g} \times \frac{6.022 \times 10^{23} \text{ molecules}}{33.99 \text{ g}} = 8.42 \times 10^{22} \text{ molecules}$$

c. molar mass of $Pb(C_2H_3O_2)_2$ = 325.3 g

$$1.25 \times 10^{-2} \text{ g} \times \frac{6.022 \times 10^{23} \text{ formula units}}{325.3 \text{ g}} = 2.31 \times 10^{19} \text{ formula units}$$

d. $$1.25 \times 10^{-2} \text{ mol} \times \frac{6.022 \times 10^{23} \text{ formula units}}{1 \text{ mol}} = 7.53 \times 10^{21} \text{ formula units}$$

e. If the sample contains a total of 5.40 mol of carbon, then because each benzene contains six carbons, there must be (5.40/6) = 0.900 mol of benzene present.

$$0.900 \text{ mol} \times \frac{6.022 \times 10^{23} \text{ molecules}}{1 \text{ mol}} = 5.42 \times 10^{23} \text{ molecules}$$

40. a. $$3.54 \text{ mol } SO_2 \times \frac{6.022 \times 10^{23} \text{ molecules}}{1 \text{ mol}} = 2.13 \times 10^{24} \text{ molecules } SO_2$$

b. molar mass of SO_2 = 64.07 g

$$3.54 \text{ g} \times \frac{6.022 \times 10^{23} \text{ molecules}}{64.07 \text{ g}} = 3.33 \times 10^{22} \text{ molecules } SO_2$$

c. molar mass of NH_3 = 17.034 g

$$4.46 \times 10^{-5} \text{ g} \times \frac{6.022 \times 10^{23} \text{ molecules}}{17.034 \text{ g}} = 1.58 \times 10^{18} \text{ molecules } NH_3$$

d. $$4.46 \times 10^{-5} \text{ mol } NH_3 \times \frac{6.022 \times 10^{23} \text{ molecules}}{1 \text{ mol}} = 2.69 \times 10^{19} \text{ molecules } NH_3$$

e. molar mass of C_2H_6 = 30.068 g

$$1.96 \text{ mg} \times \frac{1 \text{ g}}{1000 \text{ mg}} \times \frac{1 \text{ mol}}{30.068 \text{ g}} \times \frac{6.022 \times 10^{23} \text{ molecules}}{1 \text{ mol}} = 3.93 \times 10^{19} \text{ molecules } C_2H_6$$

Chapter 8:   Chemical Composition

41. a. molar mass of C_2H_6O = 46.07 g

$$1.271 \text{ g} \times \frac{1 \text{ mol}}{46.07 \text{ g}} = 0.02759 \text{ mol } C_2H_6O$$

$$0.02759 \text{ mol } C_2H_6O \times \frac{2 \text{ mol C}}{1 \text{ mol } C_2H_6O} = 0.05518 \text{ mol C}$$

b. molar mass of $C_6H_4Cl_2$ = 147.0 g

$$3.982 \text{ g} \times \frac{1 \text{ mol}}{147.0 \text{ g}} = 0.027088 \text{ mol } C_6H_4Cl_2$$

$$0.027088 \text{ mol } C_6H_4Cl_2 \times \frac{6 \text{ mol C}}{1 \text{ mol } C_6H_4Cl_2} = 0.1625 \text{ mol C}$$

c. molar mass of C_3O_2 = 68.03 g

$$0.4438 \text{ g} \times \frac{1 \text{ mol}}{68.03 \text{ g}} = 0.0065236 \text{ mol } C_3O_2$$

$$0.0065236 \text{ mol } C_3O_2 \times \frac{3 \text{ mol C}}{1 \text{ mol } C_3O_2} = 0.01957 \text{ mol C}$$

d. molar mass of CH_2Cl_2 = 84.93 g

$$2.910 \text{ g} \times \frac{1 \text{ mol}}{84.93 \text{ g}} = 0.034264 \text{ mol } CH_2Cl_2$$

$$0.034264 \text{ mol } CH_2Cl_2 \times \frac{1 \text{ mol C}}{1 \text{ mol } CH_2Cl_2} = 0.03426 \text{ mol C}$$

42. a. molar mass of Na_2SO_4 = 142.1 g

$$2.01 \text{ g } Na_2SO_4 \times \frac{1 \text{ mol } Na_2SO_4}{142.1 \text{ g}} \times \frac{1 \text{ mol S}}{1 \text{ mol } Na_2SO_4} = 0.0141 \text{ mol S}$$

b. molar mass of Na_2SO_3 = 126.1 g

$$2.01 \text{ g } Na_2SO_3 \times \frac{1 \text{ mol } Na_2SO_3}{126.1 \text{ g}} \times \frac{1 \text{ mol S}}{1 \text{ mol } Na_2SO_3} = 0.0159 \text{ mol S}$$

c. molar mass of Na_2S = 78.05 g

$$2.01 \text{ g } Na_2S \times \frac{1 \text{ mol } Na_2S}{78.05 \text{ g}} \times \frac{1 \text{ mol S}}{1 \text{ mol } Na_2S} = 0.0258 \text{ mol S}$$

d. molar mass of $Na_2S_2O_3$ = 158.1 g

$$2.01 \text{ g } Na_2S_2O_3 \times \frac{1 \text{ mol } Na_2S_2O_3}{158.1 \text{ g}} \times \frac{2 \text{ mol S}}{1 \text{ } Na_2S_2O_3} = 0.0254 \text{ mol S}$$

43. molar

Chapter 8: Chemical Composition

44. The percent composition of each element in the compound does not change because a compound consists of the same percent composition by mass regardless of the starting amount in the sample.

45. a. mass of H present = 1.008 g = 1.008 g

    mass of Cl present = 35.45 g = 35.45 g

    mass of O present = 3(16.00 g) = 48.00 g

    molar mass of $HClO_3$ = 84.46 g

    % H = $\dfrac{1.008 \text{ g H}}{84.46 \text{ g}}$ × 100 = 1.193 %H

    % Cl = $\dfrac{35.45 \text{ g Cl}}{84.46 \text{ g}}$ × 100 = 41.97 %Cl

    % O = $\dfrac{48.00 \text{ g O}}{84.46 \text{ g}}$ × 100 = 56.83 %O

    b. mass of U present = 238.0 g = 238.0 g

    mass of F present = 4(19.00 g) = 76.00 g

    molar mass of UF_4 = 314.0 g

    % U = $\dfrac{238.0 \text{ g U}}{314.0 \text{ g}}$ × 100 = 75.80% U

    % F = $\dfrac{76.00 \text{ g F}}{314.0 \text{ g}}$ × 100 = 24.20% F

    c. mass of Ca present = 40.08 g = 40.08 g

    mass of H present = 2(1.008 g) = 2.016 g

    molar mass of CaH_2 = 42.10 g

    % Ca = $\dfrac{40.08 \text{ g C}}{42.10 \text{ g}}$ × 100 = 95.21% Ca

    % H = $\dfrac{2.016 \text{ g H}}{42.10 \text{ g}}$ × 100 = 4.789% H

    d. mass of Ag present = 2(107.9 g) = 215.8 g

    mass of S present = 32.07 g = 32.07 g

    molar mass of Ag_2S = 247.9 g

    % Ag = $\dfrac{215.8 \text{ g Ag}}{247.9 \text{ g}}$ × 100 = 87.06% Ag

    % S = $\dfrac{32.07 \text{ g S}}{247.9 \text{ g}}$ × 100 = 12.94% S

Chapter 8: Chemical Composition

e. mass of Na present = 22.99 g = 22.99 g
    mass of H present = 1.008 g = 1.008 g
    mass of S present = 32.07 g = 32.07 g
    mass of O present = 3(16.00 g) = 48.00 g
    molar mass of $NaHSO_3$ = 104.07 g

    $\% \, Na = \dfrac{22.99 \text{ g Na}}{104.07 \text{ g}} \times 100 = 22.09\% \text{ Na}$

    $\% \, H = \dfrac{1.008 \text{ g H}}{104.07 \text{ g}} \times 100 = 0.9686\% \text{ H}$

    $\% \, S = \dfrac{32.07 \text{ g S}}{104.07 \text{ g}} \times 100 = 30.82\% \text{ S}$

    $\% \, O = \dfrac{48.00 \text{ g O}}{104.07 \text{ g}} \times 100 = 46.12\% \text{ O}$

f. mass of Mn present = 54.94 g = 54.94 g
    mass of O present = 2(16.00 g) = 32.00 g
    molar mass of MnO_2 = 86.94 g

    $\% \, Mn = \dfrac{54.94 \text{ g Mn}}{86.94 \text{ g}} \times 100 = 63.19\% \text{ Mn}$

    $\% \, S = \dfrac{32.00 \text{ g O}}{86.94 \text{ g}} \times 100 = 36.81\% \text{ O}$

46. a. mass of Zn present = 65.38 g
    mass of O present = 16.00 g
    molar mass of ZnO = 81.38 g

    $\%Zn = \dfrac{65.38 \text{ g Zn}}{81.38 \text{ g}} \times 100 = 80.34\% \text{ Zn}$

    $\%O = \dfrac{16.00 \text{ g O}}{81.38 \text{ g}} \times 100 = 19.66\% \text{ O}$

  b. mass of Na present = 2(22.99 g) = 45.98 g
    mass of S present = 32.07 g
    molar mass of Na_2S = 78.05 g

    $\%Na = \dfrac{45.98 \text{ g Na}}{78.05 \text{ g}} \times 100 = 58.91\% \text{ Na}$

    $\%S = \dfrac{32.07 \text{ g S}}{78.05 \text{ g}} \times 100 = 41.09\% \text{ S}$

c. mass of Mg present = 24.31 g

mass of O present = 2(16.00 g) = 32.00 g

mass of H present = 2(1.008 g) = 2.016 g

molar mass of Mg(OH)$_2$ = 58.33 g

$$\%Mg = \frac{24.31 \text{ g Mg}}{58.33 \text{ g}} \times 100 = 41.68\% \text{ Mg}$$

$$\%O = \frac{32.00 \text{ g O}}{58.33 \text{ g}} \times 100 = 54.86\% \text{ O}$$

$$\%H = \frac{2.016 \text{ g H}}{58.33 \text{ g}} \times 100 = 3.456\% \text{ H}$$

d. mass of H present = 2(1.008 g) = 2.016 g

mass of O present = 2(16.00 g) = 32.00 g

molar mass of H$_2$O$_2$ = 34.02 g

$$\%H = \frac{2.016 \text{ g H}}{34.02 \text{ g}} \times 100 = 5.926\% \text{ H}$$

$$\%O = \frac{32.00 \text{ g O}}{34.02 \text{ g}} \times 100 = 94.06\% \text{ O}$$

e. mass of Ca present = 40.08 g

mass of H present = 2(1.008 g) = 2.016 g

molar mass of CaH$_2$ = 42.10 g

$$\%Ca = \frac{40.08 \text{ g Ca}}{42.10 \text{ g}} \times 100 = 95.20\% \text{ Ca}$$

$$\%H = \frac{2.016 \text{ g H}}{42.10 \text{ g}} \times 100 = 4.789\% \text{ H}$$

f. mass of K present = 2(39.10 g) = 78.20 g

mass of O present = 16.00 g

molar mass of K$_2$O = 94.20 g

$$\%K = \frac{78.20 \text{ g K}}{94.20 \text{ g}} \times 100 = 83.01\% \text{ K}$$

$$\%O = \frac{16.00 \text{ g O}}{94.20 \text{ g}} \times 100 = 16.99\% \text{ O}$$

47. a. molar mass of CH$_4$ = 16.04 g

$$\%C = \frac{12.01 \text{ g C}}{16.04 \text{ g}} \times 100 = 74.88\% \text{ C}$$

Chapter 8: Chemical Composition

    b.   molar mass $NaNO_3$ = 85.00 g

$$\% \text{ Na} = \frac{22.99 \text{ g Na}}{85.00 \text{ g}} \times 100 = 27.05\% \text{ Na}$$

    c.   molar mass of CO = 28.01 g

$$\% \text{ C} = \frac{12.01 \text{ g C}}{28.01 \text{ g}} \times 100 = 42.88\% \text{ C}$$

    d.   molar mass of NO_2 = 46.01 g

$$\% \text{N} = \frac{14.01 \text{ g N}}{46.01 \text{ g}} \times 100 = 30.45\% \text{ N}$$

    e.   molar mass of $C_8H_{18}O$ = 130.2 g

$$\% \text{ C} = \frac{96.08 \text{ g C}}{130.2 \text{ g}} \times 100 = 73.79\% \text{ C}$$

    f.   molar mass of $Ca_3(PO_4)_2$ = 310.1 g

$$\% \text{ Ca} = \frac{120.2 \text{ g Ca}}{310.1 \text{ g}} \times 100 = 38.76\% \text{ Ca}$$

    g.   molar mass of $C_{12}H_{10}O$ = 170.2 g

$$\% \text{ C} = \frac{144.1 \text{ g C}}{170.2 \text{ g}} \times 100 = 84.67\% \text{ C}$$

    h.   molar mass of $Al(C_2H_3O_2)_3$ = 204.1 g

$$\% \text{ Al} = \frac{26.98 \text{ g Al}}{204.1 \text{ g}} \times 100 = 13.22\% \text{ Al}$$

48.   a.   molar mass of $CuBr_2$ = 223.35 g

$$\% \text{ Cu} = \frac{63.55 \text{ g Cu}}{223.35 \text{ g}} \times 100 = 28.45\% \text{ Cu}$$

    b.   molar mass of CuBr = 143.45 g

$$\% \text{ Cu} = \frac{63.55 \text{ g Cu}}{143.45 \text{ g}} \times 100 = 44.30\% \text{ Cu}$$

    c.   molar mass of $FeCl_2$ = 126.75 g

$$\% \text{ Fe} = \frac{55.85 \text{ g Fe}}{126.75 \text{ g}} \times 100 = 44.06\% \text{ Fe}$$

    d.   molar mass of $FeCl_3$ = 162.2 g

$$\% \text{ Fe} = \frac{55.85 \text{ g Fe}}{162.2 \text{ g}} \times 100 = 34.43\% \text{ Fe}$$

Chapter 8: Chemical Composition

    e.   molar mass of CoI_2 = 312.73 g

$$\% \text{ Co} = \frac{58.93 \text{ g Co}}{312.73 \text{ g}} \times 100 = 18.84\% \text{ Co}$$

    f.   molar mass of CoI_3 = 439.63 g

$$\% \text{ Co} = \frac{58.93 \text{ g Co}}{439.63 \text{ g}} \times 100 = 13.40\% \text{ Co}$$

    g.   molar mass of SnO = 134.7 g

$$\% \text{ Sn} = \frac{118.7 \text{ g Sn}}{134.7 \text{ g}} \times 100 = 88.12\% \text{ Sn}$$

    h.   molar mass of SnO_2 = 150.7 g

$$\% \text{ Sn} = \frac{118.7 \text{ g Sn}}{150.7 \text{ g}} \times 100 = 78.77\% \text{ Sn}$$

49.   a.   molar mass of $C_6H_{10}O_4$ = 146.1 g

$$\% \text{ C} = \frac{72.06 \text{ g C}}{146.1 \text{ g}} \times 100 = 49.32\% \text{ C}$$

    b.   molar mass of NH_4NO_3 = 80.05 g

$$\% \text{ N} = \frac{28.02 \text{ g N}}{80.05 \text{ g}} \times 100 = 35.00\% \text{ N}$$

    c.   molar mass of $C_8H_{10}N_4O_2$ = 194.2 g

$$\% \text{ C} = \frac{96.09 \text{ g C}}{194.2 \text{ g}} \times 100 = 49.47\% \text{ C}$$

    d.   molar mass of ClO_2 = 67.45 g

$$\% \text{ Cl} = \frac{35.45 \text{ g Cl}}{67.45 \text{ g}} \times 100 = 52.56\% \text{ Cl}$$

    e.   molar mass of $C_6H_{11}OH$ = 100.2 g

$$\% \text{ C} = \frac{72.06 \text{ g C}}{100.2 \text{ g}} \times 100 = 71.92\% \text{ C}$$

    f.   molar mass of $C_6H_{12}O_6$ = 180.2 g

$$\% \text{ C} = \frac{72.06 \text{ g C}}{180.2 \text{ g}} \times 100 = 39.99\% \text{ C}$$

    g.   molar mass of $C_{20}H_{42}$ = 282.5 g

$$\% \text{ C} = \frac{240.2 \text{ g C}}{282.5 \text{ g}} \times 100 = 85.03\% \text{ C}$$

Chapter 8: Chemical Composition

    h.   molar mass of C_2H_5OH = 46.07 g

$$\% C = \frac{24.02 \text{ g C}}{46.07 \text{ g}} \times 100 = 52.14\% \text{ C}$$

50.   a.   molar mass of CO = 28.01 g

$$\% O = \frac{16.00 \text{ g O}}{28.01 \text{ g}} \times 100 = 57.12\% \text{ O}$$

    b.   molar mass of MnO_2 = 86.94 g

$$\% O = \frac{32.00 \text{ g O}}{86.94 \text{ g}} \times 100 = 36.81\% \text{ O}$$

    c.   molar mass of $KClO_3$ = 122.55 g

$$\% O = \frac{48.00 \text{ g O}}{122.55 \text{ g}} \times 100 = 39.17\% \text{ O}$$

    d.   molar mass of FeO = 71.85 g

$$\% O = \frac{16.00 \text{ g O}}{71.85 \text{ g}} \times 100 = 22.27\% \text{ O}$$

    e.   molar mass of $Ca(OH)_2$ = 74.096 g

$$\% O = \frac{32.00 \text{ g O}}{74.096 \text{ g}} \times 100 = 43.19\% \text{ O}$$

51.   a.   molar mass of NH_4I = 144.94 g

$$4.25 \text{ g} \times \frac{1 \text{ mol } NH_4I}{144.94 \text{ g}} \times \frac{1 \text{ mol } NH_4^+}{1 \text{ mol } NH_4I} = 0.0293 \text{ mol } NH_4^+$$

molar mass of NH_4^+ = 18.04 g

$$0.0293 \text{ mol } NH_4^+ \times \frac{18.04 \text{ g } NH_4^+}{1 \text{ mol } NH_4^+} = 0.529 \text{ g } NH_4^+$$

    b. $6.31 \text{ mol } (NH_4)_2S \times \frac{2 \text{ mol } NH_4^+}{1 \text{ mol } (NH_4)_2S} = 12.6 \text{ mol } NH_4^+$

molar mass of NH_4^+ = 18.04 g

$$12.6 \text{ mol } NH_4^+ \times \frac{18.04 \text{ g } NH_4^+}{1 \text{ mol } NH_4^+} = 227 \text{ g } NH_4^+ \text{ ion}$$

    c.   molar mass of Ba_3P_2 = 473.8 g

$$9.71 \text{ g} \times \frac{1 \text{ mol } Ba_3P_2}{473.8 \text{ g}} \times \frac{3 \text{ mol } Ba^{2+}}{1 \text{ mol } Ba_3P_2} = 0.0615 \text{ mol } Ba^{2+}$$

molar mass of Ba^{2+} = 137.3 g

Chapter 8: Chemical Composition

$$0.0615 \text{ mol Ba}^{2+} \times \frac{137.3 \text{ g Ba}^{2+}}{1 \text{ mol Ba}^{2+}} = 8.44 \text{ g Ba}^{2+}$$

d. $7.63 \text{ mol Ca}_3(PO_4)_2 \times \dfrac{3 \text{ mol Ca}^{2+}}{1 \text{ mol Ca}_3(PO_4)_2} = 22.9 \text{ mol Ca}^{2+}$

molar mass of Ca^{2+} = 40.08 g

$$22.9 \text{ mol Ca}^{2+} \times \frac{40.08 \text{ g Ca}^{2+}}{1 \text{ mol Ca}^{2+}} = 918 \text{ g Ca}^{2+}$$

52. a. molar mass of $(NH_4)_2S$ = 68.15 g; molar mass of S^{2-} ion = 32.07 g

$$\% S^{2-} = \frac{32.07 \text{ g S}^{2-}}{68.15 \text{ g NH}_4S} \times 100 = 47.06\% \text{ S}^{2-}$$

b. molar mass of $CaCl_2$ = 110.98 g; molar mass of Cl^- = 35.45 g

$$\% Cl^- = \frac{70.90 \text{ g Cl}^-}{110.98 \text{ g CaCl}_2} \times 100 = 63.89\% \text{ Cl}^-$$

c. molar mass of BaO = 153.3 g; molar mass of O^{2-} ion = 16.00 g

$$\% O^{2-} = \frac{16.00 \text{ g O}^{2-}}{153.3 \text{ g BaO}} \times 100 = 10.44\% \text{ O}^{2-}$$

d. molar mass of $NiSO_4$ = 154.76 g; molar mass of SO_4^{2-} ion = 96.07 g

$$\% SO_4^{2-} = \frac{96.07 \text{ g SO}_4^{2-}}{154.76 \text{ g NiSO}_4} \times 100 = 62.08\% \text{ SO}_4^{2-}$$

53. To determine the *empirical* formula of a new compound, the composition of the compound by mass must be known. To determine the *molecular* formula of the compound, the molar mass of the compound must also be known.

54. The empirical formula indicates the smallest whole number ratio of the number and type of atoms present in a molecule. For example, NO_2 and N_2O_4 both have two oxygen atoms for every nitrogen atom and therefore have the same empirical formula

55. a. NaO
    b. $C_4H_3O_2$
    c. $C_{12}H_{12}N_2O_3$ is already the empirical formula.
    d. C_2H_3Cl

56. a. yes (each of these has empirical formula CH)
    b. no (the number of hydrogen atoms is wrong)
    c. yes (both have empirical formula NO_2)
    d. no (the number of hydrogen and oxygen atoms is wrong)

Chapter 8: Chemical Composition

57. Assume we have 100.0 g of the compound so that the percentages become masses.

    $89.56 \text{ g Ba} \times \dfrac{1 \text{ mol}}{137.3 \text{ g}} = 0.6253 \text{ mol Ba}$

    $10.44 \text{ g O} \times \dfrac{1 \text{ mol}}{16.00 \text{ g}} = 0.6525 \text{ mol O}$

    Dividing both of these numbers of moles by the smaller number of moles (0.6523 mol Ba) gives

    $\dfrac{0.6523 \text{ mol Ba}}{0.6523 \text{ mol}} = 1.000 \text{ mol Ba}$

    $\dfrac{0.6525 \text{ mol O}}{0.6523 \text{ mol}} = 1.000 \text{ mol O}$

    The empirical formula is BaO.

58. Assume we have 100.0 g of the compound so that the percentages become masses.

    $11.64 \text{ g N} \times \dfrac{1 \text{ mol}}{14.01 \text{ g}} = 0.8308 \text{ mol N}$

    $88.36 \text{ g Cl} \times \dfrac{1 \text{ mol}}{35.45 \text{ g}} = 2.493 \text{ mol Cl}$

    Dividing both of these numbers of moles by the smaller number of moles gives

    $\dfrac{0.8308 \text{ mol N}}{0.8308} = 1.000 \text{ mol N}$

    $\dfrac{2.493 \text{ mol Cl}}{0.8308 \text{ mol}} = 3.001 \text{ mol Cl}$

    The empirical formula is NCl_3.

59. $0.2322 \text{ g C} \times \dfrac{1 \text{ mol}}{12.01 \text{ g}} = 0.01933 \text{ mol C}$

    $0.05848 \text{ g H} \times \dfrac{1 \text{ mol}}{1.008 \text{ g}} = 0.05802 \text{ mol H}$

    $0.3091 \text{ g O} \times \dfrac{1 \text{ mol}}{16.00 \text{ g}} = 0.01932 \text{ mol O}$

    Dividing each number of moles by the smallest number of moles (0.01932 mol C) gives

    $\dfrac{0.01933 \text{ mol C}}{0.01932 \text{ mol}} = 1.001 \text{ mol C}$

    $\dfrac{0.05802 \text{ mol H}}{0.01932 \text{ mol}} = 3.003 \text{ mol H}$

    $\dfrac{0.01932 \text{ mol O}}{0.01932 \text{ mol}} = 1.000 \text{ mol O}$

Chapter 8: Chemical Composition

The empirical formula is CH_3O.

60. Assume we have 100.0 g of the compound, so that the percentages become masses.

$$78.14 \text{ g B} \times \frac{1 \text{ mol}}{10.81 \text{ g}} = 7.228 \text{ mol B}$$

$$21.86 \text{ g H} \times \frac{1 \text{ mol}}{1.008 \text{ g}} = 21.69 \text{ mol H}$$

Dividing each number of moles by the smaller number of moles gives

$$\frac{7.228 \text{ mol B}}{7.228 \text{ mol}} = 1.000 \text{ mol B}$$

$$\frac{21.69 \text{ mol H}}{7.228 \text{ mol}} = 3.000 \text{ mol H}$$

The empirical formula is BH_3.

61. The mass of chlorine in the reaction is $6.280 - 1.271 = 5.009$ g Cl.

$$1.271 \text{ g Al} \times \frac{1 \text{ mol}}{26.98 \text{ g}} = 0.04711 \text{ mol Al}$$

$$5.009 \text{ g Cl} \times \frac{1 \text{ mol}}{35.45 \text{ g}} = 0.1413 \text{ mol Cl}$$

Dividing each of these number of moles by the smaller (0.04711 mol Al) gives

$$\frac{0.04711 \text{ mol Al}}{0.04711 \text{ mol}} = 1.000 \text{ mol Al}$$

$$\frac{0.1413 \text{ mol Cl}}{0.04711 \text{ mol}} = 3.000 \text{ mol Cl}$$

The empirical formula is $AlCl_3$.

62. Consider 100.0 g of the compound so that percentages become masses.

$$45.56 \text{ g Sn} \times \frac{1 \text{ mol}}{118.7 \text{ g}} = 0.3838 \text{ mol Sn}$$

$$54.43 \text{ g Cl} \times \frac{1 \text{ mol}}{35.45 \text{ g}} = 1.535 \text{ mol Cl}$$

Dividing each number of moles by the smaller number of moles gives

$$\frac{0.3838 \text{ mol Sn}}{0.3838 \text{ mol}} = 1.000 \text{ mol Sn}$$

$$\frac{1.535 \text{ mol Cl}}{0.3838 \text{ mol}} = 3.999 \text{ mol Cl}$$

The empirical formula is $SnCl_4$.

Chapter 8: Chemical Composition

63. $3.269 \text{ g Zn} \times \dfrac{1 \text{ mol}}{65.38 \text{ g}} = 0.05000 \text{ mol Zn}$

    $0.800 \text{ g O} \times \dfrac{1 \text{ mol}}{16.00 \text{ g}} = 0.0500 \text{ mol O}$

    Because the two components are present in equal amounts on a molar basis, the empirical formula must be simply ZnO.

64. Consider 100.0 g of the compound.

    $55.06 \text{ g Co} \times \dfrac{1 \text{ mol}}{58.93 \text{ g}} = 0.9343 \text{ mol Co}$

    If the sulfide of cobalt is 55.06% Co, then it is 44.94% S by mass.

    $44.94 \text{ g S} \times \dfrac{1 \text{ mol}}{32.07 \text{ g}} = 1.401 \text{ mol S}$

    Dividing each number of moles by the smaller (0.9343 mol Co) gives

    $\dfrac{0.09343 \text{ mol Co}}{0.9343} = 1.000 \text{ mol Co}$

    $\dfrac{1.401 \text{ mol S}}{0.9343 \text{ mol}} = 1.500 \text{ mol S}$

    Multiplying by two, to convert to whole numbers of moles, gives the empirical formula for the compound as Co_2S_3.

65. The amount of fluorine that reacted with the aluminum sample must be (3.89 g – 1.25 g) = 2.64 g of fluorine

    $1.25 \text{ g Al} \times \dfrac{1 \text{ mol}}{26.98 \text{ g}} = 0.04633 \text{ mol Al}$

    $2.64 \text{ g F} \times \dfrac{1 \text{ mol}}{19.00 \text{ g}} = 0.1389 \text{ mol F}$

    Dividing each number of moles by the smaller number of moles gives

    $\dfrac{0.04633 \text{ mol Al}}{0.04633 \text{ mol}} = 1.000 \text{ mol Al}$

    $\dfrac{0.1389 \text{ mol F}}{0.04633 \text{ mol}} = 2.999 \text{ mol F}$

    The empirical formula is just AlF_3

66. $2.50 \text{ g Al} \times \dfrac{1 \text{ mol}}{26.98 \text{ g}} = 0.09266 \text{ mol Al}$

    $5.28 \text{ g F} \times \dfrac{1 \text{ mol}}{19.00 \text{ g}} = 0.2779 \text{ mol F}$

Chapter 8: Chemical Composition

Dividing each number of moles by the smaller number of moles gives

$$\frac{0.09266 \text{ mol Al}}{0.09266 \text{ mol}} = 1.000 \text{ mol Al}$$

$$\frac{0.2779 \text{ mol F}}{0.09266 \text{ mol}} = 2.999 \text{ mol F}$$

The empirical formula is just AlF_3. Note the similarity between this problem and question 65: they differ in the way the data is given. In question 65, you were given the mass of the product, and first had to calculate how much fluorine had reacted.

67. Consider 100.0 g of the compound.

$$67.61 \text{ g U} \times \frac{1 \text{ mol}}{238.0 \text{ g}} = 0.2841 \text{ mol U}$$

$$32.39 \text{ g F} \times \frac{1 \text{ mol}}{19.00 \text{ g}} = 1.705 \text{ mol F}$$

Dividing each number of moles by the smaller number of moles (0.2841 mol U) gives

$$\frac{0.2841 \text{ mol U}}{0.2841 \text{ mol}} = 1.000 \text{ mol U}$$

$$\frac{1.705 \text{ mol F}}{0.2841 \text{ mol}} = 6.000 \text{ mol F}$$

The empirical formula is UF_6.

68. Consider 100.0 g of the compound so that percentages become masses.

$$46.46 \text{ g Li} \times \frac{1 \text{ mol}}{6.941 \text{ g}} = 6.694 \text{ mol Li}$$

$$53.54 \text{ g O} \times \frac{1 \text{ mol}}{16.00 \text{ g}} = 3.346 \text{ mol O}$$

Dividing each number of moles by the smaller number of moles gives

$$\frac{6.694 \text{ mol Li}}{3.346 \text{ mol}} = 2.001 \text{ mol Li}$$

$$\frac{3.346 \text{ mol O}}{3.346 \text{ mol}} = 1.000 \text{ mol O}$$

The empirical formula is Li_2O

Chapter 8:   Chemical Composition

69.   Consider 100.0 g of the compound.

$$33.88 \text{ g Cu} \times \frac{1 \text{ mol}}{63.55 \text{ g}} = 0.5331 \text{ mol Cu}$$

$$14.94 \text{ g N} \times \frac{1 \text{ mol}}{14.01 \text{ g}} = 1.066 \text{ mol N}$$

$$51.18 \text{ g O} \times \frac{1 \text{ mol}}{16.00 \text{ g}} = 3.199 \text{ mol O}$$

Dividing each number of moles by the smallest number of moles (0.5331 mol Cu) gives

$$\frac{0.5331 \text{ mol Cu}}{0.5331 \text{ mol}} = 1.000 \text{ mol Cu}$$

$$\frac{1.066 \text{ mol N}}{0.5331 \text{ mol}} = 2.000 \text{ mol N}$$

$$\frac{3.199 \text{ mol O}}{0.5331 \text{ mol}} = 6.001 \text{ mol O}$$

The empirical formula is CuN_2O_6 [i.e., $Cu(NO_3)_2$].

70.   Consider 100.0 g of the compound.

$$59.78 \text{ g Li} \times \frac{1 \text{ mol}}{6.941 \text{ g}} = 8.613 \text{ mol Li}$$

$$40.22 \text{ g N} \times \frac{1 \text{ mol}}{14.01 \text{ g}} = 2.871 \text{ mol N}$$

Dividing each number of moles by the smaller number of moles (2.871 mol N) gives

$$\frac{8.613 \text{ mol Li}}{2.871 \text{ mol}} = 3.000 \text{ mol Li}$$

$$\frac{2.871 \text{ mol N}}{2.871 \text{ mol}} = 1.000 \text{ mol N}$$

The empirical formula is Li_3N.

71.   Consider 100.0 g of the compound.

$$66.75 \text{ g Cu} \times \frac{1 \text{ mol}}{63.55 \text{ g}} = 1.050 \text{ mol Cu}$$

$$10.84 \text{ g P} \times \frac{1 \text{ mol}}{30.97 \text{ g}} = 0.3500 \text{ mol P}$$

$$22.41 \text{ g O} \times \frac{1 \text{ mol}}{16.00 \text{ g}} = 1.401 \text{ mol O}$$

Dividing each number of moles by the smallest number of moles (0.3500 mol P) gives

Chapter 8: Chemical Composition

$$\frac{1.050 \text{ mol Cu}}{0.3500 \text{ mol}} = 3.000 \text{ mol Cu}$$

$$\frac{0.3500 \text{ mol P}}{0.3500} = 1.000 \text{ mol P}$$

$$\frac{1.401 \text{ mol O}}{0.3500 \text{ mol}} = 4.003 \text{ mol O}$$

The empirical formula is thus Cu_3PO_4.

72. Consider 100.0 g of the compound so that percentages become masses.

$$48.64 \text{ g C} \times \frac{1 \text{ mol C}}{12.01 \text{ g}} = 4.050 \text{ mol C}$$

$$8.16 \text{ g H} \times \frac{1 \text{ mol H}}{1.008 \text{ g}} = 8.095 \text{ mol H}$$

$$(100.0 \text{ g} - 48.64 \text{ g} - 8.16 \text{ g} = 43.2 \text{ g O}) \times \frac{1 \text{ mol}}{16.00 \text{ g}} = 2.70 \text{ mol O}$$

Dividing each number of moles by the smaller number of moles gives

$$\frac{4.050 \text{ mol C}}{2.70 \text{ mol}} = 1.500 \text{ mol C}$$

$$\frac{8.095 \text{ mol H}}{2.70 \text{ mol}} = 2.998 \text{ mol H}$$

$$\frac{2.70 \text{ mol O}}{2.70 \text{ mol}} = 1.000 \text{ mol O}$$

Multiplying these relative numbers of moles by 2 to give whole numbers gives the empirical formula as $C_3H_6O_2$.

73. The compound must contain 1.00 mg of lithium and 2.73 mg of fluorine.

$$1.00 \text{ mg Li} \times \frac{1 \text{ mmol}}{6.941 \text{ mg}} = 0.144 \text{ mmol Li}$$

$$2.73 \text{ mg F} \times \frac{1 \text{ mmol}}{19.00 \text{ mg}} = 0.144 \text{ mol F}$$

The empirical formula of the compound is LiF.

74. Compound 1: Assume 100.0 g of the compound.

$$22.55 \text{ g P} \times \frac{1 \text{ mol}}{30.97 \text{ g}} = 0.7281 \text{ mol P}$$

$$77.45 \text{ g Cl} \times \frac{1 \text{ mol}}{35.45 \text{ g}} = 2.185 \text{ mol Cl}$$

Dividing each number of moles by the smaller (0.7281 mol P) indicates that the formula of Compound 1 is PCl_3.

Compound 2: Assume 100.0 g of the compound.

$$14.87 \text{ g P} \times \frac{1 \text{ mol}}{30.97 \text{ g}} = 0.4801 \text{ mol P}$$

$$85.13 \text{ g Cl} \times \frac{1 \text{ mol}}{35.45 \text{ g}} = 2.401 \text{ mol Cl}$$

Dividing each number of moles by the smaller (0.4801 mol P) indicates that the formula of Compound 2 is PCl_5.

75. The *empirical formula* of a compound represents only the smallest whole number relationship between the number and type of atoms in a compound, whereas the *molecular formula* represents the actual number of atoms of each type in a true molecule of the substance. Many compounds (for example, H_2O) may have the same empirical and molecular formulas.

76. molecular formula: $C_6H_{12}O_6$ (count the number of each type of element in the molecule represented); empirical formula: CH_2O (simplest whole-number ratio of $C_6H_{12}O_6$; divisible by 6)

77. Assume that we have 100.00 g of the compound: then 78.14 g will be boron and 21.86 g will be hydrogen.

$$78.14 \text{ g B} \times \frac{1 \text{ mol}}{10.81 \text{ g}} = 7.228 \text{ mol B}$$

$$21.86 \text{ g H} \times \frac{1 \text{ mol}}{1.008 \text{ g}} = 21.69 \text{ mol H}$$

Dividing each number of moles by the smaller number (7.228 mol B) gives the empirical formula as BH_3. The empirical molar mass of BH_3 would be [10.81 g + 3(1.008 g)] = 13.83 g. This is approximately half of the indicated actual molar mass, and therefore the molecular formula must be B_2H_6.

78. empirical formula mass of CH = 13 g

$$n = \frac{\text{molar mass}}{\text{empirical formula mass}} = \frac{78 \text{ g}}{13 \text{ g}} = 6$$

The molecular formula is $(CH)_6$ or C_6H_6.

79. empirical formula mass of CH_2 = 14

$$n = \frac{\text{molar mass}}{\text{empirical formula mass}} = \frac{84 \text{ g}}{14 \text{ g}} = 6$$

molecular formula is $(CH_2)_6 = C_6H_{12}$.

80. empirical formula mass of C_2H_5O = 46 g

$$n = \frac{\text{molar mass}}{\text{empirical formula mass}} = \frac{90\text{ g}}{46\text{ g}} = \sim 2$$

molecular formula is $(C_2H_5O)_2 = C_4H_{10}O_2$

81. Consider 100.0 g of the compound.

$$42.87\text{ g C} \times \frac{1\text{ mol}}{12.01\text{ g}} = 3.570\text{ mol C}$$

$$3.598\text{ g H} \times \frac{1\text{ mol}}{1.008\text{ g}} = 3.569\text{ mol H}$$

$$28.55\text{ g O} \times \frac{1\text{ mol}}{16.00\text{ g}} = 1.784\text{ mol O}$$

$$25.00\text{ g N} \times \frac{1\text{ mol}}{14.01\text{ g}} = 1.784\text{ mol N}$$

Dividing each number of moles by the smallest number of moles (1.784 mol O or N) gives

$$\frac{3.570\text{ mol C}}{1.784\text{ mol}} = 2.001\text{ mol C}$$

$$\frac{3.569\text{ mol H}}{1.784\text{ mol}} = 2.001\text{ mol H}$$

$$\frac{1.784\text{ mol O}}{1.784\text{ mol}} = 1.000\text{ mol O}$$

$$\frac{1.784\text{ mol N}}{1.784\text{ mol}} = 1.000\text{ mol N}$$

The empirical formula of the compound is C_2H_2ON and the empirical formula mass of C_2H_2ON = 56.

$$n = \frac{\text{molar mass}}{\text{empirical formula mass}} = \frac{168\text{ g}}{56\text{ g}} = 3$$

The molecular formula is $(C_2H_2ON)_3 = C_6H_6O_3N_3$.

82.
$$81.71\text{ g C} \times \frac{1\text{ mol C}}{12.01\text{ g C}} = \frac{6.803\text{ mol C}}{6.803\text{ mol}} = 1 \times 3 = 3$$

$$(100 - 81.71)\text{ g H} \times \frac{1\text{ mol H}}{1.008\text{ g H}} = \frac{18.14\text{ mol H}}{6.803\text{ mol}} = 2.6667 \times 3 = 8$$

The empirical formula is therefore C_3H_8, with a molar mass of 44.094 g/mol. Since the molar mass of the compound is 44.1 g/mol, the molecular formula and the empirical formula are the same. The molecular formula is C_3H_8.

83. [1] c   [6] d

Chapter 8: Chemical Composition

[2]	e	[7]	a
[3]	j	[8]	g
[4]	h	[9]	i
[5]	b	[10]	f

84.

5.00 g Al	0.185 mol	1.12×10^{23} atoms
0.140 g Fe	0.00250 mol	1.51×10^{21} atoms
2.7×10^2 g Cu	4.3 mol	2.6×10^{24} atoms
0.00250 g Mg	1.03×10^{-4} mol	6.19×10^{19} atoms
0.062 g Na	2.7×10^{-3} mol	1.6×10^{21} atoms
3.95×10^{-18} g U	1.66×10^{-20} mol	1.00×10^4 atoms

85.

4.24 g	0.0543 mol	3.27×10^{22} molec.	3.92×10^{23} atoms
4.04 g	0.224 mol	1.35×10^{23} molec.	4.05×10^{23} atoms
1.98 g	0.0450 mol	2.71×10^{22} molec.	8.13×10^{22} atoms
45.9 g	1.26 mol	7.59×10^{23} molec.	1.52×10^{24} atoms
126 g	6.99 mol	4.21×10^{24} molec.	1.26×10^{25} atoms
0.297 g	0.00927 mol	5.58×10^{21} molec.	3.35×10^{22} atoms

86. mass of 2 mol X = 2(41.2 g) = 82.4 g

   mass of 1 mol Y = 57.7 g = 57.7 g

   mass of 3 mol Z = 3(63.9 g) = 191.7 g

   molar mass of X_2YZ_3 = 331.8 g

   $\%X = \dfrac{82.4 \text{ g}}{331.8 \text{ g}} \times 100 = 24.8\ \%X$

   $\% Y = \dfrac{57.7 \text{ g}}{331.8 \text{ g}} \times 100 = 17.4\%\ Y$

   $\% Z = \dfrac{191.7 \text{ g}}{331.8 \text{ g}} \times 100 = 57.8\%\ Z$

   If the molecular formula were actually $X_4Y_2Z_6$, the percentage composition would be the same, and the *relative* mass of each element present would not change. The molecular formula is always a whole number multiple of the empirical formula.

87. magnesium/nitrogen compound: mass of nitrogen = 1.2791 g – 0.9240 = 0.3551 g N

   $0.9240 \text{ g Mg} \times \dfrac{1 \text{ mol}}{24.31 \text{ g}} = 0.03801 \text{ mol Mg}$

   $0.3551 \text{ g N} \times \dfrac{1 \text{ mol}}{14.01 \text{ g}} = 0.02535 \text{ mol N}$

   Dividing each number of moles by the smaller number of moles gives

$$\frac{0.03801 \text{ mol Mg}}{0.02535 \text{ mol}} = 1.499 \text{ mol Mg}$$

$$\frac{0.02535 \text{ mol N}}{0.02535 \text{ mol}} = 1.000 \text{ mol N}$$

Multiplying by two, to convert to whole numbers, gives the empirical formula as Mg_3N_2.

magnesium/oxygen compound: Consider 100.0 g of the compound.

$$60.31 \text{ g Mg} \times \frac{1 \text{ mol}}{24.31 \text{ g}} = 2.481 \text{ mol Mg}$$

$$39.69 \text{ g O} \times \frac{1 \text{ mol}}{16.00 \text{ g}} = 2.481 \text{ mol O}$$

The empirical formula is MgO.

88. For the first compound (*restricted* amount of oxygen)

$$2.118 \text{ g Cu} \times \frac{1 \text{ mol}}{63.55 \text{ g}} = 0.03333 \text{ mol Cu}$$

$$0.2666 \text{ g O} \times \frac{1 \text{ mol}}{16.00 \text{ g}} = 0.01666 \text{ mol O}$$

Since the number of moles of Cu (0.03333 mol) is twice the number of moles of O (0.01666 mol), the empirical formula is Cu_2O.

For the second compound (stream of pure oxygen)

$$2.118 \text{ g Cu} \times \frac{1 \text{ mol}}{63.55 \text{ g}} = 0.03333 \text{ mol Cu}$$

$$0.5332 \text{ g O} \times \frac{1 \text{ mol}}{16.00 \text{ g}} = 0.03333 \text{ mol O}$$

Since the numbers of moles are the same, the empirical formula is CuO.

89.

Compound	Molar Mass	% H
HF	20.01 g	5.037%
HCl	36.46 g	2.765%
HBr	80.91 g	1.246%
HI	127.9 g	0.7881%

90. a. molar mass H_2O = 18.02 g

$$4.21 \text{ g} \times \frac{1 \text{ mol}}{18.02 \text{ g}} \times \frac{6.022 \times 10^{23} \text{ molecules}}{1 \text{ mol}} = 1.41 \times 10^{23} \text{ molecules}$$

The sample contains 1.41×10^{23} oxygen atoms and $2(1.41 \times 10^{23}) = 2.82 \times 10^{23}$ hydrogen atoms.

Chapter 8: Chemical Composition

    b.   molar mass CO_2 = 44.01 g

$$6.81 \text{ g} \times \frac{1 \text{ mol}}{44.01 \text{ g}} \times \frac{6.022 \times 10^{23} \text{ molecules}}{1 \text{ mol}} = 9.32 \times 10^{22} \text{ molecules}$$

        The sample contains 9.32×10^{22} carbon atoms and $2(9.32 \times 10^{22}) = 1.86 \times 10^{23}$ oxygen atoms.

    c.   molar mass C_6H_6 = 78.11 g

$$0.000221 \text{ g} \times \frac{1 \text{ mol}}{78.11 \text{ g}} \times \frac{6.022 \times 10^{23} \text{ molecules}}{1 \text{ mol}} = 1.70 \times 10^{18} \text{ molec.}$$

        The sample contains $6(1.70 \times 10^{18}) = 1.02 \times 10^{19}$ atoms of each element.

    d. $2.26 \text{ mol} \times \dfrac{6.022 \times 10^{23} \text{ molecules}}{1 \text{ mol}} = 1.36 \times 10^{24}$ molecules

        atoms C = $12(1.36 \times 10^{24}) = 1.63 \times 10^{25}$ atoms

        atoms H = $22(1.36 \times 10^{24}) = 2.99 \times 10^{25}$ atoms

        atoms O = $11(1.36 \times 10^{24}) = 1.50 \times 10^{25}$ atoms

91.   a.   Assuming $10{,}000{,}000{,}000 = 1.0 \times 10^{10}$

$$1.0 \times 10^{10} \text{ molecules} \times \frac{28.02 \text{ g N}_2}{6.022 \times 10^{23} \text{ molecules}} = 4.7 \times 10^{-13} \text{ g N}_2$$

    b. 2.49×10^{20} molecules $\times \dfrac{44.01 \text{ g CO}_2}{6.022 \times 10^{23} \text{ molecules}} = 0.0182$ g CO_2

    c.   7.0983 mol NaCl $\times \dfrac{58.44 \text{ g}}{1 \text{ mol}} = 414.8$ g NaCl

    d. 9.012×10^{-6} mol $C_2H_4Cl_2 \times \dfrac{98.95 \text{ g}}{1 \text{ mol}} = 8.918 \times 10^{-4}$ g $C_2H_4Cl_2$

92.   a.   molar mass of C_3O_2 = 3(12.01 g) + 2(16.00 g) = 68.03 g

        % C = $\dfrac{36.03 \text{ g C}}{68.03 \text{ g}} \times 100 = 52.96\%$ C

        7.819 g $C_3O_2 \times \dfrac{52.96 \text{ g C}}{100.0 \text{ g C}_3\text{O}_2} = 4.141$ g C

        4.141 g C $\times \dfrac{6.022 \times 10^{23} \text{ molecules}}{12.01 \text{ g C}} = 2.076 \times 10^{23}$ C atoms

    b.   molar mass of CO = 12.01 g + 16.00 g = 28.01 g

        % C = $\dfrac{12.01 \text{ g C}}{28.01 \text{ g}} \times 100 = 42.88\%$ C

        1.53×10^{21} molecules CO $\times \dfrac{1 \text{ C atom}}{1 \text{ molecule CO}} = 1.53 \times 10^{21}$ C atoms

Chapter 8: Chemical Composition

$$1.53 \times 10^{21} \text{ C atoms} \times \frac{12.01 \text{ g C}}{6.022 \times 10^{23} \text{ C atoms}} = 0.0305 \text{ g C}$$

c. molar mass of C_6H_6O = 6(12.01 g) + 6(1.008 g) + 16.00 g = 94.11 g

$$\% \text{ C} = \frac{72.06 \text{ g C}}{94.11 \text{ g}} \times 100 = 76.57\% \text{ C}$$

$$0.200 \text{ mol } C_6H_6O \times \frac{6 \text{ mol C}}{1 \text{ mol } C_6H_6O} = 1.20 \text{ mol C}$$

$$1.20 \text{ mol C} \times \frac{12.01 \text{ g}}{1 \text{ mol}} = 14.4 \text{ g C}$$

$$14.4 \text{ g C} \times \frac{6.022 \times 10^{23} \text{ C atoms}}{12.01 \text{ g C}} = 7.22 \times 10^{23} \text{ C atoms}$$

93. [1] g   [6] i
    [2] c   [7] f
    [3] b   [8] h
    [4] a   [9] e
    [5] j   [10] d

94. All are true (a, b, c, d, e). 6.022×10^{23} = 1 mol. 18.016 g/mol is the molar mass of H_2O. There are 3 atoms in every one molecule of H_2O, thus $3 \times (6.022 \times 10^{23}) = 1.807 \times 10^{24}$ atoms. There are 2 moles of H in every 1 mole of H_2O.

95. $2.24 \text{ g Fe} \times \dfrac{58.93 \text{ g Co}}{55.85 \text{ g Fe}} = 2.36 \text{ g Co}$

96. (a); Na has the smallest molar mass (22.99 g/mol). Since the same mass is used for each sample, Na contains the largest number of moles and thus atoms (because the mass is divided by the molar mass to get moles).

97. 1.00 kg = 1.00×10^3 g

$$1.00 \times 10^3 \text{ g Zr} \times \frac{6.941 \text{ g Li}}{91.22 \text{ g Zr}} = 76.1 \text{ g Li}$$

98. 153.8 g CCl_4 = 6.022×10^{23} molecules CCl_4

$$1 \text{ molecule} \times \frac{153.8 \text{ g } CCl_4}{6.022 \times 10^{23} \text{ molecules}} = 2.554 \times 10^{-22} \text{ g}$$

Chapter 8: Chemical Composition

99. a.   molar mass of C_6H_6 = 78.11 g

$$2.500 \text{ g} \times \frac{6.048 \text{ g H}}{78.11 \text{ g}} = 0.1936 \text{ g H}$$

b.   molar mass of CaH_2 = 42.10 g

$$2.500 \text{ g} \times \frac{2.016 \text{ g H}}{42.10 \text{ g}} = 0.1197 \text{ g H}$$

c.   molar mass of C_2H_5OH = 46.07 g

$$2.500 \text{ g} \times \frac{6.048 \text{ g H}}{46.07 \text{ g}} = 0.3282 \text{ g H}$$

d.   molar mass of $C_3H_7O_3N$ = 105.1 g

$$2.500 \text{ g} \times \frac{7.056 \text{ g H}}{105.1 \text{ g}} = 0.1678 \text{ g H}$$

100. (a); Both NO_2 and F_2 are molecules and since you have equal moles of each (but not necessarily 1 mole of each), the number of molecules must be the same (6.022×10^{23} = 1 mol). However, the number of atoms is different since NO_2 contains 3 atoms and F_2 contains 2. Furthermore, the masses are different since NO_2 and F_2 have different molar masses.

101. Consider 100.0 g of the compound.

$$25.45 \text{ g Cu} \times \frac{1 \text{ mol}}{63.55 \text{ g}} = 0.4005 \text{ mol Cu}$$

$$12.84 \text{ g S} \times \frac{1 \text{ mol}}{32.07 \text{ g}} = 0.4004 \text{ mol S}$$

$$4.036 \text{ g H} \times \frac{1 \text{ mol}}{1.008 \text{ g}} = 4.004 \text{ mol H}$$

$$57.67 \text{ g O} \times \frac{1 \text{ mol}}{16.00 \text{ g}} = 3.604 \text{ mol O}$$

Dividing each number of moles by the smallest number of moles (0.4004 mol) gives

$$\frac{0.4005 \text{ mol Cu}}{0.4004 \text{ mol}} = 1.000 \text{ mol Cu}$$

$$\frac{0.4004 \text{ mol S}}{0.4004 \text{ mol}} = 1.000 \text{ mol S}$$

$$\frac{4.004 \text{ mol H}}{0.4004 \text{ mol}} = 10.00 \text{ mol H}$$

$$\frac{3.604 \text{ mol O}}{0.4004 \text{ mol}} = 9.001 \text{ mol O}$$

The empirical formula is $CuSH_{10}O_9$ (which is usually written as $CuSO_4 \cdot 5H_2O$).

102. Consider 100.0 g of the compound so that percentages become masses.

$$60.87 \text{ g C} \times \frac{1 \text{ mol C}}{12.01 \text{ g}} = 5.068 \text{ mol C}$$

$$4.38 \text{ g H} \times \frac{1 \text{ mol H}}{1.008 \text{ g}} = 4.345 \text{ mol H}$$

$$(100.0 \text{ g} - 60.87 \text{ g} - 4.38 \text{ g} = 34.75 \text{ g O}) \times \frac{1 \text{ mol}}{16.00 \text{ g}} = 2.172 \text{ mol O}$$

Dividing each number of moles by the smaller number of moles gives

$$\frac{5.068 \text{ mol C}}{2.172 \text{ mol}} = 2.333 \text{ mol C}$$

$$\frac{4.345 \text{ mol H}}{2.172 \text{ mol}} = 2.000 \text{ mol H}$$

$$\frac{2.172 \text{ mol O}}{2.172 \text{ mol}} = 1.000 \text{ mol O}$$

Multiplying these relative numbers of moles by 3 to give whole numbers gives the empirical formula as $C_7H_6O_3$.

103. atomic mass unit (amu)

104. We use the *average* mass because this average is a *weighted average* and takes into account both the masses and the relative abundances of the various isotopes.

105. a. $160{,}000 \text{ amu} \times \dfrac{1 \text{ O atom}}{16.00 \text{ amu}} = 1.0 \times 10^4$ O atoms (assuming exact)

   b. $8139.81 \text{ amu} \times \dfrac{1 \text{ N atom}}{14.01 \text{ amu}} = 581$ N atoms

   c. $13{,}490 \text{ amu} \times \dfrac{1 \text{ Al atom}}{26.98 \text{ amu}} = 500$ Al atoms

   d. $5040 \text{ amu} \times \dfrac{1 \text{ H atom}}{1.008 \text{ amu}} = 5.00 \times 10^3$ H atoms

   e. $367{,}495.15 \text{ amu} \times \dfrac{1 \text{ Na atom}}{22.99 \text{ amu}} = 1.599 \times 10^4$ Na atoms

106. $1.98 \times 10^{13} \text{ amu} \times \dfrac{1 \text{ Na atom}}{22.99 \text{ amu}} = 8.61 \times 10^{11}$ Na atoms

$3.01 \times 10^{23} \text{ Na atoms} \times \dfrac{22.99 \text{ amu}}{1 \text{ Na atom}} = 6.92 \times 10^{24}$ amu

Chapter 8: Chemical Composition

107. a. 1.5 mg = 0.0015 g

$$0.0015 \text{ g Cr} \times \frac{1 \text{ mol}}{52.00 \text{ g}} = 2.9 \times 10^{-5} \text{ mol Cr}$$

b. $2.0 \times 10^{-3} \text{ g Sr} \times \frac{1 \text{ mol}}{87.62 \text{ g}} = 2.3 \times 10^{-5} \text{ mol Sr}$

c. $4.84 \times 10^4 \text{ g B} \times \frac{1 \text{ mol}}{10.81 \text{ g}} = 4.48 \times 10^3 \text{ mol B}$

d. $3.6 \times 10^{-6} \text{ μg} = 3.6 \times 10^{-12} \text{ g}$

$3.6 \times 10^{-12} \text{ g Cf} \times \frac{1 \text{ mol}}{251 \text{ g}} = 1.4 \times 10^{-14} \text{ mol Cf}$

e. $2000 \text{ lb} \times \frac{453.59 \text{ g}}{1 \text{ lb}} = 9.1 \times 10^5 \text{ g}$

$9.1 \times 10^5 \text{ g Fe} \times \frac{1 \text{ mol}}{55.85 \text{ g}} = 1.6 \times 10^4 \text{ mol Fe}$

f. $20.4 \text{ g Ba} \times \frac{1 \text{ mol}}{137.3 \text{ g}} = 0.149 \text{ mol Ba}$

g. $62.8 \text{ g Co} \times \frac{1 \text{ mol}}{58.93 \text{ g}} = 1.07 \text{ mol Co}$

108. a. $5.0 \text{ mol K} \times \frac{39.10 \text{ g}}{1 \text{ mol}} = 195 \text{ g} = 2.0 \times 10^2 \text{ g K}$

b. $0.000305 \text{ mol Hg} \times \frac{200.6 \text{ g}}{1 \text{ mol}} = 0.0612 \text{ g Hg}$

c. $2.31 \times 10^{-5} \text{ mol Mn} \times \frac{54.94 \text{ g}}{1 \text{ mol}} = 1.27 \times 10^{-3} \text{ g Mn}$

d. $10.5 \text{ mol P} \times \frac{30.97 \text{ g}}{1 \text{ mol}} = 325 \text{ g P}$

e. $4.9 \times 10^4 \text{ mol Fe} \times \frac{55.85 \text{ g}}{1 \text{ mol}} = 2.7 \times 10^6 \text{ g Fe}$

f. $125 \text{ mol Li} \times \frac{6.941 \text{ g}}{1 \text{ mol}} = 868 \text{ g Li}$

g. $0.01205 \text{ mol F} \times \frac{19.00 \text{ g}}{1 \text{ mol}} = 0.2290 \text{ g F}$

109. a. $2.89 \text{ g Au} \times \frac{6.022 \times 10^{23} \text{ Au atoms}}{197.0 \text{ g Au}} = 8.83 \times 10^{21} \text{ Au atoms}$

b. $0.000259 \text{ mol Pt} \times \dfrac{6.022 \times 10^{23} \text{ Pt atoms}}{1 \text{ mol}} = 1.56 \times 10^{20}$ Pt atoms

c. $0.000259 \text{ g Pt} \times \dfrac{6.022 \times 10^{23} \text{ Pt atoms}}{195.1 \text{ g Pt}} = 7.99 \times 10^{17}$ Pt atoms

d. $2.0 \text{ lb} \times \dfrac{453.59 \text{ g}}{1 \text{ lb}} = 908$ g

$908 \text{ g Mg} \times \dfrac{6.022 \times 10^{23} \text{ Mg atoms}}{24.31 \text{ g Mg}} = 2.2 \times 10^{25}$ Mg atoms

e. $1.90 \text{ mL} \times \dfrac{13.6 \text{ g}}{1 \text{ mL}} = 25.8$ g Hg

$25.8 \text{ g Hg} \times \dfrac{6.022 \times 10^{23} \text{ Hg atoms}}{200.6 \text{ g Hg}} = 7.75 \times 10^{22}$ Hg atoms

f. $4.30 \text{ mol W} \times \dfrac{6.022 \times 10^{23} \text{ W atoms}}{1 \text{ mol}} = 2.59 \times 10^{24}$ W atoms

g. $4.30 \text{ g W} \times \dfrac{6.022 \times 10^{23} \text{ W atoms}}{183.9 \text{ g W}} = 1.41 \times 10^{22}$ W atoms

110. a.   mass of 1 mol Fe = 1(55.85 g) = 55.85 g

mass of 1 mol S = 1(32.07 g) = 32.07 g

mass of 4 mol O = 4(16.00 g) = 64.00 g

molar mass of $FeSO_4$ = 151.92 g

b.   mass of 1 mol Hg = 1(200.6 g) = 200.6 g

mass of 2 mol I = 2(126.9 g) = 253.8 g

molar mass of HgI_2 =   454.4 g

c.   mass of 1 mol Sn = 1(118.7 g) = 118.7 g

mass of 2 mol O = 2(16.00 g) = 32.00 g

molar mass of SnO_2 = 150.7 g

d.   mass of 1 mol Co = 1(58.93 g) = 58.93 g

mass of 2 mol Cl = 2(35.45 g) = 70.90 g

molar mass of $CoCl_2$ = 129.83 g

e.   mass of 1 mol Cu = 1(63.55 g) = 63.55 g

mass of 2 mol N = 2(14.01 g) = 28.02 g

mass of 6 mol O = 6(16.00 g) = 96.00 g

molar mass of $Cu(NO_3)_2$ = 187.57 g

Chapter 8: Chemical Composition

111. a. mass of 6 mol C = 6(12.01 g) = 72.06 g
mass of 10 mol H = 10(1.008 g) = 10.08 g
mass of 4 mol O = 4(16.00 g) = 64.00 g
molar mass of $C_6H_{10}O_4$ = 146.14 g

b. mass of 8 mol C = 8(12.01 g) = 96.08 g
mass of 10 mol H = 10(1.008 g) = 10.08 g
mass of 4 mol N = 4(14.01 g) = 56.04 g
mass of 2 mol O = 1(16.00 g) = 32.00 g
molar mass of $C_8H_{10}N_4O_2$ = 194.20 g

c. mass of 20 mol C = 20(12.01 g) = 240.2 g
mass of 42 mol H = 42(1.008 g) = 42.34 g
molar mass of $C_{20}H_{42}$ = 282.5 g

d. mass of 6 mol C = 6(12.01 g) = 72.06 g
mass of 12 mol H = 12(1.008 g) = 12.10 g
mass of 1 mol O = 16.00 g = 16.00 g
molar mass of $C_6H_{11}OH$ = 100.16 g

e. mass of 4 mol C = 4(12.01 g) = 48.04 g
mass of 6 mol H = 6(1.008 g) = 6.048 g
mass of 2 mol O = 2(16.00 g) = 32.00 g
molar mass of $C_4H_6O_2$ = 86.09 g

f. mass of 6 mol C = 6(12.01 g) = 72.06 g
mass of 12 mol H = 12(1.008 g) = 12.10 g
mass of 6 mol O = 6(16.00 g) = 96.00 g
molar mass of $C_6H_{12}O_6$ = 180.16 g

112. a. molar mass of $(NH_4)_2S$ = 68.15 g

$21.2 \text{ g} \times \dfrac{1 \text{ mol}}{68.15 \text{ g}} = 0.311 \text{ mol } (NH_4)_2S$

b. molar mass of $Ca(NO_3)_2$ = 164.1 g

$44.3 \text{ g} \times \dfrac{1 \text{ mol}}{164.1 \text{ g}} = 0.270 \text{ mol } Ca(NO_3)_2$

c. molar mass of Cl_2O = 86.9 g

$4.35 \text{ g} \times \dfrac{1 \text{ mol}}{86.9 \text{ g}} = 0.0501 \text{ mol } Cl_2O$

Chapter 8: Chemical Composition

  d. 1.0 lb = 454 g; molar mass of $FeCl_3$ = 162.2

$$454 \text{ g} \times \frac{1 \text{ mol}}{162.2 \text{ g}} = 2.8 \text{ mol } FeCl_3$$

  e. 1.0 kg = 1.0×10^3 g; molar mass of $FeCl_3$ = 162.2 g

$$1.0 \times 10^3 \text{ g} \times \frac{1 \text{ mol}}{162.2 \text{ g}} = 6.2 \text{ mol } FeCl_3$$

113. a. molar mass of $FeSO_4$ = 151.92 g

$$1.28 \text{ g} \times \frac{1 \text{ mol}}{151.92 \text{ g}} = 8.43 \times 10^{-3} \text{ mol } FeSO_4$$

  b. 5.14 mg = 0.00514 g; molar mass of HgI_2 = 454.4 g

$$0.00514 \text{ g} \times \frac{1 \text{ mol}}{454.4 \text{ g}} = 1.13 \times 10^{-5} \text{ mol } HgI_2$$

  c. 9.21 μg = 9.21×10^{-6} g; molar mass of SnO_2 = 150.7 g

$$9.21 \times 10^{-6} \text{ g} \times \frac{1 \text{ mol}}{150.7 \text{ g}} = 6.11 \times 10^{-8} \text{ mol } SnO_2$$

  d. 1.26 lb = 1.26(453.59 g) = 572 g; molar mass of $CoCl_2$ = 129.83 g

$$572 \text{ g} \times \frac{1 \text{ mol}}{129.83 \text{ g}} = 4.41 \text{ mol } CoCl_2$$

  e. molar mass of $Cu(NO_3)_2$ = 187.57 g

$$4.25 \text{ g} \times \frac{1 \text{ mol}}{187.57 \text{ g}} = 2.27 \times 10^{-2} \text{ mol } Cu(NO_3)_2$$

114. $CuCO_3$, Na_3PO_4, P_4O_{10}; Assuming 1 mole of each, rank in order of increasing molar mass: 123.56 g/mol for $CuCO_3$, 163.94 g/mol for Na_3PO_4, and 283.88 g/mol for P_4O_{10}.

115. a. molar mass of $(NH_4)_2CO_3$ = 96.09 g

$$3.09 \text{ mol} \times \frac{96.09 \text{ g}}{1 \text{ mol}} = 297 \text{ g } (NH_4)_2CO_3$$

  b. molar mass of $NaHCO_3$ = 84.01 g

$$4.01 \times 10^{-6} \text{ mol} \times \frac{84.01 \text{ g}}{1 \text{ mol}} = 3.37 \times 10^{-4} \text{ g } NaHCO_3$$

  c. molar mass of CO_2 = 44.01 g

$$88.02 \text{ mol} \times \frac{44.01 \text{ g}}{1 \text{ mol}} = 3874 \text{ g } CO_2$$

Chapter 8: Chemical Composition

    d.   1.29 mmol = 0.00129 mol; molar mass of $AgNO_3$ = 169.9 g

$$0.00129 \text{ mol} \times \frac{169.9 \text{ g}}{1 \text{ mol}} = 0.219 \text{ g } AgNO_3$$

    e.   molar mass of $CrCl_3$ = 158.4 g

$$0.0024 \text{ mol} \times \frac{158.4 \text{ g}}{1 \text{ mol}} = 0.38 \text{ g } CrCl_3$$

116.   a.   molar mass of $C_6H_{12}O_6$ = 180.2 g

$$3.45 \text{ g} \times \frac{6.022 \times 10^{23} \text{ molecules}}{180.2 \text{ g}} = 1.15 \times 10^{22} \text{ molecules } C_6H_{12}O_6$$

    b. $3.45 \text{ mol} \times \frac{6.022 \times 10^{23} \text{ molecules}}{1 \text{ mol}} = 2.08 \times 10^{24} \text{ molecules } C_6H_{12}O_6$

    c.   molar mass of ICl_5 = 304.2 g

$$25.0 \text{ g} \times \frac{6.022 \times 10^{23} \text{ molecules}}{304.2 \text{ g}} = 4.95 \times 10^{22} \text{ molecules } ICl_5$$

    d.   molar mass of B_2H_6 = 27.67 g

$$1.00 \text{ g} \times \frac{6.022 \times 10^{23} \text{ molecules}}{27.67 \text{ g}} = 2.18 \times 10^{22} \text{ molecules } B_2H_6$$

    e.   1.05 mmol = 0.00105 mol

$$0.00105 \text{ mol} \times \frac{6.022 \times 10^{23} \text{ formula units}}{1 \text{ mol}} = 6.32 \times 10^{20} \text{ formula units}$$

117.   a.   molar mass of NH_3 = 17.03 g

$$2.71 \text{ g} \times \frac{1 \text{ mol}}{17.03 \text{ g}} = 0.159 \text{ mol } NH_3$$

        mol H = 3(0.159 mol) = 0.477 mol H

    b.   mol H = 2(0.824 mol) = 1.648 mol H = 1.65 mol H

    c.   6.25 mg = 0.00625 g; molar mass of H_2SO_4 = 98.09 g

$$0.00625 \text{ g} \times \frac{1 \text{ mol}}{98.09 \text{ g}} = 6.37 \times 10^{-5} \text{ mol } H_2SO_4$$

        mol H = 2(6.37 × 10^{-5} mol) = 1.27 × 10^{-4} mol H

    d.   molar mass of $(NH_4)_2CO_3$ = 96.09 g

$$451 \text{ g} \times \frac{1 \text{ mol}}{96.09 \text{ g}} = 4.69 \text{ mol } (NH_4)_2CO_3$$

        mol H = 8(4.69 mol) = 37.5 mol H

Chapter 8: Chemical Composition

118. molar mass of Mg_3N_2 = 100.95 g

$$5.00 \text{ g } Mg_3N_2 \times \frac{1 \text{ mol } Mg_3N_2}{100.95 \text{ g } Mg_3N_2} \times \frac{2 \text{ mol N}}{1 \text{ mol } Mg_3N_2} \times \frac{6.022 \times 10^{23} \text{ N atoms}}{1 \text{ mol N}} = 5.97 \times 10^{22} \text{ N atoms}$$

119. a. molar mass of NaN_3 = 65.02 g

$$\% \text{ Na} = \frac{22.99 \text{ g Na}}{65.02 \text{ g}} \times 100 = 35.36\% \text{ Na}$$

b. molar mass of $CuSO_4$ = 159.62 g

$$\% \text{ Cu} = \frac{63.55 \text{ g Cu}}{159.62 \text{ g}} \times 100 = 39.81\% \text{ Cu}$$

c. molar mass of $AuCl_3$ = 303.4 g

$$\% \text{ Au} = \frac{197.0 \text{ g Au}}{303.4 \text{ g}} \times 100 = 64.93\% \text{ Au}$$

d. molar mass of $AgNO_3$ = 169.9 g

$$\% \text{ Ag} = \frac{107.9 \text{ g Ag}}{169.9 \text{ g}} \times 100 = 63.51\% \text{ Ag}$$

e. molar mass of Rb_2SO_4 = 267.0 g

$$\% \text{ Rb} = \frac{170.9 \text{ g Rb}}{267.0 \text{ g}} \times 100 = 64.01\% \text{ Rb}$$

f. molar mass of $NaClO_3$ = 106.44 g

$$\% \text{ Na} = \frac{22.99 \text{ g Na}}{106.44 \text{ g}} \times 100 = 21.60\% \text{ Na}$$

g. molar mass of NI_3 = 394.7 g

$$\% \text{ N} = \frac{14.01 \text{ g N}}{394.7 \text{ g}} \times 100 = 3.550\% \text{ N}$$

h. molar mass of CsBr = 212.8 g

$$\% \text{ Cs} = \frac{132.9 \text{ g Cs}}{212.8 \text{ g}} \times 100 = 62.45\% \text{ Cs}$$

120. (d); The percent mass of an element in a compound is independent of the amount of compound present.

121. $0.7238 \text{ g C} \times \dfrac{1 \text{ mol}}{12.01 \text{ g}} = 0.06027 \text{ mol C}$

$0.07088 \text{ g H} \times \dfrac{1 \text{ mol}}{1.008 \text{ g}} = 0.07032 \text{ mol H}$

Chapter 8: Chemical Composition

$$0.1407 \text{ g N} \times \frac{1 \text{ mol}}{14.01 \text{ g}} = 0.01004 \text{ mol N}$$

$$0.3214 \text{ g O} \times \frac{1 \text{ mol}}{16.00 \text{ g}} = 0.02009 \text{ mol O}$$

Dividing each number of moles by the smallest number of moles (0.01004 mol) gives

$$\frac{0.06027 \text{ mol C}}{0.01004} = 6.003 \text{ mol C} \qquad \frac{0.07032 \text{ mol H}}{0.01004} = 7.004 \text{ mol H}$$

$$\frac{0.01004 \text{ mol N}}{0.01004} = 1.000 \text{ mol N} \qquad \frac{0.02009 \text{ mol O}}{0.01004} = 2.001 \text{ mol O}$$

The empirical formula is $C_6H_7NO_2$.

122. $$100.0 \text{ g (NH}_4)_2\text{CO}_3 \times \frac{1 \text{ mol (NH}_4)_2\text{CO}_3}{96.09 \text{ g (NH}_4)_2\text{CO}_3} \times \frac{3 \text{ mol O}}{1 \text{ mol (NH}_4)_2\text{CO}_3} \times \frac{1 \text{ mol NaOH}}{1 \text{ mol O}} \times \frac{40.00 \text{ g NaOH}}{1 \text{ mol NaOH}} = 124.9 \text{ g NaOH}$$

123. $$2.004 \text{ g Ca} \times \frac{1 \text{ mol}}{40.08 \text{ g}} = 0.05000 \text{ mol Ca}$$

$$0.4670 \text{ g N} \times \frac{1 \text{ mol}}{14.01 \text{ g}} = 0.03333 \text{ mol N}$$

Dividing each number of moles by the smaller number of moles gives

$$\frac{0.05000 \text{ mol Ca}}{0.03333 \text{ mol}} = 1.500 \text{ mol Ca} \qquad \frac{0.03333 \text{ mol N}}{0.03333 \text{ mol}} = 1.000 \text{ mol N}$$

Multiplying these relative numbers of moles by 2 to give whole numbers gives the empirical formula as Ca_3N_2.

124. Assume 100.000 grams. 100.000 g − 13.102 g = 86.898 g Cl

$$86.898 \text{ g Cl} \times \frac{1 \text{ mol Cl}}{35.45 \text{ g Cl}} = 2.451 \text{ mol Cl}$$

$$2.451 \text{ mol Cl} \times \frac{1 \text{ mol X}}{6 \text{ mol Cl}} = 0.4085 \text{ mol X}$$

$$\text{Molar Mass} = \frac{13.102 \text{ g X}}{0.4085 \text{ mol X}} = 32.07 \text{ g/mol} = \text{Sulfur (S)}$$

125. Mass of chlorine in compound = 3.045 g − 1.00 g = 2.045 g Cl

$$1.00 \text{ g Cr} \times \frac{1 \text{ mol}}{52.00 \text{ g}} = 0.0192 \text{ mol Cr}$$

$$2.045 \text{ g Cl} \times \frac{1 \text{ mol}}{35.45 \text{ g}} = 0.0577 \text{ mol Cl}$$

Chapter 8: Chemical Composition

Dividing each number of moles by the smaller number of moles

$$\frac{0.0192 \text{ mol Cr}}{0.0192} = 1.00 \text{ mol Cr} \qquad \frac{0.0577 \text{ mol Cl}}{0.0192 \text{ mol}} = 3.01 \text{ mol Cl}$$

The empirical formula is $CrCl_3$.

126. Assume we have 100.0 g of the compound.

$$65.95 \text{ g Ba} \times \frac{1 \text{ mol}}{137.3 \text{ g}} = 0.4803 \text{ mol Ba}$$

$$34.05 \text{ g Cl} \times \frac{1 \text{ mol}}{35.45 \text{ g}} = 0.9605 \text{ mol Cl}$$

Dividing each of these number of moles by the smaller number gives

$$\frac{0.4803 \text{ mol Ba}}{0.4803 \text{ mol}} = 1.000 \text{ mol Ba} \qquad \frac{0.9605 \text{ mol Cl}}{0.4803 \text{ mol}} = 2.000 \text{ mol Cl}$$

The empirical formula is then $BaCl_2$.

127. a. H_2O water

mass of 2 mol H = 2(1.008 g) = 2.016 g

mass of 1 mol O = 16.00 g

molar mass of H_2O = (2.016 g + 16.00 g) = 18.02 g

b. $FeCl_3$ iron(III) chloride

mass of 1 mol Fe = 55.85 g

mass of 3 mol Cl = 3(35.45 g) = 106.35 g

molar mass of $FeCl_3$ = (55.85 g + 106.35 g) = 162.2 g

c. KBr potassium bromide

mass of 1 mol K = 39.1 g

mass of 1 mol Br = 79.90 g

molar mass of KBr = (39.1 g + 79.90 g) = 119.0 g

d. NH_4NO_3 ammonium nitrate

mass of 2 mol N = 2(14.01 g) = 28.02 g

mass of 4 mol H = 4(1.008 g) = 4.032 g

mass of 3 mol O = 3(16.00 g) = 48.00 g

molar mass of NH_4NO_3 = (28.02 g + 4.032 g + 48.00 g) = 80.05 g

Chapter 8: Chemical Composition

  e. NaOH sodium hydroxide

   mass of 1 mol Na = 22.99 g

   mass of 1 mol O = 16.00 g

   mass of 1 mol H = 1.008 g

   molar mass of NaOH = (22.99 g + 16.00 g + 1.008 g) = 40.00 g

128. a. $C_{63}H_{88}CoN_{14}O_{14}P$

   mass of 63 mol C = 63(12.01 g) = 756.63 g

   mass of 88 mol H = 88(1.008 g) = 88.704 g

   mass of 1 mol Co = 58.93 g

   mass of 14 mol N = 14(14.01 g) = 196.14 g

   mass of 14 mol O = 14(16.00 g) = 224.00 g

   mass of 1 mol P = 30.97 g

  molar mass of $C_{63}H_{88}CoN_{14}O_{14}P$ = (756.63+88.704+58.93+196.14+224.00+30.97) = 1355.37 g

  b. $250 \text{ mg } C_{63}H_{88}CoN_{14}O_{14}P \times \dfrac{1 \text{ g}}{1000 \text{ mg}} \times \dfrac{1 \text{ mol}}{1355.37 \text{ g}} = 1.8 \times 10^{-4}$ mol

  c. $0.60 \text{ mol } C_{63}H_{88}CoN_{14}O_{14}P \times \dfrac{1355.37 \text{ g}}{1 \text{ mol}} = 810$ g

  d. $1.0 \text{ mol } C_{63}H_{88}CoN_{14}O_{14}P \times \dfrac{88 \text{ mol H}}{1 \text{ mol } C_{63}H_{88}CoN_{14}O_{14}P} \times \dfrac{6.022 \times 10^{23} \text{ H atoms}}{1 \text{ mol H}} =$

   $= 5.3 \times 10^{25}$ H atoms

  e. $1.0 \times 10^7 \text{ molecules } C_{63}H_{88}CoN_{14}O_{14}P \times \dfrac{1 \text{ mol}}{6.022 \times 10^{23}} \times \dfrac{1355.37 \text{ g}}{1 \text{ mol}} = 2.3 \times 10^{-14}$ g

  f. $1.0 \text{ molecule } C_{63}H_{88}CoN_{14}O_{14}P \times \dfrac{1 \text{ mol}}{6.022 \times 10^{23}} \times \dfrac{1355.37 \text{ g}}{1 \text{ mol}} = 2.3 \times 10^{-21}$ g

129. $326.4 \text{ g } Mg_3(PO_4)_2 \times \dfrac{1 \text{ mol}}{262.87 \text{ g}} = 1.242$ mol

  $303.0 \text{ g } Ca(NO_3)_2 \times \dfrac{1 \text{ mol}}{164.1 \text{ g}} = 1.846$ mol

  $141.6 \text{ g } K_2CrO_4 \times \dfrac{1 \text{ mol}}{194.2 \text{ g}} = 0.7291$ mol

  $406.3 \text{ g } N_2O_5 \times \dfrac{1 \text{ mol}}{108.02 \text{ g}} = 3.761$ mol

Chapter 8: Chemical Composition

130. a. $1.0 \text{ g CH}_4\text{O} \times \dfrac{1 \text{ mol CH}_4\text{O}}{32.042 \text{ g CH}_4\text{O}} \times \dfrac{1 \text{ mol C}}{1 \text{ mol CH}_4\text{O}} \times \dfrac{6.022 \times 10^{23} \text{ C atoms}}{1 \text{ mol C}} = 1.9 \times 10^{22}$ C atoms

b. $1.0 \text{ g CH}_3\text{CH}_2\text{OH} \times \dfrac{1 \text{ mol CH}_3\text{CH}_2\text{OH}}{46.068 \text{ g CH}_3\text{CH}_2\text{OH}} \times \dfrac{2 \text{ mol C}}{1 \text{ mol CH}_3\text{CH}_2\text{OH}} \times \dfrac{6.022 \times 10^{23} \text{ C atoms}}{1 \text{ mol C}} =$

$= 2.6 \times 10^{22}$ C atoms

c. $25.0 \text{ g CO(NH}_2)_2 \times \dfrac{1 \text{ mol CO(NH}_2)_2}{60.062 \text{ g CO(NH}_2)_2} \times \dfrac{2 \text{ mol N}}{1 \text{ mol CO(NH}_2)_2} \times \dfrac{6.022 \times 10^{23} \text{ N atoms}}{1 \text{ mol N}} =$

$= 5.01 \times 10^{23}$ N atoms

131. Least to greatest number of H atoms: HF, H$_2$S, PH$_3$, H$_2$O

$119 \text{ g HF} \times \dfrac{1 \text{ mol HF}}{20.008 \text{ g HF}} \times \dfrac{1 \text{ mol H}}{1 \text{ mol HF}} \times \dfrac{6.022 \times 10^{23} \text{ H atoms}}{1 \text{ mol H}} = 3.58 \times 10^{24}$ H atoms

$119 \text{ g H}_2\text{S} \times \dfrac{1 \text{ mol H}_2\text{S}}{34.086 \text{ g H}_2\text{S}} \times \dfrac{2 \text{ mol H}}{1 \text{ mol H}_2\text{S}} \times \dfrac{6.022 \times 10^{23} \text{ H atoms}}{1 \text{ mol H}} = 4.20 \times 10^{24}$ H atoms

$119 \text{ g PH}_3 \times \dfrac{1 \text{ mol PH}_3}{33.994 \text{ g PH}_3} \times \dfrac{3 \text{ mol H}}{1 \text{ mol PH}_3} \times \dfrac{6.022 \times 10^{23} \text{ H atoms}}{1 \text{ mol H}} = 6.32 \times 10^{24}$ H atoms

$119 \text{ g H}_2\text{O} \times \dfrac{1 \text{ mol H}_2\text{O}}{18.016 \text{ g H}_2\text{O}} \times \dfrac{2 \text{ mol H}}{1 \text{ mol H}_2\text{O}} \times \dfrac{6.022 \times 10^{23} \text{ H atoms}}{1 \text{ mol H}} = 7.96 \times 10^{24}$ H atoms

132. molar mass of C$_9$H$_8$O$_4$ = 180.154 g

$\% \text{ C} = \dfrac{108.09 \text{ g C}}{180.154 \text{ g}} \times 100 = 60.00\%$ C

$\% \text{ H} = \dfrac{8.064 \text{ g H}}{180.154 \text{ g}} \times 100 = 4.476\%$ H

$\% \text{ O} = \dfrac{64.00 \text{ g O}}{180.154 \text{ g}} \times 100 = 35.53\%$ O

133. SNH, NO, N$_2$O, NH$_3$

SNH: $\% \text{ N} = \dfrac{14.01 \text{ g N}}{47.088 \text{ g}} \times 100 = 29.75\%$ N

NO: $\% \text{ N} = \dfrac{14.01 \text{ g N}}{30.01 \text{ g}} \times 100 = 46.68\%$ N

N$_2$O: $\% \text{ N} = \dfrac{28.02 \text{ g N}}{44.02 \text{ g}} \times 100 = 63.65\%$ N

NH$_3$: $\% \text{ N} = \dfrac{14.01 \text{ g N}}{17.034 \text{ g}} \times 100 = 82.25\%$ N

Chapter 8: Chemical Composition

134. Consider 100.0 g of the compound so that percentages become masses.

$$40.0 \text{ g C} \times \frac{1 \text{ mol C}}{12.01 \text{ g}} = 3.33 \text{ mol C}$$

$$6.70 \text{ g H} \times \frac{1 \text{ mol H}}{1.008 \text{ g}} = 6.65 \text{ mol H}$$

$$53.3 \text{ g O} \times \frac{1 \text{ mol}}{16.00 \text{ g}} = 3.33 \text{ mol O}$$

Dividing each number of moles by the smaller number of moles gives

$$\frac{3.33 \text{ mol C}}{3.33 \text{ mol}} = 1.00 \text{ mol C}$$

$$\frac{6.65 \text{ mol H}}{3.33 \text{ mol}} = 2.00 \text{ mol H}$$

$$\frac{3.33 \text{ mol O}}{3.33 \text{ mol}} = 1.000 \text{ mol O}$$

The empirical formula is therefore CH_2O (with a molar mass of 30.026 g/mol). The molar mass of the compound is 180.1 g/mol, thus the empirical formula mass goes into the molecular formula mass 6 times (180.1/30.026 = 6). The molecular formula is $6 \times (CH_2O) = C_6H_{12}O_6$.

CHAPTER 9

Chemical Quantities

1. The coefficients of the balanced chemical equation for a reaction give the *relative numbers of molecules* of reactants and products that are involved in the reaction.

2. The coefficients of this balanced chemical equation indicate the relative numbers of molecules (or moles) of each reactant that combine, as well as the number of molecules (or moles) of the product formed.

3. Although we define mass as the "amount of matter in a substance," the *units* in which we measure mass are a human invention. Atoms and molecules react on an individual particle-by-particle basis, and we have to count individual particles when doing chemical calculations.

4. (e); Subscripts cannot be changed to balance the equation or else the identities of the reactants and/or product are changed. The balanced equation is: $N_2(g) + 3H_2(g) \rightarrow 2NH_3(g)$. Nitrogen and hydrogen will react regardless of how much is initially present (just one might be used up before the other).

5. a. $PCl_3(l) + 3H_2O(l) \rightarrow H_3PO_3(aq) + 3HCl(g)$

   One molecule of liquid phosphorus trichloride reacts with three molecules of liquid water, producing one molecule of aqueous phosphorous acid and molecules of gaseous hydrogen chloride. One mole of phosphorus trichloride reacts with three moles of water to produce one mole of phosphorous acid and three moles of hydrogen chloride.

   b. $2XeF_2(g) + 2H_2O(l) \rightarrow 2Xe(g) + 4HF(g) + O_2(g)$

   Two molecules of gaseous xenon difluoride react with two molecules of liquid water, producing two gaseous xenon atoms, four molecules of gaseous hydrogen fluoride, and one molecule of oxygen gas. Two moles of xenon difluoride reacts with two moles of water, to produce two moles of xenon, four moles of hydrogen fluoride, and one mole of oxygen.

   c. $S(s) + 6HNO_3(aq) \rightarrow H_2SO_4(aq) + 2H_2O(l) + 6NO_2(g)$

   One sulfur atom reacts with six molecules of aqueous nitric acid, producing one molecule of aqueous sulfuric acid, two molecules of water, and six molecules of nitrogen dioxide gas. One mole of sulfur reacts with six moles of nitric acid, to produce one mole of sulfuric acid, two moles of water, and six moles of nitrogen dioxide.

   d. $2NaHSO_3(s) \rightarrow Na_2SO_3(s) + SO_2(g) + H_2O(l)$

   Two formula units of solid sodium hydrogen sulfite react to produce one formula unit of solid sodium sulfite, one molecule of gaseous sulfur dioxide, and one molecule of liquid water. Two moles of sodium hydrogen sulfite react to produce one mole of sodium sulfite, one mole of sulfur dioxide, and one mole of water.

Chapter 9: Chemical Quantities

6. a. $3MnO_2(s) + 4Al(s) \rightarrow 3Mn(s) + 2Al_2O_3(s)$

    Three formula units (or three moles) of manganese(IV) oxide react with four atoms (or four moles) of aluminum, producing three atoms (or three moles) of manganese and two formula units (or two moles) of aluminum oxide.

   b. $B_2O_3(s) + 3CaF_2(s) \rightarrow 2BF_3(g) + 3CaO(s)$

    One molecule (or one mole) of diboron trioxide reacts with three formula units (or three moles) of calcium fluoride, producing two molecules (or two moles) of boron trifluoride and three formula units (three moles) of calcium oxide.

   c. $3NO_2(g) + H_2O(l) \rightarrow 2HNO_3(aq) + NO(g)$

    Three molecules (or three moles) of nitrogen dioxide react with one molecule (or one mole) of water, producing two molecules (or two moles) of nitric acid and one molecule (or one mole) of nitrogen monoxide.

   d. $C_6H_6(g) + 3H_2(g) \rightarrow C_6H_{12}(g)$

    One molecule (or one mole) of benzene (C_6H_6) reacts with three molecules (or three moles) of hydrogen gas, producing one molecule (or one mole) of cyclohexane (C_6H_{12}).

7. False. The coefficients of the balanced chemical equation represent the ratios on a *mole* basis by which carbon combines with oxygen.

8. Balanced chemical equations tell us in what molar ratios substances combine to form products, not in what mass proportions they combine. How could a total of 3 g of reactants produce 2 g of product?

9. $4Al(s) + 3O_2(g) \rightarrow 2Al_2O_3(s)$

    For converting from a given number of moles of aluminum metal to the number of moles of oxygen needed for reaction, the correct mole ratio is

    $$\left(\frac{3 \text{ mol } O_2}{4 \text{ mol Al}}\right)$$

    For converting from a given number of moles of aluminum metal to the number of moles of product produced, the mole ratio is

    $$\left(\frac{2 \text{ mol } Al_2O_3}{4 \text{ mol Al}}\right)$$

10. $CH_4(g) + 2O_2(g) \rightarrow CO_2(g) + 2H_2O(g)$

    For converting from a given number of moles of methane to the number of moles of oxygen required, the mole ratio is

    $$\left(\frac{2 \text{ mol } O_2}{1 \text{ mol } CH_4}\right)$$

    For a given number of moles of methane reacting completely, the mole ratios used to calculate the number of moles of each product are

    $$\text{For } CO_2: \left(\frac{1 \text{ mol } CO_2}{1 \text{ mol } CH_4}\right) \qquad \text{For } H_2O: \left(\frac{2 \text{ mol } H_2O}{1 \text{ mol } CH_4}\right)$$

Chapter 9: Chemical Quantities

11. a. $CO_2(g) + 4H_2(g) \rightarrow CH_4(g) + 2H_2O(l)$

$$0.500 \text{ mol } CO_2 \times \frac{1 \text{ mol } CH_4}{1 \text{ mol } CO_2} = 0.500 \text{ mol } CH_4$$

$$0.500 \text{ mol } CO_2 \times \frac{2 \text{ mol } H_2O}{1 \text{ mol } CO_2} = 1.00 \text{ mol } H_2O$$

b. $BaCl_2(aq) + 2AgNO_3(aq) \rightarrow 2AgCl(s) + Ba(NO_3)_2(aq)$

$$0.500 \text{ mol } BaCl_2 \times \frac{2 \text{ mol } AgCl}{1 \text{ mol } BaCl_2} = 1.00 \text{ mol } AgCl$$

$$0.500 \text{ mol } BaCl_2 \times \frac{1 \text{ mol } Ba(NO_3)_2}{1 \text{ mol } BaCl_2} = 0.500 \text{ mol } Ba(NO_3)_2$$

c. $C_3H_8(g) + 5O_2(g) \rightarrow 4H_2O(l) + 3CO_2(g)$

$$0.500 \text{ mol } C_3H_8 \times \frac{4 \text{ mol } H_2O}{1 \text{ mol } C_3H_8} = 2.00 \text{ mol } H_2O$$

$$0.500 \text{ mol } C_3H_8 \times \frac{3 \text{ mol } CO_2}{1 \text{ mol } C_3H_8} = 1.50 \text{ mol } CO_2$$

d. $3H_2SO_4(aq) + 2Fe(s) \rightarrow Fe_2(SO_4)_3(aq) + 3H_2(g)$

$$0.500 \text{ mol } H_2SO_4 \times \frac{1 \text{ mol } Fe_2(SO_4)_3}{3 \text{ mol } H_2SO_4} = 0.167 \text{ mol } Fe_2(SO_4)_3$$

$$0.500 \text{ mol } H_2SO_4 \times \frac{3 \text{ mol } H_2}{3 \text{ mol } H_2SO_4} = 0.500 \text{ mol } H_2$$

12. Before doing the calculations, the equations must be *balanced*.

a. $2FeO(s) + C(s) \rightarrow 2Fe(l) + CO_2(g)$

$$0.125 \text{ mol } FeO \times \frac{2 \text{ mol } Fe}{2 \text{ mol } FeO} = 0.125 \text{ mol } Fe$$

$$0.125 \text{ mol } FeO \times \frac{1 \text{ mol } CO_2}{2 \text{ mol } FeO} = 0.0625 \text{ mol } CO_2$$

b. $Cl_2(g) + 2KI(aq) \rightarrow 2KCl(aq) + I_2(s)$

$$0.125 \text{ mol } KI \times \frac{2 \text{ mol } KCl}{2 \text{ mol } KI} = 0.125 \text{ mol } KCl$$

$$0.125 \text{ mol } KI \times \frac{1 \text{ mol } I_2}{2 \text{ mol } KI} = 0.0625 \text{ mol } I_2$$

Chapter 9: Chemical Quantities

c. $2Na_2B_4O_7(s) + 2H_2SO_4(aq) + 10H_2O(l) \rightarrow 8H_3BO_3(s) + 2Na_2SO_4(aq)$

$$0.125 \text{ mol } Na_2B_4O_7 \times \frac{8 \text{ mol } H_3BO_3}{2 \text{ mol } Na_2B_4O_7} = 0.500 \text{ mol } H_3BO_3$$

$$0.125 \text{ mol } Na_2B_4O_7 \times \frac{2 \text{ mol } Na_2SO_4}{2 \text{ mol } Na_2B_4O_7} = 0.125 \text{ mol } Na_2SO_4$$

d. $CaC_2(s) + 2H_2O(l) \rightarrow Ca(OH)_2(s) + C_2H_2(g)$

$$0.125 \text{ mol } CaC_2 \times \frac{1 \text{ mol } Ca(OH)_2}{1 \text{ mol } CaC_2} = 0.125 \text{ mol } Ca(OH)_2$$

$$0.125 \text{ mol } CaC_2 \times \frac{1 \text{ mol } C_2H_2}{1 \text{ mol } CaC_2} = 0.125 \text{ mol } C_2H_2$$

13. a. $AgNO_3(aq) + LiOH(aq) \rightarrow AgOH(s) + LiNO_3(aq)$

   molar masses: AgOH, 124.91 g; LiNO$_3$, 68.95 g

$$0.125 \text{ mol } AgNO_3 \times \frac{1 \text{ mol } AgOH}{1 \text{ mol } AgNO_3} = 0.125 \text{ mol } AgOH$$

$$0.125 \text{ mol } AgOH \times \frac{124.91 \text{ g } AgOH}{1 \text{ mol } AgOH} = 15.6 \text{ g } AgOH$$

$$0.125 \text{ mol } AgNO_3 \times \frac{1 \text{ mol } LiNO_3}{1 \text{ mol } AgNO_3} = 0.125 \text{ mol } LiNO_3$$

$$0.125 \text{ mol } LiNO_3 \times \frac{68.95 \text{ g } LiNO_3}{1 \text{ mol } LiNO_3} = 8.62 \text{ g } LiNO_3$$

b. $Al_2(SO_4)_3(aq) + 3CaCl_2(aq) \rightarrow 2AlCl_3(aq) + 3CaSO_4(s)$

   molar masses: AlCl$_3$, 133.33 g; CaSO$_4$, 136.15 g

$$0.125 \text{ mol } Al_2(SO_4)_3 \times \frac{2 \text{ mol } AlCl_3}{1 \text{ mol } Al_2(SO_4)_3} = 0.250 \text{ mol } AlCl_3$$

$$0.250 \text{ mol } AlCl_3 \times \frac{133.33 \text{ g } AlCl_3}{1 \text{ mol } AlCl_3} = 33.3 \text{ g } AlCl_3$$

$$0.125 \text{ mol } Al_2(SO_4)_3 \times \frac{3 \text{ mol } CaSO_4}{1 \text{ mol } Al_2(SO_4)_3} = 0.375 \text{ mol } CaSO_4$$

$$0.375 \text{ mol } CaSO_4 \times \frac{136.15 \text{ g } CaSO4}{1 \text{ mol } CaSO_4} = 51.1 \text{ g } CaSO_4$$

c. $CaCO_3(s) + 2HCl(aq) \rightarrow CaCl_2(aq) + CO_2(g) + H_2O(l)$

   molar masses: CaCl$_2$, 110.98 g; CO$_2$, 44.01 g; H$_2$O, 18.02 g

$$0.125 \text{ mol } CaCO_3 \times \frac{1 \text{ mol } CaCl_2}{1 \text{ mol } CaCO_3} = 0.125 \text{ mol } CaCl_2$$

Chapter 9: Chemical Quantities

$$0.125 \text{ mol CaCl}_2 \times \frac{110.98 \text{ g CaCl}_2}{1 \text{ mol CaCl}_2} = 13.9 \text{ g CaCl}_2$$

$$0.125 \text{ mol CaCO}_3 \times \frac{1 \text{ mol CO}_2}{1 \text{ mol CaCO}_3} = 0.125 \text{ mol CO}_2$$

$$0.125 \text{ mol CO}_2 \times \frac{44.01 \text{ g CO}_2}{1 \text{ mol CO}_2} = 5.50 \text{ g CO}_2$$

$$0.125 \text{ mol CaCO}_3 \times \frac{1 \text{ mol H}_2\text{O}}{1 \text{ mol CaCO}_3} = 0.125 \text{ mol H}_2\text{O}$$

$$0.125 \text{ mol H}_2\text{O} \times \frac{18.02 \text{ g H}_2\text{O}}{1 \text{ mol H}_2\text{O}} = 2.25 \text{ g H}_2\text{O}$$

d. $2C_4H_{10}(g) + 13O_2(g) \rightarrow 8CO_2(g) + 10H_2O(g)$

molar masses: CO_2, 44.01 g; H_2O, 18.02 g

$$0.125 \text{ mol C}_4\text{H}_{10} \times \frac{8 \text{ mol CO}_2}{2 \text{ mol C}_4\text{H}_{10}} = 0.500 \text{ mol CO}_2$$

$$0.500 \text{ mol CO}_2 \times \frac{44.01 \text{ g CO}_2}{1 \text{ mol CO}_2} = 22.0 \text{ g CO}_2$$

$$0.125 \text{ mol C}_4\text{H}_{10} \times \frac{10 \text{ mol H}_2\text{O}}{2 \text{ mol C}_4\text{H}_{10}} = 0.625 \text{ mol H}_2\text{O}$$

$$0.625 \text{ mol H}_2\text{O} \times \frac{18.02 \text{ g H}_2\text{O}}{1 \text{ mol H}_2\text{O}} = 11.3 \text{ g H}_2\text{O}$$

14. a. $NH_3(g) + HCl(g) \rightarrow NH_4Cl(s)$

molar mass: NH_4Cl, 53.492 g

$$0.50 \text{ mol NH}_3 \times \frac{1 \text{ mol NH}_4\text{Cl}}{1 \text{ mol NH}_3} = 0.50 \text{ mol NH}_4\text{Cl}$$

$$0.50 \text{ mol NH}_4\text{Cl} \times \frac{53.492 \text{ g NH}_4\text{Cl}}{1 \text{ mol NH}_3} = 27 \text{ g NH}_4\text{Cl}$$

b. $CH_4(g) + 4S(s) \rightarrow CS_2(l) + 2H_2S(g)$

molar masses: CS_2, 76.15 g; H_2S, 34.086 g

$$0.50 \text{ mol S} \times \frac{1 \text{ mol CS}_2}{4 \text{ mol S}} = 0.125 \text{ mol CS}_2$$

$$0.125 \text{ mol CS}_2 \times \frac{76.15 \text{ g CS}_2}{1 \text{ mol CS}_2} = 9.5 \text{ g CS}_2$$

Chapter 9: Chemical Quantities

$$0.50 \text{ mol S} \times \frac{2 \text{ mol H}_2\text{S}}{4 \text{ mol S}} = 0.25 \text{ mol H}_2\text{S}$$

$$0.25 \text{ mol H}_2\text{S} \times \frac{34.086 \text{ g H}_2\text{S}}{1 \text{ mol H}_2\text{S}} = 8.5 \text{ g H}_2\text{S}$$

c. $PCl_3(l) + 3H_2O(l) \rightarrow H_3PO_3(aq) + 3HCl(aq)$

molar masses: H_3PO_3, 81.994 g; HCl, 36.458 g

$$0.50 \text{ mol PCl}_3 \times \frac{1 \text{ mol H}_3\text{PO}_3}{1 \text{ mol PCl}_3} = 0.50 \text{ mol H}_3\text{PO}_3$$

$$0.50 \text{ mol H}_3\text{PO}_3 \times \frac{81.994 \text{ g H}_3\text{PO}_3}{1 \text{ mol H}_3\text{PO}_3} = 41 \text{ g H}_3\text{PO}_3$$

$$0.50 \text{ mol PCl}_3 \times \frac{3 \text{ mol HCl}}{1 \text{ mol PCl}_3} = 1.50 \text{ mol HCl}$$

$$1.50 \text{ mol HCl} \times \frac{36.458 \text{ g HCl}}{1 \text{ mol HCl}} = 55 \text{ g HCl}$$

d. $NaOH(s) + CO_2(g) \rightarrow NaHCO_3(s)$

molar masses: $NaHCO_3$, 84.008 g

$$0.50 \text{ mol NaOH} \times \frac{1 \text{ mol NaHCO}_3}{1 \text{ mol NaOH}} = 0.50 \text{ mol NaHCO}_3$$

$$0.50 \text{ mol NaHCO}_3 \times \frac{84.008 \text{ g NaHCO}_3}{1 \text{ mol NaHCO}_3} = 42 \text{ g NaHCO}_3$$

15. a. $Cl_2(g) + 2KI(aq) \rightarrow 2KCl(aq) + I_2(s)$

$$0.275 \text{ mol Cl}_2 \times \frac{2 \text{ mol KI}}{1 \text{ mol Cl}_2} = 0.550 \text{ mol KI}$$

b. $6Co(s) + P_4(s) \rightarrow 2Co_3P_2(s)$

$$0.275 \text{ mol Co} \times \frac{1 \text{ mol P}_4}{6 \text{ mol Co}} = 0.0458 \text{ mol P}_4$$

c. $Zn(s) + 2HNO_3(aq) \rightarrow Zn(NO_3)_2(aq) + H_2(g)$

$$0.275 \text{ mol Zn} \times \frac{2 \text{ mol HNO}_3}{1 \text{ mol Zn}} = 0.550 \text{ mol HNO}_3$$

d. $C_5H_{12}(l) + 8O_2(g) \rightarrow 5CO_2(g) + 6H_2O(g)$

$$0.275 \text{ mol C}_5\text{H}_{12} \times \frac{8 \text{ mol O}_2}{1 \text{ mol C}_5\text{H}_{12}} = 2.20 \text{ mol O}_2$$

16. Before doing the calculations, the equations must be *balanced*.

    a. $4KO_2(s) + 2H_2O(l) \rightarrow 3O_2(g) + 4KOH(s)$

    $$0.625 \text{ mol KOH} \times \frac{3 \text{ mol } O_2}{4 \text{ mol KOH}} = 0.469 \text{ mol } O_2$$

    b. $SeO_2(g) + 2H_2Se(g) \rightarrow 3Se(s) + 2H_2O(g)$

    $$0.625 \text{ mol } H_2O \times \frac{3 \text{ mol Se}}{2 \text{ mol } H_2O} = 0.938 \text{ mol Se}$$

    c. $2CH_3CH_2OH(l) + O_2(g) \rightarrow 2CH_3CHO(aq) + 2H_2O(l)$

    $$0.625 \text{ mol } H_2O \times \frac{2 \text{ mol } CH_3CHO}{2 \text{ mol } H_2O} = 0.625 \text{ mol } CH_3CHO$$

    d. $Fe_2O_3(s) + 2Al(s) \rightarrow 2Fe(l) + Al_2O_3(s)$

    $$0.625 \text{ mol } Al_2O_3 \times \frac{2 \text{ mol Fe}}{1 \text{ mol } Al_2O_3} = 1.25 \text{ mol Fe}$$

17. the molar mass of the substance

18. Stoichiometry is the process of using a chemical equation to calculate the relative masses (or moles) of reactants and products involved in a reaction.

19. a. molar mass Si = 28.09 g

    $$4.15 \text{ g Si} \times \frac{1 \text{ mol Si}}{28.09 \text{ g Si}} = 0.148 \text{ mol Si}$$

    b. molar mass $AuCl_3$ = 303.4 g

    $$2.72 \text{ mg } AuCl_3 \times \frac{1 \text{ g}}{1000 \text{ mg}} \times \frac{1 \text{ mol } AuCl_3}{303.4 \text{ g } AuCl_3} = 8.97 \times 10^{-6} \text{ mol } AuCl_3$$

    c. molar mass S = 32.07 g

    $$1.05 \text{ kg S} \times \frac{1000 \text{ g}}{1 \text{ kg}} \times \frac{1 \text{ mol S}}{32.07 \text{ g S}} = 32.7 \text{ mol S}$$

    d. molar mass $FeCl_3$ = 162.20 g

    $$0.000901 \text{ g } FeCl_3 \times \frac{1 \text{ mol } FeCl_3}{162.2 \text{ g } FeCl_3} = 5.55 \times 10^{-6} \text{ mol } FeCl_3$$

    e. molar mass MgO = 40.31 g

    $$5.62 \times 10^3 \text{ g MgO} \times \frac{1 \text{ mol MgO}}{40.31 \text{ g MgO}} = 139 \text{ mol MgO}$$

20. a. molar mass Ag = 107.9 g

    $$2.01 \times 10^{-2} \text{ g Ag} \times \frac{1 \text{ mol Ag}}{107.9 \text{ g Ag}} = 1.86 \times 10^{-4} \text{ mol Ag}$$

Chapter 9: Chemical Quantities

b. molar mass $(NH_4)_2S$ = 68.154 g

$$45.2 \text{ mg } (NH_4)_2S \times \frac{1 \text{ g}}{1000 \text{ mg}} \times \frac{1 \text{ mol } (NH_4)_2S}{68.154 \text{ g } (NH_4)_2S} = 6.63 \times 10^{-4} \text{ mol } (NH_4)_2S$$

c. molar mass U = 238.00 g

$$61.7 \text{ μg U} \times \frac{1 \text{ g}}{10^6 \text{ μg}} \times \frac{1 \text{ mol U}}{238.00 \text{ g U}} = 2.59 \times 10^{-7} \text{ mol U}$$

d. molar mass SO_2 = 64.07 g

$$5.23 \text{ kg } SO_2 \times \frac{1000 \text{ g}}{1 \text{ kg}} \times \frac{1 \text{ mol } SO_2}{64.07 \text{ g } SO_2} = 81.6 \text{ mol } SO_2$$

e. molar mass $Fe(NO_3)_3$ = 241.88 g

$$272 \text{ g } Fe(NO_3)_3 \times \frac{1 \text{ mol } Fe(NO_3)_3}{241.88 \text{ g } Fe(NO_3)_3} = 1.12 \text{ mol } Fe(NO_3)_3$$

21. a. molar mass Ge = 72.59 g

$$2.17 \text{ mol Ge} \times \frac{72.59 \text{ g Ge}}{1 \text{ mol Ge}} = 158 \text{ g Ge}$$

b. molar mass $PbCl_2$ = 278.1 g; 4.24 millimol = 0.00424 mol

$$0.00424 \text{ mol } PbCl_2 \times \frac{278.1 \text{ g } PbCl_2}{1 \text{ mol } PbCl_2} = 1.18 \text{ g } PbCl_2$$

c. molar mass NH_3 = 17.03 g

$$0.0971 \text{ mol } NH_3 \times \frac{17.03 \text{ g } NH_3}{1 \text{ mol } NH_3} = 1.65 \text{ g } NH_3$$

d. molar mass C_6H_{14} = 86.17 g

$$4.26 \times 10^3 \text{ mol } C_6H_{14} \times \frac{86.17 \text{ g } C_6H_{14}}{1 \text{ mol } C_6H_{14}} = 3.67 \times 10^5 \text{ g } C_6H_{14}$$

e. molar mass ICl = 162.35 g

$$1.71 \text{ mol ICl} \times \frac{162.35 \text{ g ICl}}{1 \text{ mol ICl}} = 278 \text{ g ICl}$$

22. a. molar mass of K_3N = 131.31 g

$$0.341 \text{ mol } K_3N \times \frac{131.31 \text{ g } K_3N}{1 \text{ mol } K_3N} = 44.8 \text{ g } K_3N$$

b. molar mass of Ne = 20.18 g; 2.62 millimol = 0.00262 mol

$$0.00262 \text{ mol Ne} \times \frac{20.18 \text{ g Ne}}{1 \text{ mol Ne}} = 0.0529 \text{ g Ne}$$

Chapter 9: Chemical Quantities

c. molar mass of MnO = 70.94 g

$$0.00449 \text{ mol MnO} \times \frac{70.94 \text{ g MnO}}{1 \text{ mol MnO}} = 0.319 \text{ g MnO}$$

d. molar mass of SiO_2 = 60.09 g

$$7.18 \times 10^5 \text{ mol SiO}_2 \times \frac{60.09 \text{ g SiO}_2}{1 \text{ mol SiO}_2} = 4.31 \times 10^7 \text{ g SiO}_2$$

e. molar mass of $FePO_4$ = 150.82 g

$$0.000121 \text{ mol FePO}_4 \times \frac{150.82 \text{ g FePO}_4}{1 \text{ mol FePO}_4} = 0.0182 \text{ g FePO}_4$$

23. Before any calculations are done, the equations must be *balanced*.

   a. $2Co(s) + 3F_2(g) \rightarrow 2CoF_3(s)$

   $$0.413 \text{ mol Co} \times \frac{3 \text{ mol F}_2}{2 \text{ mol Co}} = 0.620 \text{ mol F}_2$$

   b. $2Al(s) + 3H_2SO_4(aq) \rightarrow Al_2(SO_4)_3(aq) + 3H_2(g)$

   $$0.413 \text{ mol Al} \times \frac{3 \text{ mol H}_2SO_4}{2 \text{ mol Al}} = 0.620 \text{ mol H}_2SO_4$$

   c. $2K(s) + 2H_2O(l) \rightarrow 2KOH(aq) + H_2(g)$

   $$0.413 \text{ mol K} \times \frac{2 \text{ mol H}_2O}{2 \text{ mol K}} = 0.413 \text{ mol H}_2O$$

   d. $4Cu(s) + O_2(g) \rightarrow 2Cu_2O(s)$

   $$0.413 \text{ mol Cu} \times \frac{1 \text{ mol O}_2}{4 \text{ mol Cu}} = 0.103 \text{ mol O}_2$$

24. Before any calculations are done, the equations must be *balanced*.

   a. $2Al(s) + 3Br_2(l) \rightarrow 2AlBr_3(s)$

   molar mass Al = 26.98 g

   $$0.557 \text{ g Al} \times \frac{1 \text{ mol Al}}{26.98 \text{ g Al}} \times \frac{3 \text{ mol Br}_2}{2 \text{ mol Al}} = 0.0310 \text{ mol Br}_2$$

   b. $Hg(s) + 2HClO_4(aq) \rightarrow Hg(ClO_4)_2(aq) + H_2(g)$

   molar mass Hg = 200.6 g

   $$0.557 \text{ g Hg} \times \frac{1 \text{ mol Hg}}{200.6 \text{ g}} \times \frac{2 \text{ mol HClO}_4}{1 \text{ mol Hg}} = 0.00555 \text{ mol HClO}_4$$

   c. $3K(s) + P(s) \rightarrow K_3P(s)$

   molar mass K = 39.10 g

   $$0.557 \text{ g K} \times \frac{1 \text{ mol K}}{39.10 \text{ g K}} \times \frac{1 \text{ mol P}}{3 \text{ mol K}} = 0.00475 \text{ mol P}$$

Chapter 9: Chemical Quantities

d. $CH_4(g) + 4Cl_2(g) \rightarrow CCl_4(l) + 4HCl(g)$

molar mass CH_4 = 16.04 g

$0.557 \text{ g } CH_4 \times \dfrac{1 \text{ mol } CH_4}{16.04 \text{ g } CH_4} \times \dfrac{4 \text{ mol } Cl_2}{1 \text{ mol } CH_4} = 0.139 \text{ mol } Cl_2$

25. Before any calculations are done, the equations must be *balanced*.

a. $TiBr_4(g) + 2H_2(g) \rightarrow Ti(s) + 4HBr(g)$

molar mass H_2 = 2.016 g; molar mass Ti = 47.90 g; molar mass of HBr = 80.91 g

$12.5 \text{ g } H_2 \times \dfrac{1 \text{ mol } H_2}{2.016 \text{ g } H_2} = 6.20 \text{ mol } H_2$

$6.20 \text{ mol } H_2 \times \dfrac{1 \text{ mol Ti}}{2 \text{ mol } H_2} = 3.10 \text{ mol Ti}$

$3.10 \text{ mol Ti} \times \dfrac{47.90 \text{ g Ti}}{1 \text{ mol Ti}} = 148 \text{ g Ti}$

$6.20 \text{ mol } H_2 \times \dfrac{4 \text{ mol HBr}}{2 \text{ mol } H_2} = 12.4 \text{ mol HBr}$

$12.4 \text{ mol HBr} \times \dfrac{80.91 \text{ g HBr}}{1 \text{ mol HBr}} = 1.00 \times 10^3 \text{ g HBr}$

b. $3SiH_4(g) + 4NH_3(g) \rightarrow Si_3N_4(s) + 12H_2(g)$

molar mass SiH_4 = 32.12 g; molar mass Si_3N_4 = 140.3 g; molar mass H_2 = 2.016 g

$12.5 \text{ g } SiH_4 \times \dfrac{1 \text{ mol } SiH_4}{32.12 \text{ g } SiH_4} = 0.389 \text{ mol } SiH_4$

$0.389 \text{ mol } SiH_4 \times \dfrac{1 \text{ mol } Si_3N_4}{3 \text{ mol } SiH_4} = 0.130 \text{ mol } Si_3N_4$

$0.130 \text{ mol } Si_3N_4 \times \dfrac{140.3 \text{ g } Si_3N_4}{1 \text{ mol } Si_3N_4} = 18.2 \text{ g } Si_3N_4$

$0.389 \text{ mol } SiH_4 \times \dfrac{12 \text{ mol } H_2}{3 \text{ mol } SiH_4} = 1.56 \text{ mol } H_2$

$1.56 \text{ mol } H_2 \times \dfrac{2.016 \text{ g } H_2}{1 \text{ mol } H_2} = 3.14 \text{ g } H_2$

c. $2NO(g) + 2H_2(g) \rightarrow N_2(g) + 2H_2O(l)$

molar mass H_2 = 2.016 g; molar mass N_2 = 28.02 g; molar mass H_2O = 18.02 g

$12.5 \text{ g } H_2 \times \dfrac{1 \text{ mol } H_2}{2.016 \text{ g } H_2} = 6.20 \text{ mol } H_2$

Chapter 9: Chemical Quantities

$$6.20 \text{ mol } H_2 \times \frac{1 \text{ mol } N_2}{2 \text{ mol } H_2} = 3.10 \text{ mol } N_2$$

$$3.10 \text{ mol } N_2 \times \frac{28.02 \text{ g } N_2}{1 \text{ mol } N_2} = 86.9 \text{ g } N_2$$

$$6.20 \text{ mol } H_2 \times \frac{2 \text{ mol } H_2O}{2 \text{ mol } H_2} = 6.20 \text{ mol } H_2O$$

$$6.20 \text{ mol } H_2O \times \frac{18.02 \text{ g } H_2O}{1 \text{ mol } H_2O} = 112 \text{ g } H_2O$$

d. $Cu_2S(s) \rightarrow 2Cu(s) + S(g)$

molar mass Cu_2S = 159.2 g; molar mass Cu = 63.55 g; molar mass S = 32.07 g

$$12.5 \text{ g } Cu_2S \times \frac{1 \text{ mol } Cu_2S}{159.2 \text{ g } Cu_2S} = 0.0785 \text{ mol } Cu_2S$$

$$0.0785 \text{ mol } Cu_2S \times \frac{2 \text{ mol } Cu}{1 \text{ mol } Cu_2S} = 0.157 \text{ mol } Cu$$

$$0.157 \text{ mol } Cu \times \frac{63.55 \text{ g } Cu}{1 \text{ mol } Cu} = 9.98 \text{ g } Cu$$

$$0.0785 \text{ mol } Cu_2S \times \frac{1 \text{ mol } S}{1 \text{ mol } Cu_2S} = 0.0785 \text{ mol } S$$

$$0.0785 \text{ mol } S \times \frac{32.07 \text{ g } S}{1 \text{ mol } S} = 2.52 \text{ g } S$$

26. molar masses: PCl_3, 137.32 g; H_2O, 18.016 g

$$20.0 \text{ g } PCl_3 \times \frac{1 \text{ mol } PCl_3}{137.32 \text{ g } PCl_3} = 0.146 \text{ mol } PCl_3$$

$$0.146 \text{ mol } PCl_3 \times \frac{3 \text{ mol } H_2O}{1 \text{ mol } PCl_3} = 0.437 \text{ mol } H_2O$$

$$0.437 \text{ mol } H_2O \times \frac{18.016 \text{ g } H_2O}{1 \text{ mol } H_2O} = 7.87 \text{ g } H_2O$$

27. The equation must first be *balanced*.

$(NH_4)_2CO_3(s) \rightarrow 2NH_3(g) + CO_2(g) + H_2O(g)$

molar masses: $(NH_4)_2CO_3$, 96.09 g; NH_3, 17.03 g

$$1.25 \text{ g } (NH_4)_2CO_3 \times \frac{1 \text{ mol } (NH_4)_2CO_3}{96.09 \text{ g } (NH_4)_2CO_3} = 0.0130 \text{ mol } (NH_4)_2CO_3$$

Chapter 9: Chemical Quantities

$$0.0130 \text{ mol } (NH_4)_2CO_3 \times \frac{2 \text{ mol } NH_3}{1 \text{ mol } (NH_4)_2CO_3} = 0.0260 \text{ mol } NH_3$$

$$0.0260 \text{ mol } NH_3 \times \frac{17.03 \text{ g } NH_3}{1 \text{ mol } NH_3} = 0.443 \text{ g } NH_3$$

28. The balanced equation for the reaction is:

    $CaC_2(s) + 2H_2O(l) \rightarrow C_2H_2(g) + Ca(OH)_2(s)$

    molar masses: CaC_2, 64.10 g; C_2H_2, 26.04 g

    $$3.75 \text{ g } CaC_2 \times \frac{1 \text{ mol } CaC_2}{64.10 \text{ g } CaC_2} = 0.0585 \text{ mol } CaC_2$$

    $$0.0585 \text{ mol } CaC_2 \times \frac{1 \text{ mol } C_2H_2}{1 \text{ mol } CaC_2} = 0.0585 \text{ mol } C_2H_2$$

    $$0.0585 \text{ mol } C_2H_2 \times \frac{26.04 \text{ g } C_2H_2}{1 \text{ mol } C_2H_2} = 1.52 \text{ g } C_2H_2$$

29. molar masses: C, 12.01 g; CO, 28.01 g; CO_2, 44.01 g

    $$5.00 \text{ g C} \times \frac{1 \text{ mol C}}{12.01 \text{ g C}} = 0.4163 \text{ mol C}$$

    carbon dioxide: $C(s) + O_2(g) \rightarrow CO_2(g)$

    $$0.4163 \text{ mol C} \times \frac{1 \text{ mol } CO_2}{1 \text{ mol C}} = 0.4163 \text{ mol } CO_2$$

    $$0.4163 \text{ mol } CO_2 \times \frac{44.01 \text{ g } CO_2}{1 \text{ mol } CO_2} = 18.3 \text{ g } CO_2$$

    carbon monoxide: $2C(s) + O_2(g) \rightarrow 2CO(g)$

    $$0.4163 \text{ mol C} \times \frac{2 \text{ mol CO}}{2 \text{ mol C}} = 0.4163 \text{ mol CO}$$

    $$0.4163 \text{ mol CO} \times \frac{28.01 \text{ g CO}}{1 \text{ mol CO}} = 11.7 \text{ g CO}$$

30. $2NaHCO_3(s) \rightarrow Na_2CO_3(s) + H_2O(g) + CO_2(g)$

    molar masses: $NaHCO_3$, 84.01 g; Na_2CO_3, 106.0 g

    $$1.52 \text{ g } NaHCO_3 \times \frac{1 \text{ mol } NaHCO_3}{84.01 \text{ g } NaHCO_3} = 0.01809 \text{ mol } NaHCO_3$$

    $$0.01809 \text{ mol } NaHCO_3 \times \frac{1 \text{ mol } Na_2CO_3}{2 \text{ mol } NaHCO_3} = 0.009047 \text{ mol } Na_2CO_3$$

    $$0.009047 \text{ mol } Na_2CO_3 \times \frac{106.0 \text{ g } Na_2CO_3}{1 \text{ mol } Na_2CO_3} = 0.959 \text{ g } Na_2CO_3$$

Chapter 9: Chemical Quantities

31. $2Fe(s) + 3Cl_2(g) \rightarrow 2FeCl_3(s)$

    millimolar masses: iron, 55.85 mg; FeCl$_3$, 162.2 mg

    $$15.5 \text{ mg Fe} \times \frac{1 \text{ mmol Fe}}{55.85 \text{ mg Fe}} = 0.2775 \text{ mmol Fe}$$

    $$0.2775 \text{ mmol Fe} \times \frac{2 \text{ mmol FeCl}_3}{2 \text{ mmol Fe}} = 0.2775 \text{ mmol FeCl}_3$$

    $$0.2775 \text{ mmol FeCl}_3 \times \frac{162.2 \text{ mg FeCl}_3}{1 \text{ mmol FeCl}_3} = 45.0 \text{ mg FeCl}_3$$

32. $C_6H_{12}O_6(aq) \rightarrow 2C_2H_5OH(aq) + 2CO_2(g)$

    molar masses: C$_6$H$_{12}$O$_6$, 180.2 g; C$_2$H$_5$OH, 46.07 g

    $$5.25 \text{ g } C_6H_{12}O_6 \times \frac{1 \text{ mol } C_6H_{12}O_6}{180.2 \text{ g } C_6H_{12}O_6} = 0.02913 \text{ mol } C_6H_{12}O_6$$

    $$0.02913 \text{ mol } C_6H_{12}O_6 \times \frac{2 \text{ mol } C_2H_5OH}{1 \text{ mol } C_6H_{12}O_6} = 0.5826 \text{ mol } C_2H_5OH$$

    $$0.5286 \text{ mol } C_2H_5OH \times \frac{46.07 \text{ g } C_2H_5OH}{1 \text{ mol } C_2H_5OH} = 2.68 \text{ g ethyl alcohol}$$

33. $H_2SO_3(aq) \rightarrow H_2O(l) + SO_2(g)$

    molar masses: H$_2$SO$_3$, 82.09 g; SO$_2$, 64.07 g

    $$4.25 \text{ g } H_2SO_3 \times \frac{1 \text{ mol } H_2SO_3}{82.09 \text{ g } H_2SO_3} = 0.05177 \text{ mol } H_2SO_3$$

    $$0.05177 \text{ mol } H_2SO_3 \times \frac{1 \text{ mol } SO_2}{1 \text{ mol } H_2SO_3} = 0.05177 \text{ mol } SO_2$$

    $$0.05177 \text{ mol } SO_2 \times \frac{64.07 \text{ g } SO_2}{1 \text{ mol } SO_2} = 3.32 \text{ g } SO_2$$

34. The balanced equation for the reaction is:

    $2H_2O_2(aq) \rightarrow 2H_2O(l) + O_2(g)$

    molar masses: H$_2$O$_2$, 34.016 g; O$_2$, 32.00 g

    $$10.00 \text{ g } H_2O_2 \times \frac{1 \text{ mol } H_2O_2}{34.016 \text{ g } H_2O_2} = 0.2940 \text{ mol } H_2O_2$$

    $$0.2940 \text{ mol } H_2O_2 \times \frac{1 \text{ mol } O_2}{2 \text{ mol } H_2O_2} = 0.1470 \text{ mol } O_2$$

    $$0.1470 \text{ mol } O_2 \times \frac{32.00 \text{ g } O_2}{1 \text{ mol } O_2} = 4.704 \text{ g } O_2$$

Chapter 9: Chemical Quantities

35. $P_4(s) + 5O_2(g) \rightarrow 2P_2O_5(s)$

    molar masses: P_4, 123.88 g; O_2, 32.00 g

    $$4.95 \text{ g } P_4 \times \frac{1 \text{ mol } P_4}{123.88 \text{ g } P_4} = 0.03996 \text{ mol } P_4$$

    $$0.03996 \text{ mol } P_4 \times \frac{5 \text{ mol } O_2}{1 \text{ mol } P_4} = 0.1998 \text{ mol } O_2$$

    $$0.1998 \text{ mol } O_2 \times \frac{32.00 \text{ g } O_2}{1 \text{ mol } O_2} = 6.39 \text{ g } O_2$$

36. $4HgS(s) + 4CaO(s) \rightarrow 4Hg(l) + 3CaS(s) + CaSO_4(s)$

    molar masses: HgS, 232.7 g; Hg, 200.6 g; 10.0 kg = 1.00×10^4 g

    $$1.00 \times 10^4 \text{ g HgS} \times \frac{1 \text{ mol HgS}}{232.7 \text{ g HgS}} = 42.97 \text{ mol HgS}$$

    $$42.97 \text{ mol HgS} \times \frac{4 \text{ mol Hg}}{4 \text{ mol HgS}} = 42.97 \text{ mol Hg}$$

    $$42.97 \text{ mol Hg} \times \frac{200.6 \text{ g Hg}}{1 \text{ mol Hg}} = 8.62 \times 10^3 \text{ g Hg} = 8.62 \text{ kg Hg}$$

37. $2NH_4NO_3(s) \rightarrow 2N_2(g) + O_2(g) + 4H_2O(g)$

    molar masses: NH_4NO_3, 80.05 g; N_2, 28.02 g; O_2, 32.00 g; H_2O, 18.02 g

    $$1.25 \text{ g } NH_4NO_3 \times \frac{1 \text{ mol } NH_4NO_3}{80.05 \text{ g } NH_4NO_3} = 0.0156 \text{ mol } NH_4NO_3$$

    $$0.0156 \text{ mol } NH_4NO_3 \times \frac{2 \text{ mol } N_2}{2 \text{ mol } NH_4NO_3} = 0.0156 \text{ mol } N_2$$

    $$0.0156 \text{ mol } N_2 \times \frac{28.02 \text{ g } N_2}{1 \text{ mol } N_2} = 0.437 \text{ g } N_2$$

    $$0.0156 \text{ mol } NH_4NO_3 \times \frac{1 \text{ mol } O_2}{2 \text{ mol } NH_4NO_3} = 0.00780 \text{ mol } O_2$$

    $$0.00780 \text{ mol } O_2 \times \frac{32.00 \text{ g } O_2}{1 \text{ mol } O_2} = 0.250 \text{ g } O_2$$

    $$0.0156 \text{ mol } NH_4NO_3 \times \frac{4 \text{ mol } H_2O}{2 \text{ mol } NH_4NO_3} = 0.0312 \text{ mol } H_2O$$

    $$0.0312 \text{ mol } H_2O \times \frac{18.02 \text{ g } H_2O}{1 \text{ mol } H_2O} = 0.562 \text{ g } H_2O$$

    As a check, note that 0.437 g + 0.250 g + 0.562 g = 1.249 g = 1.25 g.

38. $C_{12}H_{22}O_{11}(s) \rightarrow 12C(s) + 11H_2O(g)$

    molar masses: $C_{12}H_{22}O_{11}$, 342.3 g; C, 12.01

    $1.19 \text{ g } C_{12}H_{22}O_{11} \times \dfrac{1 \text{ mol } C_{12}H_{22}O_{11}}{342.3 \text{ g } C_{12}H_{22}O_{11}} = 3.476 \times 10^{-3} \text{ mol } C_{12}H_{22}O_{11}$

    $3.476 \times 10^{-3} \text{ mol } C_{12}H_{22}O_{11} \times \dfrac{12 \text{ mol C}}{1 \text{ mol } C_{12}H_{22}O_{11}} = 0.04172 \text{ mol C}$

    $0.04172 \text{ mol C} \times \dfrac{12.01 \text{ g C}}{1 \text{ mol C}} = 0.501 \text{ g C}$

39. $SOCl_2(l) + H_2O(l) \rightarrow SO_2(g) + 2HCl(g)$

    molar masses: $SOCl_2$, 119.0 g; H_2O, 18.02 g

    $35.0 \text{ g } SOCl_2 \times \dfrac{1 \text{ mol } SOCl_2}{119.0 \text{ g } SOCl_2} = 0.294 \text{ mol } SOCl_2$

    $0.294 \text{ mol } SOCl_2 \times \dfrac{1 \text{ mol } H_2O}{1 \text{ mol } SOCl_2} = 0.294 \text{ mol } H_2O$

    $0.294 \text{ mol } H_2O \times \dfrac{18.02 \text{ g } H_2O}{1 \text{ mol } H_2O} = 5.30 \text{ g } H_2O$

40. The balanced equation is:

    $2C_8H_{18} + 25O_2 \rightarrow 16CO_2 + 18H_2O$

    molar masses: C_8H_{18}, 114.22 g; CO_2, 44.01 g   1 lb of CO_2 = 453.59 g CO_2

    $453.59 \text{ g } CO_2 \times \dfrac{1 \text{ mol } CO_2}{44.01 \text{ g } CO_2} = 10.31 \text{ mol } CO_2$

    From the balanced chemical equation, we can calculate the number of moles and number of grams of pure octane that would be required to produce 10.31 mol CO_2.

    $10.31 \text{ mol } CO_2 \times \dfrac{2 \text{ mol } C_8H_{18}}{16 \text{ mol } CO_2} = 1.288 \text{ mol } C_8H_{18}$

    $1.288 \text{ mol } C_8H_{18} \times \dfrac{114.22 \text{ g } C_8H_{18}}{1 \text{ mol } C_8H_{18}} = 147.2 \text{ g } C_8H_{18}$

    From the density of C_8H_{18} we can calculate the volume of 147.2 g C_8H_{18}.

    $147.2 \text{ g } C_8H_{18} \times \dfrac{1 \text{ mL } C_8H_{18}}{0.75 \text{ g } C_8H_{18}} = 196.3 \text{ mL } C_8H_{18}$ (2.0×10^2 mL to two significant figures)

    From the preceding, we know that to travel 1 mile, we need approximately 200 mL of octane

    $\dfrac{1 \text{ mi}}{196.3 \text{ mL}} \times \dfrac{1000 \text{ mL}}{1 \text{ L}} \times \dfrac{3.7854 \text{ L}}{1 \text{ gal}}$ = approximately 19 mi/gal

Chapter 9:   Chemical Quantities

41. To determine the limiting reactant, first calculate the number of moles of each reactant present. Then determine how these numbers of moles correspond to the stoichiometric ratio indicated by the balanced chemical equation for the reaction. Specific answer depends on student response.

42. To determine the limiting reactant, first calculate the number of moles of each reactant present. Then determine how these numbers of moles correspond to the stoichiometric ratio indicated by the balanced chemical equation for the reaction.

43. (c); The limiting reactant must be determined by the starting masses of each reactant, molar masses of each reactant, and the ratio in which they react in the balanced chemical equation.

44. (a); When checking to see if mass is conserved in a reaction, add the masses of the two reactants before the reaction takes place. This should be equal to the total mass of products after the reaction plus whatever excess reactant is leftover.

45. a. $Na_2B_4O_7(s) + H_2SO_4(aq) + 5H_2O(l) \rightarrow 4H_3BO_3(s) + Na_2SO_4(aq)$

    molar masses: $Na_2B_4O_7$, 201.2 g; H_2SO_4, 98.09 g; H_2O, 18.02 g

    $5.00 \text{ g } Na_2B_4O_7 \times \dfrac{1 \text{ mol}}{201.2 \text{ g}} = 0.0249 \text{ mol } Na_2B_4O_7$

    $5.00 \text{ g } H_2SO_4 \times \dfrac{1 \text{ mol}}{98.09 \text{ g}} = 0.0510 \text{ mol } H_2SO_4$

    $5.00 \text{ g } H_2O \times \dfrac{1 \text{ mol}}{18.02 \text{ g}} = 0.277 \text{ mol } H_2O$

    $Na_2B_4O_7$ is the limiting reactant.

    mol H_2SO_4 remaining unreacted = 0.0510 – 0.0249 = 0.0261 mol

    mol H_2O remaining unreacted = 0.277 – 5(0.0249) = 0.153 mol

    mass of H_2SO_4 remaining = $0.0261 \text{ mol} \times \dfrac{98.09 \text{ g}}{1 \text{ mol}} = 2.56 \text{ g } H_2SO_4$

    mass of H_2O remaining = $0.153 \text{ mol} \times \dfrac{18.02 \text{ g}}{1 \text{ mol}} = 2.76 \text{ g } H_2O$

    b. $CaC_2(s) + 2H_2O(l) \rightarrow Ca(OH)_2(s) + C_2H_2(g)$

    molar masses: CaC_2, 64.10 g; H_2O, 18.02 g

    $5.00 \text{ g } CaC_2 \times \dfrac{1 \text{ mol}}{64.10 \text{ g}} = 0.0780 \text{ mol } CaC_2$

    $5.00 \text{ g } H_2O \times \dfrac{1 \text{ mol}}{18.02 \text{ g}} = 0.277 \text{ mol } H_2O$

    CaC_2 is the limiting reactant; water is present in excess.

    mol of H_2O remaining = 0.277 – 2(0.0780) = 0.121 mol H_2O

    mass of H_2O remaining = $0.121 \text{ mol} \times \dfrac{18.02 \text{ g}}{1 \text{ mol}} = 2.18 \text{ g } H_2O$

c. $2NaCl(s) + H_2SO_4(l) \rightarrow 2HCl(g) + Na_2SO_4(s)$

molar masses: NaCl, 58.44 g; H_2SO_4, 98.09 g

$$5.00 \text{ g NaCl} \times \frac{1 \text{ mol}}{58.44 \text{ g}} = 0.0856 \text{ mol NaCl}$$

$$5.00 \text{ g H}_2\text{SO}_4 \times \frac{1 \text{ mol}}{98.09 \text{ g}} = 0.0510 \text{ mol H}_2\text{SO}_4$$

NaCl is the limiting reactant; H_2SO_4 is present in excess.

mol H_2SO_4 that reacts = 0.5(0.0856) = 0.0428 mol H_2SO_4

mol H_2SO_4 remaining = 0.0510 − 0.0428 = 0.0082 mol

$$\text{mass of H}_2\text{SO}_4 \text{ remaining} = 0.0082 \text{ mol} \times \frac{98.09 \text{ g}}{1 \text{ mol}} = 0.80 \text{ g H}_2\text{SO}_4$$

d. $SiO_2(s) + 2C(s) \rightarrow Si(l) + 2CO(g)$

molar masses: SiO_2, 60.09 g; C, 12.01 g

$$5.00 \text{ g SiO}_2 \times \frac{1 \text{ mol}}{60.09 \text{ g}} = 0.0832 \text{ mol SiO}_2$$

$$5.00 \text{ g C} \times \frac{1 \text{ mol}}{12.01 \text{ g}} = 0.416 \text{ mol C}$$

SiO_2 is the limiting reactant; C is present in excess.

mol C remaining = 0.416 − 2(0.0832) = 0.250 mol

$$\text{mass of C remaining} = 0.250 \text{ mol} \times \frac{12.01 \text{ g}}{1 \text{ mol}} = 3.00 \text{ g C}$$

46. a. $S(s) + 2H_2SO_4(aq) \rightarrow 3SO_2(g) + 2H_2O(l)$

Molar masses: S, 32.07 g; H_2SO_4, 98.09 g; SO_2, 64.07 g; H_2O, 18.02 g

$$5.00 \text{ g S} \times \frac{1 \text{ mol}}{32.07 \text{ g}} = 0.1559 \text{ mol S}$$

$$5.00 \text{ g H}_2\text{SO}_4 \times \frac{1 \text{ mol}}{98.09 \text{ g}} = 0.05097 \text{ mol H}_2\text{SO}_4$$

According to the balanced chemical equation, we would need twice as much sulfuric acid as sulfur for complete reaction of both reactants. We clearly have much less sulfuric acid present than sulfur: sulfuric acid is the limiting reactant. The calculation of the masses of products produced is based on the number of moles of the sulfuric acid.

$$0.05097 \text{ mol H}_2\text{SO}_4 \times \frac{3 \text{ mol SO}_2}{2 \text{ mol H}_2\text{SO}_4} \times \frac{64.07 \text{ g SO}_2}{1 \text{ mol SO}_2} = 4.90 \text{ g SO}_2$$

$$0.05097 \text{ mol H}_2\text{SO}_4 \times \frac{2 \text{ mol H}_2\text{O}}{2 \text{ mol H}_2\text{SO}_4} \times \frac{18.02 \text{ g H}_2\text{O}}{1 \text{ mol H}_2\text{O}} = 0.918 \text{ g H}_2\text{O}$$

Chapter 9: Chemical Quantities

b. $MnO_2(s) + 2H_2SO_4(aq) \rightarrow Mn(SO_4)_2 + 2H_2O(l)$

molar masses: MnO_2, 86.94 g; H_2SO_4 98.09 g; $Mn(SO_4)_2$, 247.1 g; H_2O, 18.02 g

$$5.00 \text{ g } MnO_2 \times \frac{1 \text{ mol}}{86.94 \text{ g}} = 0.05751 \text{ mol } MnO_2$$

$$5.00 \text{ g } H_2SO_4 \times \frac{1 \text{ mol}}{98.09 \text{ g}} = 0.05097 \text{ mol } H_2SO_4$$

According to the balanced chemical equation, we would need twice as much sulfuric acid as manganese(IV) oxide for complete reaction of both reactants. We do not have this much sulfuric acid, so sulfuric acid must be the limiting reactant. The amount of each product produced will be based on the sulfuric acid reacting completely.

$$0.05097 \text{ mol } H_2SO_4 \times \frac{1 \text{ mol } Mn(SO_4)_2}{2 \text{ mol } H_2SO_4} \times \frac{247.1 \text{ g } Mn(SO_4)_2}{1 \text{ mol } Mn(SO_4)_2} = 6.30 \text{ g } Mn(SO_4)_2$$

$$0.05097 \text{ mol } H_2SO_4 \times \frac{2 \text{ mol } H_2O}{2 \text{ mol } H_2SO_4} \times \frac{18.02 \text{ g } H_2O}{1 \text{ mol } H_2O} = 0.918 \text{ g } H_2O$$

c. $2H_2S(g) + 3O_2(g) \rightarrow 2SO_2(g) + 2H_2O(l)$

Molar masses: H_2S, 34.09 g; O_2, 32.00 g; SO_2, 64.07 g; H_2O, 18.02 g

$$5.00 \text{ g } H_2S \times \frac{1 \text{ mol}}{34.09 \text{ g}} = 0.1467 \text{ mol } H_2S$$

$$5.00 \text{ g } O_2 \times \frac{1 \text{ mol}}{32.00 \text{ g}} = 0.1563 \text{ mol } O_2$$

According to the balanced equation, we would need 1.5 times as much O_2 as H_2S for complete reaction of both reactants. We don't have that much O_2, so O_2 must be the limiting reactant that will control the masses of each product produced.

$$0.1563 \text{ mol } O_2 \times \frac{2 \text{ mol } SO_2}{3 \text{ mol } O_2} \times \frac{64.07 \text{ g } SO_2}{1 \text{ mol } SO_2} = 6.67 \text{ g } SO_2$$

$$0.1563 \text{ mol } O_2 \times \frac{2 \text{ mol } H_2O}{3 \text{ mol } O_2} \times \frac{18.02 \text{ g } H_2O}{1 \text{ mol } H_2O} = 1.88 \text{ g } H_2O$$

d. $3AgNO_3(aq) + Al(s) \rightarrow 3Ag(s) + Al(NO_3)_3(aq)$

Molar masses: $AgNO_3$, 169.9 g; Al, 26.98 g; Ag, 107.9 g; $Al(NO_3)_3$, 213.0 g

$$5.00 \text{ g } AgNO_3 \times \frac{1 \text{ mol}}{169.9 \text{ g}} = 0.02943 \text{ mol } AgNO_3$$

$$5.00 \text{ g } Al \times \frac{1 \text{ mol}}{26.98 \text{ g}} = 0.1853 \text{ mol } Al$$

According to the balanced chemical equation, we would need three moles of AgNO$_3$ for every mole of Al for complete reaction of both reactants. We in fact have fewer moles of AgNO$_3$ than aluminum, so AgNO$_3$ must be the limiting reactant. The amount of product produced is calculated from the number of moles of the limiting reactant present:

$$0.02943 \text{ mol AgNO}_3 \times \frac{3 \text{ mol Ag}}{3 \text{ mol AgNO}_3} \times \frac{107.9 \text{ g Ag}}{1 \text{ mol Ag}} = 3.18 \text{ g Ag}$$

$$0.02943 \text{ mol AgNO}_3 \times \frac{1 \text{ mol Al(NO}_3)_3}{3 \text{ mol AgNO}_3} \times \frac{213.0 \text{ g Al(NO}_3)_3}{1 \text{ mol Al(NO}_3)_3} = 2.09 \text{ g}$$

47. Before any calculations are attempted, the equations must be balanced.

   a. $C_3H_8(g) + 5O_2(g) \rightarrow 3CO_2(g) + 4H_2O(g)$

   molar masses: C$_3$H$_8$, 44.09 g; O$_2$, 32.00 g; CO$_2$, 44.01 g; H$_2$O, 18.02 g

   $$10.0 \text{ g C}_3\text{H}_8 \times \frac{1 \text{ mol}}{44.09 \text{ g}} = 0.2268 \text{ mol C}_3\text{H}_8$$

   $$10.0 \text{ g O}_2 \times \frac{1 \text{ mol}}{32.00 \text{ g}} = 0.3125 \text{ mol O}_2$$

   For 0.2268 mol C$_3$H$_8$, the amount of O$_2$ that would be needed is

   $$0.2268 \text{ mol C}_3\text{H}_8 \times \frac{5 \text{ mol O}_2}{1 \text{ mol C}_3\text{H}_8} = 1.134 \text{ mol O}_2$$

   Because we do not have this amount of O$_2$, then O$_2$ is the limiting reactant.

   $$0.3125 \text{ mol O}_2 \times \frac{3 \text{ mol CO}_2}{5 \text{ mol O}_2} \times \frac{44.01 \text{ g CO}_2}{1 \text{ mol CO}_2} = 8.25 \text{ g CO}_2$$

   $$0.3125 \text{ mol O}_2 \times \frac{4 \text{ mol H}_2\text{O}}{5 \text{ mol O}_2} \times \frac{18.02 \text{ g H}_2\text{O}}{1 \text{ mol H}_2\text{O}} = 4.51 \text{ g H}_2\text{O}$$

   b. $2Al(s) + 3Cl_2(g) \rightarrow 2AlCl_3(s)$

   molar masses: Al, 26.98 g; Cl$_2$, 70.90 g; AlCl$_3$, 133.3 g

   $$10.0 \text{ g Al} \times \frac{1 \text{ mol}}{26.98 \text{ g}} = 0.3706 \text{ mol Al}$$

   $$10.0 \text{ g Cl}_2 \times \frac{1 \text{ mol}}{70.90 \text{ g}} = 0.1410 \text{ mol Cl}_2$$

   For 0.1410 mol Cl$_2$(g), the amount of Al(s) required is

   $$0.1410 \text{ mol Cl}_2 \times \frac{2 \text{ mol Al}}{3 \text{ mol Cl}_2} = 0.09400 \text{ mol Al}$$

Chapter 9: Chemical Quantities

We have far more than this amount of Al(s) present, so $Cl_2(g)$ must be the limiting reactant that will control the amount of $AlCl_3$ which forms.

$$0.1410 \text{ mol } Cl_2 \times \frac{2 \text{ mol } AlCl_3}{3 \text{ mol } Cl_2} \times \frac{133.3 \text{ g } AlCl_3}{1 \text{ mol } AlCl_3} = 12.5 \text{ g } AlCl_3$$

c. $2NaOH(s) + CO_2(g) \rightarrow Na_2CO_3(s) + H_2O(l)$

molar masses: NaOH, 40.00 g; CO_2, 44.01 g; Na_2CO_3, 106.0 g; H_2O, 18.02 g

$$10.0 \text{ g NaOH} \times \frac{1 \text{ mol}}{40.00 \text{ g}} = 0.2500 \text{ mol NaOH}$$

$$10.0 \text{ g } CO_2 \times \frac{1 \text{ mol}}{44.01 \text{ g}} = 0.2272 \text{ mol } CO_2$$

Without having to calculate, according to the balanced chemical equation, we would need *twice* as many moles of NaOH as CO_2 for complete reaction. For the amounts calculated above, there is not nearly enough NaOH present for the amount of Cl_2 used: NaOH is the limiting reactant.

$$0.2500 \text{ mol NaOH} \times \frac{1 \text{ mol } Na_2CO_3}{2 \text{ mol NaOH}} \times \frac{106.0 \text{ g } Na_2CO_3}{1 \text{ mol } Na_2CO_3} = 13.3 \text{ g } Na_2CO_3$$

$$0.2500 \text{ mol NaOH} \times \frac{1 \text{ mol } H_2O}{2 \text{ mol NaOH}} \times \frac{18.02 \text{ g } H_2O}{1 \text{ mol } H_2O} = 2.25 \text{ g } H_2O$$

d. $NaHCO_3(s) + HCl(aq) \rightarrow NaCl(aq) + H_2O(l) + CO_2(g)$

molar masses: $NaHCO_3$, 84.01 g; HCl, 36.46 g; NaCl, 58.44 g; H_2O, 18.02 g; CO_2, 44.01 g

$$10.0 \text{ g } NaHCO_3 \times \frac{1 \text{ mol}}{84.01 \text{ g}} = 0.1190 \text{ mol } NaHCO_3$$

$$10.0 \text{ g HCl} \times \frac{1 \text{ mol}}{36.46 \text{ g}} = 0.2742 \text{ mol HCl}$$

Because the coefficients of $NaHCO_3(s)$ and HCl(aq) are both *one* in the balanced chemical equation for the reaction, there is not enough $NaHCO_3$ present to react with the amount of HCl present: the 0.1190 mol $NaHCO_3$ present is the limiting reactant. Because all the coefficients of the products are also each *one*, then if 0.1190 mol $NaHCO_3$ reacts completely (with 0.1190 mol HCl), 0.1190 mol of each product will form.

$$0.1190 \text{ mol NaCl} \times \frac{58.44 \text{ g}}{1 \text{ mol}} = 6.95 \text{ g NaCl}$$

$$0.1190 \text{ mol } H_2O \times \frac{18.02 \text{ g}}{1 \text{ mol}} = 2.14 \text{ g } H_2O$$

$$0.1190 \text{ mol } CO_2 \times \frac{44.01 \text{ g}}{1 \text{ mol}} = 5.24 \text{ g } CO_2$$

Chapter 9: Chemical Quantities

48. a. $CS_2(l) + 3O_2(g) \rightarrow CO_2(g) + 2SO_2(g)$

Molar masses: CS_2, 76.15 g; O_2, 32.00 g; CO_2, 44.01 g

$$1.00 \text{ g CS}_2 \times \frac{1 \text{ mol}}{76.15 \text{ g}} = 0.01313 \text{ mol CS}_2$$

$$1.00 \text{ g O}_2 \times \frac{1 \text{ mol}}{32.00 \text{ g}} = 0.03125 \text{ mol O}_2$$

From the balanced chemical equation, we would need three times as much oxygen as carbon disulfide for complete reaction of both reactants. We do not have this much oxygen, and so oxygen must be the limiting reactant.

$$0.03125 \text{ mol O}_2 \times \frac{1 \text{ mol CO}_2}{3 \text{ mol O}_2} \times \frac{44.01 \text{ g CO}_2}{1 \text{ mol CO}_2} = 0.458 \text{ g CO}_2$$

b. $2NH_3(g) + CO_2(g) \rightarrow CN_2H_4O(s) + H_2O(l)$

Molar masses: NH_3, 17.03 g; CO_2, 44.01 g; H_2O, 18.02 g

$$1.00 \text{ g NH}_3 \times \frac{1 \text{ mol}}{17.03 \text{ g}} = 0.05872 \text{ mol NH}_3$$

$$1.00 \text{ g CO}_2 \times \frac{1 \text{ mol}}{44.01 \text{ g}} = 0.02272 \text{ mol CO}_2$$

The balanced chemical equation tells us that we would need twice as many moles of ammonia as carbon dioxide for complete reaction of both reactants. We have *more* than this amount of ammonia present, so the reaction will be limited by the amount of carbon dioxide present.

$$0.02272 \text{ mol CO}_2 \times \frac{1 \text{ mol H}_2\text{O}}{1 \text{ mol CO}_2} \times \frac{18.02 \text{ g H}_2\text{O}}{1 \text{ mol H}_2\text{O}} = 0.409 \text{ g H}_2\text{O}$$

c. $H_2(g) + MnO_2(s) \rightarrow MnO(s) + H_2O(l)$

Molar masses: H_2, 2.016 g; MnO_2, 86.94 g; H_2O, 18.02 g

$$1.00 \text{ g H}_2 \times \frac{1 \text{ mol}}{2.016 \text{ g}} = 0.496 \text{ mol H}_2$$

$$1.00 \text{ g MnO}_2 \times \frac{1 \text{ mol}}{86.94 \text{ g}} = 0.0115 \text{ mol MnO}_2$$

Because the coefficients of both reactants in the balanced chemical equation are the same, we would need equal amounts of both reactants for complete reaction. Therefore, manganese(IV) oxide must be the limiting reactant and controls the amount of product obtained.

$$0.0115 \text{ mol MnO}_2 \times \frac{1 \text{ mol H}_2\text{O}}{1 \text{ mol MnO}_2} \times \frac{18.02 \text{ g H}_2\text{O}}{1 \text{ mol H}_2\text{O}} = 0.207 \text{ g H}_2\text{O}$$

Chapter 9: Chemical Quantities

d. $I_2(s) + Cl_2(g) \rightarrow 2ICl(g)$

Molar masses: I_2, 253.8 g; Cl_2, 70.90 g; ICl, 162.35 g

$$1.00 \text{ g } I_2 \times \frac{1 \text{ mol}}{253.8 \text{ g}} = 0.00394 \text{ mol } I_2$$

$$1.00 \text{ g } Cl_2 \times \frac{1 \text{ mol}}{70.90 \text{ g}} = 0.0141 \text{ mol } Cl_2$$

From the balanced chemical equation, we would need equal amounts of I_2 and Cl_2 for complete reaction of both reactants. As we have much less iodine than chlorine, iodine must be the limiting reactant.

$$0.00394 \text{ mol } I_2 \times \frac{2 \text{ mol ICl}}{1 \text{ mol } I_2} \times \frac{162.35 \text{ g ICl}}{1 \text{ mol ICl}} = 1.28 \text{ g ICl}$$

49. a. $UO_2(s) + 4HF(aq) \rightarrow UF_4(aq) + 2H_2O(l)$

UO_2 is the limiting reactant; 1.16 g UF_4, 0.133 g H_2O

b. $2NaNO_3(aq) + H_2SO_4(aq) \rightarrow Na_2SO_4(aq) + 2HNO_3(aq)$

$NaNO_3$ is the limiting reactant; 0.836 g Na_2SO_4; 0.741 g HNO_3

c. $Zn(s) + 2HCl(aq) \rightarrow ZnCl_2(aq) + H_2(g)$

HCl is the limiting reactant; 1.87 g $ZnCl_2$; 0.0276 g H_2

d. $B(OH)_3(s) + 3CH_3OH(l) \rightarrow B(OCH_3)_3(s) + 3H_2O(l)$

CH_3OH is the limiting reactant; 1.08 g $B(OCH_3)_3$; 0.562 g H_2O

50. a. $2Al(s) + 6HCl(aq) \rightarrow 2AlCl_3(aq) + 3H_2(g)$

HCl is the limiting reactant; 18.3 g $AlCl_3$; 0.415 g H_2

b. $2NaOH(aq) + CO_2(g) \rightarrow Na_2CO_3(aq) + H_2O(l)$

NaOH is the limiting reactant; 19.9 g Na_2CO_3; 3.38 g H_2O

c. $Pb(NO_3)_2(aq) + 2HCl(aq) \rightarrow PbCl_2(s) + 2HNO_3(aq)$

$Pb(NO_3)_2$ is the limiting reactant; 12.6 g $PbCl_2$; 5.71 g HNO_3

d. $2K(s) + I_2(s) \rightarrow 2KI(s)$

I_2 is the limiting reactant; 19.6 g KI

51. $Pb(C_2H_3O_2)_2(aq) + H_2O(l) + CO_2(g) \rightarrow PbCO_3(s) + 2HC_2H_3O_2(aq)$

molar masses: $Pb(C_2H_3O_2)_2$, 325.3 g; CO_2, 44.01 g; $PbCO_3$, 267.21 g

$$1.25 \text{ g } Pb(C_2H_3O_2)_2 \times \frac{1 \text{ mol}}{325.3 \text{ g}} = 0.00384 \text{ mol } Pb(C_2H_3O_2)_2$$

$$5.95 \text{ g } CO_2 \times \frac{1 \text{ mol}}{44.01 \text{ g}} = 0.135 \text{ mol } CO_2$$

Pb(C₂H₃O₂)₂ is the limiting reactant, which determines the yield of product.

$$0.00384 \text{ mol Pb(C}_2\text{H}_3\text{O}_2)_2 \times \frac{1 \text{ mol PbCO}_3}{1 \text{ mol Pb(C}_2\text{H}_3\text{O}_2)_2} \times \frac{267.21 \text{ g PbCO}_3}{1 \text{ mol PbCO}_3} = 1.03 \text{ g PbCO}_3$$

52. $CuO(s) + H_2SO_4(aq) \rightarrow CuSO_4(aq) + H_2O(l)$

    molar masses: CuO, 79.55 g; H₂SO₄, 98.09 g

    $$2.49 \text{ g CuO} \times \frac{1 \text{ mol CuO}}{79.55 \text{ g CuO}} = 0.0313 \text{ mol CuO}$$

    $$5.05 \text{ g H}_2\text{SO}_4 \times \frac{1 \text{ mol H}_2\text{SO}_4}{98.09 \text{ g H}_2\text{SO}_4} = 0.0515 \text{ mol H}_2\text{SO}_4$$

    Since the reaction is of 1:1 stoichiometry, CuO must be the limiting reactant since it is present in the lesser amount on a molar basis.

53. $PbO(s) + C(s) \rightarrow Pb(l) + CO(g)$

    molar masses: PbO, 223.2 g; C, 12.01 g; Pb, 207.2 g

    $$50.0 \times 10^3 \text{ g PbO} \times \frac{1 \text{ mol}}{223.2 \text{ g}} = 224.0 \text{ mol PbO}$$

    $$50.0 \times 10^3 \text{ g C} \times \frac{1 \text{ mol}}{12.01 \text{ g}} = 4163 \text{ mol C}$$

    PbO is the limiting reactant.

    $$224.0 \text{ mol PbO} \times \frac{1 \text{ mol Pb}}{1 \text{ mol PbO}} \times \frac{207.2 \text{ g Pb}}{1 \text{ mol Pb}} = 4.64 \times 10^4 \text{ g} = 46.4 \text{ kg Pb}$$

54. $4Fe(s) + 3O_2(g) \rightarrow 2Fe_2O_3(s)$

    Molar masses: Fe, 55.85 g; Fe₂O₃, 159.7 g

    $$1.25 \text{ g Fe} \times \frac{1 \text{ mol}}{55.85 \text{ g}} = 0.0224 \text{ mol Fe present}$$

    Calculate how many mol of O₂ are required to react with this amount of Fe

    $$0.0224 \text{ mol Fe} \times \frac{3 \text{ mol O}_2}{4 \text{ mol Fe}} = 0.0168 \text{ mol O}_2$$

    Because we have more O₂ than this, Fe must be the limiting reactant.

    $$0.0224 \text{ mol Fe} \times \frac{2 \text{ mol Fe}_2\text{O}_3}{4 \text{ mol Fe}} \times \frac{159.7 \text{ g Fe}_2\text{O}_3}{1 \text{ mol Fe}_2\text{O}_3} = 1.79 \text{ g Fe}_2\text{O}_3$$

Chapter 9: Chemical Quantities

55. $Ag^+(aq) + Cl^-(aq) \rightarrow AgCl(s)$

    molar masses: $AgNO_3$, 169.91 g; NaCl, 58.44 g

    The number of moles of silver ion present will be the same as the number of moles of silver nitrate taken because each formula unit of silver nitrate contains one silver ion.

    $$1.15 \text{ g AgNO}_3 \times \frac{1 \text{ mol AgNO}_3}{169.91 \text{ g AgNO}_3} = 0.00677 \text{ mol AgNO}_3 = 0.00677 \text{ mol Ag}^+ \text{ ion}$$

    The number of moles of chloride ion present will be the same as the number of moles of sodium chloride taken, because each formula unit of sodium chloride contains one sodium ion.

    $$5.45 \text{ g NaCl} \times \frac{1 \text{ mol NaCl}}{58.44 \text{ g NaCl}} = 0.0933 \text{ mol NaCl} = 0.0933 \text{ mol Cl}^- \text{ ion}$$

    Because the balanced chemical equation indicates a 1:1 stoichiometry for the reaction, there is not nearly enough silver ion present (0.00677 mol) to precipitate the amount of chloride ion in the sample (0.0933 mol).

56. $CaCl_2(aq) + Na_2SO_4(aq) \rightarrow CaSO_4(s) + 2NaCl(aq)$

    molar masses: $CaCl_2$, 110.98 g; Na_2SO_4, 142.05 g

    $$5.21 \text{ g CaCl}_2 \times \frac{1 \text{ mol CaCl}_2}{110.98 \text{ g CaCl}_2} = 0.0469 \text{ mol CaCl}_2 = 0.0469 \text{ mol Ca}^{2+} \text{ ion}$$

    $$4.95 \text{ g Na}_2SO_4 \times \frac{1 \text{ mol Na}_2SO_4}{142.05 \text{ g Na}_2SO_4} = 0.0348 \text{ mol Na}_2SO_4 = 0.0348 \text{ mol SO}_4^{2-} \text{ ion}$$

    Because the balanced chemical equation indicates a 1:1 stoichiometry for the reaction, there is not nearly enough sulfate ion present (0.0348 mol) to precipitate the amount of calcium ion in the sample (0.0469 mol). Sodium sulfate (sulfate ion) is the limiting reactant. Calcium chloride (calcium ion) is present in excess.

57. $BaO_2(s) + 2HCl(aq) \rightarrow H_2O_2(aq) + BaCl_2(aq)$

    molar masses: BaO_2, 169.3 g; HCl, 36.46 g; H_2O_2, 34.02 g

    $$1.50 \text{ g BaO}_2 \times \frac{1 \text{ mol}}{169.3 \text{ g}} = 8.860 \times 10^{-3} \text{ mol BaO}_2$$

    $$25.0 \text{ mL solution} \times \frac{0.0272 \text{ g HCl}}{1 \text{ mL solution}} = 0.680 \text{ g HCl}$$

    $$0.680 \text{ g HCl} \times \frac{1 \text{ mol}}{36.46 \text{ g}} = 1.865 \times 10^{-2} \text{ mol HCl}$$

    BaO_2 is the limiting reactant.

    $$8.860 \times 10^{-3} \text{ mol BaO}_2 \times \frac{1 \text{ mol H}_2O_2}{1 \text{ mol BaO2}} \times \frac{34.02 \text{ g H}_2O_2}{1 \text{ mol H}_2O_2} = 0.301 \text{ g H}_2O_2$$

Chapter 9: Chemical Quantities

58. $SiO_2(s) + 3C(s) \rightarrow 2CO(g) + SiC(s)$

    molar masses: SiO_2, 60.09 g; SiC, 40.10 g; 1.0 kg = 1.0×10^3 g

    1.0×10^3 g $SiO_2 \times \dfrac{1 \text{ mol}}{60.09 \text{ g}} = 16.64$ mol SiO_2

    From the balanced chemical equation, if 16.64 mol of SiO_2 were to react completely (an excess of carbon is present), then 16.64 mol of SiC should be produced (the coefficients of SiO_2 and SiC are the same).

    16.64 mol SiC $\times \dfrac{40.01 \text{ g}}{1 \text{ mol}} = 6.7 \times 10^2$ g SiC = 0.67 kg SiC

59. The *theoretical yield* represents the yield we calculate from the stoichiometry of the reaction and the masses of reactants taken for the experiment. The *actual yield* is what is actually obtained in an experiment. The *percent yield* is the ratio of what is actually obtained to the theoretical amount that could be obtained, converted to a percent basis.

60. If the reaction is performed in a solvent, the product may have a substantial solubility in the solvent; the reaction may come to equilibrium before the full yield of product is achieved (see Chapter 16); loss of product may occur through operator error.

61. Percent yield = $\dfrac{\text{actual yield}}{\text{theoretical yield}} \times 100 = \dfrac{1.23 \text{ g}}{1.44 \text{ g}} \times 100 = 85.4\%$

62. $2NaN_3(s) \rightarrow 2Na(s) + 3N_2(g)$

    molar mass: NaN_3, 65.02 g; Na, 22.99 g

    10.5 g $NaN_3 \times \dfrac{1 \text{ mol } NaN_3}{65.02 \text{ g } NaN_3} = 0.161$ mol NaN_3

    0.161 mol $NaN_3 \times \dfrac{2 \text{ mol Na}}{2 \text{ mol } NaN_3} = 0.161$ mol Na

    0.161 mol Na $\times \dfrac{22.99 \text{ g Na}}{1 \text{ mol Na}} = 3.71$ g Na theoretical yield

    % yield = $\dfrac{2.84 \text{ g actual yield}}{3.71 \text{ g theoretical yield}} \times 100 = 76.5\%$

63. $S_8(s) + 8Na_2SO_3(aq) + 40H_2O(l) \rightarrow 8Na_2S_2O_3 \cdot 5H_2O$

    molar masses: S_8, 256.6 g; Na_2SO_3, 126.1 g; $Na_2S_2O_3 \cdot 5H_2O$, 248.2 g

    3.25 g $S_8 \times \dfrac{1 \text{ mol}}{256.6 \text{ g}} = 0.01267$ mol S_8

    13.1 g $Na_2SO_3 \times \dfrac{1 \text{ mol}}{126.1 \text{ g}} = 0.1039$ mol Na_2SO_3

Chapter 9: Chemical Quantities

S_8 is the limiting reactant.

$$0.01267 \text{ mol } S_8 \times \frac{8 \text{ mol Na}_2\text{S}_2\text{O}_3 \cdot 5\text{H}_2\text{O}}{1 \text{ mol } S_8} = 0.1014 \text{ mol Na}_2\text{S}_2\text{O}_3 \cdot 5\text{H}_2\text{O}$$

$$0.1014 \text{ mol Na}_2\text{S}_2\text{O}_3 \cdot 5\text{H}_2\text{O} \times \frac{248.2 \text{ g Na}_2\text{S}_2\text{O}_3 \cdot 5\text{H}_2\text{O}}{1 \text{ mol Na}_2\text{S}_2\text{O}_3 \cdot 5\text{H}_2\text{O}} = 25.2 \text{ g Na}_2\text{S}_2\text{O}_3 \cdot 5\text{H}_2\text{O}$$

$$\text{Percent yield} = \frac{\text{actual yield}}{\text{theoretical yield}} \times 100 = \frac{5.26 \text{ g}}{25.2 \text{ g}} \times 100 = 20.9\%$$

64. $2\text{LiOH}(s) + \text{CO}_2(g) \rightarrow \text{Li}_2\text{CO}_3(s) + \text{H}_2\text{O}(g)$

    molar masses: LiOH, 23.95 g; CO_2, 44.01 g

    $$155 \text{ g LiOH} \times \frac{1 \text{ mol LiOH}}{23.95 \text{ g LiOH}} \times \frac{1 \text{ mol CO}_2}{2 \text{ mol LiOH}} \times \frac{44.01 \text{ g CO}_2}{1 \text{ mol CO}_2} = 142 \text{ g CO}_2$$

    As the cartridge has only absorbed 102 g CO_2 out of a total capacity of 142 g CO_2, the cartridge has absorbed

    $$\frac{102 \text{ g}}{142 \text{ g}} \times 100 = 71.8\% \text{ of its capacity.}$$

65. $\text{Xe}(g) + 2\text{F}_2(g) \rightarrow \text{XeF}_4(s)$

    molar masses: Xe, 131.3 g; F_2, 38.00 g; XeF_4, 207.3 g

    $$130. \text{ g Xe} \times \frac{1 \text{ mol}}{131.3 \text{ g}} = 0.9901 \text{ mol Xe}$$

    $$100. \text{ g F}_2 \times \frac{1 \text{ mol}}{38.00 \text{ g}} = 2.632 \text{ mol F}_2$$

    Xe is the limiting reactant.

    $$0.9901 \text{ mol Xe} \times \frac{1 \text{ mol XeF}_4}{1 \text{ mol Xe}} \times \frac{207.3 \text{ g XeF}_4}{1 \text{ mol XeF}_4} = 205 \text{ g XeF}_4$$

    $$\text{Percent yield} = \frac{\text{actual yield}}{\text{theoretical yield}} \times 100 = \frac{145 \text{ g}}{205 \text{ g}} \times 100 = 70.7 \% \text{ of theory}$$

66. $2\text{NH}_3(g) + 3\text{CuO}(s) \rightarrow \text{N}_2(g) + 3\text{Cu}(s) + 3\text{H}_2\text{O}(g)$

    molar masses: NH_3, 17.034 g; CuO, 79.55 g; Cu, 63.55 g

    $$18.1 \text{ g NH}_3 \times \frac{1 \text{ mol NH}_3}{17.034 \text{ g NH}_3} = 1.06 \text{ mol NH}_3$$

    $$90.4 \text{ g BaCl}_2 \times \frac{1 \text{ mol CuO}}{79.55 \text{ g CuO}} = 1.14 \text{ mol CuO}$$

CuO is the limiting reactant.

$$1.14 \text{ mol CuO} \times \frac{3 \text{ mol Cu}}{3 \text{ mol CuO}} \times \frac{63.55 \text{ g Cu}}{1 \text{ mol Cu}} = 72.4 \text{ g Cu}$$

$$\text{Percent yield} = \frac{\text{actual yield}}{\text{theoretical yield}} \times 100 = \frac{45.3 \text{ g}}{72.4 \text{ g}} \times 100 = 62.6\%$$

67. $Ca(HCO_3)_2(aq) \rightarrow CaCO_3(s) + CO_2(g) + H_2O(l)$

    millimolar masses: $Ca(HCO_3)_2$, 162.1 mg; $CaCO_3$, 100.1 mg

    $$2.0 \times 10^{-3} \text{ mg Ca(HCO}_3)_2 \times \frac{1 \text{ mmol}}{162.1 \text{ mg}} = 1.23 \times 10^{-5} \text{ mmol Ca(HCO}_3)_2$$

    $$1.23 \times 10^{-5} \text{ mmol Ca(HCO}_3)_2 \times \frac{1 \text{ mmol CaCO}_3}{1 \text{ mmol Ca(HCO}_3)_2} = 1.23 \times 10^{-5} \text{ mmol CaCO}_3$$

    $$1.23 \times 10^{-5} \text{ mmol} \times \frac{100.1 \text{ mg}}{1 \text{ mmol}} = 1.2 \times 10^{-3} \text{ mg} = 1.2 \times 10^{-6} \text{ g CaCO}_3$$

68. $NaCl(aq) + NH_3(aq) + H_2O(l) + CO_2(s) \rightarrow NH_4Cl(aq) + NaHCO_3(s)$

    molar masses: NH_3, 17.03 g; CO_2, 44.01 g; $NaHCO_3$, 84.01 g

    $$10.0 \text{ g NH}_3 \times \frac{1 \text{ mol}}{17.03 \text{ g}} = 0.5872 \text{ mol NH}_3$$

    $$15.0 \text{ g CO}_2 \times \frac{1 \text{ mol}}{44.01 \text{ g}} = 0.3408 \text{ mol CO}_2$$

    CO_2 is the limiting reactant.

    $$0.3408 \text{ mol CO}_2 \times \frac{1 \text{ mol NaHCO}_3}{1 \text{ mol CO}_2} = 0.3408 \text{ mol NaHCO}_3$$

    $$0.3408 \text{ mol NaHCO}_3 \times \frac{84.01 \text{ g}}{1 \text{ mol}} = 28.6 \text{ g NaHCO}_3$$

69. $Fe(s) + S(s) \rightarrow FeS(s)$

    molar masses: Fe, 55.85 g; S, 32.07 g; FeS, 87.92 g

    $$5.25 \text{ g Fe} \times \frac{1 \text{ mol}}{55.85 \text{ g}} = 0.0940 \text{ mol Fe}$$

    $$12.7 \text{ g S} \times \frac{1 \text{ mol}}{32.07 \text{ g}} = 0.396 \text{ mol S}$$

    Fe is the limiting reactant.

    $$0.0940 \text{ mol Fe} \times \frac{1 \text{ mol FeS}}{1 \text{ mol Fe}} \times \frac{87.92 \text{ g FeS}}{1 \text{ mol FeS}} = 8.26 \text{ g FeS produced}$$

Chapter 9: Chemical Quantities

70. $C_6H_{12}O_6(s) + 6O_2(g) \rightarrow 6CO_2(g) + 6H_2O(g)$

    molar masses: glucose, 180.2 g; CO_2, 44.01 g

    1.00 g glucose × = 5.549×10^{-3} mol glucose

    5.549×10^{-3} mol glucose × $\dfrac{6 \text{ mol } CO_2}{1 \text{ mol glucose}}$ = 3.33×10^{-2} mol CO_2

    3.33×10^{-2} mol CO_2 × $\dfrac{44.01 \text{ g}}{1 \text{ mol}}$ = 1.47 g CO_2

71. $Cu(s) + S(s) \rightarrow CuS(s)$

    molar masses: Cu, 63.55 g; S, 32.07 g; CuS, 95.62 g

    31.8 g Cu × $\dfrac{1 \text{ mol}}{63.55 \text{ g}}$ = 0.5004 mol Cu

    50.0 g S × $\dfrac{1 \text{ mol}}{32.07 \text{ g}}$ = 1.559 mol S

    Cu is the limiting reactant.

    0.5004 mol Cu × $\dfrac{1 \text{ mol CuS}}{1 \text{ mol Cu}}$ = 0.5004 mol CuS

    0.5004 mol CuS × $\dfrac{95.62 \text{ g}}{1 \text{ mol}}$ = 47.8 g CuS

    % yield = $\dfrac{40.0 \text{ g}}{47.8 \text{ g}} \times 100$ = 83.7%

72. $Ba^{2+}(aq) + SO_4^{2-}(aq) \rightarrow BaSO_4(s)$

    millimolar ionic masses: Ba^{2+}, 137.3 mg; SO_4^{2-}, 96.07 mg; $BaCl_2$, 208.2 mg

    150 mg SO_4^{2-} × $\dfrac{1 \text{ mmol}}{96.07 \text{ mg}}$ = 1.56 millimol SO_4^{2-}

    As barium ion and sulfate ion react on a 1:1 stoichiometric basis, then 1.56 millimol of barium ion is needed, which corresponds to 1.56 millimol of $BaCl_2$

    1.56 millimol $BaCl_2$ × $\dfrac{208.2 \text{ mg}}{1 \text{ mmol}}$ = 325 milligrams $BaCl_2$ needed

73. mass of Cl^- present = 1.054 g sample × $\dfrac{10.3 \text{ g } Cl^-}{100.0 \text{ g sample}}$ = 0.1086 g Cl^-

    molar masses: Cl^-, 35.45 g; $AgNO_3$, 169.9 g; AgCl, 143.4 g

    0.1086 g Cl^- × $\dfrac{1 \text{ mol}}{35.45 \text{ g}}$ = 3.063×10^{-3} mol Cl^-

    3.063×10^{-3} mol Cl^- × $\dfrac{1 \text{ mol } AgNO_3}{1 \text{ mol } Cl^-}$ = 3.063×10^{-3} mol $AgNO_3$

Chapter 9: Chemical Quantities

$$3.063 \times 10^{-3} \text{ mol AgNO}_3 \times \frac{169.9 \text{ g}}{1 \text{ mol}} = 0.520 \text{ g AgNO}_3 \text{ required}$$

$$3.063 \times 10^{-3} \text{ mol Cl}^- \times \frac{1 \text{ mol AgCl}}{1 \text{ mol Cl}^-} = 3.063 \times 10^{-3} \text{ mol AgCl}$$

$$3.063 \times 10^{-3} \text{ mol AgCl} \times \frac{143.4 \text{ g}}{1 \text{ mol}} = 0.439 \text{ g AgCl produced}$$

74. a. $UO_2(s) + 4HF(aq) \rightarrow UF_4(aq) + 2H_2O(l)$

   One molecule (formula unit) of uranium(IV) oxide will combine with four molecules of hydrofluoric acid, producing one uranium(IV) fluoride molecule and two water molecules. One mole of uranium(IV) oxide will combine with four moles of hydrofluoric acid to produce one mole of uranium(IV) fluoride and two moles of water.

   b. $2NaC_2H_3O_2(aq) + H_2SO_4(aq) \rightarrow Na_2SO_4(aq) + 2HC_2H_3O_2(aq)$

   Two molecules (formula units) of sodium acetate react exactly with one molecule of sulfuric acid, producing one molecule (formula unit) of sodium sulfate and two molecules of acetic acid. Two moles of sodium acetate will combine with one mole of sulfuric acid, producing one mole of sodium sulfate and two moles of acetic acid.

   c. $Mg(s) + 2HCl(aq) \rightarrow MgCl_2(aq) + H_2(g)$

   One magnesium atom will react with two hydrochloric acid molecules (formula units) to produce one molecule (formula unit) of magnesium chloride and one molecule of hydrogen gas. One mole of magnesium will combine with two moles of hydrochloric acid, producing one mole of magnesium chloride and one mole of gaseous hydrogen.

   d. $B_2O_3(s) + 3H_2O(l) \rightarrow 2B(OH)_3(aq)$

   One molecule of diboron trioxide will react exactly with three molecules of water, producing two molecules of boron trihydroxide (boric acid). One mole of diboron trioxide will combine with three moles of water to produce two moles of boron trihydroxide (boric acid).

75. False. For 0.40 mol of $Mg(OH)_2$ to react, 0.80 mol of HCl will be needed. According to the balanced equation, for a given amount of $Mg(OH)_2$, *twice* as many moles of HCl is needed.

76. For O_2: $\left(\dfrac{5 \text{ mol O}_2}{1 \text{ mol C}_3H_8}\right)$ For CO_2: $\left(\dfrac{3 \text{ mol CO}_2}{1 \text{ mol C}_3H_8}\right)$ For H_2O: $\left(\dfrac{4 \text{ mol H}_2O}{1 \text{ mol C}_3H_8}\right)$

77. a. $2H_2O_2(l) \rightarrow 2H_2O(l) + O_2(g)$

   $$0.50 \text{ mol H}_2O_2 \times \frac{2 \text{ mol H}_2O}{2 \text{ mol H}_2O_2} = 0.50 \text{ mol H}_2O$$

   $$0.50 \text{ mol H}_2O_2 \times \frac{1 \text{ mol O}_2}{2 \text{ mol H}_2O_2} = 0.25 \text{ mol O}_2$$

Chapter 9: Chemical Quantities

b. $2KClO_3(s) \rightarrow 2KCl(s) + 3O_2(g)$

$0.50 \text{ mol } KClO_3 \times \dfrac{2 \text{ mol } KCl}{2 \text{ mol } KClO_3} = 0.50 \text{ mol } KCl$

$0.50 \text{ mol } KClO_3 \times \dfrac{3 \text{ mol } O_2}{2 \text{ mol } KClO_3} = 0.75 \text{ mol } O_2$

c. $2Al(s) + 6HCl(aq) \rightarrow 2AlCl_3(aq) + 3H_2(g)$

$0.50 \text{ mol } Al \times \dfrac{2 \text{ mol } AlCl_3}{2 \text{ mol } Al} = 0.50 \text{ mol } AlCl_3$

$0.50 \text{ mol } Al \times \dfrac{3 \text{ mol } H_2}{2 \text{ mol } Al} = 0.75 \text{ mol } H_2$

d. $C_3H_8(g) + 5O_2(g) \rightarrow 3CO_2(g) + 4H_2O(l)$

$0.50 \text{ mol } C_3H_8 \times \dfrac{3 \text{ mol } CO_2}{1 \text{ mol } C_3H_8} = 1.5 \text{ mol } CO_2$

$0.50 \text{ mol } C_3H_8 \times \dfrac{4 \text{ mol } H_2O}{1 \text{ mol } C_3H_8} = 2.0 \text{ mol } H_2O$

78. a. $NH_3(g) + HCl(g) \rightarrow NH_4Cl(s)$

molar mass of NH_3 = 17.01 g

$1.00 \text{ g } NH_3 \times \dfrac{1 \text{ mol}}{17.01 \text{ g}} = 0.0588 \text{ mol } NH_3$

$0.0588 \text{ mol } NH_3 \times \dfrac{1 \text{ mol } NH_4Cl}{1 \text{ mol } NH_3} = 0.0588 \text{ mol } NH_4Cl$

b. $CaO(s) + CO_2(g) \rightarrow CaCO_3(s)$

molar mass CaO = 56.08 g

$1.00 \text{ g } CaO \times \dfrac{1 \text{ mol}}{56.08 \text{ g}} = 0.0178 \text{ mol } CaO$

$0.0178 \text{ mol } CaO \times \dfrac{1 \text{ mol } CaCO_3}{1 \text{ mol } CaO} = 0.0178 \text{ mol } CaCO_3$

c. $4Na(s) + O_2(g) \rightarrow 2Na_2O(s)$

molar mass Na = 22.99 g

$1.00 \text{ g } Na \times \dfrac{1 \text{ mol}}{22.99 \text{ g}} = 0.0435 \text{ mol } Na$

$0.0435 \text{ mol } Na \times \dfrac{2 \text{ mol } Na_2O}{4 \text{ mol } Na} = 0.0217 \text{ mol } Na_2O$

d. $2P(s) + 3Cl_2(g) \rightarrow 2PCl_3(l)$

molar mass P = 30.97 g

$1.00 \text{ g P} \times \dfrac{1 \text{ mol}}{30.97 \text{ g}} = 0.0323 \text{ mol P}$

$0.0323 \text{ mol P} \times \dfrac{2 \text{ mol PCl}_3}{2 \text{ mol P}} = 0.0323 \text{ mol PCl}_3$

79. a. molar mass $CuSO_4$ = 159.6 g

$4.21 \text{ g CuSO}_4 \times \dfrac{1 \text{ mol}}{159.6 \text{ g}} = 0.0264 \text{ mol CuSO}_4$

b. molar mass $Ba(NO_3)_2$ = 261.3 g

$7.94 \text{ g Ba(NO}_3)_2 \times \dfrac{1 \text{ mol}}{261.3 \text{ g}} = 0.0304 \text{ mol Ba(NO}_3)_2$

c. molar mass water = 18.02 g; 1.24 mg = 0.00124 g

$0.00124 \text{ g} \times \dfrac{1 \text{ mol}}{18.02 \text{ g}} = 6.88 \times 10^{-5} \text{ mol H}_2\text{O}$

d. molar mass W = 183.9 g

$9.79 \text{ g W} \times \dfrac{1 \text{ mol}}{183.9 \text{ g}} = 5.32 \times 10^{-2} \text{ mol W}$

e. molar mass S = 32.07 g; 1.45 lb = 1.45(454) = 658 g

$658 \text{ g S} \times \dfrac{1 \text{ mol}}{32.07 \text{ g}} = 20.5 \text{ mol S}$

f. molar mass C_2H_5OH = 46.07 g

$4.65 \text{ g C}_2\text{H}_5\text{OH} \times \dfrac{1 \text{ mol}}{46.07 \text{ g}} = 0.101 \text{ mol C}_2\text{H}_5\text{OH}$

g. molar mass C = 12.01 g

$12.01 \text{ g C} \times \dfrac{1 \text{ mol}}{12.01 \text{ g}} = 1.00 \text{ mol C}$

80. a. molar mass HNO_3 = 63.0 g

$5.0 \text{ mol HNO}_3 \times \dfrac{63.0 \text{ g}}{1 \text{ mol}} = 3.2 \times 10^2 \text{ g HNO}_3$

b. molar mass Hg = 200.6 g

$0.000305 \text{ mol Hg} \times \dfrac{200.6 \text{ g}}{1 \text{ mol}} = 0.0612 \text{ g Hg}$

Chapter 9: Chemical Quantities

c. molar mass K_2CrO_4 = 194.2 g

$$2.31 \times 10^{-5} \text{ mol } K_2CrO_4 \times \frac{194.2 \text{ g}}{1 \text{ mol}} = 4.49 \times 10^{-3} \text{ g } K_2CrO_4$$

d. molar mass $AlCl_3$ = 133.3 g

$$10.5 \text{ mol } AlCl_3 \times \frac{133.3 \text{ g}}{1 \text{ mol}} = 1.40 \times 10^3 \text{ g } AlCl_3$$

e. molar mass SF_6 = 146.1 g

$$4.9 \times 10^4 \text{ mol } SF_6 \times \frac{146.1 \text{ g}}{1 \text{ mol}} = 7.2 \times 10^6 \text{ g } SF_6$$

f. molar mass NH_3 = 17.01 g

$$125 \text{ mol } NH_3 \times \frac{17.01 \text{ g}}{1 \text{ mol}} = 2.13 \times 10^3 \text{ g } NH_3$$

g. molar mass Na_2O_2 = 77.98 g

$$0.01205 \text{ mol } Na_2O_2 \times \frac{77.98 \text{ g}}{1 \text{ mol}} = 0.9397 \text{ g } Na_2O_2$$

81. Before any calculations are done, the equations must be *balanced*.

   a. $BaCl_2(aq) + H_2SO_4(aq) \rightarrow BaSO_4(s) + 2HCl(aq)$

   $$0.145 \text{ mol } BaCl_2 \times \frac{1 \text{ mol } H_2SO_4}{1 \text{ mol } BaCl_2} = 0.145 \text{ mol } H_2SO_4$$

   b. $AgNO_3(aq) + NaCl(aq) \rightarrow AgCl(s) + NaNO_3(aq)$

   $$0.145 \text{ mol } AgNO_3 \times \frac{1 \text{ mol } NaCl}{1 \text{ mol } AgNO_3} = 0.145 \text{ mol } NaCl$$

   c. $Pb(NO_3)_2(aq) + Na_2CO_3(aq) \rightarrow PbCO_3(s) + 2NaNO_3(aq)$

   $$0.145 \text{ mol } Pb(NO_3)_2 \times \frac{1 \text{ mol } Na_2CO_3}{1 \text{ mol } Pb(NO_3)_2} = 0.145 \text{ mol } Na_2CO_3$$

   d. $C_3H_8(g) + 5O_2(g) \rightarrow 3CO_2(g) + 4H_2O(g)$

   $$0.145 \text{ mol } C_3H_8 \times \frac{5 \text{ mol } O_2}{1 \text{ mol } C_3H_8} = 0.725 \text{ mol } O_2$$

82. $2SO_2(g) + O_2(g) \rightarrow 2SO_3(g)$

   molar masses: SO_2, 64.07 g; SO_3, 80.07 g; 150 kg = 1.5×10^5 g

   $$1.5 \times 10^5 \text{ g } SO_2 \times \frac{1 \text{ mol}}{64.07 \text{ g}} = 2.34 \times 10^3 \text{ mol } SO_2$$

Chapter 9: Chemical Quantities

$$2.34 \times 10^3 \text{ mol SO}_2 \times \frac{2 \text{ mol SO}_3}{2 \text{ mol SO}_2} = 2.34 \times 10^3 \text{ mol SO}_3$$

$$2.34 \times 10^3 \text{ mol SO}_3 \times \frac{80.07 \text{ g}}{1 \text{ mol}} = 1.9 \times 10^5 \text{ g SO}_3 = 1.9 \times 10^2 \text{ kg SO}_3$$

83. $2\text{ZnS}(s) + 3\text{O}_2(g) \rightarrow 2\text{ZnO}(s) + 2\text{SO}_2(g)$

    molar masses: ZnS, 97.45 g; SO$_2$, 64.07 g; 1.0×10^2 kg = 1.0×10^5 g

    $$1.0 \times 10^5 \text{ g ZnS} \times \frac{1 \text{ mol}}{97.45 \text{ g}} = 1.026 \times 10^3 \text{ mol ZnS}$$

    $$1.026 \times 10^3 \text{ mol ZnS} \times \frac{2 \text{ mol SO}_2}{2 \text{ mol ZnS}} = 1.026 \times 10^3 \text{ mol SO}_2$$

    $$1.026 \times 10^3 \text{ mol SO}_2 \times \frac{64.07 \text{ g}}{1 \text{ mol}} = 6.6 \times 10^4 \text{ g SO}_2 = 66 \text{ kg SO}_2$$

84. $2\text{Na}_2\text{O}_2(s) + 2\text{H}_2\text{O}(l) \rightarrow 4\text{NaOH}(aq) + \text{O}_2(g)$

    molar masses: Na$_2$O$_2$, 77.98 g; O$_2$, 32.00 g

    $$3.25 \text{ g Na}_2\text{O}_2 \times \frac{1 \text{ mol}}{77.98 \text{ g}} = 0.0417 \text{ mol Na}_2\text{O}_2$$

    $$0.0417 \text{ mol Na}_2\text{O}_2 \times \frac{1 \text{ mol O}_2}{2 \text{ mol Na}_2\text{O}_2} = 0.0209 \text{ mol O}_2$$

    $$0.0209 \text{ mol O}_2 \times \frac{32.00 \text{ g}}{1 \text{ mol}} = 0.667 \text{ g O}_2$$

85. $\text{Cu}(s) + 2\text{AgNO}_3(aq) \rightarrow \text{Cu(NO}_3)_2(aq) + 2\text{Ag}(s)$

    millimolar masses: Cu, 63.55 mg; AgNO$_3$, 169.9 mg

    $$1.95 \text{ mg AgNO}_3 \times \frac{1 \text{ mmol}}{169.9 \text{ mg}} = 0.01148 \text{ mmol AgNO}_3$$

    $$0.01148 \text{ mmol AgNO}_3 \times \frac{1 \text{ mmol Cu}}{2 \text{ mmol AgNO}_3} = 0.005740 \text{ mmol Cu}$$

    $$0.005740 \text{ mmol Cu} \times \frac{63.55 \text{ g}}{1 \text{ mol}} = 0.365 \text{ mg Cu}$$

86. $\text{Zn}(s) + 2\text{HCl}(aq) \rightarrow \text{ZnCl}_2(aq) + \text{H}_2(g)$

    molar masses: Zn, 65.38 g; H$_2$, 2.016 g

    $$2.50 \text{ g Zn} \times \frac{1 \text{ mol}}{65.38 \text{ g}} = 0.03824 \text{ mol Zn}$$

    $$0.03824 \text{ mol Zn} \times \frac{1 \text{ mol H}_2}{1 \text{ mol Zn}} = 0.03824 \text{ mol H}_2$$

Chapter 9: Chemical Quantities

$$0.03824 \text{ mol H}_2 \times \frac{2.016 \text{ g}}{1 \text{ mol}} = 0.0771 \text{ g H}_2$$

87. $2C_2H_2(g) + 5O_2(g) \rightarrow 4CO_2(g) + 2H_2O(g)$

molar masses: C_2H_2, 26.04 g; O_2, 32.00 g; 150 g = 1.5×10^2 g

$$1.5 \times 10^2 \text{ g C}_2\text{H}_2 \times \frac{1 \text{ mol}}{26.04 \text{ g}} = 5.760 \text{ mol C}_2\text{H}_2$$

$$5.760 \text{ mol C}_2\text{H}_2 \times \frac{5 \text{ mol O}_2}{2 \text{ mol C}_2\text{H}_2} = 14.40 \text{ mol O}_2$$

$$14.40 \text{ mol O}_2 \times \frac{32.00 \text{ g}}{1 \text{ mol}} = 4.6 \times 10^2 \text{ g O}_2$$

88. a. $2Na(s) + Br_2(l) \rightarrow 2NaBr(s)$

molar masses: Na, 22.99 g; Br_2, 159.8 g; NaBr, 102.9 g

$$5.0 \text{ g Na} \times \frac{1 \text{ mol}}{22.99 \text{ g}} = 0.2175 \text{ mol Na}$$

$$5.0 \text{ g Br}_2 \times \frac{1 \text{ mol}}{159.8 \text{ g}} = 0.03129 \text{ mol Br}_2$$

Intuitively, we would suspect that Br_2 is the limiting reactant, because there is much less Br_2 than Na on a mole basis. To *prove* that Br_2 is the limiting reactant, the following calculation is needed:

$$0.03129 \text{ mol Br}_2 \times \frac{2 \text{ mol Na}}{1 \text{ mol Br}_2} = 0.06258 \text{ mol Na}.$$

Clearly, there is more Na than this present, so Br_2 limits the reaction extent and the amount of NaBr formed.

$$0.03129 \text{ mol Br}_2 \times \frac{2 \text{ mol NaBr}}{1 \text{ mol Br}_2} = 0.06258 \text{ mol NaBr}$$

$$0.06258 \text{ mol NaBr} \times \frac{102.9 \text{ g}}{1 \text{ mol}} = 6.4 \text{ g NaBr}$$

b. $Zn(s) + CuSO_4(aq) \rightarrow ZnSO_4(aq) + Cu(s)$

molar masses: Zn, 65.38 g; Cu, 63.55 g; $ZnSO_4$, 161.5 g; $CuSO_4$, 159.6 g

$$5.0 \text{ g Zn} \times \frac{1 \text{ mol}}{65.38 \text{ g}} = 0.07648 \text{ mol Zn}$$

$$5.0 \text{ g CuSO}_4 \times \frac{1 \text{ mol}}{159.6 \text{ g}} = 0.03132 \text{ mol CuSO}_4$$

Chapter 9: Chemical Quantities

As the coefficients of Zn and $CuSO_4$ are the *same* in the balanced chemical equation, an equal number of moles of Zn and $CuSO_4$ would be needed for complete reaction. There is less $CuSO_4$ present, so $CuSO_4$ must be the limiting reactant.

$$0.03132 \text{ mol } CuSO_4 \times \frac{1 \text{ mol } ZnSO_4}{1 \text{ mol } CuSO_4} = 0.03132 \text{ mol } ZnSO_4$$

$$0.03132 \text{ mol } ZnSO_4 \times \frac{161.5 \text{ g}}{1 \text{ mol}} = 5.1 \text{ g } ZnSO_4$$

$$0.03132 \text{ mol } CuSO_4 \times \frac{1 \text{ mol } Cu}{1 \text{ mol } CuSO_4} = 0.03132 \text{ mol } Cu$$

$$0.03132 \text{ mol } Cu \times \frac{63.55 \text{ g}}{1 \text{ mol}} = 2.0 \text{ g } Cu$$

c. $NH_4Cl(aq) + NaOH(aq) \rightarrow NH_3(g) + H_2O(l) + NaCl(aq)$

molar masses: NH_4Cl, 53.49 g; NaOH, 40.00 g; NH_3, 17.03 g; H_2O, 18.02 g; NaCl, 58.44 g

$$5.0 \text{ g } NH_4Cl \times \frac{1 \text{ mol}}{53.49 \text{ g}} = 0.09348 \text{ mol } NH_4Cl$$

$$5.0 \text{ g NaOH} \times \frac{1 \text{ mol}}{40.00 \text{ g}} = 0.1250 \text{ mol NaOH}$$

As the coefficients of NH_4Cl and NaOH are both *one* in the balanced chemical equation for the reaction, an equal number of moles of NH_4Cl and NaOH would be needed for complete reaction. There is less NH_4Cl present, so NH_4Cl must be the limiting reactant.

As the coefficients of the products in the balanced chemical equation are also all *one*, if 0.09348 mol of NH_4Cl (the limiting reactant) reacts completely, then 0.09348 mol of each product will be formed.

$$0.09348 \text{ mol } NH_3 \times \frac{17.03 \text{ g}}{1 \text{ mol}} = 1.6 \text{ g } NH_3$$

$$0.09348 \text{ mol } H_2O \times \frac{18.02 \text{ g}}{1 \text{ mol}} = 1.7 \text{ g } H_2O$$

$$0.09348 \text{ mol NaCl} \times \frac{58.44 \text{ g}}{1 \text{ mol}} = 5.5 \text{ g NaCl}$$

d. $Fe_2O_3(s) + 3CO(g) \rightarrow 2Fe(s) + 3CO_2(g)$

molar masses: Fe_2O_3, 159.7 g; CO, 28.01 g; Fe, 55.85 g; CO_2, 44.01 g

$$5.0 \text{ g } Fe_2O_3 \times \frac{1 \text{ mol}}{159.7 \text{ g}} = 0.03131 \text{ mol } Fe_2O_3$$

$$5.0 \text{ g CO} \times \frac{1 \text{ mol}}{28.01 \text{ g}} = 0.1785 \text{ mol CO}$$

Chapter 9: Chemical Quantities

Because there is considerably less Fe_2O_3 than CO on a mole basis, let's see if Fe_2O_3 is the limiting reactant.

$$0.03131 \text{ mol } Fe_2O_3 \times \frac{3 \text{ mol CO}}{1 \text{ mol } Fe_2O_3} = 0.09393 \text{ mol CO}$$

As there is 0.1785 mol of CO present, but we have determined that only 0.09393 mol CO would be needed to react with all the Fe_2O_3 present, then Fe_2O_3 must be the limiting reactant. CO is present in excess.

$$0.03131 \text{ mol } Fe_2O_3 \times \frac{2 \text{ mol Fe}}{1 \text{ mol } Fe_2O_3} \times \frac{55.85 \text{ g Fe}}{1 \text{ mol Fe}} = 3.5 \text{ g Fe}$$

$$0.03131 \text{ mol } Fe_2O_3 \times \frac{3 \text{ mol } CO_2}{1 \text{ mol } Fe_2O_3} \times \frac{44.01 \text{ g } CO_2}{1 \text{ mol } CO_2} = 4.1 \text{ g } CO_2$$

89. a. $C_2H_5OH(l) + 3O_2(g) \rightarrow 2CO_2(g) + 3H_2O(l)$

    molar masses: C_2H_5OH, 46.07 g; O_2, 32.00 g; CO_2, 44.01 g

    $$25.0 \text{ g } C_2H_5OH \times \frac{1 \text{ mol}}{46.07 \text{ g}} = 0.5427 \text{ mol } C_2H_5OH$$

    $$25.0 \text{ g } O_2 \times \frac{1 \text{ mol}}{32.00 \text{ g}} = 0.7813 \text{ mol } O_2$$

    As there is less C_2H_5OH present on a mole basis, see if this substance is the limiting reactant.

    $$0.5427 \text{ mol } C_2H_5OH \times \frac{3 \text{ mol } O_2}{1 \text{ mol } C_2H_5OH} = 1.6281 \text{ mol } O_2.$$

    From the above calculation, C_2H_5OH must *not* be the limiting reactant (even though there is a smaller number of moles of C_2H_5OH present) because more oxygen than is present would be required to react completely with the C_2H_5OH present. Oxygen is the limiting reactant.

    $$0.7813 \text{ mol } O_2 \times \frac{2 \text{ mol } CO_2}{3 \text{ mol } O_2} \times \frac{44.01 \text{ g } CO_2}{1 \text{ mol } CO_2} = 22.9 \text{ g } CO_2$$

    b. $N_2(g) + O_2(g) \rightarrow 2NO(g)$

    molar masses: N_2, 28.02 g; O_2, 32.00 g; NO, 30.01 g

    $$25.0 \text{ g } N_2 \times \frac{1 \text{ mol}}{28.02 \text{ g}} = 0.8922 \text{ mol } N_2$$

    $$25.0 \text{ g } O_2 \times \frac{1 \text{ mol}}{32.00 \text{ g}} = 0.7813 \text{ mol } O_2$$

    As the coefficients of N_2 and O_2 are the *same* in the balanced chemical equation for the reaction, an equal number of moles of each substance would be necessary for complete reaction. There is less O_2 present on a mole basis, so O_2 must be the limiting reactant.

    $$0.7813 \text{ mol } O_2 \times \frac{2 \text{ mol NO}}{1 \text{ mol O2}} \times \frac{30.01 \text{ g NO}}{1 \text{ mol NO}} = 46.9 \text{ g NO}$$

Chapter 9:   Chemical Quantities

c. $2NaClO_2(aq) + Cl_2(g) \rightarrow 2ClO_2(g) + 2NaCl(aq)$

molar masses: $NaClO_2$, 90.44 g; Cl_2, 70.90 g; NaCl, 58.44 g

$$25.0 \text{ g NaClO}_2 \times \frac{1 \text{ mol}}{90.44 \text{ g}} = 0.2764 \text{ mol NaClO}_2$$

$$25.0 \text{ g Cl}_2 \times \frac{1 \text{ mol}}{70.90 \text{ g}} = 0.3526 \text{ mol Cl}_2$$

See if $NaClO_2$ is the limiting reactant.

$$0.2764 \text{ mol NaClO}_2 \times \frac{1 \text{ mol Cl}_2}{2 \text{ mol NaClO}_2} = 0.1382 \text{ mol Cl}_2$$

As 0.2764 mol of $NaClO_2$ would require only 0.1382 mol Cl_2 to react completely (and since we have more than this amount of Cl_2), then $NaClO_2$ must indeed be the limiting reactant.

$$0.2764 \text{ mol NaClO}_2 \times \frac{2 \text{ mol NaCl}}{2 \text{ mol NaClO}_2} \times \frac{58.44 \text{ g NaCl}}{1 \text{ mol NaCl}} = 16.2 \text{ g NaCl}$$

d. $3H_2(g) + N_2(g) \rightarrow 2NH_3(g)$

molar masses: H_2, 2.016 g; N_2, 28.02 g; NH_3, 17.03 g

$$25.0 \text{ g H}_2 \times \frac{1 \text{ mol}}{2.016 \text{ g}} = 12.40 \text{ mol H}_2$$

$$25.0 \text{ g N}_2 \times \frac{1 \text{ mol}}{28.02 \text{ g}} = 0.8922 \text{ mol N}_2$$

See if N_2 is the limiting reactant.

$$0.8922 \text{ mol N}_2 \times \frac{3 \text{ mol H}_2}{1 \text{ mol N}_2} = 2.677 \text{ mol H}_2$$

N_2 is clearly the limiting reactant, because there is 12.40 mol H_2 present (a large excess).

$$0.8922 \text{ mol N}_2 \times \frac{2 \text{ mol NH}_3}{1 \text{ mol N}_2} \times \frac{17.03 \text{ g NH}_3}{1 \text{ mol NH}_3} = 30.4 \text{ g NH}_3$$

90. $N_2H_4(l) + O_2(g) \rightarrow N_2(g) + 2H_2O(g)$

molar masses: N_2H_4, 32.05 g; O_2, 32.00 g; N_2, 28.02 g; H_2O, 18.02 g

$$20.0 \text{ g N}_2H_4 \times \frac{1 \text{ mol}}{32.05 \text{ g}} = 0.624 \text{ mol N}_2H_4$$

$$20.0 \text{ g O}_2 \times \frac{1 \text{ mol}}{32.00 \text{ g}} = 0.625 \text{ mol O}_2$$

The two reactants are present in nearly the required ratio for complete reaction (due to the 1:1 stoichiometry of the reaction and the very similar molar masses of the substances). We will consider N_2H_4 as the limiting reactant in the following calculations.

Chapter 9: Chemical Quantities

$$0.624 \text{ mol N}_2\text{H}_4 \times \frac{1 \text{ mol N}_2}{1 \text{ mol N}_2\text{H}_4} \times \frac{28.02 \text{ g N}_2}{1 \text{ mol N}_2} = 17.5 \text{ g N}_2$$

$$0.624 \text{ mol N}_2\text{H}_4 \times \frac{2 \text{ mol H}_2\text{O}}{1 \text{ mol N}_2\text{H}_4} \times \frac{18.02 \text{ g H}_2\text{O}}{1 \text{ mol H}_2\text{O}} = 22.5 \text{ g H}_2\text{O}$$

91. a. $3.50 \text{ mol P}_2\text{O}_5 \times \dfrac{10 \text{ mol KCl}}{6 \text{ mol P}_2\text{O}_5} = 5.83 \text{ mol KCl}$

    b. $3.50 \text{ mol P}_2\text{O}_5 \times \dfrac{3 \text{ mol P}_4}{6 \text{ mol P}_2\text{O}_5} = 1.75 \text{ mol P}_4$

92. $12.5 \text{ g theory} \times \dfrac{40 \text{ g actual}}{100 \text{ g theory}} = 5.0 \text{ g}$

93. $C_5H_{12}(l) + 8O_2(g) \rightarrow 5CO_2(g) + 6H_2O(l)$

    molar masses: C_5H_{12}, 72.146 g; H_2O, 18.016 g

    $$20.4 \text{ g } C_5H_{12} \times \frac{1 \text{ mol } C_5H_{12}}{72.146 \text{ g } C_5H_{12}} = 0.283 \text{ mol } C_5H_{12}$$

    $$0.283 \text{ mol } C_5H_{12} \times \frac{6 \text{ mol } H_2O}{1 \text{ mol } C_5H_{12}} = 1.70 \text{ mol } H_2O$$

    $$1.70 \text{ mol } H_2O \times \frac{18.016 \text{ g } H_2O}{1 \text{ mol } H_2O} = 30.6 \text{ g } H_2O$$

94. The balanced equation is: $2\text{NaNO}_3 \rightarrow 2\text{NaNO}_2 + O_2$

    $$0.2339 \text{ g NaNO}_2 \times \frac{1 \text{ mol NaNO}_2}{69.00 \text{ g NaNO}_2} = 0.003390 \text{ mol NaNO}_2$$

    $$0.003390 \text{ mol NaNO}_2 \times \frac{2 \text{ mol NaNO}_3}{2 \text{ mol NaNO}_2} = 0.003390 \text{ mol NaNO}_3$$

    $$0.003390 \text{ mol NaNO}_3 \times \frac{85.00 \text{ g NaNO}_3}{1 \text{ mol NaNO}_3} = 0.28815 \text{ g NaNO}_3$$

    $\% \text{ NaNO}_3 = \dfrac{0.28815 \text{ g}}{0.4230 \text{ g}} \times 100 = 68.12\%$

95. The balanced equation is: $2\text{LiOH}(s) + CO_2(g) \rightarrow Li_2CO_3(s) + H_2O(l)$

    $$67.4 \text{ g LiOH} \times \frac{1 \text{ mol LiOH}}{23.949 \text{ g LiOH}} = 2.81 \text{ mol LiOH}$$

Chapter 9: Chemical Quantities

$$2.81 \text{ mol LiOH} \times \frac{1 \text{ mol Li}_2\text{CO}_3}{2 \text{ mol LiOH}} = 1.41 \text{ mol Li}_2\text{CO}_3$$

$$1.41 \text{ mol Li}_2\text{CO}_3 \times \frac{73.892 \text{ g Li}_2\text{CO}_3}{1 \text{ mol Li}_2\text{CO}_3} = 104 \text{ g Li}_2\text{CO}_3$$

96. The balanced equation is: $\text{Fe}_2\text{O}_3(s) + 2\text{Al}(s) \rightarrow 2\text{Fe}(l) + \text{Al}_2\text{O}_3(s)$

   a. $25.69 \text{ g Fe} \times \dfrac{1 \text{ mol Fe}}{55.85 \text{ g Fe}} = 0.4600 \text{ mol Fe}$

   $0.4600 \text{ mol Fe} \times \dfrac{1 \text{ mol Fe}_2\text{O}_3}{2 \text{ mol Fe}} = 0.2300 \text{ mol Fe}_2\text{O}_3$

   $0.2300 \text{ mol Fe}_2\text{O}_3 \times \dfrac{159.7 \text{ g Fe}_2\text{O}_3}{1 \text{ mol Fe}_2\text{O}_3} = 36.73 \text{ g Fe}_2\text{O}_3$

   b. $0.4600 \text{ mol Fe} \times \dfrac{2 \text{ mol Al}}{2 \text{ mol Fe}} = 0.4600 \text{ mol Al}$

   $0.4600 \text{ mol Al} \times \dfrac{26.98 \text{ g Al}}{1 \text{ mol Al}} = 12.41 \text{ g Al}$

   c. $0.4600 \text{ mol Fe} \times \dfrac{1 \text{ mol Al}_2\text{O}_3}{2 \text{ mol Fe}} = 0.2300 \text{ mol Al}_2\text{O}_3$

   $0.2300 \text{ mol Al}_2\text{O}_3 \times \dfrac{101.96 \text{ g Al}_2\text{O}_3}{1 \text{ mol Al}_2\text{O}_3} = 23.45 \text{ g Al}_2\text{O}_3$

97. $2\text{H}_2\text{S}(g) + 3\text{O}_2(g) \rightarrow 2\text{SO}_2(g) + 2\text{H}_2\text{O}(g)$

    O_2 is the limiting reactant (only 2.0 mol H_2S needed).

    $$3.0 \text{ mol O}_2 \times \frac{2 \text{ mol SO}_2}{3 \text{ mol O}_2} = 2.0 \text{ mol SO}_2$$

98. The balanced equation is: $2\text{NH}_3(g) + 2\text{Na}(s) \rightarrow 2\text{NaNH}_2(s) + \text{H}_2(g)$

    $$32.8 \text{ g NH}_3 \times \frac{1 \text{ mol NH}_3}{17.034 \text{ g NH}_3} = 1.93 \text{ mol NH}_3$$

    $$16.6 \text{ g Na} \times \frac{1 \text{ mol Na}}{22.99 \text{ g Na}} = 0.722 \text{ mol Na}$$

    Na is the limiting reactant (only 0.722 mol NH_3 needed).

    $$0.722 \text{ mol Na} \times \frac{2 \text{ mol NaNH}_2}{2 \text{ mol Na}} \times \frac{39.016 \text{ g NaNH}_2}{1 \text{ mol NaNH}_2} = 28.2 \text{ g NaNH}_2$$

    $$0.722 \text{ mol Na} \times \frac{1 \text{ mol H}_2}{2 \text{ mol Na}} \times \frac{2.016 \text{ g H}_2}{1 \text{ mol H}_2} = 0.728 \text{ g H}_2$$

Chapter 9: Chemical Quantities

99. The balanced equation is: $SO_2(g) + 2NaOH(s) \rightarrow Na_2SO_3(s) + H_2O(l)$

$$38.3 \text{ g } SO_2 \times \frac{1 \text{ mol } SO_2}{64.07 \text{ g } SO_2} = 0.598 \text{ mol } SO_2$$

$$32.8 \text{ g NaOH} \times \frac{1 \text{ mol NaOH}}{39.998 \text{ g NaOH}} = 0.820 \text{ mol NaOH}$$

NaOH is the limiting reactant (only 0.410 mol SO_2 needed).

$$0.820 \text{ mol NaOH} \times \frac{1 \text{ mol } Na_2SO_3}{2 \text{ mol NaOH}} \times \frac{126.05 \text{ g } Na_2SO_3}{1 \text{ mol } Na_2SO_3} = 51.7 \text{ g } Na_2SO_3$$

$$0.820 \text{ mol NaOH} \times \frac{1 \text{ mol } H_2O}{2 \text{ mol NaOH}} \times \frac{18.016 \text{ g } H_2O}{1 \text{ mol } H_2O} = 7.39 \text{ g } H_2O$$

100. $2C_3H_6(g) + 2NH_3(g) + 3O_2(g) \rightarrow 2C_3H_3N(g) + 6H_2O(g)$

    a. $5.23 \times 10^2 \text{ g } C_3H_6 \times \dfrac{1 \text{ mol } C_3H_6}{42.078 \text{ g } C_3H_6} = 12.4 \text{ mol } C_3H_6$

$$12.4 \text{ mol } C_3H_6 \times \frac{2 \text{ mol } C_3H_3N}{2 \text{ mol } C_3H_6} \times \frac{53.064 \text{ g } C_3H_3N}{1 \text{ mol } C_3H_3N} = 658 \text{ g } C_3H_3N$$

$$5.00 \times 10^2 \text{ g } NH_3 \times \frac{1 \text{ mol } NH_3}{17.034 \text{ g } NH_3} = 29.4 \text{ mol } NH_3$$

$$29.4 \text{ mol } NH_3 \times \frac{2 \text{ mol } C_3H_3N}{2 \text{ mol } NH_3} \times \frac{53.064 \text{ g } C_3H_3N}{1 \text{ mol } C_3H_3N} = 1560 \text{ g } C_3H_3N$$

$$1.00 \times 10^3 \text{ g } O_2 \times \frac{1 \text{ mol } O_2}{32.00 \text{ g } O_2} = 31.25 \text{ mol } O_2$$

$$31.25 \text{ mol } O_2 \times \frac{2 \text{ mol } C_3H_3N}{3 \text{ mol } O_2} \times \frac{53.064 \text{ g } C_3H_3N}{1 \text{ mol } C_3H_3N} = 1110 \text{ g } C_3H_3N$$

Thus, C_3H_6 is the limiting reactant and 658 g C_3H_3N is produced (660. g C_3H_3N if decimals are carried through).

    b. $12.4 \text{ mol } C_3H_6 \times \dfrac{6 \text{ mol } H_2O}{2 \text{ mol } C_3H_6} \times \dfrac{18.016 \text{ g } H_2O}{1 \text{ mol } H_2O} = 670. \text{ g } H_2O$ (672 g H_2O if decimals are carried through)

Chapter 9: Chemical Quantities

c. C_3H_6: 0 g (limiting reactant and is used up)

NH_3:

$$12.4 \text{ mol } C_3H_6 \times \frac{2 \text{ mol } NH_3}{2 \text{ mol } C_3H_6} = 12.4 \text{ mol } NH_3 \text{ used up}$$

29.4 mol NH_3 – 12.4 mol NH_3 = 17.0 mol NH_3 leftover

$$17.0 \text{ mol } NH_3 \times \frac{17.034 \text{ g } NH_3}{1 \text{ mol } NH_3} = 290. \text{ g } NH_3 \text{ (289 g } NH_3 \text{ if decimals are carried through)}$$

O_2:

$$12.4 \text{ mol } C_3H_6 \times \frac{3 \text{ mol } O_2}{2 \text{ mol } C_3H_6} = 18.6 \text{ mol } O_2 \text{ used up}$$

31.25 mol O_2 – 18.6 mol O_2 = 12.65 mol O_2 leftover

$$12.65 \text{ mol } O_2 \times \frac{32.00 \text{ g } O_2}{1 \text{ mol } O_2} = 405 \text{ g } O_2.$$

CUMULATIVE REVIEW

Chapters 8 and 9

1. On the relative atomic scale, the average atomic mass of an element represents the weighted average mass of all the isotopes of an element. Average atomic masses are usually given in terms of atomic mass units (1 amu = 1.66×10^{-24} g). For example, the average atomic mass of sodium is 22.99 amu, which represents the average mass of all the sodium atoms in the world (including all the various isotopes and their relative abundances). So that we will be able to use the mass of a sample of sodium to count the number of atoms of sodium present in the sample, we consider that every sodium atom in a sample has exactly the same mass (the *average* atomic mass). The average atomic mass of an element is typically *not* a whole number of amu because of the presence of the different isotopes of the element, each with its own relative abundance. As the relative abundance of an element can be any random number, when the weighted average atomic mass of the element is calculated, the average is unlikely to be a whole number.

2. On a microscopic basis, one mole of a substance represents Avogadro's number (6.022×10^{23}) of individual units (atoms or molecules) of the substance. On a macroscopic, more practical basis, one mole of a substance represents the amount of substance present when the molar mass of the substance in grams is taken (for example 12.01 g of carbon will be one mole of carbon). Chemists have chosen these definitions so that there will be a simple relationship between measurable amounts of substances (grams) and the actual number of atoms or molecules present, and so that the number of particles present in samples of *different* substances can easily be compared. For example, it is known that carbon and oxygen react by the reaction

    $C(s) + O_2(g) \rightarrow CO_2(g)$.

    Chemists understand this equation to mean that one carbon atom reacts with one oxygen molecule to produce one molecule of carbon dioxide, and also that one mole (12.01 g) of carbon will react with one mole (32.00 g) of oxygen to produce one mole (44.01 g) of carbon dioxide.

3. It's all relative! The mass of each substance mentioned in this question happens to be the molar mass of that substance. Each of the three samples of elemental substances mentioned (O, C, and Na) contains Avogadro's number (6.022×10^{23}) of atoms of its respective element. For the compound Na_2CO_3 given, because each unit of Na_2CO_3 contains two sodium atoms, one carbon atom, and three oxygen atoms, then it's not surprising that a sample having a mass equal to the molar mass of Na_2CO_3 should contain one molar mass of carbon, two molar masses of sodium, and three molar masses of oxygen: 12.01 g + 2(22.99 g) + 3(16.00 g) = 106.0 g.

4. The molar mass of a compound is the mass in grams of one mole of the compound (6.022×10^{23} molecules of the compound), and is calculated by summing the average atomic masses of all the atoms present in a molecule of the compound. For example, a molecule of the compound H_3PO_4 contains three hydrogen atoms, one phosphorus atom, and four oxygen atoms: the molar mass is obtained by adding up the average atomic masses of these atoms: molar mass H_3PO_4 = 3(1.008 g) + 1(30.97 g) + 4(16.00 g) = 97.99 g

Cumulative Review: Chapters 8 and 9

5. The percentage composition (by mass) of a compound shows the relative amount of each element present in the compound on a mass basis. For compounds whose formulas are known (and whose molar masses are therefore known), the percentage of a given element present in the compound is given by

$$\frac{\text{mass of the element present in 1 mol of the compound}}{\text{mass of 1 mol of the compound}} \times 100$$

When a new compound is prepared, whose formula is not known, the percentage composition must be determined on an experimental basis. An elemental analysis must be done of a sample of the new compound to see what mass of each element is present in the sample. For example, if a 1.000 g sample of a hydrocarbon is analyzed, and it is found that the sample contains 0.7487 g of C, then the percentage by mass of carbon present in the compound is

$$\frac{0.7487 \text{ g C}}{1.000 \text{ g sample}} \times 100 = 74.87 \text{ \%C}$$

We can use the formula of a known compound to calculate the percentage composition by mass of the compound, so it is perhaps not surprising that, from experimentally-determined percentage compositions for an unknown compound, we can calculate the formula of the compound (see questions 7 and 8).

6. The empirical formula of a compound represents the lowest ratio of the relative number of atoms of each type present in a molecule of the compound, whereas the molecular formula represents the actual number of atoms of each type present in a real molecule of the compound. For example, both acetylene (molecular formula C_2H_2) and benzene (molecular formula C_6H_6) have the same relative number of carbon and hydrogen atoms (one hydrogen for each carbon atom), and so have the same empirical formula (CH). Once the empirical formula of a compound has been determined, it is also necessary to determine the molar mass of the compound before the actual molecular formula can be calculated. As real molecules cannot contain fractional parts of atoms, the molecular formula is always a whole number multiple of the empirical formula. For the examples above, the molecular formula of acetylene is twice the empirical formula, and the molecular formula of benzene is six times the empirical formula (both factors are integers).

7. For our example, let's use the "known" compound phosphoric acid, H_3PO_4. First we will calculate the percentage composition (by mass) for H_3PO_4, and then we will use our results to calculate back the empirical formula!

Molar mass = 3(1.008 g) + 1(30.97 g) + 4(16.00 g) = 97.99 g

$$\%H = \frac{3(1.008 \text{ g H})}{97.99 \text{ g}} \times 100 = 3.086 \text{ \%H}$$

$$\%P = \frac{30.97 \text{ g P}}{97.99 \text{ g}} \times 100 = 31.60 \text{ \%P}$$

$$\%O = \frac{4(16.00 \text{ g O})}{97.99 \text{ g}} \times 100 = 65.31 \text{ \%O}$$

Cumulative Review: Chapters 8 and 9

Now we will use this percentage composition data to calculate back the empirical formula. We will pretend that we did not know the formula, and were just presented with a question of the type: "a compound contains 3.086 % hydrogen, 31.60 % phosphorus, and 65.31 % oxygen; calculate the empirical formula".

Let's go! First, we will assume, as usual, that we have 100.0 g of the compound, so that the percentages turn into masses in grams. Our sample, therefore will contain 3.086 g H, 31.60 g P, and 65.31 g O. Next, we can calculate the number of moles of each element these masses represent.

$$\text{mol H} = 3.086 \text{ g H} \times \frac{1 \text{ mol H}}{1.008 \text{ g H}} = 3.062 \text{ mol H}$$

$$\text{mol P} = 31.60 \text{ g P} \times \frac{1 \text{ mol P}}{30.97 \text{ g P}} = 1.020 \text{ mol P}$$

$$\text{mol O} = 65.31 \text{ g O} \times \frac{1 \text{ mol O}}{16.00 \text{ g O}} = 4.082 \text{ mol O}$$

To get the empirical formula, divide each of these numbers of moles by the smallest number of moles: this puts things on a relative basis.

$$\frac{3.062 \text{ mol H}}{1.020} = 3.002 \text{ mol H}$$

$$\frac{1.020 \text{ mol P}}{1.020} = 1.000 \text{ mol P}$$

$$\frac{4.082 \text{ mol O}}{1.020} = 4.002 \text{ mol O}$$

This gives us H_3PO_4 as the empirical formula!

8. In question 7, we chose to calculate the percentage composition of phosphoric acid, H_3PO_4: 3.086% H, 31.60% P, 65.31% O. We could convert this percentage composition data into "experimental" data by first choosing a mass of sample to be "analyzed", and then calculating what mass of each element is present in this size sample using the percentage of each element. For example, suppose we choose our sample to have a mass of 2.417 g. Then the masses of H, P, and O present in this sample would be given by the following:

$$\text{g H} = (2.417 \text{ g sample}) \times \frac{3.086 \text{ g H}}{100.0 \text{ g sample}} = 0.07459 \text{ g H}$$

$$\text{g P} = (2.417 \text{ g sample}) \times \frac{31.60 \text{ g P}}{100.0 \text{ g sample}} = 0.7638 \text{ g P}$$

$$\text{g O} = (2.417 \text{ g sample}) \times \frac{65.31 \text{ g O}}{100.0 \text{ g sample}} = 1.579 \text{ g O}$$

Note that (0.07459 g + 0.7638 g + 1.579 g) = 2.41739 = 2.417 g.

So our new problem could be worded as follows: "A 2.417 g sample of a compound has been analyzed and was found to contain 0.07459 g H, 0.7638 g of P, and 1.579 g of oxygen. Calculate the empirical formula of the compound".

$$\text{mol H} = (0.07459 \text{ g H}) \times \frac{1 \text{ mol H}}{1.008 \text{ g H}} = 0.07400 \text{ mol H}$$

$$\text{mol P} = (0.7638 \text{ g P}) \times \frac{1 \text{ mol P}}{30.97 \text{ g P}} = 0.02466 \text{ mol P}$$

$$\text{mol O} = (1.579 \text{ g O}) \times \frac{1 \text{ mol O}}{16.00 \text{ g O}} = 0.09869 \text{ mol O}$$

Dividing each of these numbers of moles by the smallest number of moles (0.02466 mol P) gives the following:

$$\frac{0.07400 \text{ mol H}}{0.02466} = 3.001 \text{ mol H}$$

$$\frac{0.02466 \text{ mol P}}{0.02466} = 1.000 \text{ mol P}$$

$$\frac{0.09869 \text{ mol O}}{0.02466} = 4.002 \text{ mol O}$$

The empirical formula is (not surprisingly) just H_3PO_4!

9. You could, of course, have chosen practically any reaction as your example. For this discussion, let's keep it simple and use the equation

$$2H_2O_2(aq) \rightarrow 2H_2O(l) + O_2(g),$$

which describes the decomposition reaction of hydrogen peroxide.

Microscopic: Two molecules of hydrogen peroxide (in aqueous solution) decompose to produce two molecules of liquid water and one molecule of oxygen gas.

Macroscopic: Two moles of hydrogen peroxide (present in aqueous solution) decompose to produce two moles of liquid water and one mole of oxygen gas.

10. The mole ratios for a reaction are based on the *coefficients* of the balanced chemical equation for the reaction: these coefficients show in what proportions molecules (or moles of molecules) combine. For a given amount of propane, the following mole ratios could be constructed, which would enable you to calculate the number of moles of each product, or of the second reactant, that would be involved.

$$C_3H_8(g) + 5O_2(g) \rightarrow 3CO_2(g) + 4H_2O(g)$$

for O_2: $\dfrac{5 \text{ mol } O_2}{1 \text{ mol } C_3H_8}$; $0.55 \text{ mol } C_3H_8 \times \dfrac{5 \text{ mol } O_2}{1 \text{ mol } C_3H_8} = 2.8 \ (2.75) \text{ mol } O_2$

for CO_2: $\dfrac{3 \text{ mol } CO_2}{1 \text{ mol } C_3H_8}$; $0.55 \text{ mol } C_3H_8 \times \dfrac{3 \text{ mol } CO_2}{1 \text{ mol } C_3H_8} = 1.7 \ (1.65) \text{ mol } CO_2$

for H_2O: $\dfrac{4 \text{ mol } H_2O}{1 \text{ mol } C_3H_8}$; $0.55 \text{ mol } C_3H_8 \times \dfrac{4 \text{ mol } H_2O}{1 \text{ mol } C_3H_8} = 2.2 \text{ mol } H_2O$

Cumulative Review: Chapters 8 and 9

11. Let's consider this simple reaction showing the decomposition of calcium carbonate, producing calcium oxide and carbon dioxide.

    $$CaCO_3(s) \rightarrow CaO(s) + CO_2(g)$$

    Let's suppose that 50.0 g of $CaCO_3$ is to be decomposed.

    In any calculation of yield for a reaction, the molar masses of the reactants and products are almost certainly going to be needed, so let's write these down now.

    Molar masses: $CaCO_3$, 100.09 g; CaO, 56.08 g; CO_2, 44.01 g

    $$\text{mol } CaCO_3 = 50.0 \text{ g} \times \frac{1 \text{ mol}}{100.09 \text{ g}} = 0.4995 \text{ mol } CaCO_3$$

    $$\text{mol CaO} = 0.4995 \text{ mol } CaCO_3 \times \frac{1 \text{ mol CaO}}{1 \text{ mol } CaCO_3} = 0.4995 \text{ mol CaO}$$

    $$\text{mass CaO} = 0.4995 \text{ mol CaO} \times \frac{56.08 \text{ g}}{1 \text{ mol}} = 28.0 \text{ g CaO}$$

    $$\text{mol } CO_2 = 0.4995 \text{ mol } CaCO_3 \times \frac{1 \text{ mol } CO_2}{1 \text{ mol } CaCO_3} = 0.4995 \text{ mol } CO_2$$

    $$\text{mass } CO_2 = 0.4995 \text{ mol } CO_2 \times \frac{44.01 \text{ g}}{1 \text{ mol}} = 22.0 \text{ g } CO_2$$

    Although this example was especially simple, because the coefficients of the balanced chemical equation were all unity, the results illustrate an important point: the sum of the masses of the two products (28.0 g + 22.0 g) equals the mass of the reactant (50.0 g).

12. Although we can calculate specifically the exact amounts of each reactant needed for a chemical reaction, oftentimes reaction mixtures are prepared using more or less arbitrary amounts of the reagents. However, regardless of how much of each reagent may be used for a reaction, the substances still react stoichiometrically, according to the mole ratios derived from the balanced chemical equation for the reaction. When arbitrary amounts of reactants are used, there will be one reactant which, stoichiometrically, is present in the least amount. This substance is called the *limiting reactant* for the experiment. It is the limiting reactant that controls how much product is formed, regardless of how much of the other reactants are present. The limiting reactant limits the amount of product that can form in the experiment, because once the limiting reactant has reacted completely, the reaction must stop. We say that the other reactants in the experiment are present in excess, which means that a portion of these reactants will still be present unchanged after the reaction has ended and the limiting reactant has been used up completely.

13. For our discussion, let's use the simple balanced chemical equation

    $$2H_2(g) + O_2(g) \rightarrow 2H_2O(l).$$

    Molar masses: H_2, 2.016 g; O_2, 32.00 g

    To determine which reactant is limiting, we first need to realize that the masses of the reactants (25.0 g of each) tell us nothing: we need to calculate how many *moles* of each reactant is present.

    $$\text{mol } H_2 = 25.0 \text{ g } H_2 \times \frac{1 \text{ mol } H_2}{2.016 \text{ g } H_2} = 12.40 \text{ mol } H_2$$

$$\text{mol } O_2 = 25.0 \text{ g } O_2 \times \frac{1 \text{ mol } O_2}{32.00 \text{ g } O_2} = 0.7813 \text{ mol } O_2$$

Considering these numbers of moles, it is clear that there is considerably more hydrogen present than oxygen. Chances are, the hydrogen is present in excess, and oxygen is the limiting reactant. We need to *prove* this, however, by calculation. If we consider that the 0.7813 mol of oxygen may be the limiting reactant, we can calculate how much hydrogen would be needed for complete reaction: this requires the mole ratio as determined by the coefficients of the balanced chemical equation.

$$0.7813 \text{ mol } O_2 \times \frac{2 \text{ mol } H_2}{1 \text{ mol } O_2} = 1.56 \text{ mol } H_2 \text{ required for reaction}$$

As only 1.56 mol of H_2 is required to react with 0.7813 mol of O_2, and since we have considerably more hydrogen present in our sample than this amount, then clearly hydrogen is present in excess and oxygen is, indeed, the limiting reactant.

Suppose we had not initially considered that oxygen was the limiting reactant (because there is so much less oxygen present on a mole basis) and had wondered if H_2 were the limiting reactant. For the given amount of H_2 (12.40 mol), we could calculate how much oxygen would be required to react

$$12.40 \text{ mol } H_2 \times \frac{1 \text{ mol } O_2}{2 \text{ mol } H_2} = 6.20 \text{ mol } O_2 \text{ would be required}$$

As we do not have 6.20 mol of O_2 (we have only 0.7813 mol O_2), then clearly there is not enough oxygen present to react with all the hydrogen, and we would conclude again that oxygen must be the limiting reactant.

14. The *theoretical yield* for an experiment is the mass of product calculated based on the limiting reactant for the experiment being completely consumed. The *actual yield* for an experiment is the mass of product actually collected by the experimenter. Obviously, any experiment is restricted by the skills of the experimenter and by the inherent limitations of the experimental method being used. For these reasons, the actual yield is often *less* than the theoretical yield (most scientific writers report the actual or percentage yield for their experiments as an indication of the usefulness of their experiments). Although one would expect that the actual yield should never be more than the theoretical yield, in real experiments, sometimes this happens: however, an actual yield greater than a theoretical yield is usually taken to mean that something is *wrong* in either the experiment (for example, impurities may be present, or the reaction may not occur as envisioned) or in the calculations.

15. a. molar mass Fe_2O_3 = 159.7 g

   $$2.45 \text{ g} \times \frac{1 \text{ mol}}{159.7 \text{ g}} = 0.0153 \text{ mol}$$

   b. molar mass P_4 = 123.88 g

   $$2.45 \text{ g} \times \frac{1 \text{ mol}}{123.88 \text{ g}} = 0.0198 \text{ mol}$$

c. molar mass Cl_2 = 70.90 g

$$2.45 \text{ g} \times \frac{1 \text{ mol}}{70.90 \text{ g}} = 0.0346 \text{ mol}$$

d. molar mass Hg_2O = 417.2 g

$$2.45 \text{ g} \times \frac{1 \text{ mol}}{417.2 \text{ g}} = 0.00587 \text{ mol}$$

e. molar mass HgO = 216.6 g

$$2.45 \text{ g} \times \frac{1 \text{ mol}}{216.6 \text{ g}} = 0.0113 \text{ mol}$$

f. molar mass $Ca(NO_3)_2$ = 164.1 g

$$2.45 \text{ g} \times \frac{1 \text{ mol}}{164.1 \text{ g}} = 0.0149 \text{ mol}$$

g. molar mass C_3H_8 = 44.09 g

$$2.45 \text{ g} \times \frac{1 \text{ mol}}{44.09 \text{ g}} = 0.0556 \text{ mol}$$

h. molar mass $Al_2(SO_4)_3$ = 342.2 g

$$2.45 \text{ g} \times \frac{1 \text{ mol}}{342.2 \text{ g}} = 0.00716 \text{ mol}$$

16. % element X = $\dfrac{\text{mass of element X in compound}}{\text{molar mass of compound}} \times 100$

   a. 92.26% C   b. 32.37% Na
   c. 15.77% C   d. 20.24% Al
   e. 88.82% Cu   f. 79.89% Cu
   g. 71.06% Co   h. 40.00% C

17. Assume 100.0 g of compound so that the percentages may be expressed in grams

$$43.38 \text{ g Na} \times \frac{1 \text{ mol Na}}{22.99 \text{ g Na}} = 1.887 \text{ mol Na}$$

$$11.33 \text{ g C} \times \frac{1 \text{ mol C}}{12.01 \text{ g C}} = 0.9434 \text{ mol C}$$

$$45.29 \text{ g O} \times \frac{1 \text{ mol O}}{16.00 \text{ g O}} = 2.831 \text{ mol O}$$

Dividing each number of moles by the smallest number of moles (0.9434) gives

$$\frac{1.887 \text{ mol Na}}{0.9434} = 2.000 \text{ mol Na}$$

$$\frac{0.9434 \text{ mol C}}{0.9434 \text{ mol}} = 1.000 \text{ mol C}$$

$$\frac{2.831 \text{ mol O}}{0.9434} = 3.001 \text{ mol O}$$

The empirical formula is Na_2CO_3.

18. a. molar masses: SiC, 40.10 g; $SiCl_4$, 169.9 g

$$12.5 \text{ g SiC} \times \frac{1 \text{ mol}}{40.10 \text{ g}} = 0.3117 \text{ mol SiC}$$

for $SiCl_4$: $0.3117 \text{ mol SiC} \times \frac{1 \text{ mol SiCl}_4}{1 \text{ mol SiC}} \times \frac{169.9 \text{ g SiCl}_4}{1 \text{ mol SiCl}_4} = 53.0 \text{ g SiCl}_4$

for C: $0.3117 \text{ mol SiC} \times \frac{1 \text{ mol C}}{1 \text{ mol SiC}} \times \frac{12.01 \text{ g C}}{1 \text{ mol C}} = 3.75 \text{ g C}$

b. molar masses: Li_2O, 29.88 g; $LiOH$, 23.95 g

$$12.5 \text{ g Li}_2\text{O} \times \frac{1 \text{ mol}}{29.88 \text{ g}} = 0.4183 \text{ mol Li}_2\text{O}$$

$$0.4183 \text{ mol Li}_2\text{O} \times \frac{2 \text{ mol LiOH}}{1 \text{ mol Li}_2\text{O}} \times \frac{23.95 \text{ g LiOH}}{1 \text{ mol LiOH}} = 20.0 \text{ g LiOH}$$

c. molar masses: Na_2O_2, 77.98 g; $NaOH$, 40.00 g; O_2, 32.00 g

$$12.5 \text{ g} \times \frac{1 \text{ mol}}{77.98 \text{ g}} = 0.1603 \text{ mol Na}_2\text{O}_2$$

for NaOH: $0.1603 \text{ mol Na}_2\text{O}_2 \times \frac{4 \text{ mol NaOH}}{2 \text{ mol Na}_2\text{O}_2} \times \frac{40.00 \text{ g NaOH}}{1 \text{ mol NaOH}} = 12.8 \text{ g NaOH}$

for O_2: $0.1603 \text{ mol Na}_2\text{O}_2 \times \frac{1 \text{ mol O}_2}{2 \text{ mol Na}_2\text{O}_2} \times \frac{32.00 \text{ g O}_2}{1 \text{ mol O}_2} = 2.56 \text{ g O}_2$

d. molar masses: SnO_2, 150.7 g; Sn, 118.7 g; H_2O, 18.02 g

$$12.5 \text{ g SnO}_2 \times \frac{1 \text{ mol}}{150.7 \text{ g}} = 0.08295 \text{ mol SnO}_2$$

for Sn: $0.08295 \text{ mol SnO}_2 \times \frac{1 \text{ mol Sn}}{1 \text{ mol SnO}_2} \times \frac{118.7 \text{ g Sn}}{1 \text{ mol Sn}} = 9.84 \text{ g Sn}$

for H_2O: $0.08295 \text{ mol SnO}_2 \times \frac{2 \text{ mol H}_2\text{O}}{1 \text{ mol SnO}_2} \times \frac{18.02 \text{ g H}_2\text{O}}{1 \text{ mol H}_2\text{O}} = 2.99 \text{ g H}_2\text{O}$

Cumulative Review: Chapters 8 and 9

19. a. Cl_2 (0.007052 mol Cl_2 compared to 0.3117 mol SiC)
    b. H_2O (0.2775 mol H_2O compared to 0.4183 mol Li_2O)
    c. Na_2O_2 (0.1603 mol Na_2O_2 compared to 0.2775 mol H_2O)
    d. SnO_2 (0.08295 mol SnO_2 compared to 2.480 mol H_2)

20. molar masses: C, 12.01 g; CO, 28.01 g; CO_2, 44.01 g

$$5.00 \text{ g C} \times \frac{1 \text{ mol C}}{12.01 \text{ g C}} = 0.416 \text{ mol C}$$

for CO: $0.416 \text{ mol C} \times \frac{2 \text{ mol CO}}{2 \text{ mol C}} \times \frac{28.01 \text{ g CO}}{1 \text{ mol CO}} = 11.7 \text{ g CO}$

for CO_2: $0.416 \text{ mol C} \times \frac{1 \text{ mol } CO_2}{1 \text{ mol C}} \times \frac{44.01 \text{ g } CO_2}{1 \text{ mol } CO_2} = 18.3 \text{ g } CO_2$

21. molar masses: Ca, 40.08 g; CaC_2O_4, 128.1 g

theoretical yield: $0.1014 \text{ g } Ca^{2+} \times \frac{128.1 \text{ g } CaC_2O_4}{40.08 \text{ g } Ca^{2+}} = 0.3241 \text{ g } CaC_2O_4$

percent yield: $\frac{0.2995 \text{ g obtained}}{0.3241 \text{ g theoretical}} \times 100\% = 92.41\%$ of theory

CHAPTER 10

Energy

1. energy

2. Potential energy is energy due to position or composition. A stone at the top of a hill possesses potential energy since the stone may eventually roll down the hill. A gallon of gasoline possesses potential energy since heat will be released when the gasoline is burned.

3. $KE = \frac{1}{2}mv^2$

4. The total energy of the universe is *constant*.

5. A state function is a property of a system which changes independently of the pathway taken to make the change in the system. For example, the actual vertical distance between corresponding points on the 2nd floor and on the 3rd floor in your dormitory is a property which does not change. You might walk a different distance depending on whether you take the stairs or the elevator, but the actual distance between the floors is a fixed distance.

6. Ball A initially possesses potential energy by virtue of its position at the top of the hill. As Ball A rolls down the hill, its potential energy is converted to kinetic energy and frictional (heat) energy. When Ball A reaches the bottom of the hill and hits Ball B, it transfers its kinetic energy to Ball B. Ball A then has only the potential energy corresponding to its new position.

7. *Temperature* is a measure of the random motions of the particles in a substance, that is, temperature is a measure of the average kinetic energies of the particles. *Heat* is the energy that flows because of a temperature difference.

8. The hot tea is at a higher temperature, which means the particles in the hot tea have higher average kinetic energies. When the tea spills on the skin, energy flows from the hot tea to the skin, until the tea and skin are at the same temperature. This sudden inflow of energy causes the burn.

9. The thermal energy of an object represents the random motions of the particles of matter which constitute the object.

10. Temperature is the concept by which we express the thermal energy contained in a sample. We cannot measure the motions of the particles/kinetic energy in a sample of matter directly. We know, however, that if two objects are at different temperatures, the one with the higher temperature has molecules that have higher average kinetic energies than the object at the lower temperature.

11. The *system* is the part of the universe upon which we want to focus attention. In a chemical reaction, the system represents the reactants and products of the chemical reaction. The *surroundings* include everything else in the universe.

Chapter 10: Energy

12. When the chemical system evolves energy, the energy evolved from the reacting chemicals is transferred to the surroundings.

13. An exothermic reaction is one which releases energy: therefore the products will have a lower potential energy than did the reactants.

14.
    a. Endothermic; Energy is required to dissolve the solid KBr in water.
    b. Exothermic; Energy is released in combustions reactions.
    c. Exothermic; Energy is released when concentrated sulfuric acid dissociates in water.
    d. Endothermic; Energy is required to change the water from a liquid to a gas.

15. Thermodynamics is the study of energy and energy transfers. The first law of thermodynamics is the same as the law of conservation of energy, which is usually worded as "the energy of the universe is constant".

16. internal

17. $\Delta E = q + w$
    a. $\Delta E = 51 \text{ kJ} + (-15 \text{ kJ}) = 36 \text{ kJ}$
    b. $\Delta E = 100. \text{ kJ} + (-65 \text{ kJ}) = 35 \text{ kJ}$
    c. $\Delta E = -65 \text{ kJ} + (-20 \text{ kJ}) = -85 \text{ kJ}$

18. losing

19. positive

20. $\Delta E = q + w = -125 \text{ kJ} + 104 \text{ kJ} = -21 \text{ kJ}$

21. The calorie represents the amount of energy required to warm one gram of water by one Celsius degree. The "Calorie" or "nutritional" calorie represents 1000 calories (one kilocalorie). The Joule is the SI unit of energy and 1 calorie is equivalent to 4.184 Joules. The Joule has a specific SI definition as the energy required to exert a force of 1 Newton over a distance of 1 meter, but we tend to use the experimental definition above in terms of heating water.

22.
    a. $\dfrac{1 \text{ J}}{4.184 \text{ cal}}$
    b. $\dfrac{4.184 \text{ cal}}{1 \text{ J}}$
    c. $\dfrac{1 \text{ kcal}}{1000 \text{ cal}}$
    d. $\dfrac{1000 \text{ J}}{1 \text{ kJ}}$

23. $3 \times 8.40 \text{ kJ} = 25.2 \text{ kJ}$ for three times the increase in temperature (a temperature increase of 15° as opposed to 5°).

24. 6540 J = 6.54 kJ for 10 times more water

25. a. $75.2 \text{ kcal} \times \dfrac{4.184 \text{ kJ}}{1 \text{ kcal}} = 315 \text{ kJ} = 3.15 \times 10^3 \text{ J}$

    b. $75.2 \text{ cal} \times \dfrac{4.184 \text{ J}}{1 \text{ cal}} = 315 \text{ J} = 0.315 \text{ kJ}$

    c. $1.41 \times 10^3 \text{ cal} \times \dfrac{4.184 \text{ J}}{1 \text{ cal}} = 5.90 \times 10^3 \text{ J} = 5.90 \text{ kJ}$

    d. $1.41 \text{ kcal} \times \dfrac{4.184 \text{ kJ}}{1 \text{ kcal}} = 5.90 \text{ kJ} = 5.90 \times 10^3 \text{ J}$

26. a. $8254 \text{ cal} \times \dfrac{1 \text{ kcal}}{1000 \text{ cal}} = 8.254 \text{ kcal}$

    b. $41.5 \text{ cal} \times \dfrac{1 \text{ kcal}}{1000 \text{ cal}} = 0.0415 \text{ kcal}$

    c. $8.231 \times 10^3 \times \dfrac{1 \text{ kcal}}{1000 \text{ cal}} = 8.231 \text{ kcal}$

    d. $752{,}900 \text{ cal} \times \dfrac{1 \text{ kcal}}{1000 \text{ cal}} = 752.9 \text{ kcal}$

27. a. $652.1 \text{ kJ} \times \dfrac{1 \text{ kcal}}{4.184 \text{ kJ}} = 155.9 \text{ kcal}$

    b. $1.00 \text{ kJ} \times \dfrac{1 \text{ kcal}}{4.184 \text{ kJ}} = 0.239 \text{ kcal}$

    c. $4.184 \text{ kJ} \times \dfrac{1 \text{ kcal}}{4.184 \text{ kJ}} = 1.000 \text{ kcal}$

    d. $4.351 \times 10^3 \text{ kJ} \times \dfrac{1 \text{ kcal}}{4.184 \text{ kJ}} = 1040. \text{ kcal}$

28. a. $7845 \text{ cal} \times \dfrac{4.184 \text{ J}}{1 \text{ cal}} = 32820 \text{ J}$

    $32820 \text{ J} \times \dfrac{1 \text{ kJ}}{1000 \text{ J}} = 32.82 \text{ kJ}$

    b. $4.55 \times 10^4 \text{ cal} \times \dfrac{4.184 \text{ J}}{1 \text{ cal}} = 1.90 \times 10^5 \text{ J}$

    $1.90 \times 10^5 \text{ J} \times \dfrac{1 \text{ kJ}}{1000 \text{ J}} = 190. \text{ kJ}$

    c. $62.142 \text{ kcal} \times \dfrac{4.184 \text{ kJ}}{1 \text{ kcal}} = 260.0 \text{ kJ}$

Chapter 10: **Energy**

$$260.0 \text{ kJ} \times \frac{1000 \text{ J}}{1 \text{ kJ}} = 2.600 \times 10^5 \text{ J}$$

d. $\quad 43024 \text{ cal} \times \dfrac{4.184 \text{ J}}{1 \text{ cal}} = 1.800 \times 10^5 \text{ J}$

$$1.800 \times 10^5 \text{ J} \times \frac{1 \text{ kJ}}{1000 \text{ J}} = 180.0 \text{ kJ}$$

29. a. $\quad 625.2 \text{ cal} \times \dfrac{4.184 \text{ J}}{1 \text{ cal}} \times \dfrac{1 \text{ kJ}}{1000 \text{ J}} = 2.616 \text{ kJ}$

b. $\quad 82.41 \text{ kJ} \times \dfrac{1000 \text{ J}}{1 \text{ kJ}} = 8.241 \times 10^4 \text{ J}$

c. $\quad 52.61 \text{ kcal} \times \dfrac{1000 \text{ cal}}{1 \text{ kcal}} \times \dfrac{4.184 \text{ J}}{1 \text{ cal}} = 2.201 \times 10^5 \text{ J}$

d. $\quad 124.2 \text{ kJ} \times \dfrac{1 \text{ kcal}}{4.184 \text{ kJ}} = 29.68 \text{ kcal}$

30. a. $\quad 45.62 \text{ kcal} \times \dfrac{4.184 \text{ kJ}}{1 \text{ kcal}} = 190.9 \text{ kJ}$

b. $\quad 72.94 \text{ kJ} \times \dfrac{1 \text{ kcal}}{4.184 \text{ kJ}} = 17.43 \text{ kcal}$

c. $\quad 2.751 \text{ kJ} \times \dfrac{1000 \text{ J}}{1 \text{ kJ}} \times \dfrac{1 \text{ cal}}{4.184 \text{ J}} = 657.5 \text{ cal}$

d. $\quad 5.721 \text{ kcal} \times \dfrac{1000 \text{ cal}}{1 \text{ kcal}} \times \dfrac{4.184 \text{ J}}{1 \text{ cal}} = 2.394 \times 10^4 \text{ J}$

31. $Q = s \times m \times \Delta T$

    $69{,}500 \text{ J} = s \times (1012 \text{ g}) \times (11.4°C)$

    $s = 6.02 \text{ J/g °C}$

32. $Q = s \times m \times \Delta T$

    Specific heat capacity of aluminum is 0.89 J/g°C from Table 10.1.

    $Q = (0.89 \text{ J/g°C}) \times (42.7 \text{ g}) \times (15.2°C) = 5.8 \times 10^2 \text{ J}$ (only two significant figures are justified).

33. $Q = s \times m \times \Delta T$

    $\Delta T = \dfrac{Q}{s \times m}$

Chapter 10: Energy

Specific heat capacity of silver is 0.24 J/g°C.

$$\Delta T = \frac{125 \text{ J}}{0.24 \frac{\text{J}}{\text{g °C}} \times 29.3 \text{ g}} = 18°C \text{ (only two significant figures are justified)}$$

34. $Q = s \times m \times \Delta T$

   Specific heat capacity of mercury is 0.14 J/g°C

   100. J = (0.14 J/g°C) × 25 g × ΔT

   $\Delta T = 28.6°C = 29°C$

35. $Q = s \times m \times \Delta T$

   Specific heat capacity of gold is 0.13 J/g°C

   $Q = (0.13 \text{ J/g°C}) \times (55.5 \text{ g}) \times (25°C) = 180 \text{ J} = 1.8 \times 10^2 \text{ J}$

36. The reaction is exothermic: the chemical reaction liberates heat energy and warms up the beverage.

37. $Q = s \times m \times \Delta T$

   specific heat capacity of water = 4.184 J/g°C

   Let's assume that the density of water is exactly 1.00 g/mL during the measurement.

   $Q = 4.184 \text{ J/g°C} \times 1000 \text{ g} \times 15°C = 6.3 \times 10^4 \text{ J} = 63 \text{ kJ}$ (~60 kJ to one significant figure)

38. $Q = s \times m \times \Delta T$

   specific heat capacity of water = 4.184 J/g°C

   $Q = 4.184 \text{ J/g°C} \times 100.0 \text{ g} \times 35°C = 1.46 \times 10^4 \text{ J}$ (~15 kJ to two significant figures)

39. the same as

40. calorimeter

41. a. The balanced equation as written represents the formation of *two* moles of HF. The enthalpy change *per mole* of HF will be $\frac{-542 \text{ kJ}}{2 \text{ mol}} = -271 \text{ kJ/mol}$

   b. The sign of the enthalpy of reaction is negative: therefore the reaction releases heat energy and is exothermic.

   c. The conservation of energy principle requires that the heat of a reverse process be the same in magnitude but opposite in sign. The enthalpy of reaction for the given equation is –542 kJ. Therefore the enthalpy change of the reverse reaction will be +542 kJ.

42. a. molar mass of S = 32.07 g

   $1.00 \text{ g S} \times \frac{1 \text{ mol S}}{32.07 \text{ g S}} = 0.0312 \text{ mol S}$

Chapter 10: Energy

From the balanced chemical equation, combustion of 0.0312 mol S would produce 0.0312 mol SO_2

$$0.0312 \text{ mol} \times \frac{-296 \text{ kJ}}{\text{mol}} = -9.23 \text{ kJ}$$

b. From the balanced chemical equation, combustion of 0.0312 mol S would produce 0.0312 mol SO_2

$$0.501 \text{ mol} \times \frac{-296 \text{ kJ}}{\text{mol}} = -148 \text{ kJ}$$

c. The enthalpy change would be the same in magnitude, but opposite in sign = +296 kJ/mol

43. $350 \text{ cal} \times \dfrac{4.184 \text{ J}}{1 \text{ cal}} = 1.5 \times 10^3$ J (to two significant figures)

44. a. molar mass of ethanol = 46.07 g; we will assume that 1360 has only 3 significant figures.

$$-\frac{1360 \text{ kJ}}{1 \text{ mol}} \times \frac{1 \text{ mol}}{46.07 \text{ g}} = -29.5 \text{ kJ/g}$$

b. Since energy is released by the combustion, ΔH is negative: $\Delta H = -1360$ kJ

c. In the reaction as written, three moles of water vapor are produced when one mole of ethanol reacts.

$$\frac{1360 \text{ kJ}}{1 \text{ mol C}_2\text{H}_5\text{OH}} \times \frac{1 \text{ mol C}_2\text{H}_5\text{OH}}{3 \text{ mol H}_2\text{O}} = 453 \text{ kJ/mol H}_2\text{O}$$

45. If the second equation is *reversed* and then added to the first equation, the desired equation can be generated:

$X(g) + Y(g) \rightarrow XY(g)$	ΔH = a kJ (as given)
$XZ(g) \rightarrow X(g) + Z(g)$	ΔH = –b kJ (the equation was reversed)
$Y(g) + XZ(g) \rightarrow XY(g) + Z(g)$	ΔH = a + [–b] kJ or (a – b) kJ

46. The desired equation $2C(s) + O_2(g) \rightarrow 2CO(g)$ can be generated by taking twice the first equation and adding it to the reverse of the second equation:

$2 \times [C(s) + O_2(g) \rightarrow CO_2(g)]$	$\Delta H = 2 \times -393$ kJ = –786 kJ (equation doubled)
$2CO_2(g) \rightarrow 2CO(g) + O_2(g)$	$\Delta H = -(-566$ kJ$) = +566$ kJ (equation reversed)
$2C(s) + O_2(g) \rightarrow 2CO(g)$	$\Delta H = (-786) + (+566) = -220$ kJ

47. The desired equation $S(s) + O_2(g) \rightarrow SO_2(g)$ can be generated by reversing the second equation and dividing the equation by 2, and then adding this to the first equation:

$S(s) + \frac{3}{2} O_2(g) \rightarrow SO_3(g)$	$\Delta H = -395.2$ kJ (as given)
$2SO_3(g) \rightarrow 2SO_2(g) + O_2(g)$	$\Delta H = -(-198.2$ kJ$)/2 = +99$ kJ
$S(s) + O_2(g) \rightarrow SO_2(g)$	$\Delta H = (-395.2$ kJ$) + (+99$ kJ$) = -296.1$ kJ

48. The desired equation can be generated as follows:

$2CO_2(g) + H_2O(l) \rightarrow C_2H_2(g) + 5/2 O_2(g)$ $\quad \Delta H = -(-1300.\text{ kJ}) = 1300.\text{ kJ}$

$2 \times [C(s) + O_2(g) \rightarrow CO_2(g)]$ $\quad \Delta H = 2 \times -394 \text{ kJ} = -788 \text{ kJ}$

$H_2(g) + 1/2 O_2(g) \rightarrow H_2O(l)$ $\quad \Delta H = -286 \text{ kJ}$

$2C(s) + H_2(g) \rightarrow C_2H_2(g)$ $\quad \Delta H = (1300.) + (-788 \text{ kJ}) + (-286 \text{ kJ}) = 226 \text{ kJ}$

49. The energy is converted to the energy of motion of the car, and to friction as the car's tires interact with the road. Once the potential energy has been dispersed, it cannot be reused.

50. Once everything in the universe is at the same temperature, no further thermodynamic work can be done. Even though the total energy of the universe will be the same, the energy will have been dispersed evenly making it effectively useless.

51. Petroleum is especially useful because it is a concentrated, easy to transport and use, source of energy.

52. Concentrated sources of energy, such as petroleum, are being used so as to disperse the energy they contain, making it unavailable for further use.

53. These sources of energy originally came from living plants and animals, which used their metabolic processes to store energy.

54. Petroleum consists mainly of hydrocarbons, which are molecules containing chains of carbon atoms with hydrogen atoms attached to the chains. The fractions are based on the number of carbon atoms in the chains: for example, gasoline is a mixture of hydrocarbons with 5–10 carbon atoms in the chains, whereas asphalt is a mixture of hydrocarbons with 25 or more carbon atoms in the chains. Different fractions have different physical properties and uses, but all can be combusted to produce energy. See Table 10.3

55. Natural gas consists primarily of methane, with small amounts of ethane, propane and butane. It is generally found in association with petroleum deposits.

56. Tetraethyl lead was used as an additive for gasoline to promote smoother running of engines. It is no longer widely used because of concerns about the lead being released to the environment as the leaded gasoline is burned.

57. Coal matures through four stages: lignite, sub-bituminous, bituminous, and anthracite. The four types of coal differ in the ratio of carbon to the other elements. Anthracite has the highest fraction of carbon in it, and when burned, releases a larger amount of heat for a given mass.

58. The greenhouse effect is a warming effect due to the presence of gases in the atmosphere which absorb infrared radiation that has reached the earth from the sun, and do not allow it to pass back into space. A limited greenhouse effect is desirable because it moderates the temperature changes in the atmosphere that would otherwise be more drastic between daytime when the sun is shining and nighttime. Having too high a concentration of greenhouse gases, however, will elevate the temperature of the earth too much, affecting climate, crops, the polar ice caps, temperature of the oceans, and so on. Carbon dioxide produced by combustion reactions is our greatest concern as a greenhouse gas.

Chapter 10: Energy

59. driving force

60. The second law of thermodynamics says that the entropy of the universe is always increasing. Energy spread and matter spread lead to greater entropy (greater disorder) in the universe.

61. an increase in entropy

62. Formation of a solid precipitate represents a concentration of matter.

63. Entropy is a measure of the randomness or disorder in a system. The entropy of the universe increases because the natural tendency is for things to become more disordered.

64. The molecules in liquid water are moving around freely, and are therefore more "disordered" than when the molecules are held rigidly in a solid lattice in ice. The entropy increases during melting.

65. the reactants; Energy is required (input) in an endothermic reaction, thus the products are at a higher energy state than the reactants.

66. $1.00 \text{ g CH}_4 \times \dfrac{1 \text{ mol CH}_4}{16.042 \text{ g CH}_4} = 0.0623 \text{ mol CH}_4$

   $0.0623 \text{ mol CH}_4 \times \dfrac{-891 \text{ kJ}}{\text{mol}} = -55.5 \text{ kJ}$

67. a. $85.21 \text{ cal} \times \dfrac{4.184 \text{ J}}{1 \text{ cal}} = 356.5 \text{ J}$

   b. $672.1 \text{ J} \times \dfrac{1 \text{ cal}}{4.184 \text{ J}} = 160.6 \text{ cal}$

   c. $8.921 \text{ kJ} \times \dfrac{1000 \text{ J}}{1 \text{ kJ}} = 8921 \text{ J}$

   d. $556.3 \text{ cal} \times \dfrac{4.184 \text{ J}}{1 \text{ cal}} \times \dfrac{1 \text{ kJ}}{1000 \text{ J}} = 2.328 \text{ kJ}$

68. Temperature increase = 75.0 − 22.3 = 52.7°C

   $145 \text{ g} \times 4.184 \dfrac{\text{J}}{\text{g °C}} \times 52.7°\text{C} \times \dfrac{1 \text{ cal}}{4.184 \text{ J}} = 7641.5 \text{ cal} = 7.65 \text{ kcal}$

69. Specific heat of silver = 0.24 J/g°C; 1.25 kJ = 1250 J

   Temperature increase = 15.2 − 12.0 = 3.2°C

   $Q = s \times m \times \Delta T$

   1250 J = (0.24 J/g°C) × (mass of silver) × (3.2°C)

   mass of silver = 1627 g = 1.6×10^3 g

70. From Table 10.1, the specific heat capacity of iron is 0.45 J/g°C.

$Q = s \times m \times \Delta T$

$Q = 0.45 \dfrac{J}{g°C} \times 25.1 \text{ g} \times 17.5°C = 197.7 \text{ J} = 2.0 \times 10^2 \text{ J}$ (two significant figures)

71. $0.13 \dfrac{J}{g°C} \times \dfrac{1 \text{ cal}}{4.184 \text{ J}} = 0.031 \dfrac{\text{cal}}{g°C}$

72. 2.5 kg water = 2500 g

    Temperature change = 55.0 – 18.5 = 36.5°C

    $Q = s \times m \times \Delta T$

    $Q = 4.184 \text{ J/g°C} \times 2500 \text{ g} \times 36.5°C = 3.8 \times 10^5 \text{ J}$

73. For a given mass of substance, the substance with the *smallest* specific heat capacity (gold, 0.13J/g°C) will undergo the *largest* increase in temperature. Conversely, the substance with the largest specific heat capacity (water, 4.184 J/g°C) will undergo the smallest increase in temperature.

74. Let T_f represent the final temperature reached by the system.

    For the hot water, heat lost = $50.0 \text{ g} \times 4.184 \text{ J/g°C} \times (100. - T_f°C)$

    For the cold water, heat gained = $50.0 \text{ g} \times 4.184 \text{ J/g°C} \times (T_f - 25°C)$

    The heat lost by the hot water must *equal* the heat gained by the cold water; therefore

    $50.0 \text{ g} \times 4.184 \text{ J/g°C} \times (100. - T_f°C) = 50.0 \text{ g} \times 4.184 \text{ J/g°C} \times (T_f - 25°C)$

    Solving this equation for T_f gives $T_f = 62.5°C = 63°C$

75. Let T_f be the final temperature reached.

    Heat gained by water = $75 \text{ g} \times 4.184 \text{ J/g°C} \times (T_f - 20°C)$

    Heat lost by iron = $25.0 \text{ g} \times 0.45 \text{ J/g°C} \times (85 - T_f°C)$

    The heat lost by the iron must equal the heat gained by the water.

    $75 \text{ g} \times 4.184 \text{ J/g°C} \times (T_f - 20°C) = 25.0 \text{ g} \times 0.45 \text{ J/g°C} \times (85 - T_f°C)$

    Solving for T_f gives $T_f = 22.3°C = 22°C$

76. $Q = m \times s \times \Delta T$

    $Q = 7.24 \text{ kJ} \times \dfrac{1000 \text{ J}}{1 \text{ kJ}} = 7240 \text{ J}$

    $s = \dfrac{Q}{m \times \Delta T} = \dfrac{7240 \text{ J}}{(952 \text{ g})(10.7°C)} = 0.711 \text{ J/g°C}$

Chapter 10: Energy

77. For any substance, $Q = s \times m \times \Delta T$. The basic calculation for each of the substances is the same: Heat required = 150. g × (specific heat capacity) × 11.2 °C

Substance	Specific Heat Capacity	Heat Required
water (l)	4.184 J/g °C	7.03×10^3 J
water (s)	2.03 J/g °C	3.41×10^3 J
water (g)	2.0 J/g °C	3.4×10^3 J
aluminum	0.89 J/g °C	1.5×10^3 J
iron	0.45 J/g °C	7.6×10^2 J
mercury	0.14 J/g °C	2.4×10^2 J
carbon	0.71 J/g °C	1.2×10^3 J
silver	0.24 J/g °C	4.0×10^2 J
gold	0.13 J/g °C	2.2×10^2 J

78. $q = -213$ kJ (heat is released) and $\Delta E = -45$ kJ

    $\Delta E = q + w$

    -45 kJ $= -213$ kJ $+ w$

    $w = -45$ kJ $+ 213$ kJ $= +168$ kJ

79. a. $\Delta E = q + w = -47$ kJ $+ 88$ kJ $= 41$ kJ

    b. $\Delta E = 82 + 47 = 129$ kJ

    c. $\Delta E = 47 + 0 = 47$ kJ

    d. When the surroundings deliver work to the system, $w > 0$. This is the case for a and b.

80. $5.00 \text{ g C}_3\text{H}_8 \times \dfrac{1 \text{ mol C}_3\text{H}_8}{44.09 \text{ g C}_3\text{H}_8} \times \dfrac{-2221 \text{ kJ}}{\text{mol}} = -252 \text{ kJ}$

81. $4\text{Fe}(s) + 3\text{O}_2(g) \rightarrow 2\text{Fe}_2\text{O}_3(s) \quad \Delta H = -1652$ kJ;

    Note that 1652 kJ of heat are released when 4 mol Fe reacts with 3 mol O_2 to produce 2 mol Fe_2O_3.

    a. $4.00 \text{ mol Fe} \times \dfrac{-1652 \text{ kJ}}{4 \text{ mol Fe}} = -1652 \text{ kJ}$

    b. $1.00 \text{ mol Fe}_2\text{O}_3 \times \dfrac{-1652 \text{ kJ}}{2 \text{ mol Fe}_2\text{O}_3} = -826 \text{ kJ}$

    c. $1.00 \text{ g Fe} \times \dfrac{1 \text{ mol Fe}}{55.85 \text{ g Fe}} \times \dfrac{-1652 \text{ kJ}}{4 \text{ mol Fe}} = -7.39 \text{ kJ}$

    d. 10.0 g Fe = 0.179 mol Fe; 2.00 g O_2 = 0.0625 mol O_2; O_2 is the limiting reactant.

    $0.0625 \text{ mol O}_2 \times \dfrac{-1652 \text{ kJ}}{3 \text{ mol O}_2} = -34.4 \text{ kJ heat released.}$

82. Reversing the first equation and dividing by 6 we get

$$\tfrac{3}{6}D \rightarrow \tfrac{3}{6}A + B \qquad \Delta H = +403 \text{ kJ}/6$$

$$\text{or } \tfrac{1}{2}D \rightarrow \tfrac{1}{2}A + B \qquad \Delta H = +67.2 \text{ kJ}$$

Dividing the second equation by 2 we get

$$\tfrac{1}{2}E + F \rightarrow \tfrac{1}{2}A \qquad \Delta H = -105.2 \text{ kJ}/2 = -52.6 \text{ kJ}$$

Dividing the third equation by 2 we get

$$\tfrac{1}{2}C \rightarrow \tfrac{1}{2}E + \tfrac{3}{2}D \qquad \Delta H = +64.8 \text{ kJ}/2 = +32.4 \text{ kJ}$$

Adding these equations together we get

$$\tfrac{1}{2}C + F \rightarrow A + B + D \quad \Delta H = 47.0 \text{ kJ}$$

83. During exercise, the body generates about 5500 kJ/hr, or about 11,000 kJ in 2 hours. Assuming all of this heat is lost through evaporation, we can calculate the volume of the perspiration. Water has a heat of vaporization of 40.7 kJ/mol. So,

$$11{,}000 \text{ kJ} \times \frac{1 \text{ mol H}_2\text{O}}{40.7 \text{ kJ}} = 270 \text{ mol H}_2\text{O}$$

$$270 \text{ mol H}_2\text{O} \times \frac{18.02 \text{ g H}_2\text{O}}{1 \text{ mol H}_2\text{O}} = 4900 \text{ g H}_2\text{O}$$

Assuming a density of 1 g/mL for water, 4900 g of H_2O would occupy a volume of 4900 mL or 4.9 L.

84. (c)

85. (a) and (e); Changing from a solid to a gas requires energy. Splitting a molecule up into its separate atoms also requires energy. Changing from a gas to a liquid or changing from a liquid to a solid releases energy. The reaction of hydrogen and oxygen to make water is also exothermic.

86. $Q = m \times s \times \Delta T$

$$Q = 2.4 \text{ mol C} \times \frac{12.01 \text{ g C}}{1 \text{ mol C}} \times \frac{0.71 \text{ J}}{\text{g}^\circ\text{C}} \times 25.0^\circ\text{C} = 5.1 \times 10^2 \text{ J}$$

87. *Volume of water* = $10.0 \text{ m} \times 4.0 \text{ m} \times 3.0 \text{ m} = 120 \text{ m}^3$

$$120 \text{ m}^3 \times \left(\frac{100 \text{ cm}}{1 \text{ m}}\right)^3 = 1.2 \times 10^8 \text{ cm}^3$$

mass of water = $1.2 \times 10^8 \text{ cm}^3 \times \dfrac{1 \text{ g}}{\text{cm}^3} = 1.2 \times 10^8 \text{ g}$

$$Q = (1.2 \times 10^8 \text{ g}) \times \left(\frac{4.18 \text{ J}}{\text{g}^\circ\text{C}}\right) \times (24.6^\circ\text{C} - 20.2^\circ\text{C}) = 2.2 \times 10^9 \text{ J}$$

Chapter 10: Energy

88. $54.0 \text{ g B}_2\text{H}_6 \times \dfrac{1 \text{ mol B}_2\text{H}_6}{27.668 \text{ g B}_2\text{H}_6} \times \dfrac{2035 \text{ kJ}}{\text{mol}} = 3.97 \times 10^3 \text{ kJ}$

89. The desired equation can be generated as follows:

$¼ \times [2\text{NH}_3(g) + 3\text{N}_2\text{O}(g) \rightarrow 4\text{N}_2(g) + 3\text{H}_2\text{O}(l)]$ $\quad \Delta H = ¼ \times (-1010 \text{ kJ}) = -252.5 \text{ kJ}$

$¾ \times [\text{N}_2\text{H}_4(l) + \text{H}_2\text{O}(l) \rightarrow \text{N}_2\text{O}(g) + 3\text{H}_2(g)]$ $\quad \Delta H = ¾ \times -(-317 \text{ kJ}) = 237.75 \text{ kJ}$

$\tfrac{9}{4} \times [\text{H}_2(g) + ½\text{O}_2(g) \rightarrow \text{H}_2\text{O}(l)]$ $\quad \Delta H = \tfrac{9}{4} \times (-286 \text{ kJ}) = -643.5 \text{ kJ}$

$¼ \times [\text{N}_2\text{H}_4(l) + \text{H}_2\text{O}(l) \rightarrow 2\text{NH}_3(g) + ½\text{O}_2(g)]$ $\quad \Delta H = ¼ \times -(-143 \text{ kJ}) = 35.75 \text{ kJ}$

$\text{N}_2\text{H}_4(l) + \text{O}_2(g) \rightarrow \text{N}_2(g) + 2\text{H}_2\text{O}(l) \quad \Delta H = (-252.5) + (237.75) + (-643.5) + 35.75 = -623 \text{ kJ}$

CHAPTER 11

Modern Atomic Theory

1. positively; negatively

2. Rutherford was not able to determine where the electrons were in the atom or what they were doing.

3. Electromagnetic radiation is radiant energy that travels through space with wavelike behavior at the speed of light.

4. The different forms of electromagnetic radiation are similar in that they all exhibit the same type of wave-like behavior and are propagated through space at the same speed (the speed of light). The types of electromagnetic radiation differ in their frequency (and wavelength) and in the resulting amount of energy carried per photon.

5. The wavelength represents the distance between corresponding points on two successive waves. The energy of a photon is inversely proportional to the wavelength ($E = hc/\lambda$)

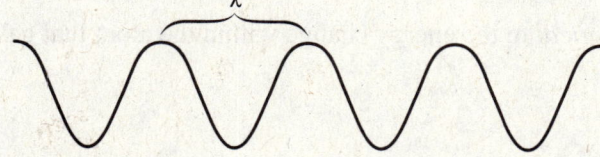

6. The *speed* of electromagnetic radiation represents how fast a given wave moves through space. The *frequency* of electromagnetic radiation represents how many complete cycles of the wave pass a given point per second. These two concepts are not the same.

7. Ultraviolet radiation is of shorter wavelength than visible light, and therefore is of higher energy than visible light.

8. The greenhouse gases do not absorb light in the visible wavelengths, enabling this light to pass through the atmosphere and continue to warm the earth, keeping the earth much warmer than it would be without these gases. The earth, in turn, emits infrared radiation which is absorbed by the greenhouse gases and which is re-emitted in all directions. As we increase our use of fossil fuels, the level of CO_2 in the atmosphere is increasing gradually, but significantly. An increase in the level of CO_2 will warm the earth further, eventually changing the weather patterns on the earth's surface and melting the polar ice caps.

9. The atoms in salts emit light of characteristic wavelengths because of how their internal electron structures interact with the applied energy. Each element's atoms have a different electron structure, so each atom emits light differently from all other atoms.

10. exactly equal to

11. The ground state represents the lowest-energy state of the atom.

Chapter 11: Modern Atomic Theory

12. A photon having an energy corresponding to the energy difference between the two states is emitted by an atom in an excited state when it returns to its ground state.

13. The energy of a photon is given by $E = hc/\lambda$ and therefore short wavelength light carries more energy per photon.

14. absorbs

15. The emission of light by excited atoms has been the key interconnection between the macroscopic world we can observe and measure, and with what is happening on a microscopic basis within an atom. Excited atoms emit light (which we can measure) because of changes in the microscopic structure of the atom. By studying the emissions of atoms we can trace back to what happened inside the atom.

16. When excited hydrogen atoms emit their excess energy, the photons of radiation emitted are always of exactly the same wavelength and energy. We consider this to mean that the hydrogen atom possesses only certain allowed energy states, and that the photons emitted correspond to the atom changing from one of these allowed energy states to another of the allowed energy state. The energy of the photon emitted corresponds to the energy difference in the allowed states. If the hydrogen atom did not possess discrete energy levels, then we would expect the photons emitted to have random wavelengths and energies.

17. transitions of electrons

18. The energy of an emitted photon is *identical* to the energy change within the atom that gave rise to the emitted photon.

19. quantized

20. Energy is emitted only at wavelengths corresponding to the specific transitions for the electron among the energy levels of hydrogen.

21. Bohr envisioned electrons as moving in circular orbits corresponding to the various allowed energy levels. He suggested that the electron could jump to a different orbit by absorbing or emitting a photon of light with exactly the correct energy content (corresponding to the difference in energy between the orbits).

22. The electron moves to an orbit farther from the nucleus of the atom.

23. Bohr suggested that the electron could jump to a different orbit by absorbing or emitting a photon of light with exactly the correct energy content (corresponding to the difference in energy between the orbits).

    As the energy levels of a given atom were fixed and definite, then the atom should always emit energy at the same discrete wavelengths.

24. Bohr's theory *explained* the experimentally *observed* line spectrum of hydrogen *exactly*. Bohr's theory was ultimately discarded because when attempts were made to extend the theory to atoms other than hydrogen, the calculated properties did *not* correspond closely to experimental measurements.

25. Schrödinger and de Broglie reasoned that, because light seems to have both wave and particle characteristics (it behaves simultaneously as a wave and as if it were a stream of particles), that perhaps the electron might exhibit both of these characteristics. That is, although the electron behaves as a discrete particle, perhaps the properties of the electron in the atom could be treated as if they were wavelike.

26. An orbit represents a definite, exact circular pathway around the nucleus in which an electron can be found. An orbital represents a region of space in which there is a high probability of finding the electron.

27. Schrödinger's mathematical treatment could only describe the movement of the electron through the atom in terms of the *probability* of finding the electron in given regions of space within the atom, but not at a particular point within the atom at a particular time. Any attempt to determine the exact position of the electron within an atom experimentally would, in fact, disturb the electron from wherever it had been.

28. The firefly analogy is intended to demonstrate the concept of a probability map for electron density. In the wave mechanical model of the atom, we cannot say specifically where the electron is in the atom, we can only say where there is a high probability of finding the electron. The analogy is to imagine a time-exposure photograph of a firefly in a closed room. Most of the time, the firefly will be found near the center of the room.

29. Although the probability of finding the electron decreases at greater distances from the nucleus, the probability of finding it even at great distances from the nucleus never becomes exactly zero. The probability becomes less and less as you move away from the nucleus, similar to the way the atmosphere becomes thinner and thinner as you move away from the surface of the earth.

30. (b)

31. The 2s orbital is similar in shape to the 1s orbital, but is larger.

32. The *p* orbitals, in general, have two lobes and are sometimes described as having a "dumbbell" shape. The 2p and 3p orbitals are similar in shape, and in fact there are three equivalent 2p or 3p orbitals in the 2p or 3p subshell. The orbitals differ in size, mean distance from the nucleus, and energy.

33. farther from

34. excited

35. The other orbitals serve as the excited states of the hydrogen atom. When energy of the right frequency is applied to the hydrogen atom, the electron can move from its normal orbital (ground state) to one of the other orbitals (excited states). Later on, the electron can move back to its normal orbital and release the absorbed energy as light.

Chapter 11: Modern Atomic Theory

36. | Value of n | Possible subshells |
    |---|---|
    | 1 | 1s |
    | 2 | 2s, 2p |
    | 3 | 3s, 3p, 3d |
    | 4 | 4s, 4p, 4d, 4f |

37. electron spin

38. Electrons have an intrinsic spin (they spin on their own axes). Geometrically, there are only two senses possible for spin (clockwise or counter-clockwise). This means only two electrons can occupy an orbital, with the opposite sense or direction of spin. This idea is called the Pauli Exclusion Principle.

39. The higher the value of the principal quantum number, n, the higher the energy of the principal energy level.

40. increases; as you move out from the nucleus, there is more space and room for more sublevels.

41. two

42. opposite

43. Choices a, b, and d are possible; choice c is *not* possible because the d subshells do not begin until the $n = 3$ orbit.

44. (a) and (c); f orbitals begin at the fourth energy level and a d orbitals begin at the third energy level

45. The $1s$ orbital is closest to the nucleus and lowest in energy, so it is always filled first.

46. When a hydrogen atom is in its ground state, the electron is found in the $1s$ orbital. The $1s$ orbital has the lowest energy of all the possible hydrogen orbitals.

47. Valence electrons are those in the outermost (highest) principal energy level of an atom. These electrons are especially important because they are at the "outside edge" of an atom, and are those electrons that are "seen" by other atoms and can interact with the electrons of another atom in a chemical reaction.

48. The elements in a given vertical column of the periodic table have the same valence electron configuration. Having the same valence electron configuration causes the elements in a given group to have similar chemical properties.

49. a. $1s^2 2s^2 2p^6 3s^2$

    b. $1s^2 2s^1$

    c. $1s^2 2s^2 2p^4$

    d. $1s^2 2s^2 2p^6 3s^2 3p^4$

Chapter 11: Modern Atomic Theory

50. Just count the electrons to get the atomic number of the element.

   a. silicon
   b. beryllium
   c. neon
   d. argon

51. a. $1s^2 2s^2 2p^6 3s^2 3p^3$
    b. $1s^2 2s^2 2p^6 3s^2 3p^6 4s^2$
    c. $1s^2 2s^2 2p^6 3s^2 3p^6 4s^1$
    d. $1s^2 2s^2 2p^1$

52. Just count the electrons to get the atomic number of the element.

   a. selenium
   b. scandium
   c. sulfur
   d. iodine

53. a. $1s(\uparrow\downarrow)$
    b. $1s(\uparrow\downarrow)\ 2s(\uparrow\downarrow)\ 2p(\uparrow\downarrow)(\uparrow\downarrow)(\uparrow\downarrow)$
    c. $1s(\uparrow\downarrow)\ 2s(\uparrow\downarrow)\ 2p(\uparrow\downarrow)(\uparrow\downarrow)(\uparrow\downarrow)\ 3s(\uparrow\downarrow)\ 3p(\uparrow\downarrow)(\uparrow\downarrow)(\uparrow\downarrow)\ 4s(\uparrow\downarrow)$
       $3d(\uparrow\downarrow)(\uparrow\downarrow)(\uparrow\downarrow)(\uparrow\downarrow)(\uparrow\downarrow)\ 4p(\uparrow\downarrow)(\uparrow\downarrow)(\uparrow\downarrow)$
    d. $1s(\uparrow\downarrow)\ 2s(\uparrow\downarrow)\ 2p(\uparrow\downarrow)(\uparrow\downarrow)(\uparrow\downarrow)\ 3s(\uparrow\downarrow)\ 3p(\uparrow\downarrow)(\uparrow\downarrow)(\uparrow\downarrow)\ 4s(\uparrow\downarrow)$
       $3d(\uparrow\downarrow)(\uparrow\downarrow)(\uparrow\downarrow)(\uparrow\downarrow)(\uparrow\downarrow)\ 4p(\uparrow\downarrow)(\uparrow\downarrow)(\uparrow\downarrow)\ 5s(\uparrow\downarrow)$
       $4d(\uparrow\downarrow)(\uparrow\downarrow)(\uparrow\downarrow)(\uparrow\downarrow)(\uparrow\downarrow)\ 5p(\uparrow\downarrow)(\uparrow\downarrow)(\uparrow\downarrow)$

54. a. $1s(\uparrow\downarrow)\ 2s(\uparrow\downarrow)\ 2p(\uparrow\downarrow)(\uparrow\downarrow)(\uparrow\downarrow)\ 3s(\uparrow\downarrow)\ 3p(\uparrow)(\ )(\ )$
    b. $1s(\uparrow\downarrow)\ 2s(\uparrow\downarrow)\ 2p(\uparrow\downarrow)(\uparrow\downarrow)(\uparrow\downarrow)\ 3s(\uparrow\downarrow)\ 3p(\uparrow)(\uparrow)(\uparrow)$
    c. $1s(\uparrow\downarrow)\ 2s(\uparrow\downarrow)\ 2p(\uparrow\downarrow)(\uparrow\downarrow)(\uparrow\downarrow)\ 3s(\uparrow\downarrow)\ 3p(\uparrow\downarrow)(\uparrow\downarrow)(\uparrow\downarrow)\ 4s(\uparrow\downarrow)$
       $3d(\uparrow\downarrow)(\uparrow\downarrow)(\uparrow\downarrow)(\uparrow\downarrow)(\uparrow\downarrow)\ 4p(\uparrow\downarrow)(\uparrow\downarrow)(\uparrow\ )$
    d. $1s(\uparrow\downarrow)\ 2s(\uparrow\downarrow)\ 2p(\uparrow\downarrow)(\uparrow\downarrow)(\uparrow\downarrow)\ 3s(\uparrow\downarrow)\ 3p(\uparrow\downarrow)(\uparrow\downarrow)(\uparrow\downarrow)$

55. Group 2 (ns^2) and Group 8 ($ns^2\ np^6$) are the only groups where all electrons are paired.

56. Specific answers depend on student choice of elements. Any Group 1 element would have one valence electron. Any Group 3 element would have three valence electrons. Any Group 5 element would have five valence electrons. Any Group 7 element would have seven valence electrons.

57. This belief is based on the *experimental properties* of K and Ca. The physical and chemical properties of K are like those of the other Group 1 elements; Ca's properties are similar to the other Group 2 elements.

Chapter 11: Modern Atomic Theory

58. The properties of Rb and Sr suggest that they are members of Groups 1 and 2, respectively, and so must be filling the 5s orbital. The 5s orbital is lower in energy (and fills before) the 4d orbitals.

59. a. [Ar] $4s^2\ 3d^{10}\ 4p^3$
    b. [Ar] $4s^2\ 3d^2$
    c. [Kr] $5s^2$
    d. [Ne] $3s^2\ 3p^5$

60. a. aluminum
    b. potassium
    c. bromine
    d. tin

61. a. [Ar] $4s^2\ 3d^1$
    b. [Kr] $5s^2\ 4d^1$
    c. [Xe] $6s^2\ 5d^1$
    d. [Rn] $7s^2\ 6d^1$

62. a. [Kr] $5s^1$: 1 valence electron
    b. [Ar] $4s^2\ 3d^{10}\ 4p^3$: 5 valence electrons (d electrons are not counted as valence electrons)
    c. [Ne] $3s^2\ 3p^1$: 3 valence electrons
    d. [Ar] $4s^2\ 3d^8$: 2 valence electrons (d electrons are not counted as valence electrons)

63. a. eight
    b. three
    c. five
    d. six

64. a. [Kr] $5s^2\ 4d^6$: 6 4d electrons
    b. [Kr] $5s^2\ 4d^8$: 8 4d electrons
    c. [Kr] $5s^2\ 4d^{10}\ 5p^2$: 10 4d electrons
    d. [Ar] $4s^2\ 3d^6$: 0 4d electrons

65. The *position* of the element (both in terms of vertical column and horizontal row) indicates which set of orbitals is being filled last.
    a. 7s
    b. 5p
    c. 5d
    d. 6p

66.   a. $[Rn]7s^2 6d^1 5f^3$

      b. $[Ar]4s^2 3d^6$

      c. $[Xe]6s^2 4f^{14} 5d^{10}$

      d. $[Rn]7s^1$

67.   The valence electrons are those beyond the noble gas "core."

      a. $[Kr] 5s^1$

      b. $[Xe] 6s^2$

      c. $[Ar] 4s^2 3d^2$

      d. $[Ar] 4s^2 3d^{10} 4p^2$

68. $[Rn] 7s^2 5f^{14} 6d^6$

69.   Some typical properties of metals are: a lustrous appearance, the ability to be pounded into sheets (malleability) or pulled into wires (ductility), and the ability to conduct heat and electricity. Nonmetals typically have a non-shiny appearance, are brittle, and do not conduct heat or electricity well. There are exceptions to these general properties: for example, graphite (a form of the nonmetal carbon) conducts electricity well.

70.   The metallic elements *lose* electrons and form *positive* ions (cations); the nonmetallic elements *gain* electrons and form *negative* ions (anions). Remember that the electron itself is *negatively* charged.

71.   The Group 1 metals are all highly reactive, and all form 1+ ions almost exclusively when they react. Physically, these metals are soft (they can be cut with a knife) and very low in density. Because of their high reactivity, these metals tend to be found with a coating of the metal oxide that hides their metallic luster (this luster can be seen, however, if a fresh surface of the metal is exposed).

72.   All exist as *diatomic* molecules (F_2, Cl_2, Br_2, I_2); all are *non*metals; all have relatively high electronegativities; all form 1- ions in reacting with metallic elements.

73.   Cs: The valence electron is farther from the nucleus than in Li or K.

74.   Elements at the *left* of a period (horizontal row) lose electrons more readily; at the left of a period (given principal energy level) the nuclear charge is the smallest and the electrons are least tightly held.

75.   The nonmetallic elements are clustered at the upper right side of the periodic table. These elements are effective at pulling electrons from metallic elements for several reasons. First, these elements have little tendency to lose electrons themselves (they have high ionization energies). Secondly, the atoms of these elements tend to be small in size, which means that electrons can be pulled in strongly since they can get closer to the nucleus. Finally, if these atoms gain electrons, they can approach the electronic configuration of the following noble gas elements (see Chapter 12 for why the electronic configuration of the noble gases are desirable for other atoms to attain).

Chapter 11: Modern Atomic Theory

76. The elements of a given period (horizontal row) have valence electrons in the same principal energy level. Nuclear charge, however, increases across a period going from left to right. Atoms at the left side have smaller nuclear charges and hold onto their valence electrons less tightly.

77. Metalloids are elements that have both metallic and non-metallic properties. They are located along either side of the "stair-step" towards the right side of the periodic table.

78. When substances absorb energy the electrons become excited (move to higher energy levels). Upon returning to the ground state, energy is released, some of which is in the visible spectrum. Since we see colors, this tells us that only *certain wavelengths* of light are released, which means that only *certain transitions* are allowed. This is what is meant by quantized energy levels. If all wavelengths of light were emitted we would see white light.

79. For most elements, the chemical activity is reflected in the ease with which the element gains or loses electrons
    a. Li (the less reactive metals are further up in a group)
    b. At (the less reactive nonmetals are at the bottom of a group)
    c. Be (the less reactive metals are further up in a group)
    d. Po (the less reactive elements are at the bottom of a group)

80. Ionization energies decrease in going from top to bottom within a vertical group; ionization energies increase in going from left to right within a horizontal period.
    a. Li
    b. Ca
    c. Cl
    d. S

81. Atomic size increases in going from top to bottom within a vertical group; atomic size decreases in going from left to right within a horizontal period.
    a. Xe < Sn < Sr < Rb
    b. He < Kr < Xe < Rn
    c. At < Pb < Ba < Cs

82. Atomic size increases in going from top to bottom within a vertical group; atomic size decreases in going from left to right within a horizontal period.
    a. Na
    b. S
    c. N
    d. F

83. The highest-energy photons are responsible for the line at 410 nm; the lowest-energy photons give rise to the line at 656 nm.

84. speed of light

Chapter 11:   Modern Atomic Theory

85. visible

86. photons

87. ground

88. quantized

89. orbits

90. orbital

91. valence

92. transition metal

93. frequency

94. spins

95. a. [Ne] $3s$(↑↓) $3p$(↑)(↑)(↑)

   P is expected to be paramagnetic; three unpaired $3p$ electrons

   b. [Kr] $5s$(↑↓) $4d$(↑↓)(↑↓)(↑↓)(↑↓)(↑↓) $5p$(↑↓)(↑↓)(↑)

   I is expected to be paramagnetic; one unpaired $5p$ electron

   c. [Ar] $4s$(↑↓) $3d$(↑↓)(↑↓)(↑↓)(↑↓)(↑↓) $4p$(↑)(↑)( )

   Ge is expected to be paramagnetic; two unpaired $4p$ electrons

96. a. $1s^2\,2s^2\,2p^6\,3s^2\,3p^6\,4s^1$ [Ar] $4s^1$

   $1s$(↑↓) $2s$(↑↓) $2p$(↑↓)(↑↓)(↑↓) $3s$(↑↓) $3p$(↑↓)(↑↓)(↑↓) $4s$(↑)

   b. $1s^2\,2s^2\,2p^6\,3s^2\,3p^6\,4s^2\,3d^2$ [Ar] $4s^2\,3d^2$

   $1s$(↑↓) $2s$(↑↓) $2p$(↑↓)(↑↓)(↑↓) $3s$(↑↓) $3p$(↑↓)(↑↓)(↑↓) $4s$(↑↓)
   $3d$(↑)(↑)( )( )( )

   c. $1s^2\,2s^2\,2p^6\,3s^2\,3p^2$ [Ne] $3s^2\,3p^2$

   $1s$(↑↓) $2s$(↑↓) $2p$(↑↓)(↑↓)(↑↓) $3s$(↑↓) $3p$(↑)(↑)( )

   d. $1s^2\,2s^2\,2p^6\,3s^2\,3p^6\,4s^2\,3d^6$ [Ar] $4s^2\,3d^6$

   $1s$(↑↓) $2s$(↑↓) $2p$(↑↓)(↑↓)(↑↓) $3s$(↑↓) $3p$(↑↓)(↑↓)(↑↓) $4s$(↑↓)
   $3d$(↑↓)(↑)(↑)(↑)(↑)

   e. $1s^2\,2s^2\,2p^6\,3s^2\,3p^6\,4s^2\,3d^{10}$ [Ar] $4s^2\,3d^{10}$

   $1s$(↑↓) $2s$(↑↓) $2p$(↑↓)(↑↓)(↑↓) $3s$(↑↓) $3p$(↑↓)(↑↓)(↑↓) $4s$(↑↓)
   $3d$(↑↓)(↑↓)(↑↓)(↑↓)(↑↓)

Chapter 11: Modern Atomic Theory

97. a. $1s^2\ 2s^2\ 2p^6\ 3s^2\ 3p^6\ 4s^2\ 3d^1$

   $1s(↑↓)\ 2s(↑↓)\ 2p(↑↓)(↑↓)(↑↓)\ 3s(↑↓)\ 3p(↑↓)(↑↓)(↑↓)\ 4s(↑↓)$
   $3d(↑\ )(\ )(\ )(\ )(\ )$

   [Ar] $4s^2\ 3d^1$

   b. $1s^2\ 2s^2\ 2p^6\ 3s^2\ 3p^3$

   $1s(↑↓)\ 2s(↑↓)\ 2p(↑↓)(↑↓)(↑↓)\ 3s(↑↓)\ 3p(↑\ )(↑\ )(↑\ )$

   [Ne] $3s^2\ 3p^3$

   c. $1s^2\ 2s^2\ 2p^6\ 3s^2\ 3p^6\ 4s^2\ 3d^{10}\ 4p^6$

   $1s(↑↓)\ 2s(↑↓)\ 2p(↑↓)(↑↓)(↑↓)\ 3s(↑↓)\ 3p(↑↓)(↑↓)(↑↓)\ 4s(↑↓)$
   $3d(↑↓)(↑↓)(↑↓)(↑↓)(↑↓)\ 4p(↑↓)(↑↓)(↑↓)$

   [Kr] is itself a noble gas

   d. $1s^2\ 2s^2\ 2p^6\ 3s^2\ 3p^6\ 4s^2\ 3d^{10}\ 4p^6\ 5s^2$

   $1s(↑↓)\ 2s(↑↓)\ 2p(↑↓)(↑↓)(↑↓)\ 3s(↑↓)\ 3p(↑↓)(↑↓)(↑↓)\ 4s(↑↓)$
   $3d(↑↓)(↑↓)(↑↓)(↑↓)(↑↓)\ 4p(↑↓)(↑↓)(↑↓)\ 5s(↑↓)$

   [Kr] $5s^2$

   e. $1s^2\ 2s^2\ 2p^6\ 3s^2\ 3p^6\ 4s^2\ 3d^{10}$

   $1s(↑↓)\ 2s(↑↓)\ 2p(↑↓)(↑↓)(↑↓)\ 3s(↑↓)\ 3p(↑↓)(↑↓)(↑↓)\ 4s(↑↓)$
   $3d(↑↓)(↑↓)(↑↓)(↑↓)(↑↓)$

   [Ar] $4s^2\ 3d^{10}$

98. a. ns^2
   b. $ns^2\ np^5$
   c. $ns^2\ np^4$
   d. ns^1
   e. $ns^2\ np^4$

99. a. four (two if the d electrons are not counted.)
   b. seven
   c. two
   d. seven (two if the d electrons are not counted.)

100. a. $\lambda = \dfrac{h}{mv}$

   $\lambda = \dfrac{6.63 \times 10^{-34}\ \text{J s}}{(9.1 \times 10^{-31}\text{kg})[0.90 \times (3.00 \times 10^8\ \text{m s}^{-1})]}$

   $\lambda = 2.7 \times 10^{-12}$ m (0.0027 nm)

Chapter 11: Modern Atomic Theory

    b. 4.4×10^{-34} m

    c. 2×10^{-35} m

    The wavelengths for the ball and the person are *infinitesimally small*, whereas the wavelength for the electron is nearly the same order of magnitude as the diameter of a typical atom.

101. 3.00×10^8 m/sec

102.   Light is emitted from the hydrogen atom only at certain fixed wavelengths. If the energy levels of hydrogen were *continuous*, a hydrogen atom would emit energy at all possible wavelengths.

103.   As an electron moves to a higher-number principal energy level, the electron's mean distance from the nucleus increases, thereby decreasing the attractive force between the electron and the nucleus.

104. $[Ar]4s^2 3d^7$: …$3d(↑↓)(↑↓)(↑)(↑)(↑)$; 3 unpaired electrons

105.   Orbitals in the 1s and 2s subshells can only contain two electrons.

106.   a. $[Ne] 3s^2 3p^4$; 16 electrons in this ground state atom which corresponds to sulfur.

        b. $[He] 2s^1 2p^4$; 7 electrons in this excited state atom which corresponds to nitrogen. One electron was excited from the 2s to 2p.

        c. $[Ar] 4s^2 3d^{10} 4p^5$; 35 electrons in this ground state ion. A charge of –1 means an electron was gained, thus making the ion Se⁻ which corresponds to 34 protons.

107.   Electrons are negatively charged particles that repel each other. By placing the three electrons in separate 2p orbitals (oriented at 90° to each other in space), the repulsion among the electrons is minimized. The electrons can also have the same spin if they are in separate orbitals but this is not discussed at length in the text.

108.   a. $1s^2 2s^2 2p^6 3s^2 3p^6 4s^2 3d^{10} 4p^5$

        b. $1s^2 2s^2 2p^6 3s^2 3p^6 4s^2 3d^{10} 4p^6 5s^2 4d^{10} 5p^6$

        c. $1s^2 2s^2 2p^6 3s^2 3p^6 4s^2 3d^{10} 4p^6 5s^2 4d^{10} 5p^6 6s^2$

        d. $1s^2 2s^2 2p^6 3s^2 3p^6 4s^2 3d^{10} 4p^4$

109.   a. $1s(↑↓)$ $2s(↑↓)$ $2p(↑↓)(↑↓)(↑↓)$ $3s(↑↓)$ $3p(↑↓)(↑↓)(↑↓)$ $4s(↑↓)$ $3d(↑)()()()()$

        b. $1s(↑↓)$ $2s(↑↓)$ $2p(↑↓)(↑↓)(↑↓)$ $3s(↑↓)$ $3p(↑↓)(↑)(↑)$

        c. $1s(↑↓)$ $2s(↑↓)$ $2p(↑↓)(↑↓)(↑↓)$ $3s(↑↓)$ $3p(↑↓)(↑↓)(↑↓)$ $4s(↑ \square)$

        d. $1s(↑↓)$ $2s(↑↓)$ $2p(↑)(↑)(↑)$

Chapter 11:   Modern Atomic Theory

110. a. five (2s, 2p)
      b. seven (3s, 3p)
      c. one (3s)
      d. three (3s, 3p)

111. transition metals

112. a. $ns^2 np^3$
      b. ns^1
      c. $ns^2 np^5$
      d. $ns^2 np^4$
      e. ns^2

113. a. [Ar] $4s^2 3d^2$
      b. [Ar] $4s^2 3d^{10} 4p^4$
      c. [Kr] $5s^2 4d^{10} 5p^3$
      d. [Kr] $5s^2$

114. (a) < (c) < (b); (a) corresponds to the element Xe. (c) corresponds to the element Sb. (b) corresponds to the element In. Atomic size increases from right to left across a row on the periodic table. Xe < Sb < In

115. a. [Ar] $4s^2 3d^8$
      b. [Kr] $5s^2 4d^3$ (actually [Kr] $5s^1 4d^4$ for reasons beyond text)
      c. [Xe] $6s^2 4f^{14} 5d^2$
      d. [Xe] $6s^2 4f^{14} 5d^{10} 6p^5$

116. metals, low; nonmetals, high

117. a. B and Al are both very reactive
      b. Na
      c. F

118. Atomic size increases in going from top to bottom within a vertical group; atomic size decreases in going from left to right within a horizontal period.
      a. Ca
      b. P
      c. K

Chapter 11: Modern Atomic Theory

119. $2f$: 0 electrons

   $2d_{xy}$: 0 electrons

   $3p$: 6 electrons

   $5d_{yz}$: 2 electrons

   $4p$: 6 electrons

120. (b), (c), and (e)

121. Ca: $1s^2 2s^2 2p^6 3s^2 3p^6 4s^2$

   B: $1s^2 2s^2 2p^1$

   H: $1s^1$

   S: $1s^2 2s^2 2p^6 3s^2 3p^4$

   Be: $1s^2 2s^2$

122. a. Te

   b. Ge

   c. F (9 total electrons)

123. K: [Ar] $4s^1$

   Be: [He] $2s^2$

   Zr: [Kr] $5s^2 4d^2$

   Se: [Ar] $4s^2 3d^{10} 4p^4$

   C: [He] $2s^2 2p^2$

124. F and B: B is the larger atom.

   C and N: C is the larger atom.

   B and Al: Al is the larger atom.

125. He and Kr: He has the larger ionization energy.

   Na and Al: Al has the larger ionization energy.

   Cl and I: Cl has the larger ionization energy.

126.

Electron configuration	Symbol	IE	AR
$1s^2 2s^2 2p^6 3s^2$	Mg	0.738	160
$1s^2 2s^2 2p^6 3s^2 3p^4$	S	0.999	104
$1s^2 2s^2 2p^6 3s^2 3p^6 4s^2$	Ca	0.590	197

CHAPTER 12

Chemical Bonding

1. A *chemical bond* is a force that holds groups of two or more atoms together and makes them function as a unit.

2. The *bond energy* represents the energy required to break a chemical bond.

3. An ionic bond is created when a metallic element reacts with a non-metallic element.

4. A covalent bond represents the *sharing* of pairs of electrons between nuclei.

5. In Cl_2, the bonding is pure covalent with the electron pair shared equally between the two chlorine atoms. In HCl there is a shared pair of electrons, but the shared pair is drawn more closely to the chlorine atom, making the bond is polar.

6. In H_2 and HF, the bonding is covalent in nature, with an electron pair being shared between the atoms. In H_2, the two atoms are identical (the sharing is equal); in HF, the two atoms are different (the sharing is unequal) and as a result the bond is polar. Both of these are in marked contrast to the situation in NaF: NaF is an ionic compound—an electron has been completely transferred from sodium to fluorine, producing separate ions.

7. electronegativity

8. A bond is polar if the centers of positive and negative charge do not coincide at the same point. The bond has a negative end and a positive end. Polar bonds will exist in any molecule with nonidentical bonded atoms (although the molecule, as a whole, may not be polar if the bond dipoles cancel each other). Two simple examples are HF and HCl: in both cases, the negative center of charge is closer to the halogen atom.

9. large

10. The level of polarity in a polar covalent bond is determined by the difference in electronegativity of the atoms in the bond.

11. In general, an element farther to the right in a given period or an element closer to the top of a given group is more electronegative.

    a. H is most electronegative; K is least electronegative

    b. F is most electronegative; Na is least electronegative

    c. F is most electronegative; B is least electronegative

12. a. At is most electronegative, Cs is least electronegative

    b. Sr is most electronegative, Ba and Ra have the same electronegativities

    c. O is most electronegative, Rb is least electronegative

13. Generally, covalent bonds between atoms of *different* elements are *polar*.
    a. covalent
    b. ionic
    c. polar covalent

14. Generally, covalent bonds between atoms of *different* elements are *polar*.
    a. covalent
    b. polar covalent
    c. ionic

15. For a bond to be polar covalent, the atoms involved in the bond must have different electronegativities (must be of different elements).
    a. polar covalent (atoms of different elements)
    b. polar covalent (atoms of different elements)
    c. nonpolar covalent (atoms of the same element)
    d. nonpolar covalent (atoms of the same element)

16. For a bond to be polar covalent, the atoms involved in the bond must have different electronegativities (must be of different elements).
    a. nonpolar covalent (atoms of the same element)
    b. nonpolar covalent (atoms of the same element)
    c. nonpolar covalent (atoms of the same element)
    d. polar covalent (atoms of different elements)

17. The *degree* of polarity of a polar covalent bond is indicated by the magnitude of the difference in electronegativities of the elements involved: the larger the difference in electronegativity, the more polar the bond. Electronegativity differences are given in parentheses below:
    a. H–F (1.9); H–Cl (0.9); the H–F bond is more polar
    b. H–Cl (0.9); H–I (0.4); the H–Cl bond is more polar
    c. H–Br (0.7); H–Cl (0.9); the H–Cl bond is more polar
    d. H–I (0.4); H–Br (0.7); the H–Br bond is more polar

18. The *degree* of polarity of a polar covalent bond is indicated by the magnitude of the difference in electronegativities of the elements involved: the larger the difference in electronegativity, the more polar the bond. Electronegativity differences are given in parentheses below:
    a. O–Cl (0.5); O–Br (0.7); the O–Br bond is more polar
    b. N–O (0.5); N–F (1.0); the N–F bond is more polar
    c. P–S (0.4); P–O (1.4); the P–O bond is more polar
    d. H–O (1.4); H–N (0.9); the H–O bond is more polar

Chapter 12: Chemical Bonding

19. The larger the difference in electronegativity between two atoms, the more ionic character the bond possesses. Electronegativity differences are given in parentheses below:

    a. Na–F (3.1); Na–I (1.6); the Na–F bond has more ionic character

    b. Ca–S (1.5); Ca–O (2.5); the Ca–O bond has more ionic character

    c. Li–Cl (2.0); Cs–Cl(2.3); the Cs–Cl bond has more ionic character

    d. Mg–N (1.8); Mg–P (0.9); the Mg–N bond has more ionic character

20. The greater the electronegativity difference between two atoms, the more ionic will be the bond between those two atoms.

    a. Na–N

    b. K–P

    c. Na–Cl

    d. Mg–Cl

21. A dipole moment is an electrical effect that occurs in a molecule that has separate centers of positive and negative charge. The simplest examples of molecules with dipole moments would be diatomic molecules involving two different elements. For example:

    $\delta+$ C $\rightarrow$ O $\delta-$

    $\delta+$ N $\rightarrow$ O $\delta-$

    $\delta+$ Cl $\rightarrow$ F $\delta-$

    $\delta+$ Br $\rightarrow$ Cl $\delta-$

22. The presence of strong bond dipoles and a large overall dipole moment in water make it a polar substance overall. Among the properties of water dependent on its dipole moment are its freezing point, melting point, vapor pressure, and its ability to dissolve many substances.

23. In a diatomic molecule containing two different elements, the more electronegative atom will be the negative end of the molecule, and the *less* electronegative atom will be the positive end.

    a. chlorine

    b. oxygen

    c. fluorine

24. In a diatomic molecule containing two different elements, the more electronegative atom will be the negative end of the molecule, and the *less* electronegative atom will be the positive end.

    a. H

    b. Cl

    c. I

25. In the figures, the arrow points toward the more electronegative atom.

    a. $\delta+$ C $\rightarrow$ F $\delta-$

    b. $\delta+$ Si $\rightarrow$ C $\delta-$

Chapter 12: Chemical Bonding

c. δ+ C → O δ–
d. δ+ B → C δ–

26. In the figures, the arrow points toward the more electronegative atom.
    a. δ+ P → S δ–
    b. δ+ S → F δ–
    c. δ+ S → Cl δ–
    d. δ+ S → Br δ–

27. In the figures, the arrow points toward the more electronegative atom.
    a. δ+ Si → H δ–
    b. P–H The atoms have very nearly the same electronegativity, so there is a very small, if any, dipole moment.
    c. δ+ H → S δ–
    d. δ+ H → Cl δ–

28. In the figures, the arrow points toward the more electronegative atom.
    a. δ+ H → C δ–
    b. δ+ N → O δ–
    c. δ+ S → N δ–
    d. δ+ C → N δ–

29. When forming ionic bonds, the element forming the positive ion loses enough electrons to have the same configuration as the previous noble gas. The element forming the negative ion gains enough electrons to have the same configuration as the next noble gas. When forming covalent compounds, electrons are shared in such a way that as many atoms as possible in the molecule have a configuration analogous to a noble gas

30. preceding

31. gaining

32. Atoms in covalent molecules gain a configuration like that of a noble gas by sharing one or more pairs of electrons between atoms: such shared pairs of electrons "belong" to each of the atoms of the bond at the same time. In ionic bonding, one atom completely gives over one or more electrons to another atom, and the resulting ions behave independently of one another.

33. a. Chlorine is in Group 7 and has configuration $1s^2\, 2s^2\, 2p^6\, 3s^2\, 3p^5$. Chlorine would be expected to gain one electron to become the Cl⁻ ion. The Cl⁻ ion has an electron configuration analogous to the noble gas Ar: $1s^2\, 2s^2\, 2p^6\, 3s^2\, 3p^6$

b. Strontium is in Group 2 and has configuration $1s^2\ 2s^2\ 2p^6\ 3s^2\ 3p^6\ 4s^2\ 3d^{10}\ 4p^6\ 5s^2$. Strontium would be expected to lose two electrons to become the Sr^{2+} ion. The Sr^{2+} ion has an electron configuration analogous to the noble gas Kr: $1s^2\ 2s^2\ 2p^6\ 3s^2\ 3p^6\ 4s^2\ 3d^{10}\ 4p^6$.

c. Oxygen is in Group 6 and has electron configuration $1s^2\ 2s^2\ 2p^4$. Oxygen would be expected to gain two electrons to become the O^{2-} ion. The O^{2-} ion has an electron configuration analogous to the noble gas Ne: $1s^2\ 2s^2\ 2p^6$

d. Rubidium is in Group 1 and has electron configuration $1s^2\ 2s^2\ 2p^6\ 3s^2\ 3p^6\ 4s^2\ 3d^{10}\ 4p^6\ 5s^1$. Rubidium would be expected to lose one electron to become the Rb^+ ion. The Rb^+ ion has an electron configuration analogous to the noble gas Kr: $1s^2\ 2s^2\ 2p^6\ 3s^2\ 3p^6\ 4s^2\ 3d^{10}\ 4p^6$.

34. a. Br^-, Kr (Br has one electron less than Kr)

b. Cs^+, Xe (Cs has one electron more than Xe)

c. P^{3-}, Ar (P has three fewer electrons than Ar)

d. S^{2-}, Ar (S has two fewer electrons than Ar)

35. a. Na^+, Mg^{2+}, Al^{3+}; $1s^2\ 2s^2\ 2p^6$

b. Li^+, Be^{2+}; $1s^2$

c. K^+, Ca^{2+} from the A group elements; $1s^2\ 2s^2\ 2p^6\ 3s^2\ 3p^6$

d. Rb^+, Sr^{2+} from the A group elements; $1s^2\ 2s^2\ 2p^6\ 3s^2\ 3p^6\ 4s^2\ 3d^{10}\ 4p^6$

36. Atoms or ions with the same number of electrons are said to be *isoelectronic*.

a. Cl^-, S^{2-}, P^{3-}

b. F^-, O^{2-}, N^{3-}

c. Br^-, Se^{2-}, As^{3-}

d. I^-, Te^{2-}

37. a. Al_2S_3: Al has three electrons more than a noble gas; S has two electrons fewer than a noble gas.

b. RaO: Ra has two electrons more than a noble gas; O has two electrons fewer than a noble gas.

c. CaF_2: Ca has two electrons more than a noble gas; F has one electron less than a noble gas.

d. Cs_3N: Cs has one electron more than a noble gas; N has three electrons fewer than a noble gas.

e. RbP_3 : Rb has one electron more than a noble gas; P has three electrons fewer than a noble gas.

38. a. $AlBr_3$: Al has three electrons more than a noble gas; Br has one electron less than a noble gas.

b. Al_2O_3: Al has three electrons more than a noble gas; O has two fewer electrons than a noble gas.

c. AlP: Al has three electrons more than a noble gas; P has three fewer electrons than a noble gas.

d. AlH$_3$: Al has three electrons more than a noble gas; H has one electron less than a noble gas.

39.
a. Ba^{2+}, [Xe]; S^{2-}, [Ar]
b. Sr^{2+}, [Kr]; F$^-$, [Ne]
c. Mg^{2+}, [Ne]; O^{2-}, [Ne]
d. Al^{3+}, [Ne]; S^{2-}, [Ar]

40. There are many examples possible. Listed below are a few compounds that fit each situation.

a. LiF: Li$^+$, [He]; F$^-$, [Ne]
b. NaF: Na$^+$, [Ne]; F$^-$, [Ne]
c. LiCl: Li$^+$, [He]; Cl$^-$, [Ar]
d. NaCl: Na$^+$, [Ne]; Cl$^-$, [Ar]

41. The formula of an ionic compound represents only the smallest whole number ratio of positive and negative ions present (the empirical formula).

42. An ionic solid such as NaCl consists of an array of alternating positively– and negatively–charged ions: that is, each positive ion has as its nearest neighbors a group of negative ions, and each negative ion has a group of positive ions surrounding it. In most ionic solids, the ions are packed as tightly as possible.

43. Positive ions are always smaller than the atoms from which they are formed, because in forming the ion, the valence electron shell (or part of it) is "removed" from the atom.

44. In forming an anion, an atom gains additional electrons in its outermost (valence) shell. Additional electrons in the valence shell increases the repulsive forces between electrons, so the outermost shell becomes larger to accommodate this.

45. Relative ionic sizes are given in Figure 12.9. Within a given horizontal row of the periodic chart, negative ions tend to be larger than positive ions because the negative ions contain a larger number of electrons in the valence shell. Within a vertical group of the periodic table, ionic size increases from top to bottom. In general, positive ions are smaller than the atoms they come from, whereas negative ions are larger than the atoms they come from.

a. H
b. N
c. Al^{3+}
d. F

46. Relative ionic sizes are given in Figure 12.9. Within a given horizontal row of the periodic chart, negative ions tend to be larger than positive ions because the negative ions contain a larger number of electrons in the valence shell. Within a vertical group of the periodic table, ionic size increases from top to bottom. In general, positive ions are smaller than the atoms they come from,

Chapter 12: Chemical Bonding

whereas negative ions are larger than the atoms they come from. If two ions contain the same number of electrons, look at the nuclear charge. In general, the larger the nuclear charge, the smaller the ion (electrons drawn closer to the nucleus).

a. Mg
b. K^+
c. Br^-
d. Se^{2-}

47. Relative ionic sizes are given in Figure 12.9. Within a given horizontal row of the periodic chart, negative ions tend to be larger than positive ions because the negative ions contain a larger number of electrons in the valence shell. Within a vertical group of the periodic table, ionic size increases from top to bottom. In general, positive ions are smaller than the atoms they come from, whereas negative ions are larger than the atoms they come from.

a. Fe^{3+}
b. Cl
c. Al^{3+}

48. Relative ionic sizes are given in Figure 12.9. Within a given horizontal row of the periodic chart, negative ions tend to be larger than positive ions because the negative ions contain a larger number of electrons in the valence shell. Within a vertical group of the periodic table, ionic size increases from top to bottom. In general, positive ions are smaller than the atoms they come from, whereas negative ions are larger than the atoms they come from.

a. I
b. F^-
c. F^-

49. Valence electrons are those found in the outermost principal energy level of the atom. The valence electrons effectively represent the outside edge of the atom and are the electrons most influenced by the electrons of another atom.

50. When atoms form covalent bonds, they try to attain a valence electronic configuration similar to that of the following noble gas element. When the elements in the first few horizontal rows of the periodic table form covalent bonds, they will attempt to gain configurations similar to the noble gases helium (2 valence electrons, duet rule), and neon and argon (8 valence electrons, octet rule).

51. noble gas electronic configuration

52. These elements attain a total of eight valence electrons, making the valence electron configurations similar to those of the noble gases Ne and Ar.

53. When two atoms in a molecule are connected by a double bond, the atoms share two pairs of electrons (4 electrons) in completing their outermost shells. A simple molecule containing a double bond is ethene (ethylene), C_2H_4 ($H_2C::CH_2$).

Chapter 12: Chemical Bonding

54. When two atoms in a molecule are connected by a triple bond, the atoms share three pairs of electrons (6 electrons) in completing their outermost shells. A simple molecule containing a triple bond is acetylene, C_2H_2 (H:C:::C:H).

55. The Group in which a representative element is found indicates the number of valence electrons.

    a. 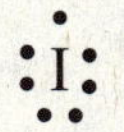 :Ï:

    b. 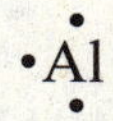 ·Al

    c.  :Xe:

    d. 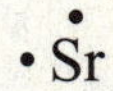 ·Sr

56. The Group in which a representative element is found indicates the number of valence electrons.

    a. 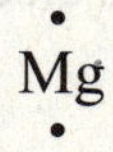 Mg

    b. 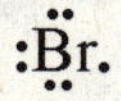 :B̈r·

    c.  :S̈·

    d. 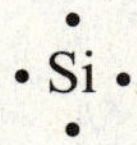 ·Si·

57. a. each N provides 5; O provides 6; total valence electrons = 16

     b. each B provides 3; each H provides 1: total valence electrons = 12

     c. each C provides 4; each H provides 1; total valence electrons = 20

     d. N provides 5; each Cl provides 7; total valence electrons = 26

58. a. each boron provides 3; each oxygen provides 6; total valence electrons = 24

     b. carbon provides 4; each oxygen provides 6; total valence electrons = 16

     c. each carbon provides 4; each hydrogen provides 1; oxygen provides 6; total valence electrons = 20

     d. N provides 5; each oxygen provides 6; total valence electrons = 17

59. a. Nitrogen provides 5 valence electrons; each bromine provides 7; total valence electrons = 26.

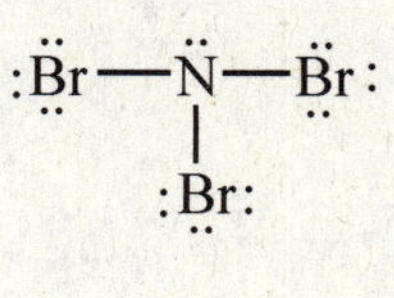

    :B̈r—N—B̈r:
          |
         :B̈r:

Chapter 12: Chemical Bonding

b. Hydrogen provides 1 valence electron; fluorine provides 7; total valence electrons = 8.

H—F̈:

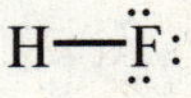

c. Carbon provides 4 valence electrons; each bromine provides 7;
total valence electrons = 32.

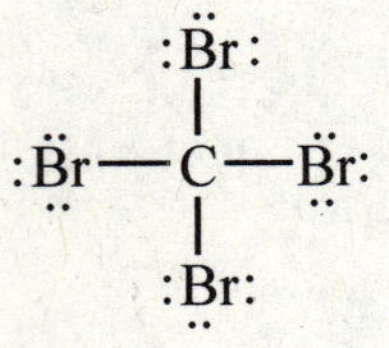

:B̈r:
|
:B̈r—C—B̈r:
|
:B̈r:

d. Each carbon provides 4 valence electrons; each hydrogen provides 1;
total valence electrons = 10.

H—C≡C—H

60. a. Each hydrogen provides 1 valence electron; sulfur provides 6 valence electrons; total valence electrons = 8

H—S̈—H

b. Each fluorine provides 7 valence electrons; silicon provides 4 valence electrons; total valence electrons = 32

:F̈:
|
:F̈—Si—F̈:
|
:F̈:

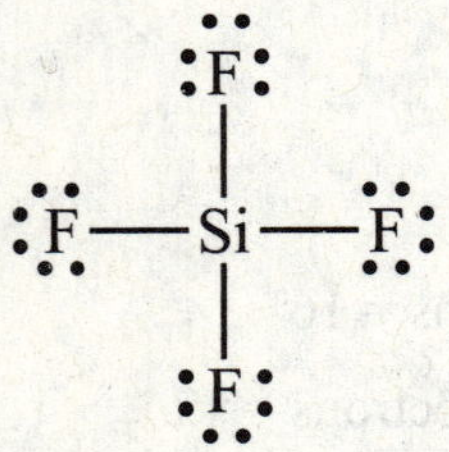

c. Each carbon provides 4 valence electrons; each hydrogen provides 1 valence electron; total valence electrons = 12

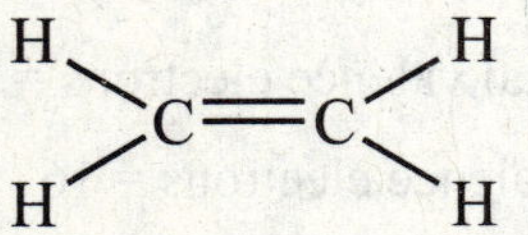

H₂C=CH₂

d. Each carbon provides 4 valence electrons; each hydrogen provides 1 valence electron; total valence electrons = 20

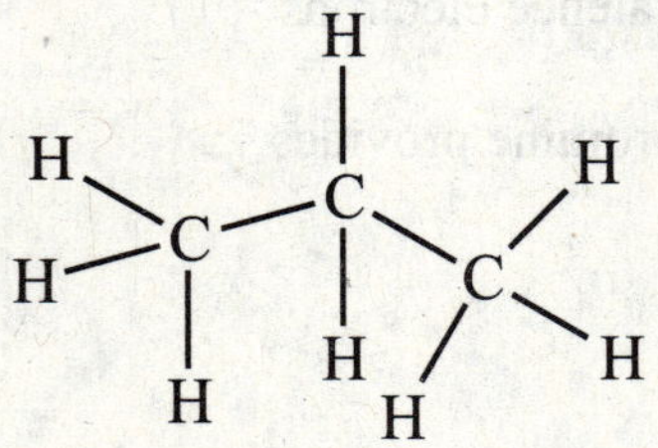

(CH₃-CH₂-CH₃ structure, propane)

Chapter 12: Chemical Bonding

61. a. Each C provides 4 valence electrons. Each H provides 1 valence electron.
Total valence electrons = 14

$$\begin{array}{c} H \quad H \\ | \quad | \\ H-C-C-H \\ | \quad | \\ H \quad H \end{array}$$

b. N provides 5 valence electrons. Each F provides 7 valence electrons.
Total valence electrons = 26

$$:\ddot{F}-\ddot{N}-\ddot{F}:$$
$$\quad \; |$$
$$\quad :\ddot{F}:$$

c. Each C provides 4 valence electrons. Each H provides 1 valence electron.
Total valence electrons = 26

$$\begin{array}{c} H \quad H \quad H \quad H \\ | \quad | \quad | \quad | \\ H-C-C-C-C-H \\ | \quad | \quad | \quad | \\ H \quad H \quad H \quad H \end{array}$$

d. Si provides 4 valence electrons. Each Cl provides 7 valence electrons.
Total valence electrons = 32

$$\begin{array}{c} :\ddot{C}l: \\ | \\ :\ddot{C}l-Si-\ddot{C}l: \\ | \\ :\ddot{C}l: \end{array}$$

62. a. P provides 5 valence electrons. Each Cl provides 7 valence electrons.
Total valence electrons = 26

$$:\ddot{C}l-\ddot{P}-\ddot{C}l:$$
$$\quad \; |$$
$$\quad :\ddot{C}l:$$

b. C provides 4 valence electrons. Each Cl provides 7 valence electrons. H provides 1 valence electron.
Total valence electrons = 26

$$\begin{array}{c} H \\ | \\ :\ddot{C}l-C-\ddot{C}l: \\ | \\ :\ddot{C}l: \end{array}$$

Chapter 12:   Chemical Bonding

c. Each C provides 4 valence electrons. Each H provides 1 valence electron. Each Cl provides 7 valence electrons
Total valence electrons = 26

$$:\ddot{\underset{..}{Cl}}-\underset{\underset{H}{|}}{\overset{\overset{H}{|}}{C}}-\underset{\underset{H}{|}}{\overset{\overset{H}{|}}{C}}-\ddot{\underset{..}{Cl}}:$$

d. Each N provides 5 valence electrons. Each H provides 1 valence electron.
Total valence electrons = 14

$$H-\underset{\underset{H}{|}}{\overset{..}{N}}-\underset{\underset{H}{|}}{\overset{..}{N}}-H$$

63. Here are three possible Lewis structures.

$$CH_3-\underset{\underset{:\ddot{O}:}{\|}}{S}-(CH_2)_4-\ddot{N}=C=\ddot{\underset{..}{S}}$$

$$CH_3-\underset{\underset{:\ddot{O}:}{\|}}{S}-(CH_2)_4-N\equiv C-\ddot{\underset{..}{S}}:$$

$$CH_3-\underset{\underset{:\ddot{O}:}{\|}}{S}-(CH_2)_4-\ddot{\underset{..}{N}}-C\equiv S:$$

64. C provides 4 valence electrons. Each oxygen provides 6 valence electrons. Having only 16 total valence electrons requires multiple bonding in the molecule.

$$:O\equiv C-\ddot{\underset{..}{O}}: \longleftrightarrow \ddot{\underset{..}{O}}=C=\ddot{\underset{..}{O}} \longleftrightarrow :\ddot{\underset{..}{O}}-C\equiv O:$$

65. a. S provides 6 valence electrons. Each O provides 6 valence electrons. The 2– charge means two additional valence electrons. Total valence electrons = 32

$$\left[\begin{array}{c} :\ddot{O}: \\ | \\ :\ddot{O}-S-\ddot{O}: \\ | \\ :\ddot{O}: \end{array} \right]^{2-}$$

Chapter 12: Chemical Bonding

b. P provides 5 valence electrons. Each O provides 6 valence electrons. The 3– charge means three additional valence electrons. Total valence electrons = 32.

$$\left[\begin{array}{c} :\ddot{O}: \\ | \\ :\ddot{O}-P-\ddot{O}: \\ | \\ :\ddot{O}: \end{array} \right]^{3-}$$

c. S provides 6 valence electrons. Each O provides 6 valence electrons. The 2– charge means two additional valence electrons. Total valence electrons = 26

$$\left[\begin{array}{c} :\ddot{O}: \\ | \\ :\ddot{O}-S-\ddot{O}: \end{array} \right]^{2-}$$

66. a. Cl provides 7 valence electrons. Each O provides 6 valence electrons. The 1– charge means 1 additional electron. Total valence electrons = 26

$$\left[\begin{array}{c} :\ddot{O}: \\ | \\ :\ddot{O}-\ddot{C}l-\ddot{O}: \end{array} \right]^{1-}$$

b. Each O provides 6 valence electrons. The 2– charge means two additional valence electrons. Total valence electrons = 14

$$\left[:\ddot{O}-\ddot{O}: \right]^{2-}$$

c. Each C provides 4 valence electrons. Each H provides 1 valence electron. Each O provides 6 valence electrons. The 1– charge means 1 additional valence electron. Total valence electrons = 24

$$\left[\begin{array}{c} H \quad :\ddot{O}: \\ | \quad \| \\ H-C-C-\ddot{O}: \\ | \\ H \end{array} \right]^{1-} \quad \left[\begin{array}{c} H \quad :\ddot{O}: \\ | \quad | \\ H-C-C=\ddot{O} \\ | \\ H \end{array} \right]^{1-}$$

67. a. Cl provides 7 valence electrons. Each O provides 6 valence electrons. The 1– charge means one additional valence electron.
Total valence electrons = 20

$$\left[:\ddot{O}-\ddot{C}l-\ddot{O}: \right]^{-}$$

Chapter 12: Chemical Bonding

b. Br provides 7 valence electrons. Each O provides 6 valence electrons. The 1– charge means one additional valence electron.
Total valence electrons = 32

$$\left[\begin{array}{c} :\!\ddot{\mathrm{O}}\!: \\ | \\ :\!\ddot{\mathrm{O}}\!-\!\mathrm{Br}\!-\!\ddot{\mathrm{O}}\!: \\ | \\ :\!\ddot{\mathrm{O}}\!: \end{array}\right]^{-}$$

c. C provides 4 valence electrons. N provides 5 valence electron. The 1– charge means one additional valence electron. Total valence electrons = 10

$$[:\!\mathrm{C}\!\equiv\!\mathrm{N}\!:]^{-}$$

68. a. C provides 4 valence electrons. Each O provides 6 valence electrons. The 2– charge means two additional valence electrons.
Total valence electrons = 24

$$\left[\begin{array}{c}:\!\mathrm{O}\!: \\ \| \\ \mathrm{C} \\ / \ \backslash \\ :\!\ddot{\mathrm{O}}\!: \quad :\!\ddot{\mathrm{O}}\!: \end{array}\right] \leftrightarrow \left[\begin{array}{c}:\!\ddot{\mathrm{O}}\!: \\ | \\ \mathrm{C} \\ \| \ \backslash \\ :\!\mathrm{O} \quad :\!\ddot{\mathrm{O}}\!: \end{array}\right] \leftrightarrow \left[\begin{array}{c}:\!\ddot{\mathrm{O}}\!: \\ | \\ \mathrm{C} \\ / \ \| \\ :\!\ddot{\mathrm{O}}\!: \quad \mathrm{O}\!: \end{array}\right]^{2-}$$

b. Each H provides 1 valence electron. N provides 5 valence electrons. The 1+ charge means one less valence electron.
Total valence electrons = 8

$$\left[\begin{array}{c} \mathrm{H} \\ | \\ \mathrm{H}\!-\!\mathrm{N}\!-\!\mathrm{H} \\ | \\ \mathrm{H} \end{array}\right]^{+}$$

c. Cl provides 7 valence electrons. O provides 6 valence electrons. The 1– charge means one additional valence electron. Total valence electrons = 14

$$[:\!\ddot{\mathrm{Cl}}\!-\!\ddot{\mathrm{O}}\!:]^{-}$$

69. The geometric structure of the water molecule is bent (or *V*-shaped). There are four pairs of valence electrons on the oxygen atom of water (two pairs are bonding pairs, two pairs are non-bonding lone pairs). The H–O–H bond angle in water is approximately 105°.

70. The geometric structure of NH_3 is that of a trigonal pyramid. The nitrogen atom of NH_3 is surrounded by four electron pairs (three are bonding, one is a lone pair). The H–N–H bond angle is somewhat less than 109.5° (due to the presence of the lone pair).

71. BF$_3$ is described as having a trigonal planar geometric structure. The boron atom of BF$_3$ is surrounded by only three pairs of valence electrons. The F–B–F bond angle in BF$_3$ is 120°.

72. The geometric structure of SiF$_4$ is that of a tetrahedron. The silicon atom of SiF$_4$ is surrounded by four bonding electron pairs. The F–Si–F bond angle is the characteristic angle of the tetrahedron, 109.5°.

73. The geometric structure of a molecule plays a very important part in its chemistry. For biological molecules, a slight change in the geometric structure of the molecule can completely destroy the molecule's usefulness to a cell, or can cause a destructive change in the cell.

74. The general molecular structure of a molecule is determined by (1) *how many electron pairs* surround the central atom in the molecule, and (2) which of those electron pairs are used for *bonding* to the other atoms of the molecule. Nonbonding electron pairs on the central atom do, however, cause minor changes in the bond angles, compared to the ideal regular geometric structure.

75. For a given atom, the positions of the other atoms bonded to it are determined by maximizing the angular separation of the valence electron pairs on the given atom to minimize the electron repulsions.

76. You will remember from high school geometry, that two points in space are all that is needed to define a straight line. A diatomic molecule represents two points (the nuclei of the atoms) in space.

77. One of the valence electron pairs of the ammonia molecule is a *lone* pair with no hydrogen atom attached. This lone pair is part of the nitrogen atom and does not enter the description of the molecule's overall shape (aside from influencing the H–N–H bond angles between the remaining valence electron pairs).

78. In NF$_3$, the nitrogen atom has *four* pairs of valence electrons, whereas in BF$_3$, there are only *three* pairs of valence electrons around the boron atom. The nonbonding electron pair on nitrogen in NF$_3$ pushes the three F atoms out of the plane of the N atom.

79. If you draw the Lewis structures for these molecules, you will see that each of the indicated atoms is surrounded by *four pairs* of electrons with a *tetrahedral* orientation of the electron pairs (the arrangement in H$_2$S is distorted because some electron pairs on the S atom are bonding and some are nonbonding).

80. a. four electron pairs in a tetrahedral arrangement with some lone-pair distortion

    b. four electron pairs in a tetrahedral arrangement with some lone-pair distortion

    c. four electron pairs in a tetrahedral arrangement

81. a. trigonal pyramidal (there is a lone pair on N)

    b. nonlinear, *V*-shaped (four electron pairs on Se, but only two atoms are attached to Se)

    c. tetrahedral (four electron pairs on Si, and four atoms attached)

Chapter 12: Chemical Bonding

82. a. tetrahedral

    b. trigonal pyramidal (there is a lone pair on P)

    c. non-linear, V-shaped (four electron pairs on O, but only two atoms are attached to O)

83. a. tetrahedral

    b. tetrahedral

    c. tetrahedral

84. a. basically tetrahedral around the P atom (the hydrogen atoms are attached to two of the oxygen atoms and do not affect greatly the geometrical arrangement of the oxygen atoms around the phosphorus)

    b. tetrahedral (4 electron pairs on Cl, and 4 atoms attached)

    c. trigonal pyramidal (4 electron pairs on S, and 3 atoms attached)

85. a. <109.5°

    b. <109.5°

    c. 109.5°

    d. 109.5°

86. a. approximately 109.5° (the molecule is V-shaped or nonlinear)

    b. approximately 109.5° (the molecule is trigonal pyramidal)

    c. 109.5°

    d. approximately 120° (the double bond makes the molecule flat)

87. The bond angles would be expected to be slightly less than the 109.5° angle expected for four electron pairs due to one of the electron pairs being non-bonding.

88. a. V-shaped; 120° (one lone pair on Se with two S atoms attached, one with a double bond)

    b. trigonal planar; 120° (three S atoms attached to Se, one with a double bond)

    c. V-shaped; 120° (one lone pair on S with two O atoms attached, one with a double bond)

    d. linear; 180° (two S atoms attached to C, each with a double bond)

89. Resonance is said to exist when more than one valid Lewis structure can be drawn for a molecule. The actual bonding and structure of such molecules is thought to be somewhere "in between" the various possible Lewis structures. In such molecules, certain electrons are delocalized over more than one bond.

90. double

91. repulsion

92. Letter *c* is the correct answer. N–N contains equal electron sharing so it is nonpolar covalent.

Chapter 12: Chemical Bonding

93. The bond with the larger electronegativity difference will be the more polar bond. See Figure 12.3 for electronegativities.

    a.   Br–F

    b.   As–O

    c.   Pb–C

94. The bond energy of a chemical bond is the quantity of energy required to break the bond and separate the atoms.

95. covalent

96. In each case, the element *higher up* within a group on the periodic table has the higher electronegativity.

    a.   Be

    b.   N

    c.   F

97. For a bond to be polar covalent, the atoms involved in the bond must have different electronegativities (must be of different elements).

    a.   polar covalent

    b.   covalent

    c.   covalent

    d.   polar covalent

98. For a bond to be polar covalent, the atoms involved in the bond must have different electronegativities (must be of different elements).

    a.   polar covalent (different elements)

    b.   *non*polar covalent (two atoms of the same element)

    c.   polar covalent (different elements)

    d.   *non*polar covalent (atoms of the same element)

99. Electronegativity differences are given in parentheses.

    a.   N–P (0.9); N–O (0.5); the N–P bond is more polar.

    b.   N–C (0.5); N–O (0.5); the bonds are of the same polarity.

    c.   N–S (0.5); N–C (0.5); the bonds are of the same polarity.

    d.   N–F (1.0); N–S (0.5); the N–F bond is more polar.

Chapter 12: Chemical Bonding

100. In a diatomic molecule containing two different elements, the more electronegative atom will be the negative end of the molecule, and the *less* electronegative atom will be the positive end.

    a. oxygen

    b. bromine

    c. iodine

101. In the figures, the arrow points toward the more electronegative atom.

    a. N–Cl: The atoms have very nearly the same electronegativity, so there is a very small, if any, dipole moment.

    b. $\delta+$ P→N $\delta-$

    c. $\delta+$ S→N $\delta-$

    d. $\delta+$ C→N $\delta-$

102. (d); The formula for calcium nitride is Ca_3N_2 consisting of Ca^{2+} and N^{3-}. The electron configuration for Ca^{2+} is: $1s^2\ 2s^2\ 2p^6\ 3s^2\ 3p^6$. Argon (Ar) is the noble gas with this same electron configuration. The electron configuration for N^{3-} is: $1s^2\ 2s^2\ 2p^6$. Neon (Ne) is the noble gas with this electron configuration.

103.
    a. Na^+
    b. I^-
    c. K^+
    d. Ca^{2+}
    e. S^{2-}
    f. Mg^{2+}
    g. Al^{3+}
    h. N^{3-}

104.
    a. Na_2Se: Na has one electron more than a noble gas; Se has two electrons fewer than a noble gas.

    b. RbF: Rb has one electron more than a noble gas; F has one electron less than a noble gas.

    c. K_2Te: K has one electron more than a noble gas; Te has two electrons fewer than a noble gas.

    d. BaSe: Ba has two electrons more than a noble gas; Se has two electrons fewer than a noble gas.

    e. KAt: K has one electron more than a noble gas; At has one electron less than a noble gas.

    f. FrCl: Fr has one electron more than a noble gas; Cl has one electron less than a noble gas.

Chapter 12: Chemical Bonding

105.   a. Ca^{2+} [Ar]; Br^- [Kr]
       b. Al^{3+} [Ne]; Se^{2-} [Kr]
       c. Sr^{2+} [Kr]; O^{2-} [Ne]
       d. K^+ [Ar]; S^{2-} [Ar]

106.   Relative ionic sizes are indicated in Figure 12.9.
       a. Na^+
       b. Al^{3+}
       c. F^-
       d. Na^+

107.   a.   He
            He:
       b.   Br
            :Br̈·
       c.   Sr
            ·
            Sr
            ·
       d.   Ne
            :N̈e:
            ··
       e.   I
            :Ï·
       f.   Ra
            ·
            Ra
            ·

108.   a.   H provides 1; N provides 5; each O provides 6; total valence electrons = 24
       b.   each H provides 1; S provides 6; each O provides 6; total valence electrons = 32
       c.   each H provides 1; P provides 5; each O provides 6; total valence electrons = 32
       d.   H provides 1; Cl provides 7; each O provides 6; total valence electrons = 32

109.   a. GeH_4: Ge provides 4 valence electrons. Each H provides 1 valence electron. Total valence electrons = 8

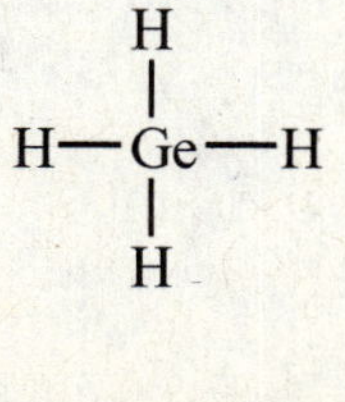

```
      H
      |
  H—Ge—H
      |
      H
```

255

Chapter 12: Chemical Bonding

    b.   ICl: I provides 7 valence electrons. Cl provides 7 valence electrons.
          Total valence electrons = 14

$$:\ddot{\underset{..}{I}}-\ddot{\underset{..}{Cl}}:$$

    c.   NI$_3$: N provides 5 valence electrons. Each I provides 7 valence electrons.
          Total valence electrons = 26

    d.   PF$_3$: P provides 5 valence electrons. Each F provides 7 valence electrons.
          Total valence electrons = 26

110.   a.   N$_2$H$_4$: Each N provides 5 valence electrons. Each H provides 1 valence electron.
          Total valence electrons = 14

    b.   C$_2$H$_6$: Each C provides 4 valence electrons. Each H provides 1 valence electron.
          Total valence electrons = 14

    c.   NCl$_3$: N provides 5 valence electrons. Each Cl provides 7 valence electrons.
          Total valence electrons = 26

    d.   SiCl$_4$: Si provides 4 valence electrons. Each Cl provides 7 valence electrons.
          Total valence electrons = 32

111. a. SO_2: S provides 6 valence electrons. Each O provides 6 valence electrons. Total valence electrons = 18

$$\ddot{\underset{}{O}}=\ddot{S}-\ddot{\underset{..}{O}}: \longleftrightarrow :\ddot{\underset{..}{O}}-\ddot{S}=\ddot{\underset{}{O}}$$

b. N_2O: Each N provides 5 valence electrons. O provides 6 valence electrons. Total valence electrons = 16

$$:N\equiv N-\ddot{\underset{..}{O}}: \longleftrightarrow :\ddot{N}=N=\ddot{O}: \longleftrightarrow :\ddot{\underset{..}{N}}-N\equiv O:$$

c. O_3: Each O provides 6 valence electrons. Total valence electrons = 18

$$\ddot{\underset{}{O}}=\ddot{O}-\ddot{\underset{..}{O}}: \longleftrightarrow :\ddot{\underset{..}{O}}-\ddot{O}=\ddot{\underset{}{O}}$$

112. a. NO_3^-: N provides 5 valence electrons. Each O provides 6 valence electrons. The 1– charge means one additional valence electron. Total valence electrons = 24

$$\left[\ddot{\underset{}{O}}=\underset{\underset{:\ddot{O}:}{|}}{N}-\ddot{\underset{..}{O}}:\right]^{1-} \longleftrightarrow \left[:\ddot{\underset{..}{O}}-\underset{\underset{:\ddot{O}:}{|}}{N}=\ddot{\underset{}{O}}\right]^{1-} \longleftrightarrow \left[\ddot{\underset{..}{O}}-\underset{\underset{:\ddot{O}:}{\|}}{N}-\ddot{\underset{..}{O}}:\right]^{1-}$$

b. CO_3^{2-}: C provides 4 valence electrons. Each O provides 6 valence electrons. The 2– charge means two additional valence electrons. Total valence electrons = 24

$$\left[\ddot{\underset{}{O}}=\underset{\underset{:\ddot{O}:}{|}}{C}-\ddot{\underset{..}{O}}:\right]^{2-} \longleftrightarrow \left[:\ddot{\underset{..}{O}}-\underset{\underset{:\ddot{O}:}{|}}{C}=\ddot{\underset{}{O}}\right]^{2-} \longleftrightarrow \left[\ddot{\underset{..}{O}}-\underset{\underset{:\ddot{O}:}{\|}}{C}-\ddot{\underset{..}{O}}:\right]^{2-}$$

c. NH_4^+: N provides 5 valence electrons. Each H provides 1 valence electron. The 1+ charge means one *less* valence electron. Total valence electrons = 8

$$\left[\begin{array}{c} H \\ | \\ H-N-H \\ | \\ H \end{array}\right]^+$$

113. Beryllium atoms only have *two* valence electrons. In BeF_2, there are single bonds between the beryllium atom and each fluorine atom. The beryllium atom of BeF_2 thus has two electron pairs around it, that lie 180° apart from one another. For the water molecule, in addition to the bonding pairs of electrons that attach the hydrogen atoms to the oxygen atoms, there are two nonbonding pairs of electrons that affect the H–O–H bond angle.

114. a. four electron pairs arranged tetrahedrally about C

b. four electron pairs arranged tetrahedrally about Ge

c. three electron pairs arranged trigonally (planar) around B

115. a. nonlinear (*V*-shaped, due to lone pairs on O)

b. nonlinear (*V*-shaped, due to lone pairs on O)

c. tetrahedral

Chapter 12: Chemical Bonding

116.   a. ClO_3^-, trigonal pyramid (lone pair on Cl)

      b. ClO_2^-, nonlinear (*V*-shaped, two lone pairs on Cl)

      c. ClO_4^-, tetrahedral (all pairs on Cl are bonding)

117.   a.   < 109.5° (molecule is nonlinear, *V*-shaped)

      b.   109.5° (molecule is tetrahedral)

      c.   180° (molecule is linear)

      d.   120° (molecule is trigonal planar)

118.   a.   nonlinear (*V*–shaped)

      b.   trigonal planar

      c.   basically trigonal planar around the C (the H is attached to one of the O atoms, and distorts the shape around the carbon only slightly)

      d.   linear

119.   a.   trigonal planar

      b.   basically trigonal planar around the N (the H is attached to one of the O atoms, and distorts the shape around the nitrogen only slightly)

      c.   nonlinear (*V*-shaped)

      d.   linear

120. Ionic compounds tend to be hard, crystalline substances with relatively high melting and boiling points. Covalently bonded substances tend to be gases, liquids, or relatively soft solids, with much lower melting and boiling points.

121. In a covalent bond between two atoms of the same element, the electron pair is shared equally and the bond is nonpolar; with a bond between atoms of different elements, the electron pair is unequally shared and the bond is polar (assuming the elements have different electronegativities).

122. Generally, covalent bonds between atoms of *different* elements are *polar*.

      a.   nonpolar covalent

      b.   ionic

      c.   ionic

      d.   polar covalent

      e.   polar covalent

      f.   nonpolar covalent

      g.   nonpolar covalent

      h.   ionic

123. In general, an element farther to the right in a given period or an element closer to the top of a given group is more electronegative.

      P and Cl: Cl

Ca and N: N

N and As: N

124. O–F, P–Cl, P–F, Si–F; The greater the electronegativity difference between two atoms, the more polar the bond between those two atoms.

125. Relative ionic sizes are given in Figure 12.9. Within a given horizontal row of the periodic chart, negative ions tend to be larger than positive ions because the negative ions contain a larger number of electrons in the valence shell. Within a vertical group of the periodic table, ionic size increases from top to bottom. In general, positive ions are smaller than the atoms they come from, whereas negative ions are larger than the atoms they come from. If two ions contain the same number of electrons, look at the nuclear charge. In general, the larger the nuclear charge, the smaller the ion (electrons drawn closer to the nucleus).

   a. $O^{2-} > O^- > O$

   b. $Fe^{2+} > Ni^{2+} > Zn^{2+}$

   c. $Cl^- > K^+ > Ca^{2+}$

126. $Na^+: 1s^2 2s^2 2p^6$

   $K^+: 1s^2 2s^2 2p^6 3s^2 3p^6$

   $Li^+: 1s^2$

   $Cs^+: 1s^2 2s^2 2p^6 3s^2 3p^6 4s^2 3d^{10} 4p^6 5s^2 4d^{10} 5p^6$

127. (a), (b), and (d)

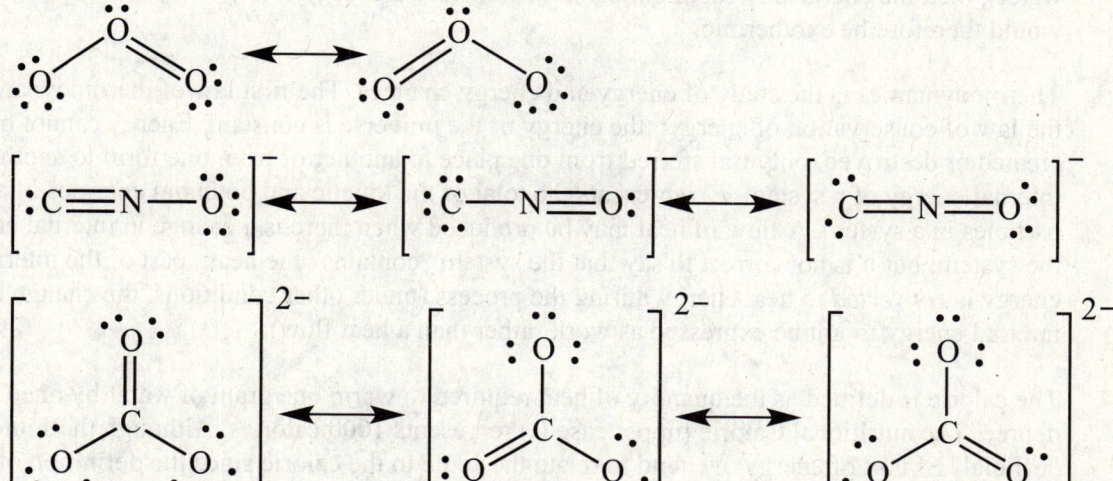

128.

Formula	Compound Name	Molecular Structure
CO_2	carbon dioxide	linear
NH_3	ammonia	trigonal pyramidal
SO_3	sulfur trioxide	trigonal planar
H_2O	water	bent or V-shaped
ClO_4^-	perchlorate ion	tetrahedral

CUMULATIVE REVIEW

Chapters 10, 11, and 12

1. Potential energy is energy due to the position of an object or its composition. Kinetic energy is the energy an object possesses due to its mass and its velocity. The law of conservation of energy states that energy cannot be created or destroyed, but can only be converted from one form to another. Work represents a force acting over a distance. A state function is a property of a system that changes independently of the pathway used to produce the change.

2. Temperature is a measure of the random motions of the components of a substance: in other words, temperature is a measure of the average kinetic energy of the particles in a sample. The molecules in warm water must be moving faster than the molecules in cold water (the molecules have the same mass, so if the temperature is higher, the average velocity of the particles must be higher in the warm water). Heat is the energy that flows because of a difference in temperature.

3. The system represents what we want to focus upon: typically, this means the chemicals in the reaction. The surroundings represent everything else in the universe, such as the water in which the chemicals of interest are dissolved. An endothermic reaction is one in which energy flows into the system from the surroundings. If the temperature of the surroundings increases (the water), then the chemicals reacting must have transferred energy to the surroundings: the reaction would therefore be exothermic.

4. Thermodynamics is the study of energy and energy changes. The first law of thermodynamics is the law of conservation of energy: the energy of the universe is constant. Energy cannot be created or destroyed, only transferred from one place to another or from one form to another. The internal energy of a system, E, represents the total of the kinetic and potential energies of all the particles in a system. A flow of heat may be produced when there is a change in internal energy in the system, but it is not correct to say that the system "contains" the heat: part of the internal energy is *converted* to heat energy during the process (under other conditions, the change in internal energy might be expressed as work rather than a heat flow).

5. The calorie is defined as the quantity of heat required to warm one gram of water by one Celsius degree. The nutritional Calorie (upper case C) represents 1000 calories. Although the Joule is the "official" SI unit of energy, we tend to relate the Joule to the calorie since the definition of the calorie is experimental and easy to comprehend: 1 calorie = 4.184 Joules. The specific heat capacity of a substance represents the quantity of heat required to warm one gram of the substance by one Celsius degree (which means that the specific heat capacity of water must be 1.00 calorie/J °C since that is also our definition of the calorie). Water's specific heat capacity is higher than most substances, and given its other properties, water is often used for cooling purposes.

6. The enthalpy change represents the heat energy that flows (at constant pressure) on a molar basis when a reaction occurs. The enthalpy change is indeed a state function (which we make great use of in Hess's Law calculations). Enthalpy changes are typically measured in insulated reaction vessels called calorimeters (a simple calorimeter is shown in Figure 10.6 in the text).

Cumulative Review: Chapters 10, 11, and 12

7. Hess's Law makes use of the fact that enthalpy is a state function and of the first law of thermodynamics. If it is possible to combine algebraically two or more chemical equations in such a way as to generate a desired chemical equation, then the heat of the desired chemical equation may be obtained from the heats of the given equations by combining their heats in the same algebraic way the chemical equations were combined. In this example, the desired chemical equation may be obtained by adding together the two given equations, combining terms, then dividing by four. Therefore the heat for the desired equation may be calculated by adding the heats of reaction for the two given equations and then dividing the sum by four.

$$P_4(s) + 6Cl_2(g) \rightarrow 4PCl_3(g)$$

$$4PCl_5(g) \rightarrow P_4(s) + 10Cl_2(g)$$

$$\cancel{P_4(s)} + \cancel{6Cl_2(g)} + 4PCl_5(g) \rightarrow 4PCl_3(g) + \cancel{P_4(s)} + \cancel{10}4Cl_2(g)$$

dividing both sides of the equation by four gives: $PCl_5(g) \rightarrow PCl_3(g) + Cl_2(g)$

$$\Delta H = \frac{[-2.44 \times 10^3 \text{ kJ} + 3.43 \times 10^3 \text{ kJ}]}{4} = 248 \text{ kJ}$$

8. Consider petroleum. A gallon of gasoline contains concentrated, stored energy. We can use that energy to make our car move, but when we do, the energy stored in the gasoline is dispersed to the environment. Although the energy is still there (it is conserved), it is no longer in a concentrated useful form. So although the energy content of the universe remains constant, the energy that is now stored in concentrated forms in oil, coal, wood, and other sources is gradually being dispersed to the universe where it can do no work.

9. Petroleum and natural gas both consist primarily of hydrocarbons: chains of carbon atoms with hydrogen atoms attached to the chains. Natural gas typically consists of molecules with 4 or fewer carbon atoms, with petroleum representing larger chains of carbon atoms. The various fractions into which petroleum is separated are based on the approximate number of carbon atoms in the chains. Gasoline, for example, consists of small chains of 5–10 carbon atoms (which makes it a volatile liquid); home heating oil consists of the fraction with 15–25 carbon atoms in the chains (which makes the oil thicker and less volatile); asphalt consists of the fraction with the longest chains, which makes it a non-volatile semi-solid useful for paving roads. "Cracking" was a process invented to produce a larger fraction of gasoline from crude petroleum: longer chains of carbon atoms are broken down during cracking to give more of the shorter chains characteristic of gasoline. Tetraethyl lead was used as an additive for gasoline to prevent engines from knocking: as engine technology has improved, the need for tetraethyl lead has decreased dramatically. The greenhouse effect considers that some gases prevent heat energy absorbed by the earth from the sun from being radiated back into space. The presence of water vapor in the atmosphere moderates the temperature of the earth, preventing heat from being radiated back into space at night. But as combustion of fossil fuels increases, the concentration of carbon dioxide in the atmosphere is increasing: carbon dioxide also prevents heat from being radiated back to space, and if there is too much carbon dioxide in the atmosphere, the earth's average temperature will increase. It is feared that an increase in the earth's average temperature may eventually produce severe climatological changes (such as changes in ocean currents, melting of the polar icecaps, etc.).

Cumulative Review: Chapters 10, 11, and 12

10. A "driving force" is an effect that tends to make a process occur. Two important driving forces are dispersion of energy during a process or dispersion of matter during a process ("energy spread" and "matter spread"). For example, a log burns in a fireplace because the energy contained in the log is dispersed to the universe when it burns. If we put a teaspoon of sugar into a glass of water, the dissolving of the sugar is a favorable process because the matter of the sugar is dispersed when it dissolves. Entropy is a measure of the randomness or disorder in a system. The entropy of the universe is constant increasing because of "matter spread" and "energy spread". A spontaneous process is one that occurs without outside intervention: the spontaneity of a reaction depends on the energy spread and matter spread if the reaction takes place. A reaction that disperses energy and also disperses matter will always be spontaneous. Reactions that require an input of energy may still be spontaneous if the matter spread is large enough.

11. $Q = s \times m \times \Delta T$ and so $\Delta T = Q/(m \times s)$; because the specific heat capacity is found in the denominator of the fraction, the substance with the smallest heat capacity will demonstrate the largest temperature change. The sample of gold will undergo a temperature change of slightly over 38°C.

$$\Delta T = Q/(m \times s) = \frac{125 \text{ J}}{25 \text{ g} \times 0.13 \text{ J/g °C}} = 38.5\text{°C} \approx 38\text{°C}$$

12. molar mass CH_4 = 16.04 g

    a. $0.521 \text{ mol} \times \dfrac{-890 \text{ kJ}}{1 \text{ mol}} = -464 \text{ kJ}$

    b. $1.25 \text{ g} \times \dfrac{1 \text{ mol}}{16.04 \text{ g}} \times \dfrac{-890 \text{ kJ}}{1 \text{ mol}} = -69.4 \text{ kJ}$

    c. $-1250 \text{ kJ} \times \dfrac{1 \text{ mol}}{-890 \text{ kJ}} = 1.40 \text{ mol } (22.5 \text{ g})$

13. Electromagnetic radiation represents the propagation of energy through space in the form of waves. Visible light, radio and television transmissions, microwaves, x-rays, and radiant heat are all examples of electromagnetic radiation. Waves of electromagnetic radiation have several characteristic properties. The wavelength (λ) represents the distance between two corresponding points (peaks or troughs) on consecutive waves and is measured in units of length (meters, centimeters, etc.). The frequency (ν) of electromagnetic radiation represents the number of complete waves that pass a given point in space per second and is measured in waves/sec (Hertz). A representative wave is depicted in Figure 11.3 in the text. The speed (c) or propagation velocity of electromagnetic radiation represents the speed at which a given wave moves through space, and is equal to 3×10^8 m/sec in a vacuum. For electromagnetic radiation, these three properties are related by the formula $\lambda \times \nu = c$.

14. An atom is said to be in its ground state when it is in its lowest possible energy state. When an atom possesses more energy than its ground state energy, the atom is said to be in an excited state. An atom is promoted from its ground state to an excited state by absorbing energy; when the atom returns from an excited state to its ground state it emits the excess energy as electromagnetic radiation. Atoms do not gain or emit radiation randomly, but rather do so only in discrete bundles of radiation called photons. The photons of radiation emitted by atoms are characterized by the wavelength (color) of the radiation: longer wavelength photons carry less

energy than shorter wavelength photons. The energy of a photon emitted by an atom corresponds exactly to the difference in energy between two allowed energy states in an atom: thus, we can use an observable phenomenon (emission of light by excited atoms), to gain insight into the energy changes taking place within the atom.

15. Excited atoms definitely do *not* emit their excess energy in a random or continuous manner. Rather, an excited atom of a given element emits only discrete photons of characteristic wavelength and energy when going back to its ground state. As an excited atom of a given element always emits photons of exactly the same energy, we take this to mean that the internal structure of the atom is such that there are only certain discrete allowed energy states for the electrons in atoms, and that the wavelengths of radiation emitted by an atom correspond to the exact energy differences between these allowed energy states. For example, excited hydrogen atoms always display the visible spectrum shown in Figure 11.11 of the text. We take this spectrum as evidence for the existence of only certain discrete energy levels within the hydrogen atom, and we describe this by saying that the energy levels of hydrogen are quantized. Previously, scientists had thought that atoms emitted energy continuously.

16. Bohr pictured the electron moving in only certain circular orbits around the nucleus. Each particular orbit (corresponding to a particular distance from the nucleus) had associated with it a particular energy (resulting from the attraction between the nucleus and the electron). When an atom absorbs energy, the electron moves from its ground state in the orbit closest to the nucleus ($n = 1$) to an orbit farther away from the nucleus ($n = 2, 3, 4, ...$). When an excited atom returns to its ground state, corresponding to the electron moving from an outer orbit to the orbit nearest the nucleus, the atom emits the excess energy as radiation. As the Bohr orbits are of fixed distances from the nucleus and from each other, when an electron moves from one fixed orbit to another, the energy change is of a definite amount. This corresponds to a photon being emitted of a particular characteristic wavelength and energy. The original Bohr theory worked very well for hydrogen: Bohr even predicted emission wavelengths for hydrogen that had not yet been seen, but were subsequently found at the exact wavelengths Bohr had calculated. However, when the simple Bohr model for the atom was applied to the emission spectra of other elements, the theory could not predict or explain the observed emission spectra.

17. De Broglie and Schrödinger took the new idea that electromagnetic radiation behaved as if it were a stream of small particles (photons), as well as a traditional wave, and basically reversed the premise. That is, if something that had previously been considered to be entirely wave-like also had a particle-like nature, then perhaps small particles also have a wave-like nature under some circumstances. The wave-mechanical theory for atomic structure developed by Schrödinger provided an exact model for the structure of the hydrogen atom which was consistent with all the observed properties of the hydrogen atom. Unlike Bohr's theory, however, which failed completely for atoms other than hydrogen, the wave-mechanical model was able to be extended to describe other atoms with considerable success. Rather than the fixed "orbits" that Bohr had postulated, the wave-mechanical model for the atom pictures the electrons of an atom as being distributed in regions of space called **orbitals**. The wave-mechanical model for the atom does not describe in classical terms the exact motion or trajectory of an electron as it moves around the nucleus, but rather predicts the probability of finding the electron in a particular location within the atom. The orbitals that constitute the solutions to the mathematical formulation of the wave-mechanical model for the atom represent probability contour maps for finding the electrons. When we draw a particular picture of a given orbital, we are saying that there is a 90% probability of finding the electron within the region indicated in the drawing.

Cumulative Review: Chapters 10, 11, and 12

18. The lowest energy hydrogen atomic orbital is called the $1s$ orbital. The $1s$ orbital is spherical in shape (that is, the electron density around the nucleus is uniform in all directions from the nucleus). The $1s$ orbital represents a probability map of electron density around the nucleus for the first principal energy level. The orbital does not have a sharp edge (it appears fuzzy) because the probability of finding the electron does not drop off suddenly with distance from the nucleus. The orbital does not represent just a spherical surface on which the electron moves (this would be similar to Bohr's original theory). When we draw a picture to represent the $1s$ orbital we are indicating that the probability of finding the electron within this region of space is greater than 90%. We know that the likelihood of finding the electron within this orbital is very high, but we still don't know exactly where in this region the electron is at a given instant in time.

19. When an atom (hydrogen, for example) absorbs energy, the electron moves to a higher energy state, which corresponds in the wave-mechanical model to a different type of orbital. The orbitals are arranged in a hierarchy of principal energy levels, as well as sub-levels of these principal levels. The principal energy levels (for hydrogen) correspond fairly well with the "orbits" of the Bohr theory and are designated by an integer, n, called the "principal quantum number ($n = 1, 2, 3, ...$). In the wave-mechanical model, however, these principal energy levels are further subdivided into sets of equivalent orbitals called subshells. For example, the $n = 2$ principal level of hydrogen is further subdivided into an s and a p subshell (indicated as the $2s$ and $2p$ subshells, respectively). These subshells, in turn, consist of the individual orbitals in which the electrons reside. The $2s$ subshell consists of the single, spherically-shaped $2s$ orbital, whereas the $2p$ subshell consists of a set of three equivalent, dumbbell-shaped $2p$ orbitals (often designated as $2p_x$, $2p_y$, and $2p_z$ to indicated their orientation in space). Similarly, the $n = 3$ principal energy level of hydrogen is subdivided into three subshells: the $3s$ subshell (1 orbital), the $3p$ subshell (a set of three orbitals), and the $3d$ subshell (a set of five orbitals).

20. The third principal energy level of hydrogen is divided into three sublevels, the $3s$, $3p$, and $3d$ sublevels. The $3s$ subshell consists of the single $3s$ orbital: like the other s orbitals, the $3s$ orbital is spherical in shape. The $3p$ subshell consists of a set of three equal-energy $3p$ orbitals: each of these $3p$ orbitals has the same shape ("dumbbell"), but each of the $3p$ orbitals is oriented in a different direction in space. The $3d$ subshell consists of a set of five $3d$ orbitals: the $3d$ orbitals have the shapes indicated in Figure 11.28, and are oriented in different directions around the nucleus (students sometimes say that the $3d$ orbitals have the shape of a 4-leaf clover). The fourth principal energy level of hydrogen is divided into four sublevels, the $4s$, $4p$, $4d$, and $4f$ orbitals. The $4s$ subshell consists of the single $4s$ orbital. The $4p$ subshell consists of a set of three $4p$ orbitals. The $4d$ subshell consists of a set of five $4d$ orbitals. The shapes of the $4s$, $4p$, and $4d$ orbitals are the same as the shapes of the orbitals of the third principal energy level (the orbitals of the fourth principal energy level are larger and further from the nucleus than the orbitals of the third level, however). The fourth principal energy level, because it is further from the nucleus, also contains a $4f$ subshell, consisting of seven $4f$ orbitals (the shapes of the $4f$ orbitals are beyond the scope of this text).

21. In simple terms, in addition to moving around the nucleus of the atom, an electron also rotates (spins) on its own axis. There are only two ways a body spinning on its own axis can rotate: in pre-digital watch days, these were often described as "clockwise" and "counter-clockwise" motions, but basically this just means to the right or to the left (for example, the earth is rotating on its axis to the right). As there are only two possible orientations for an electron's intrinsic spin, this results in a given orbital being only able to accommodate two electrons (one spinning in each direction). The Pauli exclusion principle summarizes our theory about intrinsic electron spin: a given atomic orbital can hold a maximum of two electrons, and those two electrons must have opposite spins.

Cumulative Review: Chapters 10, 11, and 12

22. Atoms have a series of principal energy levels symbolized by the letter n. The $n = 1$ level is the closest to the nucleus, and the energies of the levels increase as the value of n increases going out from the nucleus. Each principal energy level is divided into a set of sublevels of different characteristic shapes (designated by the letters s, p, d, and f). Each sublevel is further subdivided into a set of orbitals: each s subshell consists of a single s orbital; each p subshell consists of a set of three p orbitals; each d subshell consists of a set of five d orbitals; etc. A given orbital can be empty or it can contain one or two electrons, but never more than two electrons (if an orbital contains two electrons, then the electrons must have opposite intrinsic spins). The shape we picture for an orbital represents only a probability map for finding electrons: the shape does not represent a trajectory or pathway for electron movements.

23. The order in which the orbitals fill with electrons is indicated in Figure 11.32 in the text. This figure is especially useful because it also shows the specific number of orbitals of each type. As the first principal level consists only of the $1s$ orbital, the $n = 1$ level can contain only two electrons. As the second principal level consists of the $2s$ orbital and the set of three $2p$ orbitals, so the $n = 2$ level can hold a maximum of $[2 + 3(2)] = 8$ electrons. Since any p subshell consists of three orbitals, a given p subshell can hold a maximum of six electrons. A particular p orbital (like any orbital) can hold only two electrons (of opposite spin). The three orbitals within a given p subshell are of exactly the same energy and differ only in their orientation in space. Therefore when we write the electron configuration of an element like N or O that has a partially-filled p-subshell, we place the electrons in separate p-orbitals to minimize the inter-electronic repulsion. Thus, the configuration of nitrogen is written sometimes as $1s^2\ 2s^2\ 2p_x^1\ 2p_y^1\ 2p_z^1$ to emphasize this.

24. The valence electrons are the electrons in an atom's outermost shell. The valence electrons are those most likely to be involved in chemical reactions because they are at the outside edge of the atom.

25. See the inside front cover of the text. An element is placed into a particular position in the periodic table because of its electronic configuration, in particular, the number and arrangement of the valence electrons. Since all elements in a vertical group have similar valence electron configurations, they behave similarly in chemical reactions.

26. From the column and row location of an element, you should be able to determine what the valence shell of an element has for its electronic configuration. For example, the element in the third horizontal row, in the second vertical column, has $3s^2$ as its valence configuration. We know that the valence electrons are in the $n = 3$ shell because the element is in the third horizontal row. We know that the valence electrons are s electrons because the first two electrons in a horizontal row are always in an s subshell. We know that there are two electrons because the element is the second element in the horizontal row. As an additional example, the element in the seventh vertical column of the second horizontal row in the periodic table has valence configuration $2s^2 2p^5$.

27. The representative elements are the elements in Groups 1–8 of the periodic table whose valence electrons are in s and p subshells: there are two groups of representative elements at the left side of the periodic table (corresponding to the s subshells) and six groups of representative elements at the right side of the periodic table (corresponding to the p subshells). Metallic character is largest at the left-hand end of any horizontal row of the periodic table, and largest at the bottom of any vertical group. Overall, these two trends mean that the most metallic elements are at the lower left of the periodic table, and the least metallic (i.e., the nonmetals) are at the top right of

Cumulative Review: Chapters 10, 11, and 12

the periodic table. The metalloids, which have both metallic and nonmetallic properties are found in the "stairstep" region indicated on most periodic tables.

28. The ionization energy of an atom represents the energy required to remove an electron from the atom. As one goes from top to bottom in a vertical group in the periodic table, the ionization energies decrease (it becomes easier to remove an electron). As one goes down within a group, the valence electrons are farther and farther from the nucleus and are less tightly held. The ionization energies increase when going from left to right within a horizontal row within the periodic table. The left-hand side of the periodic table is where the metallic elements are found, which lose electrons relatively easily. The right-hand side of the periodic table is where the nonmetallic elements are found: rather than losing electrons, these elements tend to gain electrons. Within a given horizontal row in the periodic table, the valence electrons are all in the same principal energy shell: however, as you go from left to right in the horizontal row, the nuclear charge that holds onto the electrons is increasing one unit with each successive element, making it that much more difficult to remove an electron. The relative sizes of atoms also vary systematically with the location of an element in the periodic table. Within a given vertical group, the atoms get progressively larger when going from the top of the group to the bottom: the valence electrons of the atoms are in progressively higher principal energy shells (and are progressively further from the nucleus) as we go down in a group. In going from left to right within a horizontal row in the periodic table, the atoms get progressively smaller. Although all the elements in a given horizontal row in the periodic table have their valence electrons in the same principal energy shell, the nuclear charge is progressively increasing from left to right, making the given valence shell progressively smaller as the electrons are drawn more closely to the nucleus.

29. A chemical bond is a force that holds two or more atoms together and makes them function as a unit. The strength of a chemical bond is commonly described in terms of the bond energy, which is the quantity of energy required to break the bond. The principal types of chemical bonding are ionic bonding, pure covalent bonding, and polar covalent bonding.

30. Ionic bonding results when elements of very different electronegativities react with each other. Typically a metallic element reacts with a nonmetallic element; the metallic element losing electrons and forming positive ions and the nonmetallic element gaining electrons and forming negative ions. Sodium chloride, NaCl, is an example of a typical ionic compound. The aggregate form of such a compound consists of a crystal lattice of alternating positively and negatively charged ions. A given positive ion is attracted by several surrounding negatively charged ions, and a given negative ion is attracted by several surrounding positively charged ions. Similar electrostatic attractions go on in three dimensions throughout the crystal of ionic solid, leading to a very stable system (with very high melting and boiling points, for example). We know that ionic-bonded solids do not conduct electricity in the solid state (because the ions are held tightly in place by all the attractive forces), but such substances are strong electrolytes when melted or when dissolved in water (either process sets the ions free to move around).

31. A bond between two atoms is, in general, covalent if the atoms share a pair of electrons in mutually completing their valence electron shells. Covalent bonds can be subclassed as to whether they are pure (nonpolar) covalent or are polar covalent. In a nonpolar covalent bond, two atoms of the same electronegativity (often this means the same type of atom) equally share a pair of electrons: the electron cloud of the bond is symmetrically distributed along the bond axis. In a polar covalent bond, one of the atoms of the bond has a higher electronegativity than the other atom, and draws the shared pair of electrons more closely towards itself (pulling the electron cloud along the bond axis closer to the more electronegative atom). Because the more

electronegative atom of a polar covalent bond ends up with a higher electron density than normal, there is a center of partial negative charge at this end of the bond. Conversely, because the less electronegative element of a polar covalent bond has one of its valence electrons pulled partly away from the atom, a center of positive charge develops at this end of the bond. A polar covalent bond, thus, represents a partial transfer of an electron from one atom to another, but with the atoms still held together as a unit. This is in contrast with an ionic bond, in which one atom completely transfers an electron to a more electronegative atom, but with the resulting positive and negative ions able to separate and behave independently of one another (e.g., in solution).

32. Electronegativity represents the relative ability of an atom in a molecule to attract shared electrons towards itself. In order for a bond to be polar, one of the atoms in the bond must attract the shared electron pair towards itself and away from the other atom of the bond: this can only happen if one atom of the bond is more electronegative than the other (that is, that there is a considerable difference in electronegativity for the two atoms of the bond). The larger the difference in electronegativity between two atoms joined in a bond, the more polar is the bond. Specific examples depend on student choice of elements, but in general, a molecule like Cl_2 would be non-polar because both atoms of the bond have the same electronegativity, whereas a molecule like HCl would be polar because there is an electronegativity difference between the two atoms in the bond.

33. A molecule is said to possess an overall dipole moment if the centers of positive and negative charge in the molecule do not coincide. A distinction must be made between whether or not individual bonds within a molecule are polar and whether the molecule overall possess a net dipole moment: sometimes the geometric shape of a molecule is such that individual bond dipoles effectively cancel each other out, leaving the molecule nonpolar overall. For example, compare the two molecules H_2O and CO_2. From Chapter 12, you realize that the water molecule is nonlinear (bent or *v*-shaped), whereas the CO_2 molecule is linear because of the double bonds. Both the O–H and the C–O bonds are themselves polar (because the atoms involved do not have the same electronegativities). However, because the CO_2 molecule is linear, the two individual bond dipoles lie in opposite directions on the same axis and cancel each other out, leaving CO_2 overall as a nonpolar molecule. In water, the two individual bond dipoles lie at an angle, and combine to increase the negative charge on the oxygen atom, leading water to be a very polar molecule. Bond dipoles are in actuality vector quantities, and the overall dipole moment of the molecule represents the resultant of all the individual bond dipole vectors. The high polarity of the water molecule, combined with the existence of hydrogen bonding among water molecules, is responsible for the fact that water is a liquid at room temperature.

34. It has been observed over many experiments that when an active metal like sodium or magnesium reacts with a nonmetal, the sodium atoms always form Na^+ ions and the magnesium atoms always form Mg^{2+} ions. It has been further observed that aluminum always forms only the Al^{3+} ion. When nitrogen, oxygen, or fluorine form simple ions, the ions that are formed are always N^{3-}, O^{2-}, and F^-, respectively. Clearly the facts that these elements always form the same ions and that those ions all contain eight electrons in the outermost shell, led scientists to speculate that there must be something fundamentally stable about a species that has eight electrons in its outermost shell (like the noble gas neon). The repeated observation that so many elements, when reacting, tend to attain an electronic configuration that is isoelectronic with a noble gas led chemists to speculate that all elements try to attain such a configuration for their outermost shells. In general, when atoms of a metal react with atoms of a nonmetal, the metal atoms lose electrons until they have the configuration of the preceding noble gas, and the nonmetal atoms gain electrons until they have the configuration of the following noble gas. Covalently and polar covalently bonded molecules also strive to attain pseudo-noble gas electronic configurations. For a covalently

Cumulative Review: Chapters 10, 11, and 12

bonded molecule like F_2, in which neither fluorine atom has a greater tendency than the other to gain or lose electrons completely, each F atom provides one electron of the pair of electrons that constitutes the covalent bond. Each F atom feels also the influence of the other F atom's electron in the shared pair, and each F atom effectively fills its outermost shell. Similarly, in polar covalently bonded molecules like HF or HCl, the shared pair of electrons between the atoms effectively completes the outer electron shell of each atom simultaneously to give each atom a noble gas-like electronic configuration.

35. The properties of typical ionic substances (hardness, rigidity, high melting and boiling points, etc.) suggest that the ionic bond is a very strong one. The formulas we write for ionic substances are only empirical formulas, showing the relative numbers of each type of atom present in the substance: for example, when we write $CaCl_2$ as the formula for calcium chloride, we are only saying that there are two chloride ions for each calcium ion in the substance, and not that there are distinct molecules of $CaCl_2$. Ionic substances in the bulk consist of crystals containing an extended lattice array of positive and negative ions, in a more or less alternating pattern (that is, a given positive ion typically has several negative ions as its nearest neighbors in the crystal). A typical ionic crystal lattice is shown as Figure 12.8 in the text. When an atom forms a positive ion (cation), it sheds its outermost electron shell (the valence electrons), leaving the resulting positive ion smaller than the atom it was formed from. When an atom forms a negative ion, the atoms takes additional electrons into its outermost shell, which causes the outermost shell to increase in size because of the additional repulsive forces. This results in the negative ion being larger than the atom from which it was formed. In the case of ionic compounds involving polyatomic ions, more than one type of bonding force is involved. First of all, ionic bonding exists between the positive and negative ion. However, within the polyatomic ions themselves, the atoms that make up the polyatomic ion are held together by polar covalent bonds.

36. Bonding between atoms to form a molecule involves only the valence electrons of the atoms (not the inner core electrons). So when we draw the Lewis structure of a molecule, we show only these valence electrons (both bonding valence electrons and nonbonding valence electrons, however). The most important requisite for the formation of a stable compound (which we try to demonstrate when we write Lewis structures) is that each atom of a molecule attains a noble gas electron configuration. When we write Lewis structures, we arrange the bonding and nonbonding valence electrons to try to complete the octet (or duet) for as many atoms as is possible.

37. Both the duet and octet rules are simplified statements of our basic guiding principle when discussing how atoms bond with one another to form molecules. When atoms of one element react with atoms of another element to form a compound, as many of the atoms as possible will end up with a noble gas-like electronic configuration in the compound. The "duet" rule applies only for the element hydrogen: when a hydrogen atom forms a bond to another atom, the single electron of the hydrogen atom pairs up with an electron from the other atom to give the shared electron pair that constitutes the covalent bond. By sharing this additional electron, the hydrogen atom attains effectively the $1s^2$ configuration of the noble gas helium. The "octet" rule applies for the representative elements other than hydrogen: when one of these atoms forms bonds to other atoms, the atom shares enough electrons to end up with the ns^2np^6 (i.e., eight electrons-an octet) electron configuration of a noble gas. Bonding electrons are those electrons used in forming a covalent bond between atoms: the bonding electrons in a molecule represent the electrons that are shared between atoms. Nonbonding (or lone pair) electrons are those valence electrons that are not used in covalent bonding and that "belong" exclusively to one atom in a molecule. For example, in the molecule ammonia (NH_3), the Lewis structure shows that there are three pairs of bonding electrons (each pair constitutes a covalent bond between the nitrogen atom and one of the hydrogen atoms) as well as one nonbonding pair of electrons, which belong exclusively to the

nitrogen atom. The presence of nonbonding pairs of electrons on an atom in a molecule can have a big effect on the geometric shape of the molecule and on the molecule's properties.

38. Obviously, you could choose practically any molecule for your discussion. Let's illustrate the method for ammonia, NH_3. First count up the total number of valence electrons available in the molecule (without regard to what atom they officially come from); remember that for the representative elements, the number of valence electrons is indicated by what group the element is found in on the periodic table. For NH_3, because nitrogen is in Group 5, one nitrogen atom would contribute five valence electrons. Because hydrogen atoms only have one electron each, the three hydrogen atoms provide an additional three valence electrons, for a total of eight valence electrons overall. Next write down the symbols for the atoms in the molecule, and use one pair of electrons (represented by a line) to form a bond between each pair of bound atoms.

$$\begin{array}{c} H-N-H \\ | \\ H \end{array}$$

These three bonds use six of the eight valence electrons. Since each hydrogen already has its duet in what we have drawn so far, while the nitrogen atom only has six electrons around it so far, the final two valence electrons must represent a lone pair on the nitrogen.

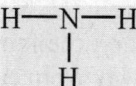

39. A double bond between two atoms represents the atoms sharing two pairs of electrons (4 electrons) between them. A triple bond represents two atoms sharing three pairs (6 electrons) between them. When writing a Lewis structure for a molecule, if you discover (after writing down a pair of bonding electrons between each of the atoms to be connected) that there do not seem to be enough valence electrons remaining to complete independently the octet (or duet) for each atom, then this strongly suggests that there must be multiple bonding present in the molecule. For example, if a molecule seems to be two electrons short to complete independently the octet, this suggests that there must be a second bond between two of the atoms (a double bond exists between those atoms). If a molecule seems to be four electrons short, this could mean either that a triple bond exists between two of the atoms, or that there are two double bonds present in the molecule. If it is possible to draw more than one valid Lewis structure for a molecule, differing only in the location of the double bonds, we say that the molecule exhibits resonance: the existence of resonance can markedly affect the geometry of the molecule and its resulting properties.

40. There are several types of exceptions to the octet rule described in the text. The octet rule is really a "rule of thumb" which we apply to molecules unless we have some evidence that a molecule does not follow the rule. There are some common molecules that, from experimental measurements, we know do not follow the octet rule. Boron and beryllium compounds sometimes do not fit the octet rule. For example, in BF_3, the boron atom only has six valence electrons, whereas in BeF_2, the beryllium atom only has four valence electrons. Other molecules that are exceptions to the octet rule include any molecule with an odd number of valence electrons (such as NO or NO_2): you can't get an octet (an even number) of electrons around each atom in a molecule with an odd number of valence electrons. Even the oxygen gas we breathe is an exception to the octet rule: although we can write a Lewis structure for O_2 satisfying the octet rule for each oxygen, we know from experiment that O_2 contains unpaired electrons (which would not be consistent with a structure in which all the electrons were paired up.)

Cumulative Review: Chapters 10, 11, and 12

41. The general geometric structure of a molecule is determined by (1) *how many electron pairs* surround the central atom in the molecule, and (2) which of those electron pairs are used for *bonding* to the other atoms of the molecule. Nonbonding electron pairs on the central atom do, however, cause minor changes in the bond angles, compared to the ideal regular geometric structure. As examples, let's show how we would determine the geometric structure of the molecules CH_4, NH_3, and H_2O. First we would draw the Lewis structures for each of these molecules.

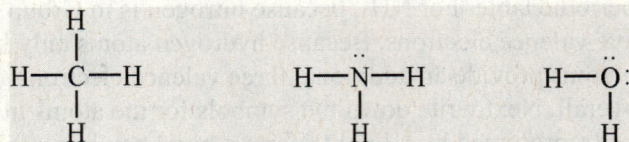

In each of these structures, the central atom (C, N, or O) is surrounded by four pairs (an octet) of electrons. According to the VSEPR theory, the four pairs of electrons repel each other and orient themselves in space so as to be as far away from each other as possible. This leads to the electron pairs being oriented tetrahedrally, separated by angles of 109.5°. For CH_4, each of the four pairs of electrons around the C atom is a bonding pair. We therefore say that CH_4 itself has a tetrahedral geometry, with H–C–H bond angles of 109.5°. For NH_3, however, although there are four tetrahedrally arranged electron pairs around the nitrogen atom, only three of these pairs are bonding pairs: there is a lone pair on the nitrogen atom. We describe the geometry of NH_3 as a trigonal pyramid: the three hydrogen atoms lie below the nitrogen atom in space as a result of the presence of the lone pair on nitrogen. The H–N–H bond angles are slightly less than the tetrahedral angle of 109.5°. Finally, in H_2O, although we have four pairs of electrons around the oxygen atom, only two of these pairs are bonding pairs: there are two lone electron pairs on the oxygen atom. We describe the geometry of H_2O as bent, V-shaped, or nonlinear: the presence of the lone pairs makes the H–O–H bond angle not 180° (linear) but somewhat less than the tetrahedral angle of 109.5°.

42.

Valence Pairs	Bond Angle	Example(s)
2	180°	BeF_2, BeH_2
3	120°	BCl_3
4	109.5°	CH_4, CCl_4, GeF_4

43. In predicting the geometric structure of a molecule, we treat a double (or triple) bond as a single entity (as if it were a single pair of electrons). This approach is reasonable, because all the electrons that bond together two particular atoms must be present in the same region of space between the atoms (whether one, two, or three electron pairs). For example, if we write the Lewis structure of acetylene, C_2H_2

H:C:::C:H

we realize that each carbon atom in the molecule has effectively only two "things" attached to it: a bonding pair of electrons (which bonds the H atom) and a triple bond (which bonds the other C atom). Since there are effectively only two things attached to each carbon atom, we would expect the bond angles for each carbon atom to be 180°, which makes the molecule linear overall.

44. a. $[Kr] 5s^2$
    b. $[Ne] 3s^2 3p^1$
    c. $[Ne] 3s^2 3p^5$
    d. $[Ar] 4s^1$

e. [Ne] $3s^2 3p^4$

f. [Ar] $4s^2 3d^{10} 4p^3$

45.   a. Al has three electrons more than a noble gas and forms Al^{3+} ions; F has one electron less than a noble gas and forms F^- ions; the simplest compound would be AlF_3.

b. Li has one electron more than a noble gas and forms Li^+ ions; N has three fewer electrons than a noble gas and forms N^{3-} ions; the simplest compound would be Li_3N.

c. Ca has two electrons more than a noble gas and forms Ca^{2+} ions; S has two fewer electrons than a noble gas and forms S^{2-} ions; the simplest compound would be CaS.

d. Mg has two electrons more than a noble gas and forms Mg^{2+} ions; P has three electrons fewer than a noble gas and forms P^{3-} ions; the simplest compound would be Mg_3P_2.

e. Al has three electrons more than a noble gas and forms Al^{3+} ions; O has two fewer electrons than a noble gas and forms O^{2-} ions; the simplest compound would be Al_2O_3.

f. K has one electron more than a noble gas and forms K^+ ions; S has two fewer electrons than a noble gas and forms S^{2-} ions; the simplest compound would be K_2S.

46.

Structure	Description
H—Ö—H	4 electron pairs tetrahedrally-oriented on O; non-linear (bent, V-shaped) geometry; H–O–H bond angle slightly less than 109.5° because of lone pairs.
H—P̈—H, H (below)	4 electron pairs tetrahedrally-oriented on P; trigonal pyramidal geometry; H–P–H bond angles slightly less than 109.5° because of lone pair.
:Br: above, :Br—C—Br:, :Br: below	4 electron pairs tetrahedrally-oriented on C; overall tetrahedral geometry; Br–C–Br bond angles 109.5°
[:Ö: above, :Ö—Cl—Ö:, :Ö: below]$^-$	4 electron pairs tetrahedrally-oriented on Cl; overall tetrahedral geometry; O–Cl–O bond angles 109.5°
:F̈—B—F̈:, :F̈ (below)	3 electron pairs trigonally-oriented on B (exception to octet rule); overall trigonal geometry; F–B–F bond angles 120°
:F̈—Be—F̈:	2 electron pairs linearly-oriented on Be (exception to octet rule); overall linear geometry; F–Be–F bond angle 180°.

CHAPTER 13

Gases

1. The atmosphere is a homogeneous mixture (a solution) of gases.

2. Solids and liquids have essentially fixed volumes and are not able to be compressed easily. Gases have volumes that depend on their conditions, and can be compressed or expanded by changes in those conditions. Although the particles of matter in solids are essentially fixed in position (the solid is rigid), the particles in liquids and gases are free to move.

3. A small amount of water is added to a metal can and then the can is heated so as to boil the water and fill the can with steam (gaseous water). The heat is then removed and the can is sealed off. As the steam in the can cools, it condenses back to a liquid. Since the gas in the can has condensed, the pressure of the atmosphere is much larger than the pressure of gas in the can, and the atmospheric pressure causes the can to collapse.

4. Figure 13.2 in the text shows a simple mercury barometer: a tube filled with mercury is inverted over a reservoir (containing mercury) that is open to the atmosphere. When the tube is inverted, the mercury falls to a level at which the pressure of the atmosphere is sufficient to support the column of mercury. One standard atmosphere of pressure is taken to be the pressure capable of supporting a column of mercury to a height of 760.0 mm above the reservoir level.

5. mix

6. Pressure units include mm Hg, torr, pascals, and psi. The unit "mm Hg" is derived from the barometer, since in a traditional mercury barometer, we measure the height of the mercury column (in millimeters) above the reservoir of mercury.

7. 1.00 atm = 760 torr = 760 mm Hg = 101.325 kPa = 14.70 psi

   a. $45.2 \text{ kPa} \times \dfrac{1 \text{ atm}}{101.325 \text{ kPa}} = 0.446 \text{ atm}$

   b. $755 \text{ mm Hg} \times \dfrac{1 \text{ atm}}{760 \text{ mm Hg}} = 0.993 \text{ atm}$

   c. $802 \text{ torr} \times \dfrac{1 \text{ atm}}{760 \text{ torr}} \times \dfrac{101.325 \text{ kPa}}{1 \text{ atm}} = 107 \text{ kPa}$

   d. $1.04 \text{ atm} \times \dfrac{760 \text{ mm Hg}}{1 \text{ atm}} = 790. \text{ mm Hg}$

8. 1.00 atm = 760 torr = 760 mm Hg = 101.325 kPa = 14.70 psi

   a. $14.9 \text{ psi} \times \dfrac{1 \text{ atm}}{14.70 \text{ psi}} = 1.01 \text{ atm}$

b. $795 \text{ torr} \times \dfrac{1 \text{ atm}}{760 \text{ torr}} = 1.05 \text{ atm}$

c. $743 \text{ mm Hg} \times \dfrac{101.325 \text{ kPa}}{760 \text{ mm Hg}} = 99.1 \text{ kPa}$

d. $99{,}436 \text{ Pa} \times \dfrac{1 \text{ kPa}}{1000 \text{ Pa}} = 99.436 \text{ kPa}$

9. $1.00 \text{ atm} = 760 \text{ torr} = 760 \text{ mm Hg} = 101.325 \text{ kPa} = 14.70 \text{ psi}$

   a. $699 \text{ mm Hg} \times \dfrac{1 \text{ atm}}{760 \text{ mm Hg}} = 0.920 \text{ atm}$

   b. $18.2 \text{ psi} \times \dfrac{760 \text{ mm Hg}}{14.70 \text{ psi}} = 941 \text{ mm Hg}$

   c. $862 \text{ mm Hg} = 862 \text{ torr}$

   d. $795 \text{ mm Hg} \times \dfrac{14.70 \text{ psi}}{760 \text{ mm Hg}} = 15.4 \text{ psi}$

10. $1.00 \text{ atm} = 760 \text{ torr} = 760 \text{ mm Hg} = 101.325 \text{ kPa} = 14.70 \text{ psi}$

    a. $17.3 \text{ psi} \times \dfrac{101.325 \text{ kPa}}{14.70 \text{ psi}} = 119 \text{ kPa}$

    b. $1.15 \text{ atm} \times \dfrac{14.70 \text{ psi}}{1 \text{ atm}} = 16.9 \text{ psi}$

    c. $4.25 \text{ atm} \times \dfrac{760 \text{ mm Hg}}{1 \text{ atm}} = 3.23 \times 10^3 \text{ mm Hg}$

    d. $224 \text{ psi} \times \dfrac{1 \text{ atm}}{14.70 \text{ psi}} = 15.2 \text{ atm}$

11. $1.00 \text{ atm} = 760 \text{ torr} = 760 \text{ mm Hg} = 101.325 \text{ kPa} = 14.70 \text{ psi}$

    a. $1.54 \times 10^5 \text{ Pa} \times \dfrac{1 \text{ atm}}{101{,}325 \text{ Pa}} = 1.52 \text{ atm}$

    b. $1.21 \text{ atm} \times \dfrac{101{,}325 \text{ Pa}}{1 \text{ atm}} = 1.23 \times 10^5 \text{ Pa}$

    c. $97{,}325 \text{ Pa} \times \dfrac{760 \text{ mm Hg}}{101{,}325 \text{ Pa}} = 730.14 \text{ mm Hg}$

    d. $1.32 \text{ kPa} \times \dfrac{1000 \text{ Pa}}{1 \text{ kPa}} = 1.32 \times 10^3 \text{ Pa}$

Chapter 13: Gases

12. 1.00 atm = 760 torr = 760 mm Hg = 101,325 Pa

   a. $774 \text{ torr} \times \dfrac{1 \text{ atm}}{760 \text{ torr}} \times \dfrac{101{,}325 \text{ Pa}}{1 \text{ atm}} = 1.03 \times 10^5 \text{ Pa}$

   b. $0.965 \text{ atm} \times \dfrac{101{,}325 \text{ Pa}}{1 \text{ atm}} = 9.78 \times 10^4 \text{ Pa}$

   c. $112.5 \text{ kPa} \times \dfrac{1000 \text{ Pa}}{1 \text{ kPa}} = 1.125 \times 10^5 \text{ Pa}$

   d. $801 \text{ mm Hg} \times \dfrac{1 \text{ atm}}{760 \text{ mm Hg}} \times \dfrac{101{,}325 \text{ Pa}}{1 \text{ atm}} = 1.07 \times 10^5 \text{ Pa}$

13. The volume of a sample of an ideal gas at constant temperature will decrease if the pressure on the gas is increased.

14. Additional mercury increases the pressure on the gas sample, causing the volume of the gas upon which the pressure is exerted to decrease (Boyle's Law)

15. pressure

16. $PV = k$; $P_1V_1 = P_2V_2$

17. a. $P_1 = 755$ mm Hg $\qquad\qquad P_2 = 780$ mm Hg
       $V_1 = 125$ mL $\qquad\qquad\qquad V_2 = ?$

       $V_2 = \dfrac{P_1V_1}{P_2} = \dfrac{(755 \text{ mm Hg})(125 \text{ mL})}{(780 \text{ mm Hg})} = 121$ mL

   b. $P_1 = 1.08$ atm $\qquad\qquad P_2 = 0.951$ atm
       $V_1 = 223$ mL $\qquad\qquad\quad V_2 = ?$

       $V_2 = \dfrac{P_1V_1}{P_2} = \dfrac{(1.08 \text{ atm})(223 \text{ mL})}{(0.951 \text{ atm})} = 253$ mL

   c. $P_1 = 103$ kPa $\qquad\qquad P_2 = 121$ kPa
       $V_1 = 3.02$ L $\qquad\qquad\quad V_2 = ?$

       $V_2 = \dfrac{P_1V_1}{P_2} = \dfrac{(103 \text{ kPa})(3.02 \text{ L})}{(121 \text{ kPa})} = 2.57$ L

18. a. $P_1 = 1.15$ atm $\qquad\qquad P_2 = 775$ mm Hg $= 1.020$ atm
       $V_1 = 375$ mL $\qquad\qquad\quad V_2 = ?$

       $V_2 = \dfrac{P_1V_1}{P_2} = \dfrac{(1.15 \text{ atm})(375 \text{ mL})}{(1.020 \text{ atm})} = 423$ mL

Chapter 13:   Gases

    b. $P_1 = 1.08$ atm $P_2 = 135$ kPa $= 1.33$ atm

          $V_1 = 195$ mL $V_2 = ?$

$$V_2 = \frac{P_1 V_1}{P_2} = \frac{(1.08 \text{ atm})(195 \text{ mL})}{(1.33 \text{ atm})} = 158 \text{ mL}$$

    c. $P_1 = 131$ kPa $= 982.6$ mm Hg $P_2 = 765$ mm Hg

          $V_1 = 6.75$ L $V_2 = ?$

$$V_2 = \frac{P_1 V_1}{P_2} = \frac{(982.6 \text{ mm Hg})(6.75 \text{ L})}{(765 \text{ mm Hg})} = 8.67 \text{ L}$$

19.   a. $P_1 = 102.1$ kPa $P_2 = ?$ kPa

              $V_1 = 19.3$ L $V_2 = 10.0$ L

$$P_2 = \frac{P_1 V_1}{V_2} = \frac{(102.1 \text{ kPa})(19.3 \text{ L})}{(10.0 \text{ L})} = 197 \text{ kPa}$$

    b. $P_1 = 755$ torr $= 755$ mm Hg $P_2 = 761$ mm Hg

          $V_1 = 25.7$ mL $V_2 = ?$ mL

$$V_2 = \frac{P_1 V_1}{P_2} = \frac{(755 \text{ mm Hg})(25.7 \text{ mL})}{(761 \text{ mm Hg})} = 25.5 \text{ mL}$$

    c. $P_1 = 1.05$ atm $P_2 = 112.2$ kPa $= 1.107$ atm

          $V_1 = 51.2$ L $V_2 = ?$

$$V_2 = \frac{P_1 V_1}{P_2} = \frac{(1.05 \text{ atm})(51.2 \text{ L})}{(1.107 \text{ atm})} = 48.6 \text{ L}$$

20.   a. $P_1 = 785$ mm Hg $P_2 = 700.$ mm Hg

             $V_1 = 53.2$ mL $V_2 = ?$

$$V_2 = \frac{P_1 V_1}{P_2} = \frac{(785 \text{ mm Hg})(53.2 \text{ mL})}{700. \text{ mm Hg}} = 59.7 \text{ mL}$$

    b. $P_1 = 1.67$ atm $P_2 = ?$

          $V_1 = 2.25$ L $V_2 = 2.00$ L

$$P_2 = \frac{P_1 V_1}{V_2} = \frac{(1.67 \text{ atm})(2.25 \text{ L})}{2.00 \text{ L}} = 1.88 \text{ atm}$$

    c. $P_1 = 695$ mm Hg $P_2 = 1.51$ atm

          $V_1 = 5.62$ L $V_2 = ?$

$$V_2 = \frac{P_1 V_1}{P_2} = \frac{(695 \text{ mm Hg})(5.62 \text{ L})}{1.51 \text{ atm} \times \left(\frac{760 \text{ mm Hg}}{1 \text{ atm}}\right)} = 3.40 \text{ L}$$

Chapter 13: Gases

21. $P_1 = 1.02$ atm $P_2 = 2.99$ atm
     $V_1 = 225$ mL $V_2 = ?$

   $$V_2 = \frac{P_1 V_1}{P_2} = \frac{(1.02 \text{ atm})(225 \text{ mL})}{(2.99 \text{ atm})} = 76.8 \text{ mL}$$

22. $P_1 = P_1$ $P_2 = 2 \times P_1$
     $V_1 = 1.04$ L $V_2 = ?$ L

   $$V_2 = \frac{P_1 V_1}{P_2} = \frac{(P_1)(1.04 \text{ L})}{(2 \times P_1)} = \frac{1.04 \text{ L}}{2} = 0.520 \text{ L}$$

23. $P_1 = 785$ mm Hg $P_2 = ?$
     $V_1 = 29.2$ mL $V_2 = 15.1$ mL

   $$P_2 = \frac{P_1 V_1}{V_2} = \frac{(785 \text{ mm Hg})(29.2 \text{ mL})}{(15.1 \text{ mL})} = 1.52 \times 10^3 \text{ mm Hg}$$

24. $$P_2 = \frac{P_1 V_1}{V_2} = \frac{\left(760. \text{ mm Hg} \times \frac{1 \text{ atm}}{760 \text{ mm Hg}}\right)(1.00 \text{ L})}{\left(50.0 \text{ mL} \times \frac{1 \text{ L}}{1000 \text{ mL}}\right)} = 20.0 \text{ atm}$$

25. Absolute zero is the lowest temperature that can exist. Absolute zero is the temperature at which the volume of an ideal gas sample would be predicted to become zero. Absolute zero is the zero-point on the Kelvin temperature scale (and corresponds to –273°C).

26. Charles's Law indicates that an ideal gas decreases by 1/273 of its volume for every degree Celsius its temperature is lowered. This means an ideal gas would approach a volume of zero at –273°C.

27. directly

28. $V = kT$; $V_1/T_1 = V_2/T_2$

29. $V_1 = 1340$ L $V_2 = ?$ mL
     $T_1 = 18°C = 291$ K $T_2 = 87°C = 360.$ K

   $$V_2 = \frac{V_1 T_2}{T_1} = \frac{(1340 \text{ L})(360. \text{ K})}{(291 \text{ K})} = 1660 \text{ L}$$

30. $V_1 = 375$ mL $V_2 = ?$ mL
     $T_1 = 78°C = 351$ K $T_2 = 22°C = 295$ K

   $$V_2 = \frac{V_1 T_2}{T_1} = \frac{(375 \text{ mL})(295 \text{ K})}{(351 \text{ K})} = 315 \text{ mL}$$

31. a. $V_1 = 2.03$ L $\qquad\qquad V_2 = 3.01$ L

$T_1 = 24°C = 297$ K $\qquad T_2 = ?$

$$T_2 = \frac{V_2 T_1}{V_1} = \frac{(3.01 \text{ L})(297 \text{ K})}{(2.03 \text{ L})} = 440 \text{ K} = 167°C$$

b. $V_1 = 127$ mL $\qquad\qquad V_2 = ?$

$T_1 = 273$ K $\qquad\qquad T_2 = 373$ K

$$V_2 = \frac{V_1 T_2}{T_1} = \frac{(127 \text{ mL})(373 \text{ K})}{(273 \text{ K})} = 174 \text{ mL}$$

c. $V_1 = 49.7$ mL $\qquad\qquad V_2 = ?$

$T_1 = 34°C = 307$ K $\qquad T_2 = 350$ K

$$V_2 = \frac{V_1 T_2}{T_1} = \frac{(49.7 \text{ mL})(350 \text{ K})}{(307 \text{ K})} = 56.7 \text{ mL}$$

32. a. $V_1 = 25.0$ L $\qquad\qquad V_2 = 50.0$ L

$T_1 = 0°C = 273$ K $\qquad T_2 = ?$ °C

$$T_2 = \frac{V_2 T_1}{V_1} = \frac{(50.0 \text{ L})(273 \text{ K})}{25.0 \text{ L}} = 546 \text{ K} = 273°C$$

b. $V_1 = 247$ mL $\qquad\qquad V_2 = 255$ mL

$T_1 = 25°C = 298$ K $\qquad T_2 = ?$ °C

$$T_2 = \frac{V_2 T_1}{V_1} = \frac{(255 \text{ mL})(298 \text{ K})}{247 \text{ mL}} = 308 \text{ K} = 35°C$$

c. $V_1 = 1.00$ mL $\qquad\qquad V_2 = ?$ mL

$T_1 = 2272°C = 2545$ K $\qquad T_2 = 25°C = 298$ K

$$V_2 = \frac{V_1 T_2}{T_1} = \frac{(1.00 \text{ mL})(298 \text{ K})}{2545 \text{ K}} = 0.117 \text{ mL}$$

33. a. $V_1 = 9.14$ L $\qquad\qquad V_2 = ?$

$T_1 = 24°C = 297$ K $\qquad T_2 = 48°C = 321$ K

$$V_2 = \frac{V_1 T_2}{T_1} = \frac{(9.14 \text{ L})(321 \text{ K})}{(297 \text{ K})} = 9.88 \text{ L}$$

b. $V_1 = 24.9$ mL $\qquad\qquad V_2 = 49.9$ mL

$T_1 = -12°C = 261$ K $\qquad T_2 = ?$

$$T_2 = \frac{V_2 T_1}{V_1} = \frac{(49.9 \text{ mL})(261 \text{ K})}{(24.9 \text{ mL})} = 523 \text{ K} = 250.°C$$

Chapter 13: Gases

    c. $V_1 = 925$ mL $V_2 = ?$

         $T_1 = 25$ K $T_2 = 273$ K

$$V_2 = \frac{V_1 T_2}{T_1} = \frac{(925 \text{ mL})(273 \text{ K})}{(25 \text{ K})} = 1.01 \times 10^4 \text{ mL}$$

34.   a. $V_1 = 2.01 \times 10^2$ L $V_2 = 5.00$ L

           $T_1 = 1150°C = 1423$ K $T_2 = ?°C$

$$T_2 = \frac{V_2 T_1}{V_1} = \frac{(5.00 \text{ L})(1423 \text{ K})}{(201 \text{ L})} = 35.4 \text{ K} = -238°C$$

     b. $V_1 = 44.2$ mL $V_2 = ?$ mL

           $T_1 = 298$ K $T_2 = 0$

$$V_2 = \frac{V_1 T_2}{T_1} = \frac{(44.2 \text{ mL})(0 \text{ K})}{(298 \text{ K})} = 0 \text{ mL (0 K is absolute zero)}$$

     c. $V_1 = 44.2$ mL $V_2 = ?$ mL

           $T_1 = 298$ K $T_2 = 0°C = 273$ K

$$V_2 = \frac{V_1 T_2}{T_1} = \frac{(44.2 \text{ mL})(273 \text{ K})}{(298 \text{ K})} = 40.5 \text{ mL}$$

35. $V_1 = 1.25$ L $V_2 = ?$ mL

       $T_1 = 291$ K $T_2 = 78$ K

$$V_2 = \frac{V_1 T_2}{T_1} = \frac{(1.25 \text{ L})(78 \text{ K})}{(291 \text{ K})} = 0.335 \text{ L} = 0.34 \text{ L}$$

36. $V_2 = \dfrac{V_1 T_2}{T_1} = \dfrac{(125 \text{ mL})(250 \text{ K})}{(450 \text{ K})} = 69.4$ mL $= 69$ mL to two significant figures

37. $24°C + 273 = 297$ K $72°C + 273 = 345$ K

$$V_2 = \frac{V_1 T_2}{T_1} = \frac{(375 \text{ mL})(345 \text{ K})}{(297 \text{ K})} = 436 \text{ mL}$$

38. $V_2 = \dfrac{V_1 T_2}{T_1}$

Temp, °C	90	80	70	60	50	40	30	20
Volume, mL	124	120.	117	113	110	107	103	100

39.   directly

40. $V = an$; $V_1/n_1 = V_2/n_2$

41. $V_1/n_1 = V_2/n_2$

    $V_1 = 242$ mL $\qquad V_2 = ?$ L

    $n_1 = 0.00901$ mol $\qquad n_2 = 0.00703$ mol

    $242 \text{ mL} \times \dfrac{0.00703 \text{ mol}}{0.00901 \text{ mol}} = 189$ mL

42. $V = an$; $V_1/n_1 = V_2/n_2$

    Since 2.08 g of chlorine contains twice the number of moles of gas contained in the 1.04 g sample, the volume of the 2.08 g sample will be twice as large = 1744 (1.74×10^3) mL

43. $V_1 = 100.$ L $\qquad V_2 = ?$ L

    $n_1 = 3.25$ mol $\qquad n_2 = 14.15$ mol

    $100. \text{ L} \times \dfrac{14.15 \text{ mol}}{3.25 \text{ mol}} = 435$ L

44. molar mass of Ar = 39.95 g

    $2.71 \text{ g Ar} \times \dfrac{1 \text{ mol}}{39.95 \text{ g}} = 0.0678$ mol Ar

    $4.21 \text{ L} \times \dfrac{1.29 \text{ mol}}{0.0678 \text{ mol}} = 80.1$ L

45. Although the definition may seem a little strange, an ideal gas is one which obeys the ideal gas law, $PV = nRT$, exactly. That is, if knowledge of *three* of the properties of a gas (pressure, volume, temperature, and amount) leads to the correct value for the *fourth* property when using this equation, then the gas under study is an ideal gas.

46. Real gases most closely approach ideal gas behavior under conditions of relatively high temperatures (0°C or higher) and relatively low pressures (1 atm or lower).

47. For an ideal gas, $PV = nRT$ is true under any conditions. Consider a particular sample of gas (so that n remains constant) at a particular fixed temperature (so that T remains constant also). Suppose that at pressure P_1 the volume of the gas sample is V_1. Then for this set of conditions, the ideal gas equation would be given by

    $P_1V_1 = nRT$.

    If we then change the pressure of the gas sample to a new pressure P_2, the volume of the gas sample changes to a new volume V_2. For this new set of conditions, the ideal gas equation would be given by

    $P_2V_2 = nRT$.

    As the right-hand sides of these equations are equal to the same quantity (because we defined n and T to be constant), then the left-hand sides of the equations must also be equal, and we obtain the usual form of Boyle's law.

    $P_1V_1 = P_2V_2$

Chapter 13: Gases

48. For an ideal gas, $PV = nRT$ is true under any conditions. Consider a particular sample of gas (so that n remains constant) at a particular fixed pressure (so that P remains constant also). Suppose that at temperature T_1 the volume of the gas sample is V_1. Then for this set of conditions, the ideal gas equation would be given by

    $PV_1 = nRT_1$.

    If we then change the temperature of the gas sample to a new temperature T_2, the volume of the gas sample changes to a new volume V_2. For this new set of conditions, the ideal gas equation would be given by

    $PV_2 = nRT_2$.

    If we make a ratio of these two expressions for the ideal gas equation for this gas sample, and cancel out terms that are constant for this situation (P, n, and R) we get

    $$\frac{PV_1}{PV_2} = \frac{nRT_1}{nRT_2}$$

    $$\frac{V_1}{V_2} = \frac{T_1}{T_2}$$

    This can be rearranged to the familiar form of Charles's law

    $$\frac{V_1}{T_1} = \frac{V_2}{T_2}$$

49. a. $P = 782.4$ mm Hg $= 1.029$ atm; $T = 26.2°C = 299$ K

    $$V = \frac{nRT}{P} = \frac{(0.1021 \text{ mol})(0.08206 \text{ L atm mol}^{-1}\text{ K}^{-1})(299 \text{ K})}{(1.029 \text{ atm})} = 2.44 \text{ L}$$

    b. $V = 27.5$ mL $= 0.0275$ L; $T = 16.6°C = 289.6$ K (290. K)

    $$P = \frac{nRT}{V} = \frac{(0.007812 \text{ mol})(0.08206 \text{ L atm mol}^{-1}\text{ K}^{-1})(290 \text{ K})}{(0.0275 \text{ L})} = 6.75 \text{ atm}$$

    $$6.75 \text{ atm} \times \frac{760 \text{ mm Hg}}{1 \text{ atm}} = 5.13 \times 10^3 \text{ mm Hg}$$

    c. $V = 45.2$ mL $= 0.0452$ L

    $$T = \frac{PV}{nR} = \frac{(1.045 \text{ atm})(0.0452 \text{ L})}{(0.002241 \text{ mol})(0.08206 \text{ L atm mol}^{-1}\text{ K}^{-1})} = 257 \text{ K}$$

    $T = 257$ K $- 273$ K $= -16°C$

50. a. $P = 782$ mm Hg $= 1.03$ atm; $T = 27°C = 300$ K

    $$V = \frac{nRT}{P} = \frac{(0.210 \text{ mol})(0.08206 \text{ L atm mol}^{-1}\text{ K}^{-1})(300 \text{ K})}{(1.03 \text{ atm})} = 5.02 \text{ L}$$

b. $V = 644$ mL $= 0.644$ L

$$P = \frac{nRT}{V} = \frac{(0.0921 \text{ mol})(0.08206 \text{ L atm mol}^{-1} \text{ K}^{-1})(303 \text{ K})}{(0.644 \text{ L})} = 3.56 \text{ atm}$$

$= 2.70 \times 10^3$ mm Hg

c. $P = 745$ mm $= 0.980$ atm

$$T = \frac{PV}{nR} = \frac{(0.980 \text{ atm})(11.2 \text{ L})}{(0.401 \text{ mol})(0.08206 \text{ L atm mol}^{-1} \text{ K}^{-1})} = 334 \text{ K}$$

51. molar mass Ne $= 20.18$ g; $25°C = 298$ K

$$n = \frac{PV}{RT} = \frac{(1.02 \text{ atm})(5.00 \text{ L})}{(0.08206 \text{ L atm mol}^{-1} \text{ K}^{-1})(298 \text{ K})} = 0.2086 \text{ mol Ne}$$

0.2086 mol Ne $\times \dfrac{20.18 \text{ g Ne}}{1 \text{ mol Ne}} = 4.21$ g Ne

52. Molar mass of $O_2 = 32.00$ g; 56.2 kg $= 5.62 \times 10^4$ g

$T = 21°C = 294$ K

$$n = 5.62 \times 10^4 \text{ g O}_2 \times \frac{1 \text{ mol O}_2}{32.00 \text{ g O}_2} = 1.76 \times 10^3 \text{ mol O}_2$$

$$P = \frac{nRT}{V} = \frac{(1.76 \times 10^3 \text{ mol})(0.08206 \text{ L atm mol}^{-1} \text{K}^{-1})(294 \text{ K})}{(125 \text{ L})} = 339 \text{ atm}$$

53. molar mass of He $= 4.003$ g; $100°C = 373$ K; 785 mm Hg $= 1.033$ atm

2.04 g He $\times \dfrac{1 \text{ mol He}}{4.003 \text{ g He}} = 0.5096$ mol He

$$V = \frac{nRT}{P} = \frac{(0.5096 \text{ mol})(0.08206 \text{ L atm mol}^{-1} \text{ K}^{-1})(373 \text{ K})}{(1.033 \text{ atm})} = 15.1 \text{ L}$$

54. molar mass Ne $= 20.18$ g; 5.00 g $= 0.248$ mol

$$T = \frac{PV}{nR} = \frac{(1.10 \text{ atm})(7.00 \text{ L})}{(0.248 \text{ mol})(0.08206 \text{ L atm mol}^{-1} \text{K}^{-1})} = 379 \text{ K} = 106°C$$

55. $T = 25°C + 273 = 298$ K; molar masses: He, 4.003 g; O_2, 32.00 g

$$n = \frac{PV}{RT} = \frac{(255 \text{ atm})(100.0 \text{ L})}{(0.08206 \text{ L atm mol}^{-1} \text{ K}^{-1})(298 \text{ K})} = 1043 \text{ mol} = 1.04 \times 10^3 \text{ mol}$$

1.04×10^3 mol of either He or O_2 would be needed.

Chapter 13: Gases

$$1.04 \times 10^3 \text{ mol He} \times \frac{4.003 \text{ g He}}{1 \text{ mol He}} = 4.16 \times 10^3 \text{ g He}$$

$$1.04 \times 10^3 \text{ mol O}_2 \times \frac{32.00 \text{ g O}_2}{1 \text{ mol O}_2} = 3.33 \times 10^4 \text{ g O}_2$$

56. molar mass Ne = 20.18 g; 25°C = 298 K; 50°C = 323 K

$$1.25 \text{ g Ne} \times \frac{1 \text{ mol}}{20.18 \text{ g}} = 0.06194 \text{ mol}$$

$$P = \frac{nRT}{V} = \frac{(0.06194 \text{ mol})(0.08206 \text{ L atm mol}^{-1} \text{ K}^{-1})(298 \text{ K})}{(10.1 \text{ L})} = 0.150 \text{ atm}$$

$$P = \frac{nRT}{V} = \frac{(0.06194 \text{ mol})(0.08206 \text{ L atm mol}^{-1} \text{ K}^{-1})(323 \text{ K})}{(10.1 \text{ L})} = 0.163 \text{ atm}$$

57. molar mass Ne = 20.18 g; P = 500. torr = 0.6579 atm

$$1.0 \text{ g Ne} \times \frac{1 \text{ mol}}{20.18 \text{ g}} = 0.0496 \text{ mol Ne}$$

$$T = \frac{PV}{nR} = \frac{(0.6579 \text{ atm})(5.0 \text{ L})}{(0.0496 \text{ mol})(0.08206 \text{ L atm mol}^{-1} \text{K}^{-1})} = 809 \text{ K} \ (810 \text{ K})$$

58. molar mass O_2 = 32.00 g; 784 mm Hg = 1.032 atm

$$4.25 \text{ g O}_2 \times \frac{1 \text{ mol O}_2}{32.00 \text{ g O}_2} = 0.1328 \text{ mol}$$

$$T = \frac{PV}{nR} = \frac{(1.032 \text{ atm})(2.51 \text{ L})}{(0.1328 \text{ mol})(0.08206 \text{ L atm mol}^{-1} \text{ K}^{-1})} = 238 \text{ K} = -35°C$$

59. 5.0 kg = 5.0×10^3 g; molar mass Ne = 20.18 g

$$5.0 \times 10^3 \text{ g Ne} \times \frac{1 \text{ mol Ne}}{20.18 \text{ g Ne}} = 247.8 \text{ mol Ne}$$

$$P = \frac{nRT}{V} = \frac{(247.8 \text{ mol})(0.08206 \text{ L atm mol}^{-1}\text{K}^{-1})(300. \text{ K})}{(200 \text{ L})} = 30 \text{ atm}$$

60. Molar masses: He, 4.003 g; Ar, 39.95 g

$$4.15 \text{ g He} \times \frac{1 \text{ mol He}}{4.003 \text{ g He}} = 1.037 \text{ mol He}$$

$$56.2 \text{ g Ar} \times \frac{1 \text{ mol Ar}}{39.95 \text{ g Ar}} = 1.407 \text{ mol Ar}$$

For He, $P = \dfrac{nRT}{V} = \dfrac{(1.037 \text{ mol})(0.08206 \text{ L atm mol}^{-1} \text{ K}^{-1})(298 \text{ K})}{(5.00 \text{ L})} = 5.07$ atm

For Ar, $P = \dfrac{nRT}{V} = \dfrac{(1.407 \text{ mol})(0.08206 \text{ L atm mol}^{-1} \text{ K}^{-1})(303 \text{ K})}{(10.00 \text{ L})} = 3.50$ atm

The helium is at a higher pressure than the argon.

61. $P_1 = 1.01$ atm $\qquad\qquad\qquad\qquad\qquad$ $P_2 = ?$ atm

    $V_1 = 24.3$ mL $\qquad\qquad\qquad\qquad\qquad$ $V_2 = 15.2$ mL

    $T_1 = 25°C = 298$ K $\qquad\qquad\qquad\quad$ $T_2 = 50°C = 323$ K

    $P_2 = \dfrac{T_2 P_1 V_1}{T_1 V_2} = \dfrac{(323 \text{ K})(1.01 \text{ atm})(24.3 \text{ mL})}{(298 \text{ K})(15.2 \text{ mL})} = 1.75$ atm

62. molar mass Ar = 39.95 g; 29°C = 302 K; 42°C = 315 K

    $1.29 \text{ g Ar} \times \dfrac{1 \text{ mol Ar}}{39.95 \text{ g Ar}} = 0.03229$ mol Ar

    $P = \dfrac{nRT}{V} = \dfrac{(0.03229 \text{ mol})(0.08206 \text{ L atm mol}^{-1}\text{K}^{-1})(302 \text{ K})}{(2.41 \text{ L})} = 0.332$ atm

    $P = \dfrac{nRT}{V} = \dfrac{(0.03229 \text{ mol})(0.08206 \text{ L atm mol}^{-1}\text{K}^{-1})(315 \text{ K})}{(2.41 \text{ L})} = 0.346$ atm

63. $P_1 = 1.05$ atm $\qquad\qquad\qquad\qquad\qquad$ $P_2 = 0.997$ atm

    $V_1 = 459$ mL $\qquad\qquad\qquad\qquad\qquad$ $V_2 = ?$ mL

    $T_1 = 27°C = 300.$ K $\qquad\qquad\qquad\quad$ $T_2 = 15°C = 288$ K

    $V_2 = \dfrac{T_2 P_1 V_1}{T_1 P_2} = \dfrac{(288 \text{ K})(1.05 \text{ atm})(459 \text{ mL})}{(300 \text{ K})(0.997 \text{ atm})} = 464$ mL

64. Molar mass of H_2O = 18.02 g; 2.0 mL = 0.0020 L; 225°C = 498 K

    $0.250 \text{ g } H_2O \times \dfrac{1 \text{ mol } H_2O}{18.02 \text{ g } H_2O} = 0.01387 \text{ mol } H_2O$

    $P = \dfrac{nRT}{V} = \dfrac{(0.01387 \text{ mol})(0.08206 \text{ L atm mol}^{-1} \text{ K}^{-1})(498 \text{ K})}{(0.0020 \text{ L})} = 283 \text{ atm} = 2.8 \times 10^2$ atm

65. In deriving the ideal gas law, we assume that the molecules of gas occupy no volume, and that the molecules do not interact with each other. Under these conditions, there is no difference between gas molecules of different substances (other than their masses) as far as the bulk behavior of the gas is concerned. Each gas behaves independently of other gases present, and the overall properties of the sample are determined by the overall quantity of gas present.

    $P_{total} = P_1 + P_2 + \ldots P_n$ where n is the number of individual gases present in the mixture.

Chapter 13: Gases

66. As a gas is bubbled through water, the bubbles of gas become saturated with water vapor, thus forming a gaseous mixture. The total pressure in a sample of gas that has been collected by bubbling through water is made up of two components: the pressure of the gas of interest and the pressure of water vapor. The partial pressure of the gas of interest is then the total pressure of the sample minus the vapor pressure of water.

67. molar masses: He, 4.003 g; Ne, 20.18 g; 25°C = 298 K

    $$2.41 \text{ g He} \times \frac{1 \text{ mol He}}{4.003 \text{ g He}} = 0.602 \text{ mol He}$$

    $$2.79 \text{ g Ne} \times \frac{1 \text{ mol Ne}}{20.18 \text{ g Ne}} = 0.138 \text{ mol Ne}$$

    $$P_{helium} = \frac{n_{helium}RT}{V} = \frac{(0.602 \text{ mol})(0.08206 \text{ L atm mol}^{-1} \text{ K}^{-1})(298 \text{ K})}{(1.04 \text{ L})} = 14.2 \text{ atm}$$

    $$P_{neon} = \frac{n_{neon}RT}{V} = \frac{(0.138 \text{ mol})(0.08206 \text{ L atm mol}^{-1} \text{ K}^{-1})(298 \text{ K})}{(1.04 \text{ L})} = 3.25 \text{ atm}$$

    $$P_{total} = 14.2 \text{ atm} + 3.25 \text{ atm} = 17.5 \text{ atm}$$

68. molar masses: Ne, 20.18 g; Ar, 39.95 g; 27°C = 300 K

    $$1.28 \text{ g Ne} \times \frac{1 \text{ mol Ne}}{20.18 \text{ g Ne}} = 0.06343 \text{ mol Ne}$$

    $$2.49 \text{ g Ar} \times \frac{1 \text{ mol Ar}}{39.95 \text{ g Ar}} = 0.06233 \text{ mol Ar}$$

    $$P_{neon} = \frac{n_{neon}RT}{V} = \frac{(0.06343 \text{ mol})(0.08206 \text{ L atm mol}^{-1} \text{ K}^{-1})(300 \text{ K})}{(9.87 \text{ L})} = 0.1582 \text{ atm}$$

    $$P_{argon} = \frac{n_{argon}RT}{V} = \frac{(0.06233 \text{ mol})(0.08206 \text{ L atm mol}^{-1} \text{ K}^{-1})(300 \text{ K})}{(9.87 \text{ L})} = 0.1555 \text{ atm}$$

    $$P_{total} = 0.1582 \text{ atm} + 0.1555 \text{ atm} = 0.314 \text{ atm}$$

69. 52.5 g O_2 = 1.641 mol O_2; 65.1 g CO_2 = 1.479 mol CO_2; total moles = 3.120 mol

    $$P_{oxygen} = 9.21 \text{ atm} \times \frac{1.641 \text{ mol O}_2}{3.120 \text{ mol total}} = 4.84 \text{ atm O}_2$$

    $$P_{carbon\ dioxide} = 9.21 \text{ atm} \times \frac{1.479 \text{ mol CO}_2}{3.120 \text{ mol total}} = 4.37 \text{ atm CO}_2$$

    Once the partial pressure of O_2 had been calculated, we also could have calculated the partial pressure of CO_2 as the difference between the total pressure (9.21 atm) and the partial pressure of O_2 (4.84 atm).

70. 925 mm Hg = 1.217 atm; 26°C = 299 K; molar masses: Ne, 20.18 g; Ar, 39.95 g

$$n = \frac{PV}{RT} = \frac{(1.217 \text{ atm})(3.00 \text{ L})}{(0.08206 \text{ L atm mol}^{-1} \text{ K}^{-1})(299 \text{ K})} = 0.1488 \text{ mol}$$

The number of moles of an ideal gas required to fill a given-sized container to a particular pressure at a particular temperature does not depend on the specific identity of the gas. So 0.1488 mol of Ne gas or 0.1488 mol of Ar gas would give the same pressure in the same flask at the same temperature.

$$\text{mass Ne} = 0.1488 \text{ mol Ne} \times \frac{20.18 \text{ g Ne}}{1 \text{ mol Ne}} = 3.00 \text{ g Ne}$$

$$\text{mass Ar} = 0.1488 \text{ mol Ar} \times \frac{39.95 \text{ g Ar}}{1 \text{ mol Ar}} = 5.94 \text{ g Ar}$$

71. $P_{\text{oxygen}} = P_{\text{total}} - P_{\text{water vapor}} = 772 - 26.7 = 745$ torr

72. The total pressure of the gases inside the container is 1.00 atm. The total number of gas particles inside the container is 10. Let x equal the pressure of each gas particle.

    1.00 atm = 10x
    x = 0.100 atm
    There are 2 argon particles, 3 neon particles, and 5 helium particles present in the container.
    Therefore, $P_{\text{Total}} = P_{\text{Ar}} + P_{\text{Ne}} + P_{\text{He}} = 2x + 3x + 5x = 10x$.
    $P_{\text{Ar}} = 2(0.100 \text{ atm}) = 0.20$ atm (taking into account significant figures when these pressures are added)
    $P_{\text{Ne}} = 3(0.100 \text{ atm}) = 0.30$ atm (taking into account significant figures when these pressures are added)
    $P_{\text{He}} = 5(0.100 \text{ atm}) = 0.50$ atm (taking into account significant figures when these pressures are added)

73. $P_{\text{oxygen}} = P_{\text{total}} - P_{\text{water vapor}} = (755 - 23)$ mm Hg $= 732$ mm Hg $= 0.9632$ atm

    $T = 24°C + 273 = 297$ K; $V = 500.$ mL $= 0.500$ L

$$n = \frac{PV}{RT} = \frac{(0.9632 \text{ atm})(0.500 \text{ L})}{(0.08206 \text{ L atm mol}^{-1} \text{ K}^{-1})(297 \text{ K})} = 1.98 \times 10^{-2} \text{ mol O}_2$$

74. 1.032 atm = 784.3 mm Hg; molar mass of Zn = 65.38 g

    $P_{\text{hydrogen}} = 784.3$ mm Hg $- 32$ mm Hg $= 752.3$ mm Hg $= 0.990$ atm

    $V = 240$ mL $= 0.240$ L; $T = 30°C + 273 = 303$ K

$$n_{\text{hydrogen}} = \frac{PV}{RT} = \frac{(0.990 \text{ atm})(0.240 \text{ L})}{(0.08206 \text{ L atm mol}^{-1} \text{ K}^{-1})(303 \text{ K})} = 0.00956 \text{ mol hydrogen}$$

$$0.00956 \text{ mol H}_2 \times \frac{1 \text{ mol Zn}}{1 \text{ mol H}_2} = 0.00956 \text{ mol of Zn must have reacted}$$

$$0.00956 \text{ mol Zn} \times \frac{65.38 \text{ g Zn}}{1 \text{ mol Zn}} = 0.625 \text{ g Zn must have reacted}$$

75. A *law* is a statement that precisely expresses generally observed behavior. A *theory* consists of a set of assumptions/hypotheses that is put forth to *explain* the observed behavior of matter. Theories attempt to explain natural laws.

Chapter 13: Gases

76. A theory is successful if it explains known experimental observations. Theories that have been successful in the past may not be successful in the future (for example, as technology evolves, more sophisticated experiments may be possible in the future).

77. assume that the volume of the molecules themselves in a gas sample is negligible compared to the bulk volume of the gas sample: this helps us to explain why gases are so compressible.

78. pressure

79. kinetic energy

80. no

81. The temperature of a gas reflects, on average, how rapidly the molecules in the gas are moving. At high temperatures, the particles are moving very fast and collide with the walls of the container frequently, whereas at low temperatures, the molecules are moving more slowly and collide with the walls of the container infrequently. The Kelvin temperature is directly proportional to the average kinetic energy of the particles in a gas.

82. If the temperature of a sample of gas is increased, the average kinetic energy of the particles of gas increases. This means that the speeds of the particles increase. If the particles have a higher speed, they will hit the walls of the container more frequently and with greater force, thereby increasing the pressure.

83. The molar volume of a gas is the volume occupied by one mole of the gas under a particular set of temperature and pressure conditions (usually STP: 0°C, 1 atm). When measured under the same conditions, all ideal gases have the same molar volume (22.4 L at STP).

84. Standard Temperature and Pressure, STP = 0°C, 1 atm pressure. These conditions were chosen because they are easy to attain and reproduce *experimentally*. The barometric pressure within a laboratory is likely to be near 1 atm most days, and 0°C can be attained with a simple ice bath.

85. molar masses: CaO, 56.08 g; CO_2; 44.01 g

    $$1.25 \text{ g CaO} \times \frac{1 \text{ mol CaO}}{56.08 \text{ g CaO}} = 0.02229 \text{ mol CaO}$$

    From the balanced chemical equation, 0.02229 mol CaO would absorb 0.02229 mol CO_2

    $$0.02229 \text{ mol } CO_2 \times \frac{44.01 \text{ g } CO_2}{1 \text{ mol } CO_2} = 0.981 \text{ g } CO_2$$

    Since one mole of an idea gas occupies 22.4 L at STP, 0.02229 mol of CO_2 would occupy

    $$0.02229 \text{ mol } CO_2 \times \frac{22.4 \text{ L}}{1 \text{ mol}} = 0.499 \text{ L at STP}$$

86. Molar mass of C = 12.01 g; 25°C = 298 K

    $$1.25 \text{ g C} \times \frac{1 \text{ mol}}{12.01 \text{ g}} = 0.1041 \text{ mol C}$$

Since the balanced chemical equation shows a 1:1 stoichiometric relationship between C and O_2, then 0.1041 mol of O_2 will be needed

$$V = \frac{nRT}{P} = \frac{(0.1041 \text{ mol})(0.08206 \text{ L atm mol}^{-1} \text{ K}^{-1})(298 \text{ K})}{(1.02 \text{ atm})} = 2.50 \text{ L } O_2$$

87. $2C_8H_{18}(l) + 25O_2(g) \rightarrow 16CO_2(g) + 18H_2O(l)$

    molar mass C_8H_{18} = 114.2 g

    $10.0 \text{ g } C_8H_{18} \times \dfrac{1 \text{ mol } C_8H_{18}}{114.2 \text{ g } C_8H_{18}} = 0.08757 \text{ mol } C_8H_{18}$

    $0.08757 \text{ mol } C_8H_{18} \times \dfrac{25 \text{ mol } O_2}{2 \text{ mol } C_8H_{18}} = 1.095 \text{ mol } O_2$

    At STP, one mole of an ideal gas occupies 22.4 L of volume.

    $1.095 \text{ mol } O_2 \times \dfrac{22.4 \text{ L}}{\text{mol}} = 24.5 \text{ L } O_2 \text{ at STP}$

88. Molar mass of Mg = 24.31 g; STP: 1.00 atm, 273 K

    $1.02 \text{ g Mg} \times \dfrac{1 \text{ mol}}{24.31 \text{ g}} = 0.0420 \text{ mol Mg}$

    As the coefficients for Mg and Cl_2 in the balanced equation are the same, for 0.0420 mol of Mg reacting we will need 0.0420 mol of Cl_2.

    $V = 0.0420 \text{ mol } Cl_2 \times \dfrac{22.4 \text{ L}}{1 \text{ mol}} = 0.940 \text{ L } Cl_2 \text{ at STP.}$

89. 27°C = 300 K; 26 °C = 299 K; molar mass NH_4Cl = 53.49 g

    mol NH_3 present = $n = \dfrac{PV}{RT} = \dfrac{(1.02 \text{ atm})(4.21 \text{ L})}{(0.08206 \text{ L atm mol}^{-1} \text{ K}^{-1})(300 \text{ K})} = 0.174 \text{ mol } NH_3$

    mol HCl present = $n = \dfrac{PV}{RT} = \dfrac{(0.998 \text{ atm})(5.35 \text{ L})}{(0.08206 \text{ L atm mol}^{-1} \text{ K}^{-1})(299 \text{ K})} = 0.218 \text{ mol HCl}$

    NH_3 and HCl react on a 1:1 basis: NH_3 is the limiting reactant.

    $0.174 \text{ mol } NH_3 \dfrac{1 \text{ mol } NH_4Cl}{1 \text{ mol } NH_3} \times \dfrac{53.49 \text{ g } NH_4Cl}{1 \text{ mol } NH_4Cl} = 9.31 \text{ g } NH_4Cl \text{ produced}$

90. molar mass CaC_2 = 64.10 g; 25°C = 298 K

    $2.49 \text{ g } CaC_2 \times \dfrac{1 \text{ mol}}{64.10 \text{ g}} = 0.03885 \text{ mol } CaC_2$

Chapter 13: Gases

From the balanced chemical equation for the reaction, 0.03885 mol of CaC_2 reacting completely would generate 0.03885 mol of acetylene, C_2H_2

$$V = \frac{nRT}{P} = \frac{(0.03885 \text{ mol})(0.08206 \text{ L atm mol}^{-1} \text{ K}^{-1})(298 \text{ K})}{(1.01 \text{ atm})} = 0.941 \text{ L}$$

$$V = \frac{nRT}{P} = \frac{(0.03885 \text{ mol})(0.08206 \text{ L atm mol}^{-1} \text{ K}^{-1})(273 \text{ K})}{(1.00 \text{ atm})} = 0.870 \text{ L at STP}$$

91. $CuSO_4 \cdot 5H_2O(s) \rightarrow CuSO_4(s) + 5H_2O(g)$

    350°C = 623 K; molar mass $CuSO_4 \cdot 5H_2O$ = 249.7 g

    $$5.00 \text{ g } CuSO_4 \cdot 5H_2O \times \frac{1 \text{ mol } CuSO_4 \cdot 5H_2O}{249.7 \text{ g } CuSO_4 \cdot 5H_2O} = 0.02002 \text{ mol } CuSO_4 \cdot 5H_2O$$

    $$0.02002 \text{ mol } CuSO_4 \cdot 5H_2O \times \frac{5 \text{ mol } H_2O}{1 \text{ mol } CuSO_4 \cdot 5H_2O} = 0.1001 \text{ mol } H_2O$$

    $$V = \frac{nRT}{P} = \frac{(0.1001 \text{ mol})(0.08206 \text{ L atm mol}^{-1} \text{ K}^{-1})(623 \text{ K})}{(1.04 \text{ atm})} = 4.92 \text{ L } H_2O$$

92. Molar mass of Mg_3N_2 = 100.95 g; T = 24°C = 297 K; P = 752 mm Hg = 0.989 atm

    $$10.3 \text{ g } Mg_3N_2 \times \frac{1 \text{ mol}}{100.95 \text{ g}} = 0.102 \text{ mol } Mg_3N_2$$

    From the balanced chemical equation, the amount of NH_3 produced will be

    $$0.102 \text{ mol } Mg_3N_2 \times \frac{2 \text{ mol } NH_3}{1 \text{ mol } Mg_3N_2} = 0.204 \text{ mol } NH_3$$

    $$V = \frac{nRT}{P} = \frac{(0.204 \text{ mol})(0.08206 \text{ L atm mol}^{-1} \text{ K}^{-1})(297 \text{ K})}{(0.989 \text{ atm})} = 5.03 \text{ L}$$

    This assumes that the ammonia was collected dry.

93. Molar masses: He, 4.003 g; H_2, 2.016 g; 28°C = 301 K

    $$14.2 \text{ g He} \times \frac{1 \text{ mol He}}{4.003 \text{ g He}} = 3.55 \text{ mol He}$$

    $$21.6 \text{ g } H_2 \times \frac{1 \text{ mol } H_2}{2.016 \text{ g } H_2} = 10.7 \text{ mol } H_2$$

    total moles = 3.55 mol + 10.7 mol = 14.25 mol

    $$V = \frac{nRT}{P} = \frac{(14.25 \text{ mol})(0.08206 \text{ L atm mol}^{-1} \text{ K}^{-1})(301 \text{ K})}{(0.985 \text{ atm})} = 357 \text{ L}$$

Chapter 13: Gases

94. a. $6Na + N_2 \rightarrow 2Na_3N$

    b. $T = 28°C = 301$ K
    Determine the number of moles of each reactant present.

    $$10.0 \text{ g Na} \times \frac{1 \text{ mol Na}}{22.99 \text{ g Na}} = 0.435 \text{ mol Na}$$

    $$n = \frac{PV}{RT} = \frac{(0.976 \text{ atm})(2.50 \text{ L})}{(0.08206 \text{ L atm mol}^{-1}\text{K}^{-1})(301 \text{ K})} = 0.0988 \text{ mol N}_2$$

    Determine how many moles of Na_3N would be produced assuming each reactant is completely consumed.

    $$0.435 \text{ mol Na} \times \frac{2 \text{ mol Na}_3\text{N}}{6 \text{ mol Na}} = 0.145 \text{ mol Na}_3\text{N}$$

    $$0.0988 \text{ mol N}_2 \times \frac{2 \text{ mol Na}_3\text{N}}{1 \text{ mol N}_2} = 0.198 \text{ mol Na}_3\text{N}$$

    Once 0.145 mol Na_3N is produced, all of the Na is used up (making Na the limiting reactant). Thus,

    $$0.145 \text{ mol Na}_3\text{N} \times \frac{82.98 \text{ g Na}_3\text{N}}{1 \text{ mol Na}_3\text{N}} = 12.0 \text{ g Na}_3\text{N}$$

95. $P_1 = 892$ mm Hg $\qquad P_2 = 1.00$ atm $= 760$ mm Hg
    $V_1 = 25.2$ mL $\qquad V_2 = ?$
    $T_1 = 95°C + 273 = 368$ K $\qquad T_2 = 273$ K

    $$V_2 = \frac{T_2 P_1 V_1}{T_1 P_2} = \frac{(273 \text{ K})(892 \text{ mm Hg})(25.2 \text{ mL})}{(368 \text{ K})(760 \text{ mm Hg})} = 21.9 \text{ mL}$$

96. $P_1V_1/T_1 = P_2V_2/T_2$

    $V_1 = 50.0$ L $\qquad V_2 = ?$ L
    $T_1 = 20.°C = 293$ K $\qquad T_2 = 273$ K
    $P_1 = 742$ torr $\qquad P_2 = 760.$ torr (1 atm)

    $$V_2 = \frac{P_1 V_1 T_2}{P_2 T_1} = \frac{(742 \text{ torr})(50.0 \text{ L})(273 \text{ K})}{(760. \text{ torr})(293 \text{ K})} = 45.5 \text{ L}$$

97. molar masses: O_2, 32.00 g; N_2, 28.02 g; CO_2, 44.01 g; Ne, 20.18 g

    $$5.00 \text{ g O}_2 \times \frac{1 \text{ mol O}_2}{32.00 \text{ g O}_2} = 0.1563 \text{ mol O}_2$$

    $$5.00 \text{ g N}_2 \times \frac{1 \text{ mol N}_2}{28.02 \text{ g N}_2} = 0.1784 \text{ mol N}_2$$

Chapter 13: Gases

$$5.00 \text{ g CO}_2 \times \frac{1 \text{ mol CO}_2}{44.01 \text{ g CO}_2} = 0.1136 \text{ mol CO}_2$$

$$5.00 \text{ g Ne} \times \frac{1 \text{ mol Ne}}{20.18 \text{ g Ne}} = 0.2478 \text{ mol Ne}$$

Total moles of gas = 0.1563 + 0.1784 + 0.1136 + 0.2478 = 0.6961 mol

22.4 L is the volume occupied by one mole of any ideal gas at STP. This would apply even if the gas sample is a *mixture* of individual gases.

$$0.6961 \text{ mol} \times \frac{22.4 \text{ L}}{1 \text{ mol}} = 15.59 \text{ L} = 15.6 \text{ L}$$

The *partial pressure* of each individual gas in the mixture will be related to what *fraction* on a mole basis each gas represents in the mixture.

$$P_{\text{oxygen}} = 1.00 \text{ atm} \times \frac{0.1563 \text{ mol O}_2}{0.6961 \text{ mol total}} = 0.225 \text{ atm O}_2$$

$$P_{\text{nitrogen}} = 1.00 \text{ atm} \times \frac{0.1784 \text{ mol N}_2}{0.6961 \text{ mol total}} = 0.256 \text{ atm N}_2$$

$$P_{\text{carbon dioxide}} = 1.00 \text{ atm} \times \frac{0.1136 \text{ mol CO}_2}{0.6961 \text{ mol total}} = 0.163 \text{ atm CO}_2$$

$$P_{\text{neon}} = 1.00 \text{ atm} \times \frac{0.2478 \text{ mol Ne}}{0.6961 \text{ mol total}} = 0.356 \text{ atm Ne}$$

98. Molar masses: He, 4.003 g; Ne, 20.18 g

    $$6.25 \text{ g He} \times \frac{1 \text{ mol He}}{4.003 \text{ g He}} = 1.561 \text{ mol He}$$

    $$4.97 \text{ g Ne} \times \frac{1 \text{ mol Ne}}{20.18 \text{ g Ne}} = 0.2463 \text{ mol Ne}$$

    $n_{\text{total}} = 1.561 \text{ mol} + 0.2463 \text{ mol} = 1.807 \text{ mol}$

    As 1 mol of an ideal gas occupies 22.4 L at STP, the volume is given by

    $$1.807 \text{ mol} \times \frac{22.4 \text{ L}}{1 \text{ mol}} = 40.48 \text{ L} = 40.5 \text{ L}.$$

    The partial pressure of a given gas in a mixture will be proportional to what *fraction* of the total number of moles of gas the given gas represents

    $$P_{\text{He}} = \frac{1.561 \text{ mol He}}{1.807 \text{ mol total}} \times 1.00 \text{ atm} = 0.8639 \text{ atm} = 0.864 \text{ atm}$$

    $$P_{\text{Ne}} = \frac{0.2463 \text{ mol Ne}}{1.807 \text{ mol total}} \times 1.00 \text{ atm} = 0.1363 \text{ atm} = 0.136 \text{ atm}$$

99. $2Na(s) + Cl_2(g) \rightarrow 2NaCl(s)$

molar mass Na = 22.99 g

$4.81 \times \dfrac{1 \text{ mol Na}}{22.99 \text{ g Na}} = 0.2092 \text{ mol Na}$

$0.2092 \text{ mol Na} \times \dfrac{1 \text{ mol Cl}_2}{2 \text{ mol Na}} = 0.1046 \text{ mol Cl}_2$

$0.1046 \text{ mol Cl}_2 \times \dfrac{22.4 \text{ L}}{1 \text{ mol}} = 2.34 \text{ L Cl}_2 \text{ at STP}$

100. $2C_2H_2(g) + 5O_2(g) \rightarrow 2H_2O(g) + 4CO_2(g)$

molar mass C_2H_2 = 26.04 g

$1.00 \text{ g } C_2H_2 \times \dfrac{1 \text{ mol}}{26.04 \text{ g}} = 0.0384 \text{ mol } C_2H_2$

From the balanced chemical equation, $2 \times 0.0384 = 0.0768$ mol of CO_2 will be produced.

$0.0768 \text{ mol } CO_2 \times \dfrac{22.4 \text{ L}}{1 \text{ mol}} = 1.72 \text{ L at STP}$

101. $FeO(s) + CO(g) \rightarrow Fe(s) + CO_2(g)$

molar mass FeO = 71.85 g; 1.45 kg = 1.45×10^3 g

$1.45 \times 10^3 \text{ g FeO} \times \dfrac{1 \text{ mol FeO}}{71.85 \text{ g FeO}} = 20.18 \text{ mol FeO}$

Since the coefficients of the balanced equation are all *one*, if 20.18 mol FeO reacts, then 20.18 mol CO(g) is required and 20.18 mol of $CO_2(g)$ is produced.

$20.18 \text{ mol} \times \dfrac{22.4 \text{ L}}{1 \text{ mol}} = 452 \text{ L}$

4.52×10^4 L CO(g) is required for reaction and 4.52×10^4 L $CO_2(g)$ are produced by the reaction.

102. 125 mL = 0.125 L

$0.125 \text{ L} \times \dfrac{1 \text{ mol}}{22.4 \text{ L}} = 0.00558 \text{ mol } H_2$

From the balanced chemical equation, one mole of zinc is required for each mole of hydrogen produced. Therefore, 0.00558 mol of Zn will be required.

$0.00558 \text{ mol Zn} \times \dfrac{65.38 \text{ g Zn}}{1 \text{ mol}} = 0.365 \text{ g Zn}$

103. kelvin (absolute)

104. twice

Chapter 13: Gases

105. Gases consist of tiny particles, which are so small that the fraction of the bulk volume of the gas occupied by the particles is negligible. The particles of a gas are in constant random motion and collide with the walls of the container (giving rise to the *pressure* of the gas). The particles of a gas do not attract or repel each other. The average kinetic energy of the particles of a gas is reflected in the *temperature* of the gas sample.

106. In both cases, the gas particles will uniformly distribute throughout both flasks.

   Case 1:

   $P_1V_1 = P_2V_2$

   $P_1(2X) = P_2(3X)$

   $\left(\dfrac{2}{3}\right)P_1 = P_2$

   Case 2:

   $P_1V_1 = P_2V_2$

   $P_1(X) = P_2(2X)$

   $\left(\dfrac{1}{2}\right)P_1 = P_2$

   Case 1: The drawing will have four particles in the left and two particles in the right.

   Case 2: The drawing will have three particles in the left and three particles in the right.

107. sum

108. First determine what volume the helium in the tank would have if it were at a pressure of 755 mm Hg (corresponding to the pressure the gas will have in the balloons).

   8.40 atm = 6384 mm Hg

   $V_2 = (25.2 \text{ L}) \times \dfrac{6384 \text{ mm Hg}}{755 \text{ mm Hg}} = 213 \text{ L}$

   Allowing for the fact that 25.2 L of He will have to remain in the tank, this leaves 213 − 25.2 = 187.8 L of He for filling the balloons.

   $187.8 \text{ L He} \times \dfrac{1 \text{ balloon}}{1.50 \text{ L He}} = 125 \text{ balloons}$

109. A decrease in temperature would tend to make the volume of the weather balloon *decrease*. As the overall volume of a weather balloon *increases* when it rises to higher altitudes, the contribution to the new volume of the gas from the decrease in pressure must be more important than the decrease in temperature (the temperature change in kelvins is not as dramatic as it seems in degrees Celsius).

110. According to the balanced chemical equation, when 1 mol of $(NH_4)_2CO_3$ reacts, a total of 4 moles of gaseous substances is produced.

   molar mass $(NH_4)_2CO_3$ = 96.09 g; 453 °C = 726 K

   $52.0 \text{ g} \times \dfrac{1 \text{ mol}}{96.09 \text{ g}} = 0.541 \text{ mol}$

   As 0.541 mol of $(NH_4)_2CO_3$ reacts, 4(0.541) = 2.16 mol of gaseous products result.

   $V = \dfrac{nRT}{P} = \dfrac{(2.16 \text{ mol})(0.08206 \text{ L atm mol}^{-1} \text{ K}^{-1})(726 \text{ K})}{(1.04 \text{ atm})} = 124 \text{ L}$

111. $CaCO_3(s) \rightarrow CaO(s) + CO_2(g)$

774 torr = 1.018 atm; 55°C + 273 = 328 K; molar mass $CaCO_3$ = 100.1 g

$$10.0 \text{ g CaCO}_3 \times \frac{1 \text{ mol}}{100.1 \text{ g}} = 0.0999 \text{ mol CaCO}_3$$

From the balanced equation, 0.0999 mol CO_2 will be produced.

$$V = \frac{nRT}{P} = \frac{(0.0999 \text{ mol})(0.08206 \text{ L atm mol}^{-1} \text{ K}^{-1})(328 \text{ K})}{(1.018 \text{ atm})} = 2.64 \text{ L CO}_2$$

112. $CaCO_3(s) + 2H^+(aq) \rightarrow Ca^{2+}(aq) + H_2O(l) + CO_2(g)$

molar mass $CaCO_3$ = 100.1 g; 60°C + 273 = 333 K

$$10.0 \text{ g CaCO}_3 \times \frac{1 \text{ mol}}{100.1 \text{ g}} = 0.0999 \text{ mol CaCO}_3 = 0.0999 \text{ mol CO}_2 \text{ also}$$

$P_{\text{carbon dioxide}} = P_{\text{total}} - P_{\text{water vapor}}$

$P_{\text{carbon dioxide}} = 774$ mm Hg $- 149.4$ mm Hg $= 624.6$ mm Hg $= 0.822$ atm

$$V_{\text{wet}} = \frac{nRT}{P} = \frac{(0.0999 \text{ mol})(0.08206 \text{ L atm mol}^{-1} \text{ K}^{-1})(333 \text{ K})}{(0.822 \text{ atm})} = 3.32 \text{ L wet CO}_2$$

$$V_{\text{dry}} = 3.32 \text{ L} \times \frac{624.6 \text{ mm Hg}}{774 \text{ mm Hg}} = 2.68 \text{ L}$$

113. $2S(s) + 3O_2(g) \rightarrow 2SO_3(g)$

350.°C + 273 = 623 K; molar mass S = 32.07 g

$$5.00 \text{ g} \times \frac{1 \text{ mol S}}{32.07 \text{ g S}} = 0.1559 \text{ mol S}$$

$$0.1559 \text{ mol S} \times \frac{3 \text{ mol O}_2}{2 \text{ mol S}} = 0.2339 \text{ mol O}_2$$

$$V = \frac{nRT}{P} = \frac{(0.2339 \text{ mol})(0.08206 \text{ L atm mol}^{-1} \text{ K}^{-1})(623 \text{ K})}{(5.25 \text{ atm})} = 2.28 \text{ L O}_2$$

114. (b)

115. molar mass He = 4.003 g

$$10.0 \text{ g He} \times \frac{1 \text{ mol He}}{4.003 \text{ g He}} = 2.498 \text{ mol He}$$

$$2.498 \text{ mol} \times \frac{22.4 \text{ L}}{1 \text{ mol}} = 56.0 \text{ L He}$$

Chapter 13: Gases

116. $n_1 = 0.214$ mol $\qquad n_2 = 0.375$ mol

$V_1 = 652$ mL $\qquad V_2 = ?$

$$V_2 = \frac{V_1 n_2}{n_1} = \frac{(652 \text{ mL})(0.375 \text{ mol})}{0.214 \text{ mol}} = 1140 \text{ mL} = 1.14 \text{ L}$$

117. a. $0.903 \text{ atm} \times \dfrac{760 \text{ mm Hg}}{1 \text{ atm}} = 686$ mm Hg

b. $2.1240 \times 10^6 \text{ Pa} \times \dfrac{760 \text{ mm Hg}}{101,325 \text{ Pa}} = 1.5931 \times 10^4$ mm Hg

c. $445 \text{ kPa} \times \dfrac{760 \text{ mm}}{101.325 \text{ kPa}} = 3.34 \times 10^3$ mm Hg

d. 342 torr = 342 mm Hg

118. a. $645 \text{ mm Hg} \times \dfrac{101,325 \text{ Pa}}{760 \text{ mm Hg}} = 8.60 \times 10^4$ Pa

b. 221 kPa = 221×10^3 Pa = 2.21×10^5 Pa

c. $0.876 \text{ atm} \times \dfrac{101,325 \text{ Pa}}{1 \text{ atm}} = 8.88 \times 10^4$ Pa

d. $32 \text{ torr} \times \dfrac{101,325 \text{ Pa}}{760 \text{ torr}} = 4.3 \times 10^3$ Pa

119. a. 1002 mm Hg = 1.318 atm

$$V = 123 \text{ L} \times \frac{4.56 \text{ atm}}{1.318 \text{ atm}} = 426 \text{ L}$$

b. 25.2 mm Hg = 0.0332 atm

$$P = 0.0332 \text{ atm} \times \frac{634 \text{ mL}}{166 \text{ mL}} = 0.127 \text{ atm}$$

c. 511 torr = 6.81×10^4 Pa = 68.1 kPa

$$V = 443 \text{ L} \times \frac{68.1 \text{ kPa}}{1.05 \text{ kPa}} = 2.87 \times 10^4 \text{ L}$$

120. a. 1.00 mm Hg = 1.00 torr

$$V = 255 \text{ mL} \times \frac{1.00 \text{ torr}}{2.00 \text{ torr}} = 128 \text{ mL}$$

b. 1.0 atm = 101.325 kPa

$$V = 1.3 \text{ L} \times \frac{1.0 \text{ kPa}}{101.325 \text{ kPa}} = 1.3 \times 10^{-2} \text{ L}$$

Chapter 13:   Gases

c. 1.0 mm Hg = 0.133 kPa

$$V = 1.3 \text{ L} \times \frac{1.0 \text{ kPa}}{0.133 \text{ kPa}} = 9.8 \text{ L}$$

121. Assume the pressure at sea level to be 1 atm (760 mm Hg). Calculate the volume the balloon would have if it rose to the point where the pressure has dropped to 500 mm Hg. If this calculated volume is greater than the balloon's specified maximum volume (2.5 L), the balloon will burst.

$$2.0 \text{ L} \times \frac{760 \text{ mm Hg}}{500 \text{ mm Hg}} = 3.0 \text{ L} > 2.5 \text{ L. The balloon will burst.}$$

122. $1.52 \text{ L} = 1.52 \times 10^3 \text{ mL}$

$$755 \text{ mm Hg} \times \frac{1.52 \times 10^3 \text{ mL}}{450 \text{ mL}} = 2.55 \times 10^3 \text{ mm Hg}$$

123. 22°C + 273 = 295 K; 100°C + 273 = 373 K

$$729 \text{ mL} \times \frac{373 \text{ K}}{295 \text{ K}} = 922 \text{ mL}$$

124. a. 74°C + 273 = 347 K; −74°C + 273 = 199 K

$$100. \text{ mL} \times \frac{199 \text{ K}}{347 \text{ K}} = 57.3 \text{ mL}$$

b. 100°C + 273 = 373 K

$$373 \text{ K} \times \frac{600 \text{ mL}}{500 \text{ mL}} = 448 \text{ K } (175°\text{C})$$

c. zero (the volume of any gas sample becomes zero at 0 K)

125. a. 0°C + 273 = 273 K

$$273 \text{ K} \times \frac{44.4 \text{ L}}{22.4 \text{ L}} = 541 \text{ K } (268°\text{C})$$

b. −272°C + 273 = 1 K; 25°C + 273 = 298 K

$$1.0 \times 10^{-3} \text{ mL} \times \frac{298 \text{ K}}{1 \text{ K}} = 0.30 \text{ mL}$$

c. −40°C + 273 = 233 K

$$233 \text{ K} \times \frac{1000 \text{ L}}{32.3 \text{ L}} = 7.21 \times 10^3 \text{ K } (6940°\text{C})$$

126. 12°C + 273 = 285 K; 192°C + 273 = 465 K

$$75.2 \text{ mL} \times \frac{465 \text{ K}}{285 \text{ K}} = 123 \text{ mL}$$

Chapter 13: Gases

127. 5.12 g O_2 = 0.160 mol; 25.0 g O_2 = 0.781 mol

$$6.21 \text{ L} \times \frac{0.781 \text{ mol}}{0.160 \text{ mol}} = 30.3 \text{ L}$$

128. Three changes you can make to double the volume are:

1) *increase the temperature (double the temperature in the kelvin scale)*; If the temperature is increased, the gas particles have more kinetic energy and will hit the piston with more force (and more pressure). Therefore, the piston will move up until the pressure inside the container is the same as outside the container (causing the volume to increase).
2) *add more moles of gas to the container (double the amount)*; By adding more moles of gas to the container, gas particles will hit the walls of the container more frequently (and thus exert more pressure). The piston will move up until the pressure inside the container is the same as outside the container (causing the volume to increase).
3) *decrease the pressure outside the container (by half)*; By decreasing the pressure outside the container, the pressure inside becomes greater than the pressure outside. The gas particles inside will push the piston up until the pressure inside the container is the same as outside the container (causing the volume to increase).

129. a. $V = 142$ mL $= 0.142$ L

$$T = \frac{PV}{nR} = \frac{(21.2 \text{ atm})(0.142 \text{ L})}{(0.432 \text{ mol})(0.08206 \text{ L atm mol}^{-1} \text{ K}^{-1})} = 84.9 \text{ K}$$

b. $V = 1.23$ mL $= 0.00123$ L

$$P = \frac{nRT}{V} = \frac{(0.000115 \text{ mol})(0.08206 \text{ L atm mol}^{-1} \text{ K}^{-1})(293 \text{ K})}{(0.00123 \text{ L})} = 2.25 \text{ atm}$$

c. $P = 755$ mm Hg $= 0.993$ atm; $T = 131°C + 273 = 404$ K

$$V = \frac{nRT}{P} = \frac{(0.473 \text{ mol})(0.08206 \text{ L atm mol}^{-1} \text{ K}^{-1})(404 \text{ K})}{(0.993 \text{ atm})} = 15.8 \text{ L} = 1.58 \times 10^4 \text{ mL}$$

130. a. $V = 21.2$ mL $= 0.0212$ L

$$T = \frac{PV}{nR} = \frac{(1.034 \text{ atm})(0.0212 \text{ L})}{(0.00432 \text{ mol})(0.08206 \text{ L atm mol}^{-1} \text{ K}^{-1})} = 61.8 \text{ K}$$

b. $V = 1.73$ mL $= 0.00173$ L

$$P = \frac{nRT}{V} = \frac{(0.000115 \text{ mol})(0.08206 \text{ L atm mol}^{-1} \text{ K}^{-1})(182 \text{ K})}{(0.00173 \text{ L})} = 0.993 \text{ atm}$$

c. $P = 1.23$ mm Hg $= 0.00162$ atm; $T = 152°C + 273 = 425$ K

$$V = \frac{nRT}{P} = \frac{(0.773 \text{ mol})(0.08206 \text{ L atm mol}^{-1} \text{ K}^{-1})(425 \text{ K})}{(0.00162 \text{ atm})} = 1.66 \times 10^4 \text{ L}$$

Chapter 13: Gases

131. molar mass N_2 = 28.02 g; T = 26°C + 273 = 299 K

$$n = 14.2 \text{ g N}_2 \times \frac{1 \text{ mol N}_2}{28.02 \text{ g N}_2} = 0.507 \text{ mol N}_2$$

$$P = \frac{nRT}{V} = \frac{(0.507 \text{ mol})(0.08206 \text{ L atm mol}^{-1} \text{ K}^{-1})(299 \text{ K})}{(10.0 \text{ L})} = 1.24 \text{ atm}$$

132. 27°C + 273 = 300 K

The number of moles of gas it takes to fill the 100. L tanks to 120 atm at 27°C is independent of the identity of the gas.

$$n = \frac{PV}{RT} = \frac{(120 \text{ atm})(100. \text{ L})}{(0.08206 \text{ L atm mol}^{-1} \text{ K}^{-1})(300 \text{ K})} = 487 \text{ mol}$$

487 mol of *any* gas will fill the tanks to the required specifications.

molar masses: CH_4, 16.0 g; N_2, 28.0 g; CO_2, 44.0 g

for CH_4: (487 mol)(16.0 g/mol) = 7792 g = 7.79 kg CH_4

for N_2: (487 mol)(28.0 g/mol) = 13,636 g = 13.6 kg N_2

for CO_2: (487 mol)(44.0 g/mol) = 21,428 g = 21.4 kg CO_2

133. molar mass He = 4.003 g

$$n = 4.00 \text{ g He} \times \frac{1 \text{ mol He}}{4.003 \text{ g He}} = 0.999 \text{ mol He}$$

$$T = \frac{PV}{nR} = \frac{(1.00 \text{ atm})(22.4 \text{ L})}{(0.999 \text{ mol})(0.08206 \text{ L atm mol}^{-1} \text{ K}^{-1})} = 273 \text{ K} = 0°C$$

134. molar mass of O_2 = 32.00 g; 55 mg = 0.055 g

$$n = 0.055 \text{ g} \times \frac{1 \text{ mol O}_2}{32.00 \text{ g O}_2} = 0.0017 \text{ mol}$$

V = 100. mL = 0.100 L; T = 26°C + 273 = 299 K

$$P = \frac{nRT}{V} = \frac{(0.0017 \text{ mol})(0.08206 \text{ L atm mol}^{-1} \text{ K}^{-1})(299 \text{ K})}{(0.100 \text{ L})} = 0.42 \text{ atm}$$

135. P_1 = 1.0 atm $\qquad$ P_2 = 220 torr = 0.289 atm

V_1 = 1.0 L $\qquad$ V_2 = ?

T_1 = 23°C + 273 = 296 K $\qquad$ T_2 = −31°C = 242 K

$$V_2 = \frac{T_2 P_1 V_1}{T_1 P_2} = \frac{(242 \text{ K})(1.0 \text{ atm})(1.0 \text{ L})}{(296 \text{ K})(0.289 \text{ atm})} = 2.8 \text{ L}$$

Chapter 13: Gases

136. molar mass N_2 = 28.02 g; 3.20 g = 0.114 mol; 8.80 g = 0.314 mol

n_1 = 0.114 mol n_2 = 0.114 mol + 0.314 mol = 0.428 mol

V_1 = 1.71 L V_2 = ?

$$V_2 = \frac{V_1 n_2}{n_1} = \frac{(1.71 \text{ L})(0.428 \text{ mol})}{0.114 \text{ mol}} = 6.41 \text{ L}$$

137. molar mass of O_2 = 32.00 g; 25°C + 273 = 298 K

$$50. \text{ g } O_2 \times \frac{1 \text{ mol } O_2}{32.00 \text{ g } O_2} = 1.56 \text{ mol } O_2$$

total number of moles of gas = 1.0 mol N_2 + 1.56 mol O_2 = 2.56 mol

$$P = \frac{nRT}{V} = \frac{(2.56 \text{ mol})(0.08206 \text{ L atm mol}^{-1} \text{ K}^{-1})(298 \text{ K})}{(5.0 \text{ L})} = 13 \text{ atm}$$

138. a. $P_{Total} = P_{H_2} + P_{N_2} + P_{O_2}$

P_{Total} = 325 torr + 475 torr + 650. torr = 1450. torr

b. Since the volume and temperature are constant, there is a direct relationship between the pressure and number of moles. The gas with the highest pressure (which is O_2) must contain the greatest number of moles and collide with the walls more frequently.

139. The pressures must be expressed in the same units, either mm Hg or atm.

$P_{hydrogen} = P_{total} - P_{water\ vapor}$

1.023 atm = 777.5 mm Hg

$P_{hydrogen}$ = 777.5 mm Hg – 42.2 mm Hg = 735.3 mm Hg

42.2 mm Hg = 0.056 atm

$P_{hydrogen}$ = 1.023 atm – 0.056 atm = 0.967 atm

140. $N_2(g) + 3H_2(g) \rightarrow 2NH_3(g)$

molar mass of NH_3 = 17.03 g; 11°C + 273 = 284 K

$$5.00 \text{ g } NH_3 \times \frac{1 \text{ mol } NH_3}{17.03 \text{ g } NH_3} = 0.294 \text{ mol } NH_3 \text{ to be produced}$$

$$0.294 \text{ mol } NH_3 \times \frac{1 \text{ mol } N_2}{2 \text{ mol } NH_3} = 0.147 \text{ mol } N_2 \text{ required}$$

$$0.294 \text{ mol } NH_3 \times \frac{3 \text{ mol } H_2}{2 \text{ mol } NH_3} = 0.441 \text{ mol } H_2 \text{ required}$$

$$V_{nitrogen} = \frac{nRT}{P} = \frac{(0.147 \text{ mol})(0.08206 \text{ L atm mol}^{-1} \text{ K}^{-1})(284 \text{ K})}{(0.998 \text{ atm})} = 3.43 \text{ L N}_2$$

$$V_{hydrogen} = \frac{nRT}{P} = \frac{(0.441 \text{ mol})(0.08206 \text{ L atm mol}^{-1} \text{ K}^{-1})(284 \text{ K})}{(0.998 \text{ atm})} = 10.3 \text{ L H}_2$$

141. $C_6H_{12}O_6(s) + 6O_2(g) \rightarrow CO_2(g) + 6H_2O(g)$

    molar mass of $C_6H_{12}O_6$ = 180. g; 28°C + 273 = 301 K

    $5.00 \text{ g C}_6H_{12}O_6 \times \dfrac{1 \text{ mol C}_6H_{12}O_6}{180. \text{ g C}_6H_{12}O_6} = 0.02778 \text{ mol C}_6H_{12}O_6$

    $0.02778 \text{ mol C}_6H_{12}O_6 \times \dfrac{6 \text{ mol O}_2}{1 \text{ mol C}_6H_{12}O_6} = 0.1667 \text{ mol O}_2$

    $$V_{oxygen} = \frac{nRT}{P} = \frac{(0.1667 \text{ mol})(0.08206 \text{ L atm mol}^{-1} \text{ K}^{-1})(301 \text{ K})}{(0.976 \text{ atm})} = 4.22 \text{ L}$$

    Because the coefficients of $CO_2(g)$ and $H_2O(g)$ in the balanced chemical equation happen to be the same as the coefficient of $O_2(g)$, the calculations for the volumes of these gases produced are identical: 4.22 L of each gaseous product is produced.

142. $2Cu_2S(s) + 3O_2(g) \rightarrow 2Cu_2O(s) + 2SO_2(g)$

    molar mass Cu_2S = 159.2 g; 27.5°C + 273 = 301 K

    $25 \text{ g Cu}_2S \times \dfrac{1 \text{ mol Cu}_2S}{159.2 \text{ g Cu}_2S} = 0.1570 \text{ mol Cu}_2S$

    $0.1570 \text{ mol Cu}_2S \times \dfrac{3 \text{ mol O}_2}{2 \text{ mol Cu}_2S} = 0.2355 \text{ mol O}_2$

    $$V_{oxygen} = \frac{nRT}{P} = \frac{(0.2355 \text{ mol})(0.08206 \text{ L atm mol}^{-1} \text{ K}^{-1})(301 \text{ K})}{(0.998 \text{ atm})} = 5.8 \text{ L O}_2$$

    $0.1570 \text{ mol Cu}_2S \times \dfrac{2 \text{ mol SO}_2}{2 \text{ mol Cu}_2S} = 0.1570 \text{ mol SO}_2$

    $$V_{sulfur\ dioxide} = \frac{nRT}{P} = \frac{(0.1570 \text{ mol})(0.08206 \text{ L atm mol}^{-1} \text{ K}^{-1})(301 \text{ K})}{(0.998 \text{ atm})} = 3.9 \text{ L SO}_2$$

143. $2NaHCO_3(s) \rightarrow Na_2CO_3(s) + H_2O(g) + CO_2(g)$

    molar mass $NaHCO_3$ = 84.01 g; 29°C + 273 = 302 K; 769 torr = 1.012 atm

    $1.00 \text{ g NaHCO}_3 \times \dfrac{1 \text{ mol NaHCO}_3}{84.01 \text{ g NaHCO}_3} = 0.01190 \text{ mol NaHCO}_3$

    $0.01190 \text{ mol NaHCO}_3 \times \dfrac{1 \text{ mol H}_2O}{2 \text{ mol NaHCO}_3} = 0.00595 \text{ mol H}_2O$

Chapter 13: Gases

Because $H_2O(g)$ and $CO_2(g)$ have the same coefficients in the balanced chemical equation for the reaction, if 0.00595 mol H_2O is produced, then 0.00595 mol CO_2 must also be produced. The total number of moles of gaseous substances produced is thus $0.00595 + 0.00595 = 0.0119$ mol.

$$V_{total} = \frac{nRT}{P} = \frac{(0.0119 \text{ mol})(0.08206 \text{ L atm mol}^{-1}\text{ K}^{-1})(302 \text{ K})}{(1.012 \text{ atm})} = 0.291 \text{ L}$$

144. One mole of any ideal gas occupies 22.4 L at STP.

$$35 \text{ mol N}_2 \times \frac{22.4 \text{ L}}{1 \text{ mol}} = 7.8 \times 10^2 \text{ L}$$

145. $P_1 = 0.987$ atm $\quad\quad\quad\quad\quad\quad\quad\quad\quad P_2 = 1.00$ atm

$V_1 = 125$ L $\quad\quad\quad\quad\quad\quad\quad\quad\quad\quad V_2 = ?$

$T_1 = 25°C + 273 = 298$ K $\quad\quad\quad\quad\quad T_2 = 273$ K

$$V_2 = \frac{T_2 P_1 V_1}{T_1 P_2} = \frac{(273 \text{ K})(0.987 \text{ atm})(125 \text{ L})}{(298 \text{ K})(1.00 \text{ atm})} = 113 \text{ L}$$

146. a. Assuming the temperature is the same on both sides, the initial pressure of helium is greater because there are more gas particles colliding with the walls of its container (within the same volume).

$$\frac{P_{He}}{P_{Ne}} = \frac{n_{He}}{n_{Ne}} = \frac{6}{4} = 1.5 \text{ times greater}$$

b. When the valve is opened, the gases disperse uniformly throughout the entire apparatus (i.e., two Ne and three He in each vessel).

c. From the diagram, when the volume doubles the pressure is halved so

$$P_{f(Ne)} = \frac{1}{2} P_{o(Ne)}$$

$$P_{f(He)} = \frac{1}{2} P_{o(He)}$$

d. The original pressure of helium is 1.5 times the original pressure of neon.

$P_{o(He)} = 1.5\, P_{o(Ne)}$

The final pressure is

$$P_f = \frac{1}{2} P_{o(Ne)} + \frac{1}{2} P_{o(He)} = \frac{1}{2} P_{o(Ne)} + \frac{1.5}{2} P_{o(Ne)}$$

$$= \frac{2.5}{2} P_{o(Ne)} = 1.25 P_{o(Ne)}$$

147. $CaCO_3(s) \rightarrow CaO(s) + CO_2(g)$

molar mass of $CaCO_3$ = 100.1 g

$$27.5 \text{ g CaCO}_3 \times \frac{1 \text{ mol CaCO}_3}{100.1 \text{ g CaCO}_3} = 0.275 \text{ mol CaCO}_3$$

From the balanced chemical equation, if 0.275 mol of $CaCO_3$ reacts, then 0.275 mol of $CaCO_3$ will be produced.

$$0.275 \text{ mol H}_2 \times \frac{22.4 \text{ L}}{1 \text{ mol}} = 6.16 \text{ L}$$

148. The solution is only 50% H_2O_2. Therefore 125 g solution = 62.5 g H_2O_2

molar mass of H_2O_2 = 34.02 g; T = 27°C = 300 K; P = 764 mm Hg = 1.01 atm

$$62.5 \text{ g H}_2O_2 \times \frac{1 \text{ mol}}{34.02 \text{ g}} = 1.84 \text{ mol H}_2O_2$$

$$1.84 \text{ mol H}_2O_2 \times \frac{1 \text{ mol O}_2}{2 \text{ mol H}_2O_2} = 0.920 \text{ mol O}_2$$

$$V = \frac{nRT}{P} = \frac{(0.920 \text{ mol})(0.08206 \text{ L atm mol}^{-1} \text{ K}^{-1})(300 \text{ K})}{(1.01 \text{ atm})} = 22.4 \text{ L}$$

149. $2NaN_3(s) \rightarrow 2Na(s) + 3N_2(g)$

molar mass of NaN_3 = 65.02 g; T = 0°C = 273 K; P = 1.00 atm

$$n = \frac{PV}{RT} = \frac{(1.00 \text{ atm})(70.0 \text{ L})}{(0.08206 \text{ L atm mol}^{-1} \text{ K}^{-1})(273 \text{ K})} = 3.12 \text{ mol N}_2$$

$$3.12 \text{ mol N}_2 \times \frac{2 \text{ mol NaN}_3}{3 \text{ mol N}_2} = 2.08 \text{ mol NaN}_3$$

$$2.08 \text{ mol NaN}_3 \times \frac{65.02 \text{ g NaN}_3}{1 \text{ mol NaN}_3} = 135 \text{ g NaN}_3$$

150. $V = .04 \times 500 \text{ mL} = 20 \text{ mL CO}_2 = 0.02 \text{ L}$

molar mass of CO_2 = 44.01 g; T = 25°C = 298 K; P = 1.00 atm

$$n = \frac{PV}{RT} = \frac{(1.00 \text{ atm})(0.02 \text{ L})}{(0.08206 \text{ L atm mol}^{-1} \text{ K}^{-1})(298 \text{ K})} = 8 \times 10^{-4} \text{ mol CO}_2$$

$$8 \times 10^{-4} \text{ mol CO}_2 \times \frac{44.01 \text{ g CO}_2}{1 \text{ mol CO}_2} = 0.04 \text{ g CO}_2$$

Chapter 13: Gases

151. $PV = nRT$

P(atm)	V(L)	n(mol)	T
6.74	10.4	2.00	155°C
0.300	1.74	0.0410	155 K
4.47	25.0	2.19	349°C
140.	2.25	10.5	93°C

152. (b), (c), (d)

    molar mass of N_2 = 28.02 g; 28 g = 1.0 mol

    molar mass of O_2 = 32.00 g; 28 g = 0.88 mol; 32 g = 1.0 mol

    Double the moles of gas would need to be present in order to double the pressure. Adding 28 g of O_2 is not double the moles, as in (a), but adding 32 g of O_2 is double the moles, as in (d). ($n_1/P_1 = n_2/P_2$)

    $T = -73°C = 200.$ K; $T = 127°C = 400.$ K; $T = 30°C = 303$ K; $T = 60°C = 333$ K

    Doubling the temperature (in Kelvin) will double the pressure, as in (b), but not in (e). ($P_1/T_1 = P_2/T_2$)

    Cutting the volume by half will double the pressure, as in (c). ($P_1V_1 = P_2V_2$)

153. $n_1 = 150.0$ mol $\qquad n_2 = ?$

    $P_1 = 8.93$ MPa $\qquad P_2 = 2.00$ MPa

    $T_1 = 25°C = 298$ K $\qquad T_2 = 19°C = 292$ K

    $n_2 = \dfrac{P_2 n_1 T_1}{P_1 T_2} = \dfrac{(2.00 \text{ MPa})(150.0 \text{ mol})(298 \text{ K})}{(8.93 \text{ MPa})(292 \text{ K})} = 34.3$ mol Ar

    $34.3 \text{ mol Ar} \times \dfrac{39.95 \text{ g Ar}}{1 \text{ mol Ar}} = 1.37 \times 10^3$ g Ar

154. $V_1 = 855$ L $\qquad V_2 = ?$

    $P_1 = 730$ torr $\qquad P_2 = 605$ torr

    $T_1 = 25°C = 298$ K $\qquad T_2 = 15°C = 288$ K

    $V_2 = \dfrac{P_1 V_1 T_2}{T_1 P_2} = \dfrac{(730 \text{ torr})(855 \text{ L})(288 \text{ K})}{(298 \text{ K})(605 \text{ torr})} = 997$ L

    $\Delta V = 997$ L $- 855$ L $= 142$ L

155. $V = 936$ mL $= 0.936$ L $\qquad P = 0.967$ atm $\qquad T = 31°C = 304$ K

    mass of gas $= 135.87$ g $- 134.66$ g $= 1.21$ g

    $n = \dfrac{PV}{RT} = \dfrac{(0.967 \text{ atm})(0.936 \text{ L})}{(0.08206 \text{ L atm mol}^{-1} \text{ K}^{-1})(304 \text{ K})} = 0.0363$ mol

    molar mass of gas $= 1.21$ g$/0.0363$ mol $= 33.3$ g/mol

156. $Xe + 2F_2 \rightarrow XeF_4$

molar mass of $XeF_4 = 207.3$ g; $T = 400°C = 673$ K; $V = 20.0$ L;

$P_{Xenon} = 0.859$ atm; $P_{Fluorine} = 1.37$ atm

$$n_{Xenon} = \frac{PV}{RT} = \frac{(0.859 \text{ atm})(20.0 \text{ L})}{(0.08206 \text{ L atm mol}^{-1} \text{ K}^{-1})(673 \text{ K})} = 0.311 \text{ mol Xe}$$

$$n_{Fluorine} = \frac{PV}{RT} = \frac{(1.37 \text{ atm})(20.0 \text{ L})}{(0.08206 \text{ L atm mol}^{-1} \text{ K}^{-1})(673 \text{ K})} = 0.496 \text{ mol F}_2$$

F_2 is the limiting reactant (only 0.248 mol Xe needed).

$$0.496 \text{ mol F}_2 \times \frac{1 \text{ mol XeF}_4}{2 \text{ mol F}_2} \times \frac{207.3 \text{ g XeF}_4}{1 \text{ mol XeF}_4} = 51.4 \text{ g XeF}_4$$

157. $CaSiO_3(s) + 6HF(g) \rightarrow CaF_2(aq) + SiF_4(g) + 3H_2O(l)$

molar mass of $CaSiO_3 = 116.17$ g; molar mass of $SiF_4 = 104.09$ g; molar mass of $H_2O = 18.016$ g

$T = 27°C = 300.$ K; $V = 31.8$ L; $P = 1.00$ atm

$$32.9 \text{ g CaSiO}_3 \times \frac{1 \text{ mol CaSiO}_3}{116.17 \text{ g CaSiO}_3} = 0.283 \text{ mol CaSiO}_3$$

$$n_{HF} = \frac{PV}{RT} = \frac{(1.00 \text{ atm})(31.8 \text{ L})}{(0.08206 \text{ L atm mol}^{-1} \text{ K}^{-1})(300. \text{ K})} = 1.29 \text{ mol HF}$$

HF is the limiting reactant (only 0.215 mol $CaSiO_3$ needed).

$$1.29 \text{ mol HF} \times \frac{1 \text{ mol SiF}_4}{6 \text{ mol HF}} \times \frac{104.09 \text{ g SiF}_4}{1 \text{ mol SiF}_4} = 22.4 \text{ g SiF}_4$$

$$1.29 \text{ mol HF} \times \frac{3 \text{ mol H}_2\text{O}}{6 \text{ mol HF}} \times \frac{18.016 \text{ g H}_2\text{O}}{1 \text{ mol H}_2\text{O}} = 11.6 \text{ g H}_2\text{O}$$

158. (a), (c), and (d); Only doubling the temperature in Kelvin will double the volume, thus (b) is not true.

CHAPTER 14

Liquids and Solids

1. Gases have *lower* densities than liquids or solids.

2. Liquids and solids are *less* compressible than are gases.

3. Ice floats on liquid water and it could only do that if it were less dense than liquid water. We also know that water expands in volume when it freezes, and since the mass of the water does not change, if the water expands when it freezes then the density must decrease as the same mass becomes dispersed in a larger volume.

4. Since it requires so much more energy to vaporize water than to melt ice, this suggests that the gaseous state is significantly different from the liquid state, but that the liquid and solid state are relatively similar.

5. From –5°C to 0°C, molecules within the solid vibrate faster and faster as the temperature of the solid increases; at 0°C (the normal melting point), the solid melts as the heat being applied is used to break apart the forces between molecules; above 0°C, the molecules within the liquid move faster and faster as additional heat is applied.

6. See Figure 14.2.

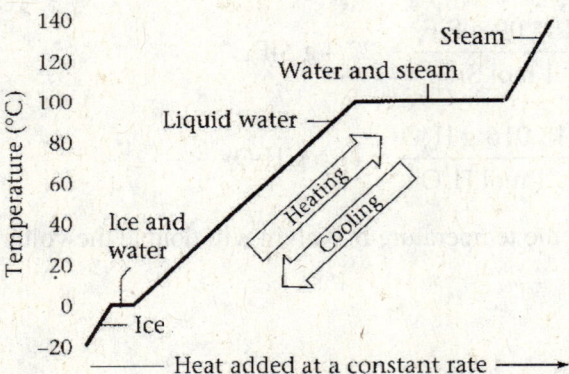

7. Changes in state for molecular solids are physical changes because no chemical bonds are broken within molecules.

8. When a solid is heated, the molecules begin to vibrate/move more quickly. When enough energy has been added to overcome the intermolecular forces that hold the molecules in a crystal lattice, the solid melts. As the liquid is heated, the molecules begin to move more quickly and more randomly. When enough energy has been added, molecules having sufficient kinetic energy will begin to escape from the surface of the liquid. Once the pressure of vapor coming from the liquid is equal to the pressure above the liquid, the liquid boils. Only intermolecular forces need to be overcome in this process: no chemical bonds are broken.

9. *Intra*molecular forces are the forces *within* a molecule itself (e.g., a covalent bond is an intramolecular force). *Inter*molecular forces are forces between or among *different* molecules. Consider liquid bromine, Br_2. *Intra*molecular forces (a covalent bond) are responsible for the fact that bromine atoms form discrete two-atom units (molecules) within the substance. *Inter*molecular forces between adjacent Br_2 molecules are responsible for the fact that the substance is a liquid at room temperature and pressure.

10. Intramolecular; intermolecular

11. In ice, water molecules are in more or less regular, fixed positions in the ice crystal; strong hydrogen bonding forces exist within the ice crystal holding the water molecules together. In liquid water, enough heat has been applied that the molecules are no longer fixed in position (but are more free to roam about in the bulk of the liquid); because the water molecules are still relatively close together, strong hydrogen bonding forces still exist, however, which keep the liquid together in one place. In steam, the water molecules possess enough kinetic energy that they have escaped from the liquid; because the water molecules are very far apart in steam, and because the molecules are moving very quickly, they do not exert any forces on each other (each water molecule in steam behaves independently).

12. The molar heat of fusion of a substance represents the quantity of energy that must be applied to melt one mole of the substance.

13. Heating curve for Substance X:

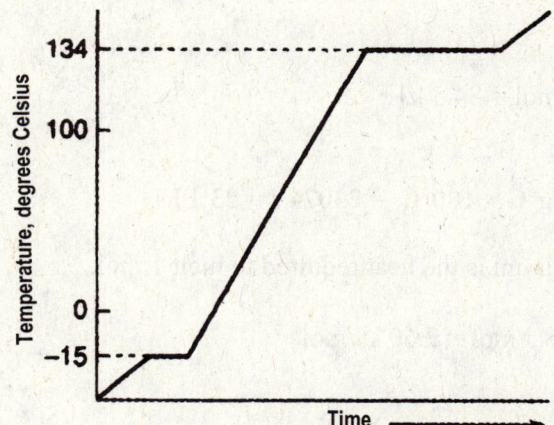

14. a. More energy is required to separate the atoms of the liquid into the freely-moving and widely-separated atoms of the vapor/gas.

   b. $1.00 \text{ g Al} \times \dfrac{1 \text{ mol Al}}{26.98 \text{ g Al}} \times \dfrac{293.4 \text{ kJ}}{1 \text{ mol Al}} = 10.9 \text{ kJ}$

   c. $5.00 \text{ g Al} \times \dfrac{1 \text{ mol Al}}{26.98 \text{ g Al}} \times \dfrac{-10.79 \text{ kJ}}{1 \text{ mol Al}} = -2.00 \text{ kJ}$ (heat is evolved)

   d. $0.105 \text{ mol Al} \times \dfrac{10.79 \text{ kJ}}{1 \text{ mol Al}} = 1.13 \text{ kJ}$

Chapter 14: Liquids and Solids

15. molar mass C_6H_6 = 78.11 g

    melt: $8.25 \text{ g } C_6H_6 \times \dfrac{1 \text{ mol } C_6H_6}{78.11 \text{ g } C_6H_6} \times \dfrac{9.92 \text{ kJ}}{1 \text{ mol } C_6H_6} = 1.05 \text{ kJ}$

    boil: $8.25 \text{ g } C_6H_6 \times \dfrac{1 \text{ mol } C_6H_6}{78.11 \text{ g } C_6H_6} \times \dfrac{30.7 \text{ kJ}}{1 \text{ mol } C_6H_6} = 3.24 \text{ kJ}$

    More energy is required to overcome the forces holding the molecules together in the liquid state (l to g) than to just allow the molecules to begin moving (s to l).

16. molar mass Ag = 107.9 g

    melt: $12.5 \text{ g Ag} \times \dfrac{1 \text{ mol Ag}}{107.9 \text{ g Ag}} \times \dfrac{11.3 \text{ kJ}}{1 \text{ mol Al}} = 1.31 \text{ kJ}$

    condense: $4.59 \text{ g Ag} \times \dfrac{1 \text{ mol Ag}}{107.9 \text{ g Ag}} \times \dfrac{-250. \text{ kJ}}{1 \text{ mol Ag}} = -10.6 \text{ kJ}$ (heat is evolved)

17. 25.0 g ice (H_2O) = 1.39 mol H_2O

    to melt the solid ice:

    1.39 mol H_2O × 6.02 kJ/mol = 8.35 kJ

    37.5 g liquid H_2O = 2.08 mol H_2O

    to vaporize the liquid water:

    2.08 mol H_2O × 40.6 kJ/mol = 84.5 kJ

    for H_2O going from 0° to 100.°C:

    Q = 55.2 g H_2O × 4.18 J/g°C × 100°C = 23074 J = 23.1 kJ

18. The *molar* heat of fusion of aluminum is the heat required to melt 1 mol.

    $\dfrac{113 \text{ J}}{1.00 \text{ g Na}} \times \dfrac{22.99 \text{ g Na}}{1 \text{ mol Na}} = 2598 \text{ J/mol} = 2.60 \text{ kJ/mol}$

19. The molecules would orient with the iodine ($\delta+$) on one molecule aimed toward the chlorine ($\delta-$) on an adjacent molecule: $^{\delta+}I \to Cl^{\delta-} \cdots ^{\delta+}I \to Cl^{\delta-}$

20. Dipole-dipole forces get *weaker* as the distance between the dipoles increases.

21. Hydrogen bonding is possible when hydrogen atoms are bonded to highly electronegative atoms such as oxygen, nitrogen, or fluorine. The small size of the hydrogen atom allows the dipoles to approach each other more closely than with other atoms.

22. Water molecules are able to form strong *hydrogen bonds* with each other. These bonds are an especially strong form of dipole-dipole forces and are only possible when hydrogen atoms are bonded to the most electronegative elements (N, O, and F). The particularly strong intermolecular forces in H_2O require much higher temperatures (higher energies) to be overcome in order to permit the liquid to boil. We take the fact that water has a much higher boiling point than the

other hydrogen compounds of the Group 6 elements as proof that a special force is at play in water (hydrogen bonding).

23. The magnitude of dipole-dipole interactions is strongly dependent on the distance between the dipoles. In the solid and liquid states, the molecular dipoles are quite close together. In the vapor phase (gaseous state), however, the molecules are too far apart from one another for dipole-dipole forces to be very strong.

24. London dispersion forces are instantaneous dipole forces that arise when the electron cloud of an atom is momentarily distorted and induces a dipole in an adjacent molecule, temporarily separating the centers of positive and negative charge in the atom.

25. a. London dispersion forces (nonpolar molecules)
    b. dipole-dipole forces (polar molecules); London dispersion forces
    c. hydrogen bonding (H bonded to O); London dispersion forces
    d. London dispersion forces (nonpolar molecules)

26. a. London dispersion forces (noble gas atoms)
    b. London dispersion forces (nonpolar molecules)
    c. dipole-dipole forces (polar molecules); London dispersion forces
    d. hydrogen bonding (H bonded to O); London dispersion forces

27. The boiling points increase with an increase in the molar mass of the noble gas. As the noble gas atoms increase in size, the valence electrons are farther from the nucleus. At greater distance from the nucleus, it is easy for another atom's electrons to momentarily distort the electron cloud of the noble gas atom. As the size of the noble gas atoms increases, so does the magnitude of the London dispersion forces.

28. An increase in the heat of fusion is observed for an increase in the size of the halogen atom involved (the electron cloud of a larger atom is more easily polarized by an approaching dipole, thus giving larger London dispersion forces).

29. Both water and ammonia molecules are capable of hydrogen bonding (H attached to O, and H attached to N, respectively). When ammonia gas is first bubbled into a sample of water, the ammonia molecules begin to hydrogen bond to the water molecules and to draw the water molecules closer together than they would ordinarily because of the increased intermolecular forces.

30. For a homogeneous mixture to be able to form at all, the forces between molecules of the two substances being mixed must be at least *comparable in magnitude* to the intermolecular forces within each *separate* substance. Apparently in the case of a water-ethanol mixture, the forces that exist when water and ethanol are mixed are stronger than water-water or ethanol-ethanol forces in the separate substances. This allows ethanol and water molecules to approach each other more closely in the mixture than either substance's molecules could approach a like molecule in separate substances. There is strong hydrogen bonding in both ethanol and water.

31. Evaporation is when a substance passes from the liquid state to the gaseous state; condensation is the reverse of this process; evaporation is endothermic, condensation is exothermic.

Chapter 14: Liquids and Solids

32. Vapor pressure is the pressure of vapor present *at equilibrium* above a liquid in a sealed container at a particular temperature. When a liquid is placed in a closed container, molecules of the liquid evaporate freely into the empty space above the liquid. As the number of molecules present in the vapor state increases with time, vapor molecules begin to rejoin the liquid state (condense). Eventually a dynamic equilibrium is reached between evaporation and condensation in which the net number of molecules present in the vapor phase becomes *constant* with time. Since vapor pressure increases with temperature, the vapor pressure of the solvent is higher on a warm day.

33. A dynamic equilibrium exists when two opposite processes are going on at the same speed, so there is no *net* change in the system. When a liquid evaporates into the empty space above it, eventually, as more molecules accumulate in the vapor state, condensation will be occurring at the same rate as further evaporation. When evaporation and condensation are going on at the same rate, a fixed pressure of vapor will have developed, and there will be no further net change in the amount of liquid present.

34. A liquid is injected at the bottom of the column of mercury and rises to the surface of the mercury, where the liquid evaporates into the vacuum above the mercury column. As the liquid evaporates, the pressure of vapor increases in the space above the mercury, and presses down on the mercury. The level of mercury, therefore, drops, and the amount by which the mercury level drops (in mm Hg) is equivalent to the vapor pressure of the liquid.

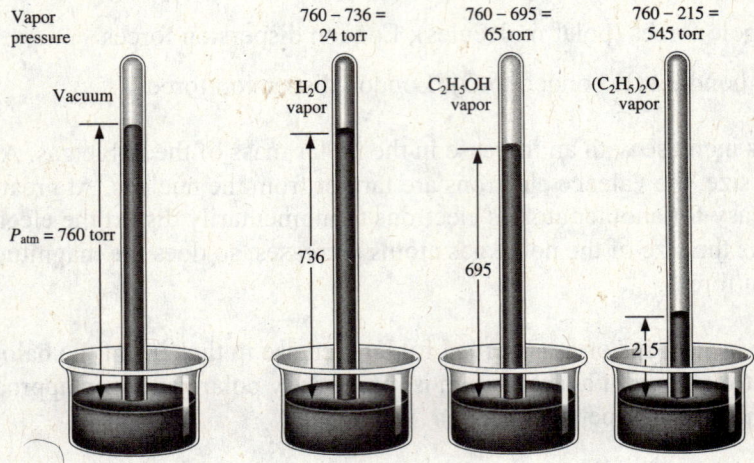

In the picture, the left tube represents a barometer—a tube of mercury inverted into a dish of mercury, with a vacuum above the mercury column: the height of the mercury column represents the pressure of the atmosphere. In the remaining 3 tubes, liquids of different volatilities are admitted to the bottom of the tube of mercury: they rise through the mercury and evaporate into the vacuum above the column of mercury. As the pressure of vapor builds up above the mercury column, the height of the mercury in the tube drops. Note that diethyl ether, $(C_2H_5)_2O$, shows the highest vapor pressure because it is the most volatile of the three liquids.

35.
   a. CH_3OH: Both substances are capable of hydrogen bonding, but CH_3OH has a much smaller molar mass. Everything else being equal, substances with lower molar masses, and thus lower London forces, tend to have lower boiling points.

   b. CH_3CH_3: The CH_3CH_2OH molecule is capable of hydrogen bonding which provides additional forces between molecules. In CH_3CH_3, only London forces are operating.

   c. CH_4: Hydrogen bonding is possible in H_2O but not in CH_4.

36. a. H_2S: H_2O exhibits hydrogen bonding, and H_2S does not. In general, substances that exhibit weaker intermolecular forces are more volatile.

    b. CH_3OH: H_2O exhibits stronger hydrogen bonding than CH_3OH because there are two locations where hydrogen bonding is possible on water.

    c. CH_3OH: Both are capable of hydrogen bonding, but generally lighter molecules are more volatile than heavier molecules.

37. Hydrogen bonding can occur in *both* molecules. Oxygen atoms are more electronegative than nitrogen atoms, however, and the polarity of the O–H bond is considerably greater than the polarity of the N–H bond. This leads to *stronger* hydrogen bonding in liquid water than in liquid NH_3.

38. Both substances have the same molar mass. However, ethyl alcohol contains a hydrogen atom directly bonded to an oxygen atom. Therefore, hydrogen bonding can exist in ethyl alcohol, whereas only weak dipole-dipole forces can exist in dimethyl ether. Dimethyl ether is more volatile; ethyl alcohol has a higher boiling point.

39. A crystalline solid is a solid with a regular, repeating microscopic arrangement of its components (ions, atoms, or molecules). This highly-ordered microscopic arrangement of the components of a crystalline solid is frequently reflected macroscopically in beautiful, regularly shaped crystals for such solids.

40. *Ionic* solids have as their fundamental particles positive and negative *ions*; a simple example is sodium chloride, in which Na^+ and Cl^- ions are held together by strong electrostatic forces.

Molecular solids have molecules as their fundamental particles, with the molecules being held together in the crystal by dipole-dipole forces, hydrogen bonding forces, or London dispersion forces (depending on the identity of the substance); simple examples of molecular solids include ice (H_2O) and ordinary table sugar (sucrose).

Atomic solids have simple atoms as their fundamental particles, with the atoms being held together in the crystal either by covalent bonding (as in graphite or diamond) or by metallic bonding (as in copper or other metals).

41. The fundamental particles in ionic solids are positive and negative ions. For example, the ionic solid sodium chloride consists of an alternating, regular array of Na^+ and Cl^- ions. Similarly, an ionic solid such as $CaBr_2$ consists of a regular array of Ca^{2+} ions and Br^- ions. In ionic crystals, each positive ion is surrounded by and attracted to a group of negative ions, and each negative ion is surrounded by and attracted to a group of positive ions. The fundamental particles in molecular solids are discrete molecules. Although the atoms in each molecule are held together by strong intramolecular forces (covalent bonds), the intermolecular forces in a molecular solid are not nearly as strong as in an ionic solid, which leads to molecular solids typically having relatively low melting points. Two examples of molecular solids are the common sugars glucose, $C_6H_{12}O_6$, and sucrose, $C_{12}H_{22}O_{11}$.

42. The interparticle forces in ionic solids (the ionic bond) are much stronger than the interparticle forces in molecular solids (dipole-dipole forces, London forces, etc.). The difference in intermolecular forces is most clearly shown in the great differences in melting points and boiling points between ionic and molecular solids. For example, table salt and ordinary sugar are both crystalline solids that appear very similar. Yet sugar can be melted easily in a saucepan during the making of candy, whereas even the full heat of a stove will not melt salt.

Chapter 14: Liquids and Solids

43. Ionic solids are held together by strong electrostatic forces between the positive and negative ions. The forces are so strong that it becomes very difficult to move or displace ions from one another when an outside force is applied to the solid: so the solid is perceived as being "hard". Molecular solids are held together by weaker dipole-dipole forces. When an outside force seeking to deform or displace the solid is applied, these weaker forces are easier to overcome. The overall magnitude of an electrostatic force is related (in part) to the magnitude of the charges involved. In ionic compounds the charges are "full" ionic charges and the forces are strong; in molecular solids the charges are "partial" and the forces are weaker.

44. Strong electrostatic forces exist between oppositely charged ions in ionic solids.

45. In molecular solids, the molecules are held together in the crystal by hydrogen-bonding, dipole-dipole, or London forces, which are weaker than the strong ion-ion forces that exist between oppositely charged ions in ionic solids (i.e., the attraction of a positive ion by several nearby ions of the opposite charge, and *vice versa*). As a result of these strong forces, ionic solids are typically much harder and have much higher melting and boiling points than molecular solids.

46. In liquid hydrogen, the only intermolecular forces are weak London dispersion forces. In ethyl alcohol and water we have hydrogen bonding possible, but the hydrogen bonding forces are weaker in ethyl alcohol because of the influence of the remainder of the molecule. In sucrose, we also have hydrogen bonding possible, but now at several places in the molecule, leading to stronger forces. In calcium chloride, we have an ionic crystal lattice with even stronger forces between the particles.

47. A network solid has strong covalent bonding among all the atoms in the solid, which results in the solid effectively being one large molecule (diamond is the best example). In a molecular solid, individual molecules are held together by covalent bonding among the atoms in the molecules, but the forces *between* the molecules are typically only dipole-dipole or London dispersion forces.

48. Although ions exist in the solid, liquid, or dissolved states, in the solid state the ions are rigidly held in place in the crystal lattice and cannot *move* so as to conduct an electrical current.

49. An alloy represents a mixture of elements that, as a whole, shows metallic properties. In a substitutional alloy, some of the host metal atoms are *replaced* by other metal atoms (e.g., brass, pewter, plumber's solder). In an interstitial alloy, other small atoms occupy the spaces between the larger host metal atoms (e.g., carbon steel).

50. Nitinol is an alloy of nickel and titanium. When nickel and titanium are heated to a sufficiently high temperature during the production of Nitinol, the atoms arrange themselves in a compact and regular pattern of the atoms.

51. m

52. j

53. g

54. f

55. i

56. d

57. e

58. a

59. c

60. l

61. ice: $1.0 \text{ g} \times \dfrac{1 \text{ cm}^3}{0.9168 \text{ g}} = 1.1 \text{ cm}^3$

   liquid: $1.0 \text{ g} \times \dfrac{1 \text{ cm}^3}{0.9971 \text{ g}} = 1.0 \text{ cm}^3$

   steam: $1.0 \text{ g} \times \dfrac{1 \text{ cm}^3}{3.26 \times 10^{-4} \text{ g}} = 3.1 \times 10^3 \text{ cm}^3$

62. Diethyl ether has the larger vapor pressure. No hydrogen bonding is possible because the O atom does not have a hydrogen atom attached. Hydrogen bonding can occur *only* when a hydrogen atom is *directly* attached to a strongly electronegative atom (such as N, O, or F). Hydrogen bonding *is* possible in 1-butanol (1-butanol contains an ÷OH group).

63. a. KBr (ionic bonding)

    b. NaCl (ionic bonding)

    c. H_2O (hydrogen bonding)

64. None of the substances listed exhibit hydrogen bonding interactions.

    CCl_2H_2: dipole-dipole forces; London dispersion forces

    BeF_2: London dispersion forces

    NO_3^-: London dispersion forces

    HCN: dipole-dipole forces; London dispersion forces

65. Evaporation of a substance is the *liquid* → *vapor* change of state. For every substance, a certain amount of energy is required to accomplish this change of state (heat of vaporization). Alcohol is a volatile liquid, with a relatively large heat of vaporization. Applying alcohol to a fever victim's skin causes internal heat from the body to be absorbed as the heat of vaporization of the alcohol.

66. substitutional; interstitial

67. molar mass of K = 39.10 g

    a. $Q = s \times m \times \Delta T$

    $0.75 \dfrac{\text{J}}{\text{g °C}} \times 5.00 \text{ g} \times (45.2°C - 25.3°C) = 75 \text{ J}$

Chapter 14: Liquids and Solids

b. $\quad 1.35 \text{ mol} \times \dfrac{2.334 \text{ kJ}}{1 \text{ mol}} = 3.15 \text{ kJ}$

c. $\quad 2.25 \text{ g} \times \dfrac{1 \text{ mol}}{39.10 \text{ g}} = 0.05754 \text{ mol}$

$\quad\quad 0.05754 \text{ mol} \times \dfrac{79.87 \text{ kJ}}{1 \text{ mol}} = 4.60 \text{ kJ}$

68. Water is the solvent in which cellular processes take place in living creatures. Water in the oceans moderates the Earth's temperature. Water is used in industry as a cooling agent. Water serves as a means of transportation on the Earth's oceans. The liquid range is 0°C to 100°C at 1 atm pressure.

69. From room temperature to the freezing point (0°C), the average kinetic energy of the molecules in liquid water decreases, and the molecules slow down. At the freezing point, the liquid freezes, with the molecules forming a crystal lattice in which there is much greater order than in the liquid state: molecules no longer move freely, but rather are only able to vibrate somewhat. Below the freezing point, the molecules' vibrations become slower as the temperature is lowered further.

70. At higher altitudes, the boiling points of liquids, such as water, are lower because there is a lower atmospheric pressure above the liquid. The temperature at which food cooks is determined by the temperature to which the water in the food can be heated before it escapes as steam. Thus, food cooks at a lower temperature at high elevations where the boiling point of water is lowered.

71. *Intra*molecular forces hold the atoms together *within* a molecule. When a molecular solid is melted, it is forces *between* molecules (not within the molecules themselves) that must be overcome. *Intra*molecular forces are typically stronger than *inter*molecular forces.

72. Heat of fusion (melt); heat of vaporization (boil).

    The heat of vaporization is always larger, because virtually all of the intermolecular forces must be overcome to form a gas. In a liquid, considerable intermolecular forces remain. Thus going from a solid to liquid requires less energy than going from the liquid to the gas.

73. molar mass CS_2 = 76.15 g

    $1.0 \text{ g } CS_2 \times \dfrac{1 \text{ mol } CS_2}{76.15 \text{ g } CS_2} = 0.0131 \text{ mol } CS_2$

    $0.131 \text{ mol} \times 28.4 \text{ kJ/mol} = 0.37 \text{ kJ required}$

    $50. \text{ g } CS_2 \times \dfrac{1 \text{ mol } CS_2}{76.15 \text{ g } CS_2} = 0.657 \text{ mol } CS_2$

    $0.657 \text{ mol } CS_2 \times 28.4 \text{ kJ/mol} = 19 \text{ kJ evolved } (-19 \text{ kJ})$

74. Dipole-dipole interactions are typically about 1% as strong as a covalent bond. Dipole-dipole interactions represent electrostatic attractions between portions of molecules that carry only a *partial* positive or negative charge. Such forces require the molecules that are interacting to come *near* enough to each other.

75. He (London forces) < CO (London forces, dipole-dipole forces) < H₂O (London forces, dipole-dipole forces, hydrogen bonding) < NaCl (ionic)

76. London dispersion forces are relatively weak forces that arise among noble gas atoms and in nonpolar molecules. London forces are due to *instantaneous dipoles* that develop when one atom (or molecule) momentarily distorts the electron cloud of another atom (or molecule). London forces are typically weaker than either permanent dipole-dipole forces or covalent bonds.

77. a. London dispersion forces (nonpolar molecules)
     b. hydrogen bonding (H attached to N); London dispersion forces
     c. London dispersion forces (nonpolar molecules)
     d. London dispersion forces (nonpolar molecules)

78. a. London dispersion forces (nonpolar atoms)
     b. hydrogen bonding (H attached to O); London dispersion forces
     c. dipole-dipole (polar molecules); London dispersion forces
     d. London dispersion forces (nonpolar molecules)

79. A *volatile* liquid is one that evaporates relatively easily. Volatile liquids typically have large vapor pressures because the intermolecular forces that would tend to prevent evaporation are small. Typically, London forces are the only intermolecular force in highly volatile liquids.

80. a. HF contains a stronger polar bond because it has a greater electronegativity difference between the two atoms versus HCl, therefore it has more unequal sharing of electrons.
     b. HF contains stronger dipole-dipole interactions because the molecule itself is more polar. It has more charge separation within the molecule, which leads to stronger dipole-dipole interactions between the molecules.
     c. HCl would boil first because the intermolecular forces are not as strong as in HF. HCl exhibits dipole-dipole interactions but HF exhibits hydrogen bonding (which is a stronger form of dipole-dipole interactions). It would take less energy to disturb the dipole-dipole interactions in HCl and make it go to the gas phase. (HF would require more energy and thus has a higher boiling point.)

81. Ionic solids typically have the highest melting points, because the ionic charges and close packing in ionic solids allow for very strong forces among a given ion and its nearest neighbors of the opposite charge.

82. In a crystal of ice, strong *hydrogen bonding* forces are present, whereas in the crystal of a nonpolar substance like oxygen, only the much weaker *London* forces exist.

83. The electron sea model envisions a metal as a cluster of positive ions through which the valence electrons are able to move freely. An electrical current represents the movement of electrons, for example through a metal wire, and is consistent with a model in which the electrons are free to roam.

84. Ice floats on liquid water; water expands when it is frozen

Chapter 14: Liquids and Solids

85. As a liquid is heated, the molecules of the liquid gain energy and begin to move faster and more randomly. At some point, some molecules of liquid will have sufficient kinetic energy and will be moving in the right direction so as to escape from the surface of the liquid. At the point where the pressure of vapor leaving the liquid is equal to the pressure above the liquid, the liquid is said to be boiling. Only intermolecular forces need to be overcome: no intramolecular bonds are broken.

86. Although they are at the same *temperature*, steam at 100°C contains a larger amount of *energy* than hot water, equal to the heat of vaporization of water.

87. A dipole-dipole interaction exists when the positive end of one polar molecule is attracted to the negative end of another polar molecule. Examples depend on student choice.

88. Hydrogen bonding is a special case of dipole-dipole interactions that occur among molecules containing hydrogen atoms bonded to highly electronegative atoms such as fluorine, oxygen, or nitrogen. The bonds are very polar, and the small size of the hydrogen atom (compared to other atoms) allows the dipoles to approach each other very closely. Examples: H_2O, NH_3, HF, etc.

89. Although the noble gas elements do not have *permanent* dipole moments, *instantaneous* dipole moments can arise in their atoms (London dispersion forces). For example, if two xenon atoms approached each other, the nuclear charge on the first atom could momentarily influence the electron cloud on the second atom. If the electron cloud on the second atom is affected so that the centers of negative and positive charge in the atom momentarily do not coincide, then the atom momentarily behaves as a weak dipole.

90. Evaporation and condensation are opposite processes. Evaporation is an endothermic process, condensation is an exothermic process. Evaporation requires an input of energy to provide the increased kinetic energy possessed by the molecules when they are in the gaseous state. Evaporation occurs when the molecules in a liquid are moving fast enough and randomly enough that molecules are able to escape from the surface of the liquid and become a gas.

91. B_2O_3

92. Diamonds are made of only one element (carbon). The very strong covalent bonds among the carbon atoms in diamond lead to a giant molecule, and these types of substances are referred to as network solids.

93. (b) and (f) only exhibit London dispersion forces; (c) and (d) exhibit hydrogen bonding interactions; (a) and (e) have dipole-dipole forces (but not hydrogen bonding)

94. (a), (d), and (e) are true; Molecules that only exhibit London dispersion forces can exist in any state of matter at room temperature depending on the molar mass of the molecule. H_2O exhibits stronger hydrogen bonding interactions because it has a greater charge separation within the molecule (greater differences in electronegativities).

95. $CF_3(CF_2CF_2)_nCF_3$: London dispersion

    CO_2: London dispersion

    NaI: ionic

    NH_4Cl: ionic

    $MgCl_2$: ionic

Chapter 14:   Liquids and Solids

96. CH_3Cl, CH_3CH_2Cl, $CH_3CH_2CH_2Cl$, $CH_3CH_2CH_2CH_2Cl$

The larger the molar mass of the molecule, the more London dispersion forces present, which leads to a higher boiling point.

97. CH_4, H_2S, H_2O, MgO

The stronger the intermolecular forces, the higher the boiling point. The strongest forces present in each are:

$\quad\quad\quad$ CH_4: London dispersion

$\quad\quad\quad$ H_2S: dipole-dipole

$\quad\quad\quad$ H_2O: hydrogen bonding

$\quad\quad\quad$ MgO: ionic

98. (b), (c), and (d) are true; LiF has stronger interactions (ionic) versus H_2S (dipole-dipole) thus it has a lower vapor pressure. Similarly, MgO has stronger interactions (ionic) versus CH_3CH_2OH (hydrogen bonding) thus it also has a lower vapor pressure. HF exhibits stronger interactions (hydrogen bonding) versus HBr (dipole-dipole) thus it has a lower vapor pressure. Cl_2 exhibits stronger interactions (London dispersion with a higher molar mass) versus Ar (London dispersion with a lower molar mass) thus it has a higher boiling point. Water is polar, thus HCl is more soluble in water than CCl_4 because HCl is also polar (whereas CCl_4 is nonpolar).

CHAPTER 15

Solutions

1. Specific answers depend on student choices.

2. A heterogeneous mixture does not have a uniform composition: the composition varies in different places within the mixture. Examples of non–homogeneous mixtures include salad dressing (mixture of oil, vinegar, water, herbs, and spices) and granite (combination of minerals).

3. Water is the solvent, sugar the solute.

4. solvent; solutes

5. When an ionic solute dissolves in water, a given ion is pulled into solution by the attractive ion–dipole force exerted by *several* water molecules. For example, in dissolving a positive ion, the ion is approached by the negatively charged end of several water molecules: if the attraction of the water molecules for the positive ion is stronger than the attraction of the negative ions near it in the crystal, the ion leaves the crystal and enters solution. After entering solution, the dissolved ion is surrounded completely by water molecules, which tends to prevent the ion from reentering the crystal.

6. "Like dissolves like." The hydrocarbons in oil have intermolecular forces that are very different from those in water, and so the oil spreads out rather than dissolving in the water.

7. Salts are composed of positive and negative ions packed closely together in a crystal lattice. The crystal is held together by the ionic forces among oppositely charged particles in the solid. When a crystal of a salt is placed in water, the polar water molecules arrange their dipoles around the ions of the crystal in such a way as to attract the ions. For example, a positive ion at a corner of the crystal will be surrounded by a group of water molecules aiming the negative end of their dipoles at the positive ion. If the forces of attraction of the water molecules for the positive ion are larger than the ionic forces holding the ion in the crystal, then the ion will leave the crystal and enter solution. See Figure 15.2

8. Carbon dioxide is somewhat soluble in water, especially if pressurized (otherwise, the soda you may be drinking while studying Chemistry would be "flat"). Carbon dioxide's solubility in water is approximately 1.5 g/L at 25°C under a pressure of approximately 1 atm. Although the carbon dioxide molecule overall is non-polar, that is because the two individual C–O bond dipoles cancel each other due to the linearity of the molecule. However these bond dipoles are able to interact with water, making CO_2 more soluble in water than non-polar molecules such as O_2 or N_2 that do not possess individual bond dipoles.

9. A saturated solution is one that contains the maximum amount of solute possible at a particular temperature. A saturated solution is one that is in equilibrium with undissolved solute.

10. unsaturated

11. variable

Chapter 15: Solutions

12. large

13. The mass percent of a particular component in a solution is the mass of that component, divided by the total mass of the solution, multiplied by 100%.

14. 100.0

15. a. $\dfrac{2.14 \text{ g KCl}}{(2.14 \text{ g KCl} + 12.5 \text{ g water})} \times 100 = 14.6\% \text{ KCl}$

    b. $\dfrac{2.14 \text{ g KCl}}{(2.14 \text{ g KCl} + 25.0 \text{ g water})} \times 100 = 7.89\% \text{ KCl}$

    c. $\dfrac{2.14 \text{ g KCl}}{(2.14 \text{ g KCl} + 37.5 \text{ g water})} \times 100 = 5.40\% \text{ KCl}$

    d. $\dfrac{2.14 \text{ g KCl}}{(2.14 \text{ g KCl} + 50.0 \text{ g water})} \times 100 = 4.10\% \text{ KCl}$

16. a. $\dfrac{5.00 \text{ g CaCl}_2}{(5.00 \text{ g CaCl}_2 + 95.0 \text{ g water})} \times 100 = 5.00\% \text{ CaCl}_2$

    b. $\dfrac{1.00 \text{ g CaCl}_2}{(1.00 \text{ g CaCl}_2 + 19.0 \text{ g water})} \times 100 = 5.00\% \text{ CaCl}_2$

    c. $\dfrac{15.0 \text{ g CaCl}_2}{(15.0 \text{ g CaCl}_2 + 285 \text{ g water})} \times 100 = 5.00\% \text{ CaCl}_2$

    d. 2.00 mg = 0.00200 g

    $\dfrac{0.00200 \text{ g CaCl}_2}{(0.00200 \text{ g CaCl}_2 + 0.0380 \text{ g water})} \times 100 = 5.00\% \text{ CaCl}_2$

17. a. $375 \text{ g} \times \dfrac{1.51 \text{ g NH}_4\text{Cl}}{100. \text{ g solution}} = 5.66 \text{ g NH}_4\text{Cl}$

    b. $125 \text{ g} \times \dfrac{2.91 \text{ g NaCl}}{100. \text{ mg solution}} = 3.64 \text{ g NaCl}$

    c. $1.31 \text{ kg} = 1.31 \times 10^3 \text{ g}$

    $1.31 \times 10^3 \text{ g} \times \dfrac{4.92 \text{ g KNO}_3}{100. \text{ g solution}} = 64.5 \text{ g KNO}_3$

    d. 478 mg = 0.478 g

    $0.478 \text{ g} \times \dfrac{12.5 \text{ g NH}_4\text{NO}_3}{100. \text{ g solution}} = 0.0598 \text{ g NH}_4\text{NO}_3$

Chapter 15: Solutions

18. a. $525 \text{ g solution} \times \dfrac{3.91 \text{ g FeCl}_3}{100. \text{ g solution}} = 20.5 \text{ g FeCl}_3$

   $525 \text{ g solution} - 20.5 \text{ g FeCl}_3 = 504.5 \text{ g water (505 g water)}$

   b. $225 \text{ g solution} \times \dfrac{11.9 \text{ g sucrose}}{100. \text{ g solution}} = 26.8 \text{ g sucrose}$

   $225 \text{ g solution} - 26.8 \text{ g sucrose} = 198.2 \text{ g water (198 g water)}$

   c. $1.45 \text{ kg} = 1.45 \times 10^3 \text{ g}$

   $1.45 \times 10^3 \text{ g solution} \times \dfrac{12.5 \text{ g NaCl}}{100. \text{ g solution}} = 181.3 \text{ g NaCl (181 g NaCl)}$

   $1.45 \times 10^3 \text{ g solution} - 181.3 \text{ g NaCl} = 1268.7 \text{ g water } (1.27 \times 10^3 \text{ g water})$

   d. $635 \text{ g solution} \times \dfrac{15.1 \text{ g KNO}_3}{100. \text{ g solution}} = 95.9 \text{ g KNO}_3$

   $635 \text{ g solution} - 95.9 \text{ g KNO}_3 = 539.1 \text{ g water (539 g water)}$

19. Total mass = 92.1 g + 2.59 g + 1.59 g = 96.28 g

   $\%\text{Fe} = \dfrac{92.1 \text{ g Fe}}{96.28 \text{ g alloy}} \times 100 = 95.7\% \text{ Fe}$

   $\%\text{C} = \dfrac{2.59 \text{ g C}}{96.28 \text{ g alloy}} \times 100 = 2.69\% \text{ C}$

   $\%\text{Cr} = \dfrac{1.59 \text{ g Cr}}{96.28 \text{ g alloy}} \times 100 = 1.65\% \text{ Cr}$

20. Percent means "per hundred". So the percentages in Question 19 represent the number of grams of the particular element present in 100. g of the alloy. Since 1.00 kg represents ten times the mass of 100. g, we would need 957 g Fe, 26.9 g C, and 16.5 g Cr to prepare 1.00 kg of the alloy.

21. To say that the solution contains 7.51% of ammonium nitrate means that every 100. grams of the solution will contain 7.51 g of ammonium nitrate.

   $1.25 \text{ kg} = 1.25 \times 10^3 \text{ g}$

   $1.25 \times 10^3 \text{ g solution} \times \dfrac{7.51 \text{ g NH}_4\text{NO}_3}{100 \text{ g solution}} = 93.9 \text{ g NH}_4\text{NO}_3$

   $1.25 \times 10^3 \text{ g solution} - 93.9 \text{ g NH}_4\text{NO}_3 = 1156.1 \text{ g water } (1.16 \times 10^3 \text{ g water})$

22. $\dfrac{67.1 \text{ g CaCl}_2}{(67.1 \text{ g CaCl}_2 + 275 \text{ g water})} \times 100 = 19.6\% \text{ CaCl}_2$

Chapter 15: Solutions

23. To say that the solution to be prepared is 4.50% by mass $CaCl_2$ means that 4.50 g of $CaCl_2$ will be contained in every 100. g of the solution.

    $$175 \text{ g solution} \times \frac{4.50 \text{ g } CaCl_2}{100. \text{ g solution}} = 7.88 \text{ g } CaCl_2$$

24. To say that the solution is 6.25% KBr by mass, means that 100. g of the solution will contain 6.25 g KBr.

    $$125 \text{ g solution} \times \frac{6.25 \text{ g KBr}}{100. \text{ g solution}} = 7.81 \text{ g KBr}$$

25. $285 \text{ g solution} \times \dfrac{5.00 \text{ g NaCl}}{100.0 \text{ g solution}} = 14.3 \text{ g NaCl}$

    $285 \text{ g solution} \times \dfrac{7.50 \text{ g } Na_2CO_3}{100.0 \text{ g solution}} = 21.4 \text{ g } Na_2CO_3$

26. molar mass O_2 = 32.00 g

    $$1.00 \text{ g } O_2 \times \frac{1 \text{ mol}}{32.00 \text{ g}} = 0.03125 \text{ mol O}$$

    from the balanced chemical equation, it will take 2(0.03125) = 0.0625 mol H_2O_2 to produce this quantity of oxygen.

    molar mass H_2O_2 = 34.02 g

    $$0.0625 \text{ mol } H_2O_2 \times \frac{34.02 \text{ g } H_2O_2}{1 \text{ mol } H_2O_2} = 2.13 \text{ g } H_2O_2$$

    $$2.13 \text{ g } H_2O_2 \times \frac{100. \text{ g solution}}{3 \text{ g } H_2O_2} = \text{approximately 71 g}$$

27. First find the mass of the solution, because the percentage is by mass.

    $1.00 \text{ L} = 1.00 \times 10^3 \text{ mL}$

    $$1.00 \times 10^3 \text{ mL solution} \times \frac{1.84 \text{ g solution}}{1 \text{ mL solution}} = 1840 \text{ g solution} = 1.84 \times 10^3 \text{ g solution}$$

    $$1.84 \times 10^3 \text{ g solution} \times \frac{98.3 \text{ g } H_2SO_4}{100. \text{ g solution}} = 1.81 \times 10^3 \text{ g } H_2SO_4$$

28. $1000 \text{ g} \times \dfrac{0.95 \text{ g stablizer}}{100. \text{ g}} = 9.5 \text{ g}$

29. Each liter of the solution contains 3 mol of hydrogen chloride.

30. 0.110 mol; 0.220 mol

Chapter 15:   Solutions

31. A standard solution is one whose concentration is known very accurately and with high precision. A standard solution is typically prepared by weighing out a precise amount of solute, and then dissolving the solute in a precise amount of solvent, or to a precise final volume using a volumetric flask.

32. The molarity represents the number of moles of solute per liter of solution: choice b is the only scenario that fulfills this definition.

33. Molarity = $\dfrac{\text{moles of solute}}{\text{liters of solution}}$

    a. 125 mL = 0.125 L

    $$M = \dfrac{0.521 \text{ mol}}{0.125 \text{ L}} = 4.17 \; M$$

    b. 250. mL = 0.250 L

    $$M = \dfrac{0.521 \text{ mol}}{0.250 \text{ L}} = 2.08 \; M$$

    c. 500. mL = 0.500 L

    $$M = \dfrac{0.521 \text{ mol}}{0.500 \text{ L}} = 1.04 \; M$$

    d. $M = \dfrac{0.521 \text{ mol}}{1.00 \text{ L}} = 0.521 \; M$

34. Molarity = $\dfrac{\text{moles of solute}}{\text{liters of solution}}$

    a. 225 mL = 0.225 L

    $$M = \dfrac{0.754 \text{ mol KNO}_3}{0.225 \text{ L}} = 3.35 \; M$$

    b. 10.2 mL = 0.0102 L

    $$M = \dfrac{0.0105 \text{ mol CaCl}_2}{0.0102 \text{ L}} = 1.03 \; M$$

    c. $M = \dfrac{3.15 \text{ mol NaCl}}{5.00 \text{ L}} = 0.630 \; M$

    d. 100. mL = 0.100 L

    $$M = \dfrac{0.499 \text{ mol NaBr}}{0.100 \text{ L}} = 4.99 \; M$$

35. Molarity = $\dfrac{\text{moles of solute}}{\text{liters of solution}}$; molar mass of NaCl = 58.44 g

    $3.51 \text{ g NaCl} \times \dfrac{1 \text{ mol NaCl}}{58.44 \text{ g NaCl}} = 0.06006 \text{ mol NaCl}$

Chapter 15: Solutions

a. 25 mL = 0.025 L

$$M = \frac{0.06006 \text{ mol NaCl}}{0.025 \text{ L}} = 2.4\ M$$

b. 50. mL = 0.050 L

$$M = \frac{0.06006 \text{ mol}}{0.050 \text{ L}} = 1.2\ M$$

c. 75 mL = 0.075 L

$$M = \frac{0.06006 \text{ mol}}{0.075 \text{ L}} = 0.80\ M$$

d. $M = \dfrac{0.06006 \text{ mol}}{1.00 \text{ L}} = 0.0601\ M$

36. Molarity = $\dfrac{\text{moles of solute}}{\text{liters of solution}}$; molar mass of $CaCl_2$ = 110.98 g; 125 mL = 0.125 L

a. $5.59 \text{ g } CaCl_2 \times \dfrac{1 \text{ mol } CaCl_2}{110.98 \text{ g } CaCl_2} = 0.05037 \text{ mol } CaCl_2$

$$M = \frac{0.05037 \text{ mol}}{0.125 \text{ L}} = 0.403\ M$$

b. $2.34 \text{ g } CaCl_2 \times \dfrac{1 \text{ mol } CaCl_2}{110.98 \text{ g } CaCl_2} = 0.02108 \text{ mol } CaCl_2$

$$M = \frac{0.02108 \text{ mol}}{0.125 \text{ L}} = 0.169\ M$$

c. $8.73 \text{ g } CaCl_2 \times \dfrac{1 \text{ mol } CaCl_2}{110.98 \text{ g } CaCl_2} = 0.07866 \text{ mol } CaCl_2$

$$M = \frac{0.07866 \text{ mol}}{0.125 \text{ L}} = 0.629\ M$$

d. $11.5 \text{ g } CaCl_2 \times \dfrac{1 \text{ mol } CaCl_2}{110.98 \text{ g } CaCl_2} = 0.1036 \text{ mol } CaCl_2$

$$M = \frac{0.1036 \text{ mol}}{0.125 \text{ L}} = 0.829\ M$$

37. molar mass of $CaCl_2$ = 110.98 g; 225 mL = 0.225 L

$$0.225 \text{ L} \times \frac{0.150 \text{ mol}}{1 \text{ L}} \times \frac{110.98 \text{ g}}{1 \text{ mol}} = 3.75 \text{ g } CaCl_2$$

38. molar mass of NH_4Cl = 53.492 g; 450. mL = 0.450 L

$$0.450 \text{ L} \times \frac{0.251 \text{ mol}}{1 \text{ L}} \times \frac{53.492 \text{ g}}{1 \text{ mol}} = 6.04 \text{ g } NH_4Cl$$

Chapter 15: Solutions

39. molar mass $CaCO_3$ = 100.1 g; 250.0 mL = 0.2500 L

    $$1.745 \text{ g CaCO}_3 \times \frac{1 \text{ mol}}{100.1 \text{ g}} = 0.01743 \text{ mol CaCO}_3$$

    One mole of $CaCO_3$ contains one mole of Ca^{2+} ion; therefore,

    $$M = \frac{0.01743 \text{ mol Ca}^{2+}}{0.2500 \text{ L solution}} = 0.06973 \text{ } M$$

40. molar mass of I_2 = 253.8 g; 225 mL = 0.225 L

    $$5.15 \text{ g I}_2 \times \frac{1 \text{ mol}}{253.8 \text{ g}} = 0.0203 \text{ mol I}_2$$

    $$M = \frac{0.0203 \text{ mol I}_2}{0.225 \text{ L solution}} = 0.0902 \text{ } M$$

41. molar mass of NaOH = 40.00 g; 225 mL = 0.225 L

    $$42.5 \text{ g NaOH} \times \frac{1 \text{ mol NaOH}}{40.00 \text{ g}} = 1.063 \text{ mol NaOH}$$

    $$M = \frac{1.063 \text{ mol NaOH}}{0.225 \text{ L solution}} = 4.72 \text{ } M$$

42. molar mass of $AgNO_3$ = 169.91 g; 250. mL = 0.2500 L

    $$0.2500 \text{ L} \times \frac{0.100 \text{ mol}}{1 \text{ L}} \times \frac{169.91 \text{ g}}{1 \text{ mol}} = 4.25 \text{ g AgNO}_3$$

43. Molarity = $\frac{\text{moles of solute}}{\text{liters of solution}}$

    a. 4.25 mL = 0.00425 L

    $$0.00425 \text{ L solution} \times \frac{0.105 \text{ mol CaCl}_2}{1.00 \text{ L solution}} = 4.46 \times 10^{-4} \text{ mol CaCl}_2$$

    b. 11.3 mL = 0.0113 L

    $$0.0113 \text{ L solution} \times \frac{0.405 \text{ mol NaOH}}{1.00 \text{ L solution}} = 4.58 \times 10^{-3} \text{ mol NaOH}$$

    c. $$1.25 \text{ L solution} \times \frac{12.1 \text{ mol HCl}}{1.00 \text{ L solution}} = 15.1 \text{ mol HCl}$$

    d. 27.5 mL = 0.0275 L

    $$0.0275 \text{ L solution} \times \frac{1.98 \text{ mol NaCl}}{1.00 \text{ L solution}} = 0.0545 \text{ mol NaCl}$$

Chapter 15: Solutions

44. a. 12.5 mL = 0.0125 L

$$0.0125 \text{ L solution} \times \frac{0.104 \text{ mol HCl}}{1.00 \text{ L solution}} = 0.00130 \text{ mol HCl}$$

b. 27.3 mL = 0.0273 L

$$0.0273 \text{ L solution} \times \frac{0.223 \text{ mol NaOH}}{1.00 \text{ L solution}} = 0.00609 \text{ mol NaOH}$$

c. 36.8 mL = 0.0368 L

$$0.0368 \text{ L solution} \times \frac{0.501 \text{ mol HNO}_3}{1.00 \text{ L solution}} = 0.0184 \text{ mol HNO}_3$$

d. 47.5 mL = 0.0475 L

$$0.0475 \text{ L solution} \times \frac{0.749 \text{ mol KOH}}{1.00 \text{ L solution}} = 0.0356 \text{ mol KOH}$$

45. a. $2.50 \text{ L solution} \times \dfrac{13.1 \text{ mol HCl}}{1.00 \text{ L solution}} = 32.75 \text{ mol HCl}$

molar mass HCl = 36.46 g

$$32.75 \text{ mol HCl} \times \frac{36.46 \text{ g HCl}}{1 \text{ mol HCl}} = 1194 \text{ g HCl} = 1.19 \times 10^3 \text{ g HCl}$$

b. 15.6 mL = 0.0156 L

$$0.0156 \text{ L solution} \times \frac{0.155 \text{ mol NaOH}}{1.00 \text{ L solution}} = 0.002418 \text{ mol NaOH}$$

molar mass NaOH = 40.00 g

$$0.002418 \text{ mol NaOH} \times \frac{40.00 \text{ g NaOH}}{1 \text{ mol}} = 0.0967 \text{ g NaOH}$$

c. 135 mL = 0.135 L

$$0.135 \text{ L solution} \times \frac{2.01 \text{ mol HNO}_3}{1.00 \text{ L solution}} = 0.2714 \text{ mol HNO}_3$$

molar mass HNO_3 = 63.02 g

$$0.2714 \text{ mol HNO}_3 \times \frac{63.02 \text{ g HNO}_3}{1 \text{ mol}} = 17.1 \text{ g HNO}_3$$

d. $4.21 \text{ L solution} \times \dfrac{0.515 \text{ mol CaCl}_2}{1.00 \text{ L solution}} = 2.168 \text{ mol CaCl}_2$

molar mass $CaCl_2$ = 111.0 g

$$2.168 \text{ mol CaCl}_2 \times \frac{111.0 \text{ g CaCl}_2}{1 \text{ mol}} = 241 \text{ g CaCl}_2$$

Chapter 15:   Solutions

46. a. molar mass of $CaCl_2$ = 110.98 g; 17.8 mL = 0.0178 L

$$0.0178 \text{ L solution} \times \frac{0.119 \text{ mol } CaCl_2}{1 \text{ L solution}} \times \frac{110.98 \text{ g } CaCl_2}{1 \text{ mol } CaCl_2} = 0.235 \text{ g } CaCl_2$$

b. molar mass of KCl = 74.55 g; 27.6 mL = 0.0276 L

$$0.0276 \text{ L solution} \times \frac{0.288 \text{ mol KCl}}{1 \text{ L solution}} \times \frac{74.55 \text{ g KCl}}{1 \text{ mol KCl}} = 0.593 \text{ g KCl}$$

c. molar mass of $FeCl_3$ = 162.2 g; 35.4 mL = 0.0354 L

$$0.0354 \text{ L solution} \times \frac{0.399 \text{ mol } FeCl_3}{1 \text{ L solution}} \times \frac{162.2 \text{ g } FeCl_3}{1 \text{ mol } FeCl_3} = 2.29 \text{ g } FeCl_3$$

d. molar mass KNO_3 = 101.11 g; 46.1 mL = 0.0461 L

$$0.0461 \text{ L solution} \times \frac{0.559 \text{ mol } KNO_3}{1 \text{ L solution}} \times \frac{101.11 \text{ g } KNO_3}{1 \text{ mol } KNO_3} = 2.61 \text{ g } KNO_3$$

47. molar mass of NaOH = 40.00 g

$$3.5 \text{ L solution} \times \frac{0.50 \text{ mol NaOH}}{1 \text{ L solution}} \times \frac{40.00 \text{ g NaOH}}{1 \text{ mol NaOH}} = 70. \text{ g NaOH}$$

48. molar mass of KBr = 119.0 g; 225 mL = 0.225 L

$$0.225 \text{ L solution} \times \frac{0.355 \text{ mol KBr}}{1 \text{ L solution}} \times \frac{119.0 \text{ g KBr}}{1 \text{ mol KBr}} = 9.51 \text{ g KBr}$$

49. a. $1.00 \text{ L solution} \times \frac{0.251 \text{ mol } Na_2SO_4}{1.00 \text{ L}} = 0.251 \text{ mol } Na_2SO_4$

$$0.251 \text{ mol } Na_2SO_4 \times \frac{2 \text{ mol } Na^+}{1 \text{ mol } Na_2SO_4} = 0.502 \text{ mol } Na^+$$

b. $5.50 \text{ L solution} \times \frac{0.1 \text{ mol } FeCl_3}{1.00 \text{ L}} = 0.550 \text{ mol } FeCl_3$

$$0.550 \text{ mol } FeCl_3 \times \frac{3 \text{ mol } Cl^-}{1 \text{ mol } FeCl_3} = 1.65 \text{ mol} = 1.7 \text{ mol } Cl^-$$

c. 100. mL = 0.100 L

$$0.100 \text{ L solution} \times \frac{0.55 \text{ mol } Ba(NO_3)_2}{1.00 \text{ L}} = 0.0550 \text{ mol } Ba(NO_3)_2$$

$$0.0550 \text{ mol } Ba(NO_3)_2 \times \frac{2 \text{ mol } NO_3^-}{1 \text{ mol } Ba(NO_3)_2} = 0.11 \text{ mol } NO_3^-$$

Chapter 15:   Solutions

    d.   250. mL = 0.250 L

$$0.250 \text{ L solution} \times \frac{0.350 \text{ mol } (NH_4)_2SO_4}{1.00 \text{ L}} = 0.0875 \text{ mol } (NH_4)_2SO_4$$

$$0.0875 \text{ mol } (NH_4)_2SO_4 \times \frac{2 \text{ mol } NH_4^+}{1 \text{ mol } (NH_4)_2SO_4} = 0.175 \text{ mol } NH_4^+$$

50.   a.   10.2 mL = 0.0102 L

$$0.0102 \text{ L} \times \frac{0.451 \text{ mol } AlCl_3}{1.00 \text{ L}} \times \frac{1 \text{ mol } Al^{3+}}{1 \text{ mol } AlCl_3} = 4.60 \times 10^{-3} \text{ mol } Al^{3+}$$

$$0.0102 \text{ L} \times \frac{0.451 \text{ mol } AlCl_3}{1.00 \text{ L}} \times \frac{3 \text{ mol } Cl^-}{1 \text{ mol } AlCl_3} = 1.38 \times 10^{-2} \text{ mol } Cl^-$$

    b. $5.51 \text{ L} \times \dfrac{0.103 \text{ mol } Na_3PO_4}{1.00 \text{ L}} \times \dfrac{3 \text{ mol } Na^+}{1 \text{ mol } Na_3PO_4} = 1.70 \text{ mol } Na^+$

$$5.51 \text{ L} \times \frac{0.103 \text{ mol } Na_3PO_4}{1.00 \text{ L}} \times \frac{1 \text{ mol } PO_4^{3-}}{1 \text{ mol } Na_3PO_4} = 0.568 \text{ mol } PO_4^{3-}$$

    c.   1.75 mL = 0.00175 L

$$0.00175 \text{ L} \times \frac{1.25 \text{ mol } CuCl_2}{1.00 \text{ L}} \times \frac{1 \text{ mol } Cu^{2+}}{1 \text{ mol } CuCl_2} = 2.19 \times 10^{-3} \text{ mol } Cu^{2+}$$

$$0.00175 \text{ L} \times \frac{1.25 \text{ mol } CuCl_2}{1.00 \text{ L}} \times \frac{2 \text{ mol } Cl^-}{1 \text{ mol } CuCl_2} = 4.38 \times 10^{-3} \text{ mol } Cl^-$$

    d.   25.2 mL = 0.0252 L

$$0.0252 \text{ L} \times \frac{0.00157 \text{ mol } Ca(OH)_2}{1.00 \text{ L}} \times \frac{1 \text{ mol } Ca^{2+}}{1 \text{ mol } Ca(OH)_2} = 3.96 \times 10^{-5} \text{ mol } Ca^{2+}$$

$$0.0252 \text{ L} \times \frac{0.00157 \text{ mol } Ca(OH)_2}{1.00 \text{ L}} \times \frac{2 \text{ mol } OH^-}{1 \text{ mol } Ca(OH)_2} = 7.91 \times 10^{-5} \text{ mol } OH^-$$

51.   molar mass NaCl = 58.44 g; 125 mL = 0.125 L

$$0.125 \text{ L} \times \frac{0.105 \text{ mol NaCl}}{1.00 \text{ L}} \times \frac{58.44 \text{ g NaCl}}{1 \text{ mol}} = 0.767 \text{ g NaCl for 125 mL of the solution}$$

$$\frac{0.105 \text{ mol NaCl}}{1.00 \text{ L}} \times \frac{58.44 \text{ g NaCl}}{1 \text{ mol}} = 6.14 \text{ g NaCl for 1.00 L of the solution}$$

52.   Molar mass of Na_2CO_3 = 106.0 g; 250 mL = 0.250 L

$$0.250 \text{ L} \times \frac{0.0500 \text{ mol } Na_2CO_3}{1.00 \text{ L}} \times \frac{106. \text{g}}{1 \text{ mol}} = 1.33 \text{ g } Na_2CO_3$$

53.   moles of solute

Chapter 15:   Solutions

54. half

55. $M_1 \times V_1 = M_2 \times V_2$

   a. $M_1 = 0.119\ M$ $M_2 = ?$
       $V_1 = 25.0\ \text{mL} = 0.0250\ \text{L}$ $V_2 = (55.0 + 25.0)\ \text{mL} = 80.0\ \text{mL} = 0.0800\ \text{L}$

       $$M_2 = \frac{(0.119\ M)(0.0250\ \text{L})}{(0.0800\ \text{L})} = 0.0372\ M$$

   b. $M_1 = 0.701\ M$ $M_2 = ?$
       $V_1 = 45.3\ \text{mL} = 0.0453\ \text{L}$ $V_2 = (45.3 + 125) = 170.3\ \text{mL} = 0.1703\ \text{L}$

       $$M_2 = \frac{(0.701\ M)(0.0453\ \text{L})}{(0.1703\ \text{L})} = 0.186\ M$$

   c. $M_1 = 3.01\ M$ $M_2 = ?$
       $V_1 = 125\ \text{mL} = 0.125\ \text{L}$ $V_2 = (125 + 550.)\ \text{mL} = 675\ \text{mL} = 0.675\ \text{L}$

       $$M_2 = \frac{(3.01\ M)(0.125\ \text{L})}{(0.675\ \text{L})} = 0.557\ M$$

   d. $M_1 = 2.07\ M$ $M_2 = ?$
       $V_1 = 75.3\ \text{mL} = 0.0753\ \text{L}$ $V_2 = (75.3 + 335)\ \text{mL} = 410.3\ \text{mL} = 0.4103\ \text{L}$

       $$M_2 = \frac{(2.07\ M)(0.0753\ \text{L})}{(0.4103\ \text{L})} = 0.380\ M$$

56. $M_1 \times V_1 = M_2 \times V_2$

   a. $M_1 = 0.251\ M$ $M_2 = ?$
       $V_1 = 125\ \text{mL}$ $V_2 = (125 + 250.) = 375\ \text{mL}$

       $$M_2 = \frac{(0.251\ M)(125\ \text{mL})}{375\ \text{mL}} = 0.0837\ M$$

   b. $M_1 = 0.499\ M$ $M_2 = ?$
       $V_1 = 445\ \text{mL}$ $V_2 = (445 + 250.) = 695\ \text{mL}$

       $$M_2 = \frac{(0.499\ M)(445\ \text{mL})}{695\ \text{mL}} = 0.320\ M$$

   c. $M_1 = 0.101\ M$ $M_2 = ?$
       $V_1 = 5.25\ \text{L}$ $V_2 = (5.25 + 0.250) = 5.50\ \text{L}$

       $$M_2 = \frac{(0.101\ M)(5.25\ \text{L})}{5.50\ \text{L}} = 0.0964\ M$$

   d. $M_1 = 14.5\ M$ $M_2 = ?$
       $V_1 = 11.2\ \text{mL}$ $V_2 = (11.2 + 250.) = 261.2\ \text{mL}$

Chapter 15: Solutions

$$M_2 = \frac{(14.5\ M)(11.2\ \text{mL})}{261.2\ \text{mL}} = 0.622\ M$$

57. $M_1 \times V_1 = M_2 \times V_2$

    HCl: $\dfrac{(3.0\ M)(225\ \text{mL})}{(12.1\ M)} = 55.8\ \text{mL} = 56\ \text{mL}$

    HNO$_3$: $\dfrac{(3.0\ M)(225\ \text{mL})}{(15.9\ M)} = 42.45\ \text{mL} = 42\ \text{mL}$

    H$_2$SO$_4$: $\dfrac{(3.0\ M)(225\ \text{mL})}{(18.0\ M)} = 37.5\ \text{mL} = 38\ \text{mL}$

    HC$_2$H$_3$O$_2$: $\dfrac{(3.0\ M)(225\ \text{mL})}{(17.5\ M)} = 38.6\ \text{mL} = 39\ \text{mL}$

    H$_3$PO$_4$: $\dfrac{(3.0\ M)(225\ \text{mL})}{(14.9\ M)} = 45.3\ \text{mL} = 45\ \text{mL}$

58. $M_1 = 19.4\ M$ $M_2 = 3.00\ M$

    $V_1 = ?\ \text{mL}$ $V_2 = 3.50\ \text{L}$

    $M_1 = \dfrac{(3.00\ M)(3.50\ \text{mL})}{(19.4\ M)} = 0.541\ \text{L}\ (541\ \text{mL})$

59. $M_1 \times V_1 = M_2 \times V_2$

    $M_1 = 2.00\ M$ $M_2 = 0.350\ M$

    $V_1 = ?$ $V_2 = 275\ \text{mL} = 0.275\ \text{L}$

    $V_1 = \dfrac{(0.350\ M)(0.275\ \text{L})}{(2.00\ M)} = 0.0481\ \text{L} = 48.1\ \text{mL}$

    Dilute 48.1 mL of the 2.00 M solution to a final volume of 275 mL.

60. $M_1 \times V_1 = M_2 \times V_2$

    $M_1 = 1.01\ M$ $M_2 = 0.150\ M$

    $V_1 = ?\ \text{mL}$ $V_2 = 325\ \text{mL}$

    $M_2 = \dfrac{(0.150\ M)(325\ \text{mL})}{(1.01\ M)} = 48.3\ \text{mL}$

    Dilute 48.3 mL of the 1.01 M solution to a final volume of 325 mL.

61. $M_1 \times V_1 = M_2 \times V_2$

    $M_1 = 0.200\ M$ $M_2 = 0.150\ M$

    $V_1 = 500.\ \text{mL} = 0.500\ \text{L}$ $V_2 = ?$

Chapter 15: Solutions

$$V_2 = \frac{(0.200\ M)(0.500\ L)}{(0.150\ M)} = 0.667\ L = 667\ mL$$

Therefore 667 – 500. = 167 mL of water must be added.

62. $\dfrac{(100.\ mL)(1.25\ M)}{(12.1\ M)} = 10.3\ mL$

63. 10.0 mL = 0.0100 L

The number of moles of Ni^{2+} ion present is given by

$$0.0100\ L \times \frac{0.103\ mol\ Ni^{2+}}{1\ L} = 0.00103\ mol\ Ni^{2+}$$

Since Ni^{2+} and DMG react on a 1:2 mole basis, then 2 × 0.00103 = 0.00206 mol of DMG are needed for the reaction

$$0.00206\ mol\ DMG \times \frac{1\ L}{0.0703\ mol\ DMG} = 0.0293\ L = 29.3\ mL\ \text{of the DMG solution is needed.}$$

64. $Na_2CO_3(aq) + CaCl_2(aq) \rightarrow CaCO_3(s) + 2NaCl(s)$

mmol Ca^{2+} ion: $37.2\ mL \times \dfrac{0.105\ mmol\ Ca^{2+}}{1.00\ mL} = 3.91\ mmol\ Ca^{2+}$

From the balanced chemical equation, 3.91 mmol CO_3^{2-} will be needed to precipitate this quantity of Ca^{2+} ion.

$$3.91\ mmol\ CO_3^{2-} \times \frac{1.00\ mL}{0.125\ mmol} = 31.2\ mL$$

65. The balanced net ionic equation for the precipitation reaction is

$Cu^{2+}(aq) + S^{2-}(aq) \rightarrow CuS(s)$

copper ion present: $27.5\ mL \times \dfrac{0.121\ mmol\ Cu^{2+}}{1.00\ mL} = 3.328\ mmol\ Cu^{2+}$

As the coefficients of the balanced equation are both 1, then 3.328 mmol of S^{2-} ion will be required to precipitate the Cu^{2+} ion.

volume sulfide solution: $3.328\ mmol\ S^{2-} \times \dfrac{1.00\ mL}{0.105\ mmol\ S^{2-}} = 31.7\ mL\ Na_2S$

66. molar mass $Na_2C_2O_4$ = 134.0 g   37.5 mL = 0.0375 L

moles Ca^{2+} ion = $0.0375\ L \times \dfrac{0.104\ mol\ Ca^{2+}}{1.00\ L} = 0.00390\ mol\ Ca^{2+}$ ion

$Ca^{2+}(aq) + C_2O_4^{2-}(aq) \rightarrow CaC_2O_4(s)$

As the precipitation reaction is of 1:1 stoichiometry, then 0.00390 mol of $C_2O_4^{2-}$ ion is needed. Moreover, each formula unit of $Na_2C_2O_4$ contains one $C_2O_4^{2-}$ ion, so 0.00390 mol of $Na_2C_2O_4$ is required.

Chapter 15: Solutions

$$0.00390 \text{ mol Na}_2\text{C}_2\text{O}_4 \times \frac{134.0 \text{ g}}{1 \text{ mol}} = 0.523 \text{ g Na}_2\text{C}_2\text{O}_4 \text{ required}$$

67. $\text{Pb(NO}_3)_2(aq) + \text{K}_2\text{CrO}_4(aq) \rightarrow \text{PbCrO}_4(s) + 2\text{KNO}_3(aq)$

    molar masses: $\text{Pb(NO}_3)_2$, 331.2 g; PbCrO_4, 323.2 g

    $$1.00 \text{ g Pb(NO}_3)_2 \times \frac{1 \text{ mol Pb(NO}_3)_2}{331.2 \text{ g Pb(NO}_3)_2} = 0.003019 \text{ mol Pb(NO}_3)_2$$

    25.0 mL = 0.0250 L

    $$0.0250 \text{ L solution} \times \frac{1.00 \text{ mol K}_2\text{CrO}_4}{1.00 \text{ L solution}} = 0.0250 \text{ mol K}_2\text{CrO}_4$$

    $\text{Pb(NO}_3)_2$ is the limiting reactant: 0.003019 mol PbCrO_4 will form.

    $$0.003019 \text{ mol PbCrO}_4 \times \frac{323.2 \text{ g PbCrO}_4}{1 \text{ mol PbCrO}_4} = 0.976 \text{ g PbCrO}_4$$

68. 10.0 mL = 0.0100 L

    $$0.0100 \text{ L} \times \frac{0.250 \text{ mol AlCl}_3}{1.00 \text{ L}} = 2.50 \times 10^{-3} \text{ mol AlCl}_3$$

    $\text{AlCl}_3(aq) + 3\text{NaOH}(s) \rightarrow \text{Al(OH)}_3(s) + 3\text{NaCl}(aq)$

    $$2.50 \times 10^{-3} \text{ mol AlCl}_3 \times \frac{3 \text{ mol NaOH}}{1 \text{ mol AlCl}_3} = 7.50 \times 10^{-3} \text{ mol NaOH}$$

    molar mass NaOH = 40.00 g

    $$7.50 \times 10^{-3} \text{ mol NaOH} \times \frac{40.00 \text{ g NaOH}}{1 \text{ mol}} = 0.300 \text{ g NaOH}$$

69. $\text{NaOH}(aq) + \text{HNO}_3(aq) \rightarrow \text{NaNO}_3(aq) + \text{H}_2\text{O}(l)$

    27.2 mL = 0.0272 L

    $$0.0272 \text{ L} \times \frac{0.491 \text{ mol HNO}_3}{1 \text{ L}} = 0.01335 \text{ mol HNO}_3$$

    $$0.01335 \text{ mol HNO}_3 \times \frac{1 \text{ mol NaOH}}{1 \text{ mol HNO}_3} = 0.01335 \text{ mol NaOH}$$

    $$0.01335 \text{ mol NaOH} \times \frac{1 \text{ L solution}}{0.502 \text{ mol NaOH}} = 0.0226 \text{ L} = 26.6 \text{ mL}$$

70. $\text{NaOH}(aq) + \text{HCl}(aq) \rightarrow \text{H}_2\text{O}(l) + \text{NaCl}(aq)$

    125 mL = 0.125 L

    $$0.125 \text{ L} \times \frac{3.01 \text{ mol NaOH}}{1 \text{ L}} = 0.3763 \text{ mol NaOH}$$

Chapter 15: Solutions

$$0.3763 \text{ mol NaOH} \times \frac{1 \text{ mol HCl}}{1 \text{ mol NaOH}} = 0.3763 \text{ mol HCl}$$

$$0.3763 \text{ mol HCl} \times \frac{1 \text{ L solution}}{0.995 \text{ mol HCl}} = 0.378 \text{ L} = 378 \text{ mL}$$

71. $HCl(aq) + NaHCO_3(s) \rightarrow NaCl(aq) + H_2O(l) + CO_2(g)$

molar mass of $NaHCO_3$ = 84.01 g; 47.21 mL = 0.04721 L

$$0.1015 \text{ g NaHCO}_3 \times \frac{1 \text{ mol}}{84.01 \text{ g}} = 0.001208 \text{ mol NaHCO}_3$$

The balanced chemical equation indicates that HCl and $NaHCO_3$ react on a 1:1 basis, so 0.001208 mol of HCl must be present. Therefore, the molarity of the HCl is given by

$$M = \frac{0.001208 \text{ mol}}{0.04721 \text{ L}} = 0.02559 \ M$$

72. 7.2 mL = 0.0072 L

$$0.0072 \text{ L} \times \frac{2.5 \times 10^{-3} \text{ mol NaOH}}{1.00 \text{ L}} = 1.8 \times 10^{-5} \text{ mol NaOH}$$

$H^+(aq) + OH^-(aq) \rightarrow H_2O(l)$

$$1.8 \times 10^{-5} \text{ mol OH}^- \times \frac{1 \text{ mol H}^+}{1 \text{ mol OH}^-} = 1.8 \times 10^{-5} \text{ mol H}^+$$

100 mL = 0.100 L

$$M = \frac{1.8 \times 10^{-5} \text{ mol H}^+}{0.100 \text{ L}} = 1.8 \times 10^{-4} \ M \text{ H}^+(aq)$$

73. a. $NaOH(aq) + HC_2H_3O_2(aq) \rightarrow NaC_2H_3O_2(aq) + H_2O(l)$

25.0 mL = 0.0250 L

$$0.0250 \text{ L} \times \frac{0.154 \text{ mol acetic acid}}{1.00 \text{ L}} = 0.00385 \text{ mol acetic acid}$$

$$0.00385 \text{ mol acetic acid} \times \frac{1 \text{ mol NaOH}}{1 \text{ mol acetic acid}} = 0.00385 \text{ mol NaOH}$$

$$0.00385 \text{ mol NaOH} \times \frac{1.00 \text{ L}}{1.00 \text{ mol NaOH}} = 0.00385 \text{ L} = 3.85 \text{ mL NaOH}$$

b. $HF(aq) + NaOH(aq) \rightarrow NaF(aq) + H_2O(l)$

35.0 mL = 0.0350 L

$$0.0350 \text{ L} \times \frac{0.102 \text{ mol HF}}{1.00 \text{ L}} = 0.00357 \text{ mol HF}$$

$$0.00357 \text{ mol HF} \times \frac{1 \text{ mol HF}}{1 \text{ mol NaOH}} = 0.00357 \text{ mol NaOH}$$

$$0.00357 \text{ mol NaOH} \times \frac{1.00 \text{ L}}{1.00 \text{ mol NaOH}} = 0.00357 \text{ L} = 3.57 \text{ mL}$$

c. $H_3PO_4(aq) + 3NaOH(aq) \rightarrow Na_3PO_4(aq) + 3H_2O(l)$

10.0 mL = 0.0100 L

$$0.0100 \text{ L} \times \frac{0.143 \text{ mol } H_3PO_4}{1.00 \text{ L}} = 0.00143 \text{ mol } H_3PO_4$$

$$0.00143 \text{ mol } H_3PO_4 \times \frac{3 \text{ mol NaOH}}{1 \text{ mol } H_3PO_4} = 0.00429 \text{ mol NaOH}$$

$$0.00429 \text{ mol NaOH} \times \frac{1.00 \text{ L}}{1.00 \text{ mol NaOH}} = 0.00429 \text{ L} = 4.29 \text{ mL}$$

d. $H_2SO_4(aq) + 2NaOH(aq) \rightarrow Na_2SO_4(aq) + 2H_2O(l)$

35.0 mL = 0.0350 L

$$0.0350 \text{ L} \times \frac{0.220 \text{ mol } H_2SO_4}{1.00 \text{ L}} = 0.00770 \text{ mol } H_2SO_4$$

$$0.00770 \text{ mol } H_2SO_4 \times \frac{2 \text{ mol NaOH}}{1 \text{ mol } H_2SO_4} = 0.0154 \text{ mol NaOH}$$

$$0.0154 \text{ mol NaOH} \times \frac{1.00 \text{ L}}{1.00 \text{ mol NaOH}} = 0.0154 \text{ L} = 15.4 \text{ mL}$$

74. Experimentally, neutralization reactions are usually performed with volumetric glassware that is calibrated in milliliters rather than liters. For convenience in calculations for such reactions, the arithmetic is often performed in terms of *milli*liters and *milli*moles, rather than in liters and moles: 1 mmol = 0.001 mol. Note that the number of moles of solute per liter of solution, the molarity, is numerically equivalent to the number of *milli*moles of solute per *milli*liter of solution.

   a. $HNO_3(aq) + NaOH(aq) \rightarrow NaNO_3(aq) + H_2O(l)$

   $$12.7 \text{ mL} \times \frac{0.501 \text{ mmol}}{1.00 \text{ mL}} = 6.36 \text{ mmol NaOH present in the sample}$$

   $$6.36 \text{ mmol NaOH} \times \frac{1 \text{ mmol } HNO_3}{1 \text{ mmol NaOH}} = 6.36 \text{ mmol } HNO_3 \text{ required to react}$$

   $$6.36 \text{ mmol } HNO_3 \times \frac{1.00 \text{ mL}}{0.101 \text{ mmol } HNO_3} = 63.0 \text{ mL } HNO_3 \text{ required}$$

   b. $2HNO_3(aq) + Ba(OH)_2 \rightarrow Ba(NO_3)_2 + 2H_2O(l)$

   $$24.9 \text{ mL} \times \frac{0.00491 \text{ mmol}}{1.00 \text{ mL}} = 0.122 \text{ mmol } Ba(OH)_2 \text{ present in sample}$$

Chapter 15: Solutions

$$0.122 \text{ mmol Ba(OH)}_2 \times \frac{2 \text{ mmol HNO}_3}{1 \text{ mmol Ba(OH)}_2} = 0.244 \text{ mmol HNO}_3 \text{ required}$$

$$0.244 \text{ mmol HNO}_3 \times \frac{1.00 \text{ mL}}{0.101 \text{ mmol HNO}_3} = 2.42 \text{ mL HNO}_3 \text{ is required}$$

c. $HNO_3(aq) + NH_3(aq) \rightarrow NH_4NO_3(aq)$

$$49.1 \text{ mL} \times \frac{0.103 \text{ mmol}}{1.00 \text{ mL}} = 5.06 \text{ mmol NH}_3 \text{ present in the sample}$$

$$5.06 \text{ mmol NH}_3 \times \frac{1 \text{ mmol HNO}_3}{1 \text{ mmol NH}_3} = 5.06 \text{ mmol HNO}_3 \text{ required}$$

$$5.06 \text{ mmol HNO}_3 \times \frac{1.00 \text{ mL}}{0.101 \text{ mmol HNO}_3} = 50.1 \text{ mL HNO}_3 \text{ required}$$

d. $KOH(aq) + HNO_3(aq) \rightarrow KNO_3(aq) + H_2O(l)$

$$1.21 \text{ L} \times \frac{0.102 \text{ mol}}{1.00 \text{ L}} = 0.123 \text{ mol KOH present in the sample}$$

$$0.123 \text{ mol KOH} \times \frac{1 \text{ mol HNO}_3}{1 \text{ mol KOH}} = 0.123 \text{ mol HNO}_3 \text{ required}$$

$$0.123 \text{ mol HNO}_3 \times \frac{1.00 \text{ L}}{0.101 \text{ mol HNO}_3} = 1.22 \text{ L HNO}_3 \text{ required}$$

75. one mole of protons (hydrogen ions)

76. 1 normal

77. When H_2SO_4 reacts with OH^-, the reaction is

$H_2SO_4(aq) + 2OH^-(aq) \rightarrow 2H_2O(l) + SO_4^{2-}(aq)$.

As each mol of H_2SO_4 provides *two* moles of H^+ ion, it is only necessary to take *half* a mole of H_2SO_4 to provide *one* mole of H^+ ion. The equivalent weight of H_2SO_4 is thus half the molar mass.

78. 1.53 equivalents OH^- ion are needed to react with 1.53 equivalents of H^+ ion. By *definition*, one equivalent of OH^- ion exactly neutralizes one equivalent of H^+ ion.

79. $N = \dfrac{\text{number of equivalents of solute}}{\text{number of liters of solution}}$

a. equivalent weight HCl = molar mass HCl = 36.46 g

$$M_2 = \frac{M_1 V_1}{V_2} = \frac{(25.2 \text{ mL})(0.105 \, M)}{(75.3 \text{ mL})} = 0.0351 \, M = 0.0351 \, N$$

Chapter 15: Solutions

b. equivalent weight $H_3PO_4 = \dfrac{\text{molar mass}}{3}$

$\dfrac{0.253 \text{ mol H3PO4}}{1.00 \text{ L}} \times \dfrac{3 \text{ equivalents } H_3PO_4}{1 \text{ mole } H_3PO_4} = 0.759 \ N$

c. equivalent weight $Ca(OH)_2 = \dfrac{\text{molar mass}}{2}$

$\dfrac{0.00103 \text{ mol Ca(OH)}_2}{1.00 \text{ L}} \times \dfrac{2 \text{ equivalents Ca(OH)}_2}{1 \text{ mol Ca(OH)}_2} = 0.00206 \ N$

80. $N = \dfrac{\text{number of equivalents of solute}}{\text{number of liters of solution}}$

   a. equivalent weight NaOH = molar mass NaOH = 40.00 g

$0.113 \text{ g NaOH} \times \dfrac{1 \text{ equiv NaOH}}{40.00 \text{ g}} = 2.83 \times 10^{-3} \text{ equiv NaOH}$

10.2 mL = 0.0102 L

$N = \dfrac{2.83 \times 10^{-3} \text{ equiv}}{0.0102 \text{ L}} = 0.277 \ N$

   b. equivalent weight $Ca(OH)_2 \ \dfrac{\text{molar mass}}{2} = \dfrac{74.10 \text{ g}}{2} = 37.05 \text{ g}$

$12.5 \text{ mg} \times \dfrac{1 \text{ g}}{10^3 \text{ mg}} \times \dfrac{1 \text{ equiv}}{37.05 \text{ g}} = 3.37 \times 10^{-4} \text{ equiv Ca(OH)}_2$

100. mL = 0.100 L

$N = \dfrac{3.37 \times 10^{-4} \text{ equiv}}{0.100 \text{ L}} = 3.37 \times 10^{-3} \ N$

   c. equivalent weight $H_2SO_4 = \dfrac{\text{molar mass}}{2} = \dfrac{98.09 \text{ g}}{2} = 49.05 \text{ g}$

$12.4 \text{ g} \times \dfrac{1 \text{ equiv}}{49.05 \text{ g}} = 0.253 \text{ equiv } H_2SO_4$

155 mL = 0.155 L

$N = \dfrac{0.253 \text{ equiv}}{0.155 \text{ L}} = 1.63 \ N$

81. a. $0.250 \ M \text{ HCl} \times \dfrac{1 \text{ equiv HCl}}{1 \text{ mol HCl}} = 0.250 \ N \text{ HCl}$

   b. $0.105 \ M \ H_2SO_4 \times \dfrac{2 \text{ equiv } H_2SO_4}{1 \text{ mol } H_2SO_4} = 0.210 \ N$

Chapter 15: Solutions

c. $5.3 \times 10^{-2} \, M \, H_3PO_4 \times \dfrac{3 \text{ equiv } H_3PO_4}{1 \text{ mol } H_3PO_4} = 0.159 \, N = 0.16 \, N$

82. a. $0.134 \, M \, NaOH \times \dfrac{1 \text{ equiv NaOH}}{1 \text{ mol NaOH}} = 0.134 \, N \, NaOH$

 b. $0.00521 \, M \, Ca(OH)_2 \times \dfrac{2 \text{ equiv } Ca(OH)_2}{1 \text{ mol } Ca(OH)_2} = 0.0104 \, N \, Ca(OH)_2$

 c. $4.42 \, M \, H_3PO_4 \times \dfrac{3 \text{ equiv } H_3PO_4}{1 \text{ mol } H_3PO_4} = 13.3 \, N \, H_3PO_4$

83. molar mass $H_3PO_4 = 98.0$ g

 $35.2 \text{ g } H_3PO_4 \times \dfrac{1 \text{ mol } H_3PO_4}{98.0 \text{ g } H_3PO_4} = 0.3592 \text{ mol } H_3PO_4$

 $M = \dfrac{0.3592 \text{ mol } H_3PO_4}{1.00 \text{ L}} = 0.3592 \, M = 0.359 \, M$

 $0.359 \, M \, H_3PO_4 \times \dfrac{3 \text{ equiv } H_3PO_4}{1 \text{ mol } H_3PO_4} = 1.08 \, N \, H_3PO_4$

84. Molar mass of $Ca(OH)_2 = 74.10$ g

 $5.21 \text{ mg } Ca(OH)_2 \times \dfrac{1 \text{ g}}{10^3 \text{ mg}} \times \dfrac{1 \text{ mol}}{74.10 \text{ g}} = 7.03 \times 10^{-5} \text{ mol } Ca(OH)_2$

 1000. mL = 1.000 L (volumetric flask volume: 4 significant figures).

 $M = \dfrac{7.03 \times 10^{-5} \text{ mol}}{1.000 \, L} = 7.03 \times 10^{-5} \, M \, Ca(OH)_2$

 $N = 7.03 \times 10^{-5} \, M \, Ca(OH)_2 \times \dfrac{2 \text{ equiv } Ca(OH)_2}{1 \text{ mol } Ca(OH)_2} = 1.41 \times 10^{-4} \, N \, Ca(OH)_2$

85. $H_2SO_4(aq) + 2NaOH(aq) \rightarrow Na_2SO_4(aq) + 2H_2O(l)$

 $N_{acid} \times V_{acid} = N_{base} \times V_{base}$

 $(0.35 \, N) \times (15.0 \text{ mL}) = (0.50 \, N) \times (V_{base})$

 $V_{base} = 10.5 \text{ mL} = 11 \text{ mL}$

86. $H_2SO_4(aq) + 2NaOH(aq) \rightarrow Na_2SO_4(aq) + 2H_2O(l)$

 $N_{acid} \times V_{acid} = N_{base} \times V_{base}$

 $(0.104 \, N)(V_{acid}) = (0.152 \, N)(15.2 \text{ mL})$

 $V_{acid} = 22.2 \text{ mL}$

Chapter 15: Solutions

The 0.104 M sulfuric acid solution is *twice as concentrated* as the 0.104 N sulfuric acid solution (1 mole = 2 equivalents), so half as much will be required to neutralize the same quantity of NaOH = 11.1 mL

87. $2NaOH(aq) + H_2SO_4(aq) \rightarrow Na_2SO_4(aq) + 2H_2O(l)$

    For the 0.125 N H_2SO_4:

    $N_{acid} \times V_{acid} = N_{base} \times V_{base}$

    $(0.125\ N) \times (24.2\ mL) = (0.151\ N) \times (V_{base})$

    V_{base} = 20.0 mL of the 0.151 N NaOH solution needed

    For the 0.125 M H_2SO_4:

    As each H_2SO_4 formula unit produces two H^+ ions, the normality of this solution will be twice its molarity.

    0.125 M H_2SO_4 = 0.250 N H_2SO_4

    $N_{acid} \times V_{acid} = N_{base} \times V_{base}$

    $(0.250\ N) \times (24.1\ mL) = (0.151\ N) \times (V_{base})$

    V_{base} = 39.9 mL of the 0.151 N NaOH solution needed

88. $2NaOH(aq) + H_2SO_4(aq) \rightarrow Na_2SO_4(aq) + 2\ H_2O(l)$

    $27.34\ mL\ NaOH \times \dfrac{0.1021\ mmol}{1.00\ mL} = 2.791\ mmol\ NaOH$

    $2.791\ mmol\ NaOH \times \dfrac{1\ mmol\ H_2SO4}{2\ mmol\ NaOH} = 1.396\ mmol\ H_2SO_4$

    $M = \dfrac{1.396\ mmol\ H_2SO4}{25.00\ mL} = 0.05583\ M\ H_2SO_4 = 0.1117\ N\ H_2SO_4$

89. total mass of solution = 50.0 g + 50.0 g + 5.0 g = 105.0 g

    % ethanol = $\dfrac{50.0\ g\ ethanol}{105.0\ g\ total} \times 100 = 47.6\%$ ethanol

    % water = $\dfrac{50.0\ g\ water}{105.0\ g\ total} \times 100 = 47.6\%$ water

    % sugar = $\dfrac{5.0\ g\ sugar}{105.0\ g\ total} \times 100 = 4.8\%$ sugar

    $1.5\ g\ sugar \times \dfrac{100.0\ g\ solution}{4.8\ g\ sugar} = 31\ g\ solution$

    $10.0\ g\ ethanol \times \dfrac{100.0\ g\ solution}{47.6\ g\ ethanol} = 21.0\ g\ solution$

Chapter 15: Solutions

90. Molarity is defined as the number of moles of solute contained in 1 liter of *total* solution volume (solute plus solvent after mixing). In the first case, where 50. g of NaCl is dissolved in 1.0 L of water, the total volume after mixing is *not* known and the molarity cannot be calculated. In the second example, the final volume after mixing is known and the molarity can be calculated simply.

91. mmol $CoCl_2$ = 50.0 mL × 0.250 M $CoCl_2$ = 12.5 mmol $CoCl_2$

   This contains 12.5 mmol Co^{2+} and 25.0 mmol Cl^-

   mmol $NiCl_2$ = 25.0 mL × 0.350 M $NiCl_2$ = 8.75 mmol $NiCl_2$

   This contains 8.75 mmol Ni^{2+} and 17.5 mmol Cl^-

   Total mmol Cl^- after mixing = 25.0 + 17.5 = 42.5 mmol Cl^-

   Total volume after mixing = 50.0 mL + 25.0 mL = 75.0 mL

   $$M_{\text{cobalt(II) ion}} = \frac{12.5 \text{ mmol Co}^{2+}}{75.0 \text{ mL}} = 0.167 \, M$$

   $$M_{\text{nickel(II) ion}} = \frac{8.75 \text{ mmol Ni}^{2+}}{75.0 \text{ mL}} = 0.117 \, M$$

   $$M_{\text{chloride ion}} = \frac{42.5 \text{ mmol Cl}^-}{75.0 \text{ mL}} = 0.567 \, M$$

92. 75 g solution × $\frac{25 \text{ g NaCl}}{100. \text{ g solution}}$ = 18.75 g NaCl

   new % = $\frac{18.75 \text{ g NaCl}}{575 \text{ g solution}} \times 100$ = 3.26 = 3.3 %

93. $AgNO_3(s) + NaCl(aq) \rightarrow AgCl(s) + NaNO_3(aq)$

   molar masses: $AgNO_3$, 169.9 g; AgCl, 143.4 g

   10.0 g $AgNO_3$ × $\frac{1 \text{ mol AgNO}_3}{169.9 \text{ g AgNO}_3}$ = 0.05886 mol $AgNO_3$

   50. mL = 0.050 L

   0.050 L × $\frac{1.0 \times 10^{-2} \text{ mol NaCl}}{1.00 \text{ L}}$ = 0.00050 mol NaCl

   NaCl is the limiting reactant. 0.00050 mol AgCl form.

   0.00050 mol AgCl × $\frac{143.4 \text{ g AgCl}}{1 \text{ mol}}$ = 0.072 g AgCl (72 mg)

   As 1 mol $AgNO_3$ contains 1 mol Ag^+, the mol Ag^+ remaining in solution = 0.05886 − 0.00050 = 0.05836 mol $AgNO_3$.

   0.05836 mol $AgNO_3$ = 0.05836 mol Ag^+

   $$M_{\text{silver ion}} = \frac{0.05836 \text{ mol Ag}^+}{0.050 \text{ L}} = 1.167 \, M = 1.2 \, M$$

Chapter 15: Solutions

94. molar mass NaHCO$_3$ = 84.01 g; 25.2 mL = 0.0252 L

   NaHCO$_3$(s) + HCl(aq) → NaCl(aq) + H$_2$O(l)

   mol HCl = mol NaHCO$_3$ required = 0.0252 L × $\dfrac{6.01 \text{ mol HCl}}{1.00 \text{ L}}$ = 0.151 mol

   0.151 mol × $\dfrac{84.01 \text{ g}}{1 \text{ mol}}$ = 12.7 g NaHCO$_3$ required

95. NiCl$_2$(aq) + H$_2$S(aq) → NiS(s) + 2HCl(aq)

   10. mL = 0.010 L

   0.010 L × $\dfrac{0.050 \text{ mol NiCl}_2}{1.00 \text{ L}}$ = 0.00050 mol NiCl$_2$

   From the balanced equation, 0.00050 mol H$_2$S will be required.

   0.00050 mol H$_2$S × $\dfrac{22.4 \text{ L}}{1 \text{ mol}}$ = 0.0112 L = 11 mL H$_2$S

96. molar mass H$_2$O = 18.0 g

   1.0 L water = 1.0 × 10^3 mL water ≅ 1.0 × 10^3 g water

   1.0 × 10^3 g H$_2$O × $\dfrac{1 \text{ mol H}_2\text{O}}{18.0 \text{ g H}_2\text{O}}$ = 56 mol H$_2$O

97. 100. mL = 0.100 L

   0.100 L × $\dfrac{14.5 \text{ mol NH}_3}{1.00 \text{ L}}$ = 1.45 mol NH$_3$

   1.45 mol NH$_3$ × $\dfrac{22.4 \text{ L}}{1 \text{ mol}}$ = 32.5 L

98. 500 mL HCl solution = 0.500 L HCl solution

   0.500 L solution × $\dfrac{0.100 \text{ mol HCl}}{1.00 \text{ L solution}}$ = 0.0500 mol HCl

   0.0500 mol HCl × $\dfrac{22.4 \text{ L}}{1 \text{ mol}}$ = 1.12 L HCl gas at STP

99. When we say "like dissolves like" we mean that two substances will be miscible if they have similar intermolecular forces, so that the forces existing in the mixture will be similar to the forces existing in each separate substance. Molecules do *not* have to be identical to be miscible, but a similarity in structure (e.g., an OH group) will aid solubility.

Chapter 15:   Solutions

100. $10.0 \text{ g HCl} \times \dfrac{100. \text{ g solution}}{33.1 \text{ g HCl}} = 30.21 \text{ g solution}$

$30.21 \text{ g solution} \times \dfrac{1.00 \text{ mL solution}}{1.147 \text{ g solution}} = 26.3 \text{ mL solution}$

101. $1.00 \text{ g AgNO}_3 \times \dfrac{100.0 \text{ g solution}}{0.50 \text{ g AgNO}_3} = 200 \text{ g solution } (2.0 \times 10^2 \text{ mL solution})$

102. 225.0 mL = 0.2250 L   150.0 mL = 0.1500 L

Determine the total number of moles of HCl in the solution and the total volume of solution.

$0.2250 \text{ L solution} \times \dfrac{2.5 \text{ mol HCl}}{1 \text{ L solution}} = 0.56 \text{ mol HCl}$

$0.1500 \text{ L solution} \times \dfrac{0.75 \text{ mol HCl}}{1 \text{ L solution}} = 0.11 \text{ mol HCl}$

Total number of moles of HCl = 0.56 mol + 0.11 mol = 0.67 mol HCl
Total volume of solution = 0.2250 L + 0.1500 L = 0.3750 L solution

$\text{Molarity} = \dfrac{\text{moles of solute}}{\text{liters of solution}} = \dfrac{0.67 \text{ mol HCl}}{0.3750 \text{ L solution}} = 1.8 \ M$

103. 0.1 g CaCl$_2$

104. a. $\dfrac{x \text{ g NaCl}}{11.5 \text{ g total}} \times 100 = 6.25\% \text{ NaCl}$

$x = 0.719 \text{ g NaCl}$

  b. $\dfrac{x \text{ g NaCl}}{6.25 \text{ g total}} \times 100 = 11.5\% \text{ NaCl}$

$x = 0.719 \text{ g NaCl}$

  c. $\dfrac{x \text{ g NaCl}}{54.3 \text{ g total}} \times 100 = 0.91\% \text{ NaCl}$

$x = 0.49 \text{ g NaCl}$

  d. $\dfrac{x \text{ g NaCl}}{452 \text{ g total}} \times 100 = 12.3\% \text{ NaCl}$

$x = 55.6 \text{ g NaCl}$

105. To say a solution is 15.0% by mass NaCl means that 100.0 g of the solution would contain 15.0 g of NaCl

  a. $10.0 \text{ g NaCl} \times \dfrac{100. \text{ g solution}}{15.0 \text{ g NaCl}} = 66.7 \text{ g solution}$

   15.0 g NaCl

Chapter 15: Solutions

b. $25.0 \text{ g NaCl} \times \dfrac{100. \text{ g solution}}{15.0 \text{ g NaCl}} = 167 \text{ g solution}$

c. $100.0 \text{ g NaCl} \times \dfrac{100. \text{ g solution}}{15.0 \text{ g NaCl}} = 667 \text{ g solution}$

d. $1.00 \text{ lb} = 453.59 \text{ g}$

$453.59 \text{ g NaCl} \times \dfrac{100. \text{ g solution}}{15.0 \text{ g NaCl}} = 3.02 \times 10^3 \text{ g solution}$

106. $\%C = \dfrac{5.0 \text{ g C}}{(5.0 \text{ g C} + 1.5 \text{ g Ni} + 100. \text{ g Fe})} \times 100 = 4.7\% \text{ C}$

$\%Ni = \dfrac{1.5 \text{ g Ni}}{(5.0 \text{ g C} + 1.5 \text{ g Ni} + 100. \text{ g Fe})} \times 100 = 1.4\% \text{ Ni}$

$\%Fe = \dfrac{100. \text{ g Fe}}{(5.0 \text{ g C} + 1.5 \text{ g Ni} + 100. \text{ g Fe})} \times 100 = 93.9\% \text{ Fe}$

107. $25 \text{ g dextrose} \times \dfrac{100. \text{ g solution}}{10. \text{ g dextrose}} = 250 \text{ g solution}$

108. To say that the solution is 5.5% by mass Na_2CO_3 means that 5.5 g of Na_2CO_3 are contained in every 100 g of the solution.

$500. \text{ g solution} \times \dfrac{5.5 \text{ g Na}_2\text{CO}_3}{100. \text{ g solution}} = 28 \text{ g Na}_2\text{CO}_3$

109. $125 \text{ g solution} \times \dfrac{1.5 \text{ g KNO}_3}{100. \text{ g solution}} = 1.9 \text{ g KNO}_3$

110. For NaCl: $125 \text{ g solution} \times \dfrac{7.5 \text{ g NaCl}}{100. \text{ g solution}} = 9.4 \text{ g NaCl}$

For KBr: $125 \text{ g solution} \times \dfrac{2.5 \text{ g KBr}}{100. \text{ g solution}} = 3.1 \text{ g KBr}$

111. $0.0117 \text{ L} \times \dfrac{0.102 \text{ mol Na}_3\text{PO}_4}{1.00 \text{ L}} = 1.19 \times 10^{-3} \text{ mol Na}_3\text{PO}_4$

The sample would contain 1.19×10^{-3} mol of PO_4^{3-} and $3(1.19 \times 10^{-3}) = 3.58 \times 10^{-3}$ mol of Na^+ ion.

112. a. $250 \text{ mL} = 0.25 \text{ L}$

$M = \dfrac{0.50 \text{ mol}}{0.25 \text{ L}} = 2.0 \ M$

Chapter 15:   Solutions

    b.   500. mL = 0.500 L

$$M = \frac{0.50 \text{ mol}}{0.500 \text{ L}} = 1.0 \; M$$

    c.   750 mL = 0.75 L

$$M = \frac{0.50 \text{ mol}}{0.75 \text{ L}} = 0.67 \; M$$

    d. $M = \dfrac{0.50 \text{ mol}}{1.0 \text{ L}} = 0.50 \; M$

113.   a.   molar mass $BaCl_2$ = 208.2 g

$$5.0 \text{ g } BaCl_2 \times \frac{1 \text{ mol}}{208.2 \text{ g } BaCl_2} = 0.0240 \text{ mol } BaCl_2$$

$$M = \frac{0.240 \text{ mol } BaCl_2}{2.5 \text{ L solution}} = 9.6 \times 10^{-3} \; M$$

    b.   molar mass KBr = 119.0 g; 75 mL = 0.075 L

$$3.5 \text{ g KBr} \times \frac{1 \text{ mol}}{119.0 \text{ g KBr}} = 0.0294 \text{ mol KBr}$$

$$M = \frac{0.0294 \text{ mol KBr}}{0.075 \text{ L solution}} = 0.39 \; M$$

    c.   molar mass Na_2CO_3 = 106.0 g; 175 mL = 0.175 L

$$21.5 \text{ g } Na_2CO_3 \times \frac{1 \text{ mol}}{106.0 \text{ g } Na_2CO_3} = 0.2028 \text{ mol } Na_2CO_3$$

$$M = \frac{0.2028 \text{ mol } Na_2CO_3}{0.175 \text{ L}} = 1.16 \; M$$

    d.   molar mass $CaCl_2$ = 111.0 g

$$55 \text{ g } CaCl_2 \times \frac{1 \text{ mol}}{111.0 \text{ g } CaCl_2} = 0.495 \text{ mol } CaCl_2$$

$$M = \frac{0.495 \text{ mol } CaCl_2}{1.2 \text{ L solution}} = 0.41 \; M$$

114.   molar mass $C_{12}H_{22}O_{11}$ = 342.3 g; 450. mL = 0.450 L

$$125 \text{ g } C_{12}H_{22}O_{11} \times \frac{1 \text{ mol}}{342.3 \text{ g}} = 0.3652 \text{ mol } C_{12}H_{22}O_{11}$$

$$M = \frac{0.3652 \text{ mol}}{0.450 \text{ L solution}} = 0.812 \; M$$

Chapter 15: Solutions

115. molar mass HCl = 36.46 g

$$439 \text{ g HCl} \times \frac{1 \text{ mol}}{36.46 \text{ g}} = 12.04 \text{ mol HCl}$$

$$M = \frac{12.04 \text{ mol HCl}}{1.00 \text{ L solution}} = 12.0 \ M$$

116. a. $1.50 \text{ L solution} \times \dfrac{2.00 \text{ mol FeCl}_3}{1 \text{ L solution}} = 3.00 \text{ mol FeCl}_3$

$3.00 \text{ mol FeCl}_3 \times \dfrac{1 \text{ mol Fe}^{3+}}{1 \text{ mol FeCl}_3} = 3.00 \text{ mol Fe}^{3+}$

$3.00 \text{ mol FeCl}_3 \times \dfrac{3 \text{ mol Cl}^-}{1 \text{ mol FeCl}_3} = 9.00 \text{ mol Cl}^-$

b. The balanced equation is:

$2\text{FeCl}_3(aq) + 3\text{Pb(NO}_3)_2(aq) \rightarrow 3\text{PbCl}_2(s) + 2\text{Fe(NO}_3)_3(aq)$

Determine the number of moles of each reactant present.

$1.50 \text{ L solution} \times \dfrac{2.00 \text{ mol FeCl}_3}{1 \text{ L solution}} = 3.00 \text{ mol FeCl}_3$ (already determined from part *a*)

$0.500 \text{ L solution} \times \dfrac{4.00 \text{ mol Pb(NO}_3)_2}{1 \text{ L solution}} = 2.00 \text{ mol Pb(NO}_3)_2$

Determine how many moles of $PbCl_2$ (the solid) would be produced assuming each reactant is completely consumed.

$3.00 \text{ mol FeCl}_3 \times \dfrac{3 \text{ mol PbCl}_2}{2 \text{ mol FeCl}_3} = 4.50 \text{ mol PbCl}_2$

$2.00 \text{ mol Pb(NO}_3)_2 \times \dfrac{3 \text{ mol PbCl}_2}{3 \text{ mol Pb(NO}_3)_2} = 2.00 \text{ mol PbCl}_2$

Once 2.00 mol $PbCl_2$ is produced, all of the Pb^{2+} ion is used up (making $Pb(NO_3)_2$ the limiting reactant). Thus,

$2.00 \text{ mol PbCl}_2 \times \dfrac{278.1 \text{ g PbCl}_2}{1 \text{ mol PbCl}_2} = 556 \text{ g PbCl}_2$

117. a. $1.5 \text{ L solution} \times \dfrac{3.0 \text{ mol H}_2\text{SO}_4}{1.00 \text{ L solution}} = 4.5 \text{ mol H}_2\text{SO}_4$

b. 35 mL = 0.035 L

$0.035 \text{ L solution} \times \dfrac{5.4 \text{ mol NaCl}}{1.00 \text{ L solution}} = 0.19 \text{ mol NaCl}$

Chapter 15:   Solutions

      c.   5.2 L solution $\times \dfrac{18 \text{ mol H}_2\text{SO}_4}{1.00 \text{ L solution}} = 94 \text{ mol H}_2\text{SO}_4$

      d.   0.050 L $\times \dfrac{1.1 \times 10^{-3} \text{ mol NaF}}{1.00 \text{ L solution}} = 5.5 \times 10^{-5} \text{ mol NaF}$

118.   a.   4.25 L solution $\times \dfrac{0.105 \text{ mol KCl}}{1.00 \text{ L solution}} = 0.446 \text{ mol KCl}$

                 molar mass KCl = 74.6 g

                 0.446 mol KCl $\times \dfrac{74.6 \text{ g KCl}}{1 \text{ mol KCl}} = 33.3 \text{ g KCl}$

       b.   15.1 mL = 0.0151 L

                 0.0151 L solution $\times \dfrac{0.225 \text{ mol NaNO}_3}{1.00 \text{ L solution}} = 3.40 \times 10^{-3} \text{ mol NaNO}_3$

                 molar mass NaNO$_3$ = 85.00 g

                 3.40×10^{-3} mol $\times \dfrac{85.00 \text{ g NaNO}_3}{1 \text{ mol NaNO}_3} = 0.289 \text{ g NaNO}_3$

       c.   25 mL = 0.025 L

                 0.025 L solution $\times \dfrac{3.0 \text{ mol HCl}}{1.00 \text{ L solution}} = 0.075 \text{ mol HCl}$

                 molar mass HCl = 36.46 g

                 0.075 mol HCl $\times \dfrac{36.46 \text{ g HCl}}{1 \text{ mol HCl}} = 2.7 \text{ g HCl}$

       d.   100. mL = 0.100 L

                 0.100 L solution $\times \dfrac{0.505 \text{ mol H}_2\text{SO}_4}{1.00 \text{ L solution}} = 0.0505 \text{ mol H}_2\text{SO}_4$

                 molar mass H$_2$SO$_4$ = 98.09 g

                 0.0505 mol H$_2$SO$_4$ $\times \dfrac{98.09 \text{ g H}_2\text{SO}_4}{1 \text{ mol H}_2\text{SO}_4} = 4.95 \text{ g H}_2\text{SO}_4$

119.   molar mass AgNO$_3$ = 169.9 g

              10. g AgNO$_3$ $\times \dfrac{1 \text{ mol AgNO}_3}{169.9 \text{ g AgNO}_3} = 0.0589 \text{ mol AgNO}_3$

              0.0589 mol AgNO$_3$ $\times \dfrac{1.00 \text{ L solution}}{0.25 \text{ mol AgNO}_3} = 0.24 \text{ L solution}$

Chapter 15: Solutions

120. a. $1.25 \text{ L} \times \dfrac{0.250 \text{ mol Na}_3\text{PO}_4}{1.00 \text{ L}} = 0.3125 \text{ mol Na}_3\text{PO}_4$

$0.3125 \text{ mol Na}_3\text{PO}_4 \times \dfrac{3 \text{ mol Na}^+}{1 \text{ mol Na}_3\text{PO}_4} = 0.938 \text{ mol Na}^+$

$0.3125 \text{ mol Na}_3\text{PO}_4 \times \dfrac{1 \text{ mol PO}_4^-}{1 \text{ mol Na}_3\text{PO}_4} = 0.313 \text{ mol PO}_4^{3-}$

b. $3.5 \text{ mL} = 0.0035 \text{ L}$

$0.0035 \text{ L} \times \dfrac{6.0 \text{ mol H}_2\text{SO}_4}{1.00 \text{ L}} = 0.021 \text{ mol H}_2\text{SO}_4$

$0.021 \text{ mol H}_2\text{SO}_4 \times \dfrac{2 \text{ mol H}^+}{1 \text{ mol H}_2\text{SO}_4} = 0.042 \text{ mol H}^+$

$0.021 \text{ mol H}_2\text{SO}_4 \times \dfrac{1 \text{ mol SO}_4^{2-}}{1 \text{ mol H}_2\text{SO}_4} = 0.021 \text{ mol SO}_4^{2-}$

c. $25 \text{ mL} = 0.025 \text{ L}$

$0.025 \text{ L} \times \dfrac{0.15 \text{ mol AlCl}_3}{1.00 \text{ L}} = 0.00375 \text{ mol AlCl}_3$

$0.00375 \text{ mol AlCl}_3 \times \dfrac{1 \text{ mol Al}^{3+}}{1 \text{ mol AlCl}_3} = 0.0038 \text{ mol Al}^{3+}$

$0.00375 \text{ mol AlCl}_3 \times \dfrac{1 \text{ mol Cl}^-}{1 \text{ mol AlCl}_3} = 0.011 \text{ mol Cl}^-$

d. $1.50 \text{ L} \times \dfrac{1.25 \text{ mol BaCl}_2}{1.00 \text{ L}} = 1.875 \text{ mol BaCl}_2$

$1.875 \text{ mol BaCl}_2 \times \dfrac{1 \text{ mol Ba}^{2+}}{1 \text{ mol BaCl}_2} = 1.88 \text{ mol Ba}^{2+}$

$1.875 \text{ mol BaCl}_2 \times \dfrac{2 \text{ mol Cl}^-}{1 \text{ mol BaCl}_2} = 3.75 \text{ mol Cl}^-$

121. molar mass $CaCO_3 = 100.1 \text{ g}$; $500.\text{ mL} = 0.500 \text{ L}$

$0.500 \text{ L} \times \dfrac{0.0200 \text{ mol CaCO}_3}{1.00 \text{ L}} = 0.0100 \text{ mol CaCO}_3 \text{ needed}$

$0.0100 \text{ mol CaCO}_3 \times \dfrac{100.1 \text{ g CaCO}_3}{1 \text{ mol CaCO}_3} = 1.00 \text{ g CaCO}_3$

Chapter 15:   Solutions

122. $M_1 \times V_1 = M_2 \times V_2$

    a. $M_1 = 0.200\ M$ $M_2 = ?$
          $V_1 = 125$ mL $V_2 = 125 + 150. = 275$ mL

$$M_2 = \frac{(0.200\ M)(125\ \text{mL})}{(275\ \text{mL})} = 0.0909\ M$$

    b. $M_1 = 0.250\ M$ $M_2 = ?$
          $V_1 = 155$ mL $V_2 = 155 + 150. = 305$ mL

$$M_2 = \frac{(0.250\ M)(155\ \text{mL})}{(305\ \text{mL})} = 0.127\ M$$

    c. $M_1 = 0.250\ M$ $M_2 = ?$
          $V_1 = 0.500$ L $= 500.$ mL $V_2 = 500. + 150. = 650.$ mL

$$M_2 = \frac{(0.250\ M)(500.\ \text{mL})}{(650\ \text{mL})} = 0.192\ M$$

    d. $M_1 = 18.0\ M$ $M_2 = ?$
          $V_1 = 15$ mL $V_2 = 15 + 150. = 165$ mL

$$M_2 = \frac{(18.0\ M)(15\ \text{mL})}{(165\ \text{mL})} = 1.6\ M$$

123. $M_1 \times V_1 = M_2 \times V_2$

    $M_1 = 18.0\ M$ $M_2 = 0.250\ M$
    $V_1 = ?$ $V_2 = 35.0$ mL

$$V_1 = \frac{(0.250\ M)(35.0\ \text{mL})}{(18.0\ \text{mL})} = 0.486\ \text{mL}$$

124. $M_1 \times V_1 = M_2 \times V_2$

    $M_1 = 5.4\ M$ $M_2 = ?$
    $V_1 = 50.$ mL $V_2 = 300.$ mL

$$M_2 = \frac{(5.4\ M)(50.\ \text{mL})}{(300.\ \text{mL})} = 0.90\ M$$

125. $M_1 \times V_1 = M_2 \times V_2$

    $M_1 = 6.0\ M$ $M_2 = ?$
    $V_1 = 3.0$ L $V_2 = 10.0 + 3.0 = 13.0$ L

$$M_2 = \frac{(6.0\ M)(3.0\ \text{L})}{(13.0\ \text{L})} = 1.4\ M$$

Chapter 15:   Solutions

126. a. $0.250\ M\ Pb(NO_3)_2 \times \dfrac{2\ M\ NO_3^-}{1\ M\ Pb(NO_3)_2} = 0.500\ M\ NO_3^-$

   b.   The balanced equation is:

   $3Pb(NO_3)_2(aq) + 2Na_3PO_4(aq) \rightarrow Pb_3(PO_4)_2(s) + 6NaNO_3(aq)$

   150.0 mL = 0.1500 L

   $0.1500\ L\ Pb(NO_3)_2 \times \dfrac{0.250\ mol\ Pb(NO_3)_2}{1\ L\ Pb(NO_3)_2} \times \dfrac{2\ mol\ Na_3PO_4}{3\ mol\ Pb(NO_3)_2} \times \dfrac{1\ L}{0.100\ M\ Na_3PO_4} = 0.250\ L\ Na_3PO_4$

   $0.250\ L\ Na_3PO_4 = 250.\ mL\ Na_3PO_4$

127.   15.3 mL = 0.0153 L; molar mass $Ba(NO_3)_2$ = 261.3 g

   $0.0153\ L \times \dfrac{0.139\ mol\ H_2SO_4}{1.00\ L} = 2.127 \times 10^{-3}\ mol\ H_2SO_4$

   $2.127 \times 10^{-3}\ mol\ H_2SO_4 \times \dfrac{1\ mol\ Ba(NO_3)_2}{1\ mol\ H_2SO_4} = 2.127 \times 10^{-3}\ mol\ Ba(NO_3)_2$

   $2.127 \times 10^{-3}\ mol\ Ba(NO_3)_2 \times \dfrac{261.3\ g\ Ba(NO_3)_2}{1\ mol\ Ba(NO_3)_2} = 0.556\ g\ Ba(NO_3)_2$

128. $M_1 \times V_1 = M_2 \times V_2$

   $M_1 = 16\ M$ $M_2 = 0.10\ M$
   $V_1 = ?$ $V_2 = 750\ mL = 0.75\ L$

   $V_1 = \dfrac{(0.10\ M)(0.75\ L)}{16\ M} = 0.0047\ L$

   0.0047 L = 4.7 mL

129. a. $HCl(aq) + NaOH(aq) \rightarrow NaCl(aq) + H_2O(l)$

   25.0 mL = 0.0250 L

   $0.0250\ L \times \dfrac{0.103\ mol\ NaOH}{1.00\ L} = 0.02575\ mol\ NaOH$

   $0.02575\ mol\ NaOH \times \dfrac{1\ mol\ HCl}{1\ mol\ NaOH} = 0.02575\ mol\ HCl$

   $0.02575\ mol\ HCl \times \dfrac{1.00\ L}{0.250\ mol\ HCl} = 0.0103\ L\ HCl = 10.3\ mL\ HCl$

   b. $2HCl(aq) + Ca(OH)_2(aq) \rightarrow CaCl_2(aq) + 2H_2O(l)$

   50.0 mL = 0.0500 L

   $0.0500\ L \times \dfrac{0.00501\ mol\ Ca(OH)_2}{1.00\ L} = 2.505 \times 10^{-4}\ mol\ Ca(OH)_2$

Chapter 15: Solutions

$$2.505 \times 10^{-4} \text{ mol Ca(OH)}_2 \times \frac{2 \text{ mol HCl}}{1 \text{ mol Ca(OH)}_2} = 5.010 \times 10^{-4} \text{ mol HCl}$$

$$5.010 \times 10^{-4} \text{ mol HCl} \times \frac{1.00 \text{ L}}{0.250 \text{ mol HCl}} = 0.00200 \text{ L} = 2.00 \text{ mL}$$

c. $HCl(aq) + NH_3(aq) \rightarrow NH_4Cl(aq)$

20.0 mL = 0.0200 L

$$0.0200 \text{ L} \times \frac{0.226 \text{ mol NH}_3}{1.00 \text{ L}} = 0.00452 \text{ mol NH}_3$$

$$0.00452 \text{ mol NH}_3 \times \frac{1 \text{ mol HCl}}{1 \text{ mol NH}_3} = 0.00452 \text{ mol HCl}$$

$$0.00452 \text{ mol HCl} \times \frac{1.00 \text{ L}}{0.250 \text{ mol HCl}} = 0.01808 \text{ L} = 18.1 \text{ mL}$$

d. $HCl(aq) + KOH(aq) \rightarrow KCl(aq) + H_2O(l)$

15.0 mL = 0.0150 L

$$0.0150 \text{ L} \times \frac{0.0991 \text{ mol KOH}}{1.00 \text{ L}} = 1.487 \times 10^{-3} \text{ mol KOH}$$

$$1.487 \times 10^{-3} \text{ mol KOH} \times \frac{1 \text{ mol HCl}}{1 \text{ mol KOH}} = 1.487 \times 10^{-3} \text{ mol HCl}$$

$$1.487 \times 10^{-3} \text{ mol HCl} \times \frac{1.00 \text{ L}}{0.250 \text{ mol HCl}} = 0.00595 \text{ L} = 5.95 \text{ mL}$$

130. a. equivalent weight HCl = molar mass HCl = 36.46 g; 500. mL = 0.500 L

$$15.0 \text{ g HCl} \times \frac{1 \text{ equiv HCl}}{36.46 \text{ g HCl}} = 0.411 \text{ equiv HCl}$$

$$N = \frac{0.411 \text{ equiv}}{0.500 \text{ L}} = 0.822 \text{ } N$$

b. equivalent weight $H_2SO_4 = \frac{\text{molar mass}}{2} = \frac{98.09 \text{ g}}{2} = 49.05$ g; 250. mL = 0.250 L

$$49.0 \text{ g H}_2\text{SO}_4 \times \frac{1 \text{ equiv H}_2\text{SO}_4}{49.05 \text{ g H}_2\text{SO}_4} = 0.999 \text{ equiv H}_2\text{SO}_4$$

$$N = \frac{0.999 \text{ equiv}}{0.250 \text{ L}} = 4.00 \text{ } N$$

c. equivalent weight $H_3PO_4 = \frac{\text{molar mass}}{3} = \frac{98.0 \text{ g}}{3} = 32.67$ g; 100. mL = 0.100 L

$$10.0 \text{ g H}_3\text{PO}_4 \times \frac{1 \text{ equiv H}_3\text{PO}_4}{32.67 \text{ g H}_3\text{PO}_4} = 0.3061 \text{ equiv H}_3\text{PO}_4$$

Chapter 15: Solutions

$$N = \frac{0.3061 \text{ equiv}}{0.100 \text{ L}} = 3.06 \, N$$

131. a. $0.50 \, M \, HC_2H_3O_2 \times \dfrac{1 \text{ equiv } HC_2H_3O_2}{1 \text{ mol } HC_2H_3O_2} = 0.50 \, N \, HC_2H_3O_2$

   b. $0.00250 \, M \, H_2SO_4 \times \dfrac{2 \text{ equiv } H_2SO_4}{1 \text{ mol } H_2SO_4} = 0.00500 \, N \, H_2SO_4$

   c. $0.10 \, M \, KOH \times \dfrac{1 \text{ equiv KOH}}{1 \text{ mol KOH}} = 0.10 \, N \, KOH$

132. molar mass NaH_2PO_4 = 120.0 g; 500. mL = 0.500 L

   $5.0 \text{ g } NaH_2PO_4 \times \dfrac{1 \text{ mol } NaH_2PO_4}{120.0 \text{ g } NaH_2PO_4} = 0.04167 \text{ mol } NaH_2PO_4$

   $M = \dfrac{0.04167 \text{ mol}}{0.500 \text{ L}} = 0.08333 \, M \, NaH_2PO_4 = 0.083 \, M \, NaH_2PO_4$

   $0.08333 \, M \, NaH_2PO_4 \times \dfrac{2 \text{ equiv } NaH_2PO_4}{1 \text{ mol } NaH_2PO_4} = 0.1667 \, N \, NaH_2PO_4 = 0.17 \, N \, NaH_2PO_4$

133. $3NaOH(aq) + H_3PO_4(aq) \rightarrow Na_3PO_4(aq) + 3H_2O(l)$

   14.2 mL = 0.0142 L

   $0.0142 \text{ L} \times \dfrac{0.141 \text{ mol } H_3PO_4}{1.00 \text{ L}} = 2.00 \times 10^{-3} \text{ mol } H_3PO_4$

   $2.00 \times 10^{-3} \text{ mol } H_3PO_4 \times \dfrac{3 \text{ mol NaOH}}{1 \text{ mol } H_3PO_4} = 6.00 \times 10^{-3} \text{ mol NaOH}$

   $6.00 \times 10^{-3} \text{ mol NaOH} \times \dfrac{1.00 \text{ L}}{0.105 \text{ mol NaOH}} = 5.72 \times 10^{-2} \text{ L} = 57.2 \text{ mL}$

134. $N_{acid} \times V_{acid} = N_{base} \times V_{base}$

   $N_{acid} \times (10.0 \text{ mL}) = (3.5 \times 10^{-2} \, N)(27.5 \text{ mL})$

   $N_{acid} = 9.6 \times 10^{-2} \, N \, HNO_3$

135. molar mass CsCl = 168.35 g; molar mass Na_2SO_4 = 142.05 g; molar mass CsF = 151.9 g; molar mass Na_3PO_4 = 163.94 g

   $\text{Molarity} = \dfrac{\text{moles of solute}}{\text{liters of solution}}$

   Soln #1: $100.0 \text{ g CsCl} \times \dfrac{1 \text{ mol CsCl}}{168.35 \text{ g CsCl}} = 0.5940 \text{ mol CsCl}/1.0 \text{ L} = 0.5940 \, M \, CsCl$

   Soln #2: $100.0 \text{ g } Na_2SO_4 \times \dfrac{1 \text{ mol } Na_2SO_4}{142.05 \text{ g } Na_2SO_4} = 0.7040 \text{ mol } Na_2SO_4/1.0 \text{ L} = 0.0.7040 \, M \, Na_2SO_4$

Chapter 15: Solutions

Soln #3: $100.0 \text{ g CsF} \times \dfrac{1 \text{ mol CsF}}{151.9 \text{ g CsF}} = 0.6583 \text{ mol CsF}/1.0 \text{ L} = 0.6583 \text{ } M \text{ CsF}$

Soln #4: $100.0 \text{ g Na}_3\text{PO}_4 \times \dfrac{1 \text{ mol Na}_3\text{PO}_4}{163.94 \text{ g Na}_3\text{PO}_4} = 0.6100 \text{ mol Na}_3\text{PO}_4/1.0 \text{ L} = 0.6100 \text{ } M \text{ Na}_3\text{PO}_4$

Therefore, Solution #2 has the highest concentration.

136. The solution with the greatest number of ions will have the greatest number of moles of ions.

Soln #1: $\text{CsCl}(aq) \rightarrow \text{Cs}^+(aq) + \text{Cl}^-(aq)$; 1 mol CsCl produces 2 moles of ions.

$0.5940 \text{ mol CsCl} \times \dfrac{2 \text{ mol ions}}{1 \text{ mol CsCl}} = 1.188 \text{ mol ions}$

Soln #2: $\text{Na}_2\text{SO}_4(aq) \rightarrow 2\text{Na}^+(aq) + \text{SO}_4^{2-}(aq)$; 1 mol Na$_2SO_4$ produces 3 moles of ions.

$0.7040 \text{ mol Na}_2\text{SO}_4 \times \dfrac{3 \text{ mol ions}}{1 \text{ mol Na}_2\text{SO}_4} = 2.112 \text{ mol ions}$

Soln #3: $\text{CsF}(aq) \rightarrow \text{Cs}^+(aq) + \text{F}^-(aq)$; 1 mol CsF produces 2 moles of ions.

$0.6583 \text{ mol CsF} \times \dfrac{2 \text{ mol ions}}{1 \text{ mol CsF}} = 1.317 \text{ mol ions}$

Soln #4: $\text{Na}_3\text{PO}_4(aq) \rightarrow 3\text{Na}^+(aq) + \text{PO}_4^{3-}(aq)$; 1 mol Na$_3PO_4$ produces 4 moles of ions.

$0.6100 \text{ mol Na}_3\text{PO}_4 \times \dfrac{4 \text{ mol ions}}{1 \text{ mol Na}_3\text{PO}_4} = 2.440 \text{ mol ions}$

Therefore, Solution #4 contains the greatest number of ions.

137. $100.0 \text{ mL} = 0.1000 \text{ L}$ molar mass MgCl$_2$ = 95.21 g

$0.160 \text{ g MgCl}_2 \times \dfrac{1 \text{ mol MgCl}_2}{95.21 \text{ g MgCl}_2} = 0.00168 \text{ mol MgCl}_2$

$M = \dfrac{0.00168 \text{ mol MgCl}_2}{0.1000 \text{ L}} = 0.0168 \text{ } M \text{ MgCl}_2$

$0.0168 \text{ } M \text{ MgCl}_2 \times \dfrac{1 \text{ } M \text{ Mg}^{2+}}{1 \text{ } M \text{ MgCl}_2} = 0.0168 \text{ } M \text{ Mg}^{2+}$

$0.0168 \text{ } M \text{ MgCl}_2 \times \dfrac{2 \text{ } M \text{ Cl}^-}{1 \text{ } M \text{ MgCl}_2} = 0.0336 \text{ } M \text{ Cl}^-$

138. molar mass H$_2$C$_2$O$_4$ = 90.036 g

100.0 mL = 0.1000 L; 10.00 mL = 0.01000 L; 250.0 mL = 0.2500 L

Chapter 15: Solutions

$$0.6706 \text{ g H}_2\text{C}_2\text{O}_4 \times \frac{1 \text{ mol H}_2\text{C}_2\text{O}_4}{90.036 \text{ g H}_2\text{C}_2\text{O}_4} = 0.007448 \text{ mol H}_2\text{C}_2\text{O}_4$$

$$M = \frac{0.007448 \text{ mol H}_2\text{C}_2\text{O}_4}{0.1000 \text{ L}} = 0.07448 \; M \text{ H}_2\text{C}_2\text{O}_4$$

$$0.01000 \text{ L} \times \frac{0.07448 \text{ mol H}_2\text{C}_2\text{O}_4}{1 \text{ L}} = 7.448 \times 10^{-4} \text{ mol H}_2\text{C}_2\text{O}_4$$

$$M = \frac{7.448 \times 10^{-4} \text{ mol H}_2\text{C}_2\text{O}_4}{0.2500 \text{ L}} = 2.979 \times 10^{-3} \; M \text{ H}_2\text{C}_2\text{O}_4$$

139. 150.0 mL = 0.1500 L

The balanced equation is: $2\text{NaOH}(aq) + \text{Ni(NO}_3)_2(aq) \rightarrow \text{Ni(OH)}_2(s) + 2\text{NaNO}_3(aq)$

$$0.1500 \text{ L} \times \frac{0.249 \text{ mol Ni(NO}_3)_2}{1 \text{ L}} = 0.03735 \text{ mol Ni(NO}_3)_2$$

$$0.03735 \text{ mol Ni(NO}_3)_2 \times \frac{2 \text{ mol NaOH}}{1 \text{ mol Ni(NO}_3)_2} = 0.0747 \text{ mol NaOH}$$

$$0.0747 \text{ mol NaOH} \times \frac{1 \text{ L}}{0.100 \text{ mol NaOH}} = 0.747 \text{ L NaOH}$$

0.747 L = 747 mL NaOH

140. 500.0 mL = 0.5000 L; 400.0 mL = 0.4000 L

molar mass $\text{Ba}_3(\text{PO}_4)_2$ = 601.84 g

The balanced equation is: $2\text{Na}_3\text{PO}_4(aq) + 3\text{BaCl}_2(aq) \rightarrow \text{Ba}_3(\text{PO}_4)_2(s) + 6\text{NaCl}(aq)$

$$0.5000 \text{ L} \times \frac{0.200 \text{ mol Na}_3\text{PO}_4}{1 \text{ L}} = 0.100 \text{ mol Na}_3\text{PO}_4$$

$$0.4000 \text{ L} \times \frac{0.289 \text{ mol BaCl}_2}{1 \text{ L}} = 0.116 \text{ mol BaCl}_2$$

BaCl_2 is the limiting reactant (only 0.0771 mol Na_3PO_4 needed).

$$0.116 \text{ mol BaCl}_2 \times \frac{1 \text{ mol Ba}_3(\text{PO}_4)_2}{3 \text{ mol BaCl}_2} = 0.0385 \text{ mol Ba}_3(\text{PO}_4)_2$$

$$0.0385 \text{ mol Ba}_3(\text{PO}_4)_2 \times \frac{601.84 \text{ g Ba}_3(\text{PO}_4)_2}{1 \text{ mol Ba}_3(\text{PO}_4)_2} = 23.2 \text{ g Ba}_3(\text{PO}_4)_2 \text{ (when decimals carried over)}$$

141. 450.0 mL = 0.4500 L; 400.0 mL = 0.4000 L

The balanced equation is: $2\text{AgNO}_3(aq) + \text{CaCl}_2(aq) \rightarrow 2\text{AgCl}(s) + \text{Ca(NO}_3)_2(aq)$

$$0.4500 \text{ L} \times \frac{0.257 \text{ mol AgNO}_3}{1 \text{ L}} = 0.116 \text{ mol AgNO}_3$$

$$0.4000 \text{ L} \times \frac{0.200 \text{ mol CaCl}_2}{1 \text{ L}} = 0.0800 \text{ mol CaCl}_2$$

Chapter 15: Solutions

AgNO₃ is the limiting reactant (only 0.0578 mol CaCl₂ needed).

$$0.116 \text{ mol AgNO}_3 \times \frac{1 \text{ mol CaCl}_2}{2 \text{ mol AgNO}_3} = 0.0578 \text{ mol CaCl}_2 \text{ used up}$$

0.0800 mol CaCl₂ to start − 0.0578 mol CaCl₂ used up = 0.0222 mol CaCl₂ leftover

$CaCl_2(aq) \rightarrow Ca^{2+}(aq) + 2Cl^-(aq)$

$$0.0222 \text{ mol CaCl}_2 \times \frac{2 \text{ mol Cl}^-}{1 \text{ mol CaCl}_2} = 0.0444 \text{ mol Cl}^-$$

$$M = \frac{0.0444 \text{ mol Cl}^-}{(0.4500 \text{ L} + 0.4000 \text{ L})} = 0.0522 \text{ } M \text{ Cl}^- \text{ (when decimals carried over)}$$

142. 34.66 mL = 0.03466 L; 50.00 mL = 0.05000 L

The balanced equation is: $Ca(OH)_2(aq) + 2HNO_3(aq) \rightarrow Ca(NO_3)_2(aq) + 2H_2O(l)$

$$0.03466 \text{ L} \times \frac{0.944 \text{ mol HNO}_3}{1 \text{ L}} = 0.0327 \text{ mol HNO}_3$$

$$0.0327 \text{ mol HNO}_3 \times \frac{1 \text{ mol Ca(OH)}_2}{2 \text{ mol HNO}_3} = 0.0164 \text{ mol Ca(OH)}_2$$

$$M = \frac{0.0164 \text{ mol Ca(OH)}_2}{0.05000 \text{ L}} = 0.327 \text{ } M \text{ Ca(OH)}_2 \text{ (when decimals carried over)}$$

143. 28.44 mL = 0.02844 L

The balanced equations are:

$H_2O_2 + SO_2 \rightarrow H_2SO_4$

$H_2SO_4 + 2NaOH \rightarrow Na_2SO_4 + 2H_2O$

$$0.02844 \text{ L} \times \frac{0.1000 \text{ mol NaOH}}{1 \text{ L}} = 0.002844 \text{ mol NaOH}$$

$$0.002844 \text{ mol NaOH} \times \frac{1 \text{ mol H}_2SO_4}{2 \text{ mol NaOH}} = 0.001422 \text{ mol H}_2SO_4$$

$$0.001422 \text{ mol H}_2SO_4 \times \frac{1 \text{ mol SO}_2}{1 \text{ mol H}_2SO_4} = 0.001422 \text{ mol SO}_2$$

$$0.001422 \text{ mol SO}_2 \times \frac{1 \text{ mol S}}{1 \text{ mol SO}_2} = 0.001422 \text{ mol S}$$

$$0.001422 \text{ mol S} \times \frac{32.07 \text{ g S}}{1 \text{ mol S}} = 0.04560 \text{ g S}$$

$$\%S = \frac{0.04560 \text{ g}}{1.302 \text{ g}} \times 100 = 3.503\% \text{ (when decimals carried over)}$$

CUMULATIVE REVIEW

Chapters 13, 14, and 15

1. Gases have no fixed volume or shape, but rather take on the shape and volume of the container in which they are confined. This is in contrast to solids and liquids: a solid has its own intrinsic volume and shape and is very incompressible; a liquid has an intrinsic volume, but takes on the shape of its container.

2. The pressure of the atmosphere represents the mass of the gases in the atmosphere pressing down on the surface of the earth. The device most commonly used to measure the pressure of the atmosphere is the mercury barometer shown in Figure 13.2 in the text.

   A simple experiment to demonstrate the pressure of the atmosphere is shown in Figure 13.1 in the text. Some water is added to a metal can, and the can heated until the water boils (boiling represents when the pressure of the vapor coming from the water is equal to the atmospheric pressure). The can is then stoppered. As the steam in the can cools, it condenses to liquid water, which lowers the pressure of gas inside the can. The pressure of the atmosphere outside the can is then much larger than the pressure inside the can, and the can collapses.

3. The SI unit of pressure is the pascal (Pa), but this unit is almost never used in everyday situations because it is too small to be practical. Rather, we tend to use units of pressure that are based on the simple instruments used to measure pressures: the mercury barometer and manometer (see Figures 13.2 and 13.3 in the text). The mercury barometer (Figure 13.2), used for measuring the pressure of the atmosphere, consists of a column of mercury that is held in a vertical glass tube by the atmosphere. The pressure of the atmosphere is then indicated in terms of the height of the surface of the mercury in the long tube (relative to the surface of the mercury in the reservoir). As the atmospheric pressure changes, the height of the mercury column changes. The height of the mercury column is given in radio and TV weather reports in inches of mercury, but most scientific applications would quote the height in millimeters of mercury (mm Hg, torr). Pressures are also quoted in standard atmospheres, where 1 atm is equivalent to a pressure of 760 mm Hg.

   Although the barometer is used to measure atmospheric pressure, a device called a mercury manometer is used to measure the pressure of samples of gas in the laboratory. A manometer consists of a U-shaped tube filled with mercury, with one arm of the U open to the atmosphere and the other arm of the U connected to the gas sample to be measured. If the pressure of the gas sample is the same as the pressure of the atmosphere, then the mercury levels will be the same in both sides of the U. If the pressure of the gas is not the same as the atmospheric pressure, then the difference in height of the mercury levels can be used to determine by how many mm Hg the pressure of the gas sample differs from atmospheric pressure.

4. In simple terms, Boyle's law states that the volume of a gas sample will decrease if you squeeze harder on it. Imagine squeezing hard on a tennis ball with your hand: the ball collapses as the gas inside is forced into a smaller volume by your hand. Of course, to be perfectly correct, the temperature and amount of gas (moles) must remain the same as you adjust the pressure for Boyle's law to hold true. There are two mathematical statements of Boyle's law you should remember. The first is

   $P \times V = \text{constant}$,

Cumulative Review: Chapters 13, 14, and 15

which basically is the definition of Boyle's law (in order for the product ($P \times V$) to remain constant, if one of these terms increases the other must decrease). The second formula is the one more commonly used in solving problems,

$$P_1 \times V_1 = P_2 \times V_2.$$

With this second formulation, we can determine pressure-volume information about a given sample under two sets of conditions. These two mathematical formulas are just two different ways of saying the same thing: if the pressure on a sample of gas is increased, the volume of the sample of gas will decrease. A graph of Boyle's law data is given as Figure 13.5: this type of graph ($xy = k$) is known to mathematicians as a hyperbola.

5. The qualification is necessary because the volume of a gas sample is dependent on all its properties. The properties of a gas are all interrelated (as shown by the ideal gas law, $PV = nRT$). If we want to use one of the derivative gas laws (Boyle's, Charles's, or Avogadro's), which isolate how the volume of a gas sample varies with only one of its properties, then we must keep all the other properties constant while that one property is studied.

6. Charles's law simply says that if you heat a sample of gas, the volume of the sample will increase. That is, when the temperature of a gas is increased, the volume of the gas also increases (assuming the pressure and amount of gas remains the same). Charles's law is a direct proportionality when the temperature is expressed in kelvins (if you increase T, this increases V), whereas Boyle's law is an inverse proportionality (if you increase P, this decreases V). There are two mathematical statements of Charles's law with which you should be familiar. The first statement is:

$$V = kT.$$

This is simply a definition (the volume of a gas sample is directly related to its Kelvin temperature: if you increase the temperature, the volume increases). The working formulation of Charles's law we use in problem solving is given as:

$$\frac{V_1}{T_1} = \frac{V_2}{T_2}$$

With this formulation, we can determine volume-temperature information for a given gas sample under two sets of conditions. Charles's law only holds true if the amount of gas remains the same (obviously the volume of a gas sample would increase if there were more gas present) and also if the pressure remains the same (a change in pressure also changes the volume of a gas sample).

7. The volume of a gas sample changes by the same factor (i.e., linearly) for each degree its temperature is changed (for a fixed amount of gas at a constant pressure). Charles realized that, if a gas were cooled, the volume of a gas sample would decrease by a constant factor for each degree the temperature was lowered. When Charles plotted his experimental data and extrapolated the linear data to very low temperatures that he could not measure experimentally, he realized that there would be an ultimate temperature where the volume of a gas sample would shrink to zero if the temperature were lowered any further. The same ultimate temperature was calculated regardless what gas sample was used for the experiment. This temperature-where the volume of an ideal gas sample would approach zero as a limit–is referred to as the absolute zero of temperature. Unlike the Fahrenheit and Celsius temperature scales, which were defined by humans with experimentally convenient reference points, the absolute zero of temperature is a fundamental, natural reference point for the measurement of temperatures. The Kelvin or absolute temperature scale is defined with absolute zero as its lowest temperature, with all temperatures positive relative to this point. The size of the Kelvin degree was chosen to be the same size as the

Celsius degree. Absolute zero (0 K) corresponds to –273°C. A graph of volume versus temperature (at constant pressure) for an ideal gas is a straight line with intercept at –273°C (See Figure 13.7 in the text).

8. Avogadro's law tells us that, with all other things being equal, two moles of gas are twice as big as one mole of gas! That is, the volume of a sample of gas is directly proportional to the number of moles or molecules of gas present (at constant temperature and pressure). If we want to compare the volumes of two samples of the same gas as an indication of the amount of gas present in the samples, we would have to make certain that the two samples of gas are at the same pressure and temperature: the volume of a sample of gas would vary with either temperature or pressure, or both. Avogadro's law holds true for comparing gas samples that are under the same conditions. Avogadro's law is a direct proportionality: the greater the number of gas molecules you have in a sample, the larger the sample's volume will be.

9. Although it may sound strange, an *ideal gas* is defined to be a gas which obeys the ideal gas law (realize that the ideal gas law is based on the experimental measurement of the properties of gases). Boyle's law tells us that the volume of a gas is inversely proportional to its pressure (at constant temperature for a fixed amount of gas):

$$V = (\text{constant})/P.$$

Charles's law indicates that the volume of a gas sample is related to its temperature (at constant pressure for a fixed amount of gas):

$$V = (\text{constant}) \times T.$$

Avogadro's law shows that the volume of a gas sample is proportional to the number of moles of gas (at constant pressure and temperature):

$$V = (\text{constant}) \times n.$$

If we combine all these relationships (and constants) to show how the volume of a gas is proportional to *all* its properties simultaneously:

$$V = (\text{constant}) \times \frac{T \times n}{P}$$

This can be arranged to the familiar form of the ideal gas law:

$$P \times V = n \times R \times T \quad \text{or just} \quad PV = nRT$$

where R is the universal gas constant, which has the value

$$0.08206 \frac{\text{L atm}}{\text{mol K}} \quad (\text{which we often write as } 0.08206 \text{ L atm mol}^{-1} \text{ K}^{-1}).$$

Although it is always important to pay attention to the units when solving a problem, this is especially important when solving gas problems involving the universal gas constant, R. The numerical value of 0.08206 for R applies only when the properties of the gas sample are given in the units specified for the constant: the volume in liters (not mL), the pressure in atmospheres (not mm Hg, torr, or Pa), the amount of gas in moles (not g), and the temperature in kelvins (not °F or °C).

Cumulative Review: Chapters 13, 14, and 15

10. The "partial" pressure of an individual gas in a mixture of gases represents the pressure the gas would have in the same container at the same temperature if it were the only gas present. The total pressure in a mixture of gases is the sum of the individual partial pressures of the gases present in the mixture. Because the partial pressures of the gases in a mixture are additive (i.e., the total pressure is the sum of the partial pressures), this suggests that the total pressure in a container is a function only of the number of molecules present in the same, and not of the identity of the molecules or any other property of the molecules (such as their inherent atomic size).

11. When a gas is bubbled through water, or is collected by displacing water, the gas becomes saturated with water vapor. A correction must be applied to the pressure of the gas to take into account the vapor pressure of water.

12. The main postulates of the kinetic-molecular theory for gases are as follows: (a) gases consist of tiny particles (atoms or molecules), and the size of these particles is negligible compared to the bulk volume of a gas sample; (b) the particles in a gas are in constant random motion, colliding with each other and with the walls of the container; (c) the particles in a gas sample do not exert any attractive or repulsive forces on one another; (d) the average kinetic energy of the particles in a sample of gas is directly related to the absolute temperature of the gas sample. The pressure exerted by a gas is a result of the molecules colliding with (and pushing on) the walls of the container. The pressure increases with temperature because at a higher temperature, the molecules are moving faster and hit the walls of the container with greater force. A gas fills whatever volume is available to it because the molecules in a gas are in constant random motion: if the motion of the molecules is random, they eventually will move out into whatever volume is available until the distribution of molecules is uniform. At constant pressure, the volume of a gas sample increases as the temperature is increased because with each collision having greater force, the container must expand so that the molecules (and therefore the collisions) are farther apart if the pressure is to remain constant.

13. The abbreviation "STP" stands for "Standard Temperature and Pressure". STP corresponds to a temperature of 0°C and a pressure of 1 atm. These conditions were chosen as STP for comparisons of gas samples because they are easy to reproduce in any laboratory (an equilibrium mixture of ice and water has a temperature of 0°C, and the pressure in most laboratories is very near to 1 atm). One mole of any ideal gas occupies a volume of 22.4 L at STP.

14. Solids and liquids are much more condensed states of matter than are gases: the molecules are much closer together in solids and liquids and interact with each other to a much greater extent. Solids and liquids have much greater densities than do gases, and are much less compressible, because there is so little room between the molecules in the solid and liquid states (solids and liquids effectively have native volumes of their own, and their volumes are not affected nearly as much by the temperature or pressure). Although solids are more rigid than liquids, the solid and liquid state have much more in common with each other than either of these states has with the gaseous state. We know this is true because it typically only takes a few kilojoules of energy to melt 1 mol of a solid (not much change has to take place in the molecules), whereas it may take 10 times more energy to vaporize a liquid (as there is a great change between the liquid and gaseous states).

15. Water is a colorless, odorless, tasteless liquid that freezes at 0°C and that boils at 100°C at 1 atm pressure. Water is one of the most important substances on Earth. Water forms the solvent for most of the biochemical processes necessary for plant and animal life. Water in the oceans moderates the temperature of the Earth. Because of its relatively large specific heat capacity and

Cumulative Review: Chapters 13, 14, and 15

its great abundance, water is used as the primary coolant in industrial machinery. Water provides a medium for transportation across vast distances on the Earth. Water provides a medium for the smallest plants and animals in many food chains.

16. The normal boiling point of water, that is, water's boiling point at a pressure of exactly 760 mm Hg, is 100°C (you will recall that the boiling point of water was used to set one of the reference temperatures of the Celsius temperature scale). Water remains at 100°C while boiling, until all the water has boiled away, because the additional heat energy being added to the sample is used to overcome attractive forces among the water molecules as they go from the condensed, liquid state to the gaseous state. The normal (760 mm Hg) freezing point of water is exactly 0°C (again, this property of water was used as one of the reference points for the Celsius temperature scale). A cooling curve for water is given in Figure 14.2. Notice how the curve shows that the amount of heat needed to boil the sample is much larger than the amount needed to melt the sample.

17. Changes in state are only physical changes: no chemical bonds are broken during the change and no new substances result (no changes in the *intra*molecular bonding forces takes place). In order to melt a solid or to boil a liquid, the *inter*molecular forces that hold the molecules together in the solid or liquid must be overcome. The quantity of energy required to melt and to boil 1 mol of a substance are called the substance's *molar heat of fusion* and *molar heat of vaporization*, respectively. The molar heat of vaporization of water (or any substance) is much larger than the molar heat of fusion because in order to form a vapor, the molecules have to be moved much farther apart, and virtually all the intermolecular forces must be overcome (when a solid melts, the intermolecular forces remaining in the liquid are still relatively strong). The boiling point of a liquid decreases with altitude because the atmospheric pressure (against which the vapor must be expanded during boiling) decreases with altitude (the atmosphere is thinner).

18. Dipole-dipole forces are a type of intermolecular force that can exist between molecules with permanent dipole moments. Molecules with permanent dipole moments try to orient themselves so that the positive end of one polar molecule can attract the negative end of another polar molecule. Dipole-dipole forces are not nearly as strong as ionic or covalent bonding forces (only about 1% as strong as covalent bonding forces) because electrostatic attraction is related to the magnitude of the charges of the attracting species. As polar molecules have only a "partial" charge at each end of the dipole, the magnitude of the attractive force is not as large. The strength of such forces also drops rapidly as molecules become farther apart and is important only in the solid and liquid states (such forces are negligible in the gaseous state because the molecules are too far apart). Hydrogen bonding is an especially strong sort of dipole-dipole attractive force that can exist when hydrogen atoms are directly bonded to the most strongly electronegative atoms (N, O, and F). Because the hydrogen atom is so small, dipoles involving N–H, O–H, and F–H bonds can approach each other much more closely than can dipoles involving other atoms. As the magnitude of dipole-dipole forces is dependent on distance, unusually strong attractive forces can exist in such molecules. We take the fact that the boiling point of water is higher than that of the other covalent hydrogen compounds of the Group 6 elements as evidence for the special strength of hydrogen bonding (it takes more energy to vaporize water because of the extra strong forces holding together the molecules in the liquid state).

19. London dispersion are the extremely weak forces that must exist to explain the fact that substances consisting of single atoms or of nonpolar molecules can be liquefied and solidified. London forces are instantaneous dipole forces, which come about as the electrons of an atom move around the nucleus. Although we usually consider that the electrons are uniformly distributed in space around the nucleus, at any given instant there may be more electronic charge on one side of the nucleus than on the other, which results in an instantaneous separation of

Cumulative Review: Chapters 13, 14, and 15

charge and a small dipole moment. Such an instantaneous dipole may induce a similar instantaneous dipole in a neighboring atom, which then results in an attractive force between the dipoles. Although an instantaneous dipole can arise in any molecule, in most cases other, much stronger intermolecular forces predominate. However, for substances that exist as single atoms (e.g., the noble gases) or exist as nonpolar molecules (e.g., H_2, O_2), London forces are the only major intermolecular forces present.

20. Vaporization of a liquid requires an input of energy because the intermolecular forces that hold the molecules together in the liquid state must be overcome. The high heat of vaporization of water is essential to life on Earth because much of the excess energy striking the Earth from the sun is dissipated in vaporizing water. Condensation is the opposite process to vaporization; that is, condensation refers to the process by which molecules in the vapor state form a liquid. In a closed container containing a liquid and some empty space above the liquid, an equilibrium is set up between vaporization and condensation. The liquid in such a sealed container never completely evaporates: when the liquid is first placed in the container, the liquid phase begins to evaporate into the empty space. As the number of molecules in the vapor phase begins to get large, however, some of these molecules begin to re-enter the liquid phase. Eventually, every time a molecule of liquid somewhere in the container enters the vapor phase, somewhere else in the container a molecule of vapor re-enters the liquid. There is no further net change in the amount of liquid phase (although molecules are continually moving between the liquid and vapor phases). The pressure of the vapor in such an equilibrium situation is characteristic for the liquid at each particular temperature (for example, the vapor pressures of water are tabulated at different temperatures in Table 13.2). A simple experiment to determine vapor pressure is shown in Figure 14.10. Samples of a liquid are injected into a sealed tube containing mercury; because mercury is so dense, the liquids float to the top of the mercury where they evaporate. As the vapor pressures of the liquids develop to the saturation point, the level of mercury in the tube changes as an index of the magnitude of the vapor pressures. Typically, liquids with strong intermolecular forces have small vapor pressures (they have more difficulty in evaporating) than do liquids with very weak intermolecular forces: for example, the components of gasoline (weak forces) have much higher vapor pressures, and evaporate more easily than does water (strong forces).

21. Crystalline solids consist of a regular lattice array, which extends in three dimensions of repeating component units (atoms, molecules, or ions). A small portion of a sodium chloride crystal lattice is show in Figure 14.11 in the text. The three important types of crystalline solids are *ionic* solids, *molecular* solids, and *atomic* solids.

    Sodium chloride is a typical ionic solid. Its crystals consists of an alternating array of positive Na^+ ions and negative Cl^- ions. Each positive ion is surrounded by several negative ions, and each negative ion is surrounded by several positive ions. The electrostatic forces that develop in such an arrangement are large. The resulting substance is very stable and has very high melting and boiling points.

    Ice represents a molecular solid. The crystals consist of polar water molecules arranged in three dimensions so as to maximize dipole-dipole interactions (and hydrogen bonding). Figure 14.14(c) shows a representation of an ice crystal and how the negative end of one water molecule is oriented towards the positive end of another water molecule. It also shows how this arrangement repeats. As dipole-dipole forces are weaker than ionic bonding forces, substances that exist as molecular solids typically have much lower melting and boiling points.

    Atomic solids vary as to how the atoms are held together in the crystal. Substances such as the noble gases are held together in the solid only by very weak London dispersion forces. Such substances have extremely low melting and boiling points because these forces are so weak. In

Cumulative Review: Chapters 13, 14, and 15

other atomic solids, such as the diamond form of carbon, adjacent atoms may actually form covalent bonds with each other, leading the entire crystal to be one giant molecule. Such atomic solids have much higher boiling and melting points than those substances held together by only London forces, because there is so much energy held in all the covalent bonds that exist in the crystal. Finally, the metallic substances are also atomic solids, in which there is strong, but nondirectional bonding that leads to the properties associated with metals. Metals are envisioned in terms of the "electron sea" model in which a regular array of metal atoms are perfused with a sea of freely moving valence electrons.

22. The simple model we use to explain many properties of metallic elements is called the electron sea model. In this model we picture a regular lattice array of metal cations in sort of a "sea" of mobile valence electrons. The electrons can move easily to conduct heat or electricity through the metal; and the lattice of cations can be deformed fairly easily, allowing the metal to be hammered into a sheet or stretched to make a wire. An alloy contains a mixture of elements, which overall has metallic properties. Substitutional alloys consist of a host metal in which some of the atoms in the metal's crystalline structure are replaced by atoms of other metallic elements of comparable size to the atoms of the host metal. For example, sterling silver consists of an alloy in which approximately 7% of the silver atoms have been replaced by copper atoms. Brass and pewter are also substitutional alloys. An interstitial alloy is formed when other smaller atoms enter the interstices (holes) between atoms in the host metal's crystal structure. Steel is an interstitial alloy in which typically carbon atoms enter the interstices of a crystal of iron atoms. The presence of the interstitial carbon atoms markedly changes the properties of the iron, making it much harder, more malleable, and more ductile. Depending on the amount of carbon introduced into the iron crystals, the properties of the resulting steel can be carefully controlled.

23. A solution is a homogeneous mixture in which the components are uniformly intermingled. When an ionic substance is dissolved in water to form a solution, the water plays an essential role in overcoming strong interparticle forces in the ionic crystal (shown in Figure 15.2 in the text). Water is a highly polar substance: one end of the water molecule dipole is strongly negative, and the other is strongly positive. Consider a crystal of sodium chloride, in which there is a negative chloride ion at one of the corners of the crystal. When this crystal is placed in water, water molecules surround the chloride ion, and orient themselves with the positive end of their dipoles aimed at the negative chloride ion. When enough water molecules have so arranged themselves, the resultant attraction of the several water molecules for the chloride ion becomes stronger than the attractive forces from the positive sodium ions in the crystal, and the chloride ion separates from the crystal and enters solution (still surrounded by the group of water molecules). Similarly, a positive sodium ion in a similar position would be attracted by a group of water molecules arranged with the negative end of their dipoles oriented toward the positive ion, and when enough water molecules had so arranged themselves so as to surpass the attractive forces from negative ions in the crystal, the sodium ion would enter solution. Once the chloride ion and the sodium ion are in solution, they remain surrounded by a layer of water molecules (called a hydration sphere), which diminishes the effective charge each ion would feel from the other and prevents them from easily recombining. For a molecular solid (such as sugar) to be able to dissolve in a solvent, there must be some portion(s) of the molecule that can be attracted by molecules of solvent. For example, common table sugar (sucrose) contains many hydroxyl groups, –OH. These hydroxyl groups are relatively polar and can be attracted by water molecule dipoles. When enough water molecules have attracted enough hydroxyl groups on a sugar molecule to overcome attractive forces from other sugar molecules within the crystal, the molecule leaves the crystal and enters solution (naturally, the –OH groups can also hydrogen bond with water molecules). In order for a substance to dissolve in water, not only must there be attractive forces possible between water molecules and solute molecules, but these forces must be strong enough to overcome the strong

Cumulative Review: Chapters 13, 14, and 15

attractive forces that water molecules have for one another. In order for a substance to dissolve, the molecules of substance must be capable of being dispersed among water molecules. If the water-solute interactions are not comparable to the water-water interactions, the substance will not dissolve.

24. A saturated solution is one that contains as much solute as can dissolve at a particular temperature. To say that a solution is saturated does not necessarily mean that the solute is present at a high concentration. For example, magnesium hydroxide only dissolves to a very small extent before the solution is saturated, whereas it takes a great deal of sugar to form a saturated solution (and the saturated solution is extremely concentrated). A saturated solution is one which is in equilibrium with undissolved solute: as molecules of solute dissolve from the solid in one place in the solution, dissolved molecules rejoin the solid phase in another place in the solution. As with the development of vapor pressure above a liquid (see Question 20 above), formation of a solution reaches a state of dynamic equilibrium: once the rates of dissolving and "undissolving" become equal, there will be no further net change in the concentration of the solution and the solution will be saturated.

25. $\text{mass percent} = \dfrac{\text{mass of solute}}{\text{total mass of solution}} \times 100 = \dfrac{5.00 \text{ g NaCl}}{(5.00 + 15.0) \text{ g total}} \times 100 = 25.0\%$

$5.00 \text{ g NaCl} = 5.00 \text{ g} \times \dfrac{1 \text{ mol}}{58.44 \text{ g}} = 0.0856 \text{ mol} \quad 16.1 \text{ mL} = 0.0161 \text{ L}$

$\text{molarity} = \dfrac{0.0856 \text{ mol}}{0.0161 \text{ L}} = 5.32 \ M$

26. Adding more solvent to a solution so as to dilute the solution *does not change* the number of moles of solute present, but only changes the volume in which the solute is dispersed. If we are using the molarity of the solution to describe its concentration, the number of liters is changed when we add solvent, and the number of moles per liter (the molarity) changes, but the actual number of moles of solute does not change. For example, 125 mL of 0.551 M NaCl contains 68.9 millimol of NaCl. The solution will still contain 68.9 millimol of NaCl after the 250 mL of water is added to it, only now the 68.9 millimol of NaCl will be dispersed in a total volume of 375 mL. This gives the new molarity as 68.9 mmol/375 mL = 0.184 M. The volume and the concentration have changed, but the number of moles of solute in the solution has not changed.

27. One equivalent of an acid is the amount of acid that can furnish one mole of H^+ ions; one equivalent of a base is the amount of base that can furnish one mole of OH^- ions. The equivalent weight of an acid or base is the mass of the substance representing one equivalent of the acid or base. The equivalent weight of a substance is determined from the molar mass of the substance, also taking into account how many H^+ or OH^- ions the substance furnishes per molecule. For example, HCl and NaOH have equivalent weights equal to their molar masses, because each of these substances furnishes one H^+ or OH^- ion per molecule

$HCl \rightarrow H^+ + Cl^-$

$NaOH \rightarrow Na^+ + OH^-$

However, sulfuric acid (H_2SO_4) has an equivalent weight that is half the molar mass, because each H_2SO_4 molecule can produce *two* H^+ ions: therefore, only half a mole of H_2SO_4 is needed to provide one mole of H^+ ion. Similarly, the equivalent weight of phosphoric acid (H_3PO_4) is one

Cumulative Review: Chapters 13, 14, and 15

third of its molar mass, because each H_3PO_4 molecule can provide three H^+ ions (and so only one-third of a mole of H_3PO_4 is needed to provide one mole of H^+ ions).

$$H_2SO_4 \rightarrow 2H^+ + SO_4^{2-}$$

$$H_3PO_4 \rightarrow 3H^+ + PO_4^{3-}$$

Similarly, bases like $Ca(OH)_2$ and $Mg(OH)_2$ have equivalent weights that are half the molar masses, because each of these substances produces two moles of OH^- ion per mole of base (and so only half a mole of base is needed to provide one mole of OH^-).

The *normality* of a solution is defined to be the number of equivalents of solute contained in one liter of the solution: a 1 N solution of an acid contains 1 mole of H^+ per liter; a 1 N solution of a base contains 1 mole of OH^- per liter. As the equivalent weight and the molar mass of a substance are related by small whole numbers (representing the number of H^+ or OH^- a molecule of the substance furnishes), the normality and molarity of a solution are also simply related by these same numbers. In fact, $N = n \times M$ for a solution, where n represents the number of H^+ or OH^- ions furnished per molecule of solute. For example, a 0.521 M HCl(aq) solution is also 0.521 N, because each HCl furnishes one H^+ ion. However, a 0.475 M H_2SO_4 solution would have a normality equal to

$$N = n \times M = 2 \times 0.475\ M = 0.950\ N$$

because each H_2SO_4 molecule furnishes two H^+ ions.

28. $P_1 \times V_1 = P_2 \times V_2$

   a. $V_2 = \dfrac{P_1 \times V_1}{P_2} = \dfrac{125\ \text{mL} \times 755\ \text{mm Hg}}{899\ \text{mm Hg}} = 105\ \text{mL}$

   b. $P_2 = \dfrac{P_1 \times V_1}{V_2} = \dfrac{455\ \text{mL} \times 755\ \text{mm Hg}}{327\ \text{mL}} = 1.05 \times 10^3\ \text{mm Hg}$

29. $\dfrac{V_1}{T_1} = \dfrac{V_2}{T_2}$

   a. $V_2 = \dfrac{V_1 \times T_2}{T_1} = \dfrac{(255\ \text{mL})(55 + 273\text{K})}{(35 + 273\text{K})} = 272\ \text{mL}$

   b. $V_2 = \dfrac{V_1 \times T_2}{T_1} = \dfrac{(325\ \text{mL})(-196 + 273\text{K})}{(25 + 273\text{K})} = 84.0\ \text{mL}$

30. a. $PV = nRT$; molar mass He = 4.003 g; 25°C = 298 K

   $1.15\ \text{g He} \times \dfrac{1\ \text{mol}}{4.003\ \text{g}} = 0.2873\ \text{mol He}$

   $V = \dfrac{nRT}{P} = \dfrac{(0.2873\ \text{mol})(0.08206\ \text{L-atm/mol-K})(298\ \text{K})}{(1.01\ \text{atm})} = 6.96\ \text{L}$

   b. molar masses: H_2, 2.016 g; He, 4.003 g; 0°C = 273 K

   $2.27\ \text{g}\ H_2 \times \dfrac{1\ \text{mol}\ H_2}{2.016\ \text{g}\ H_2} = 1.126\ \text{mol}\ H_2$

Cumulative Review: Chapters 13, 14, and 15

$$1.03 \text{ g He} \times \frac{1 \text{ mol He}}{4.003 \text{ g He}} = 0.2573 \text{ mol He}$$

$$P_{H_2} = \frac{nRT}{V} = \frac{(1.126 \text{ mol H}_2)(0.08206 \text{ L-atm/mol-K})(273 \text{ K})}{(5.00 \text{ L})} = 5.05 \text{ atm}$$

$$P_{He} = \frac{nRT}{V} = \frac{(0.2573 \text{ mol He})(0.08206 \text{ L-atm/mol-K})(273 \text{ K})}{(5.00 \text{ L})} = 1.15 \text{ atm}$$

c. molar mass of Ar = 39.95 g; 27°C = 300 K

$$42.5 \text{ g Ar} \times \frac{1 \text{ mol Ar}}{39.95 \text{ g Ar}} = 1.064 \text{ mol Ar}$$

$$P = \frac{nRT}{V} = \frac{(1.064 \text{ mol Ar})(0.08206 \text{ L-atm/mol-K})(300 \text{ K})}{(9.97 \text{ L})} = 2.63 \text{ atm}$$

31. $MnO_2(s) + 4HCl(aq) \rightarrow MnCl_2(aq) + 2H_2O(l) + Cl_2(g)$

    molar mass MnO_2 = 86.94 g

    $$4.05 \text{ g MnO}_2 \times \frac{1 \text{ mol}}{86.94 \text{ g}} = 0.04658 \text{ mol MnO}_2$$

    From the coefficients of the balanced chemical equation, if 0.04658 mol of MnO_2 react completely, then 0.04658 mol of Cl_2 will be produced.

    At STP, 0.04658 mol of Cl_2 will have a volume of $0.04658 \text{ mol} \times \frac{22.4 \text{ L}}{\text{mol}} = 1.04 \text{ L}$

32. molar masses: $CaCO_3$, 100.09 g; CO_2, 44.01 g

    $$1.25 \text{ g CaCO}_3 \times \frac{1 \text{ mol CaCO}_3}{100.09 \text{ g}} = 0.01249 \text{ mol CaCO}_3$$

    $$0.01249 \text{ mol CaCO}_3 \times \frac{1 \text{ mol CO}_2}{1 \text{ mol CaCO}_3} = 0.01249 \text{ mol CO}_2$$

    $$0.01249 \text{ mol CO}_2 \times \frac{44.01 \text{ g CO}_2}{1 \text{ mol CO}_2} = 0.550 \text{ g CO}_2$$

    $$0.01249 \text{ mol CO}_2 \times \frac{22.4 \text{ L}}{1 \text{ mol}} = 0.280 \text{ L CO}_2 \text{ at STP}$$

33. molar mass NaCl = 58.44 g; 25°C = 298 K; 767 mm Hg = 1.009 atm

    $$1.25 \text{ g NaCl} \times \frac{1 \text{ mol NaCl}}{58.44 \text{ g NaCl}} \times \frac{1 \text{ mol Cl}_2}{2 \text{ mol Na}} = 0.01069 \text{ mol Cl}_2$$

    $$V = \frac{nRT}{P} = \frac{(0.01069 \text{ mol Cl}_2)(0.08206 \text{ L atm mol}^{-1} \text{ K}^{-1})(298 \text{ K})}{1.009 \text{ atm}} = 0.259 \text{ L}$$

Cumulative Review: Chapters 13, 14, and 15

34. a. mass of solution = 2.05 g NaCl + 19.2 g water = 21.25 g solution

$$\frac{2.05 \text{ g NaCl}}{21.25 \text{ g solution}} \times 100 = 9.65\% \text{ NaCl}$$

b. $26.2 \text{ g solution} \times \dfrac{10.5 \text{ g CaCl}_2}{100 \text{ g solution}} = 2.75 \text{ g CaCl}_2$

c. $225 \text{ g solution} \times \dfrac{5.05 \text{ g NaCl}}{100 \text{ g solution}} = 11.4 \text{ g NaCl required}$

35. a. molar mass NaOH = 40.00 g; 235 mL = 0.235 L

$$0.235 \text{ L solution} \times \frac{0.251 \text{ mol NaOH}}{1 \text{ L solution}} = 0.05899 \text{ mol NaOH}$$

$$0.05899 \text{ mol NaOH} \times \frac{40.00 \text{ g NaOH}}{1 \text{ mol NaOH}} = 2.36 \text{ g NaOH}$$

b. 125 mL = 0.125 L

$$M = \frac{0.293 \text{ mol}}{0.125 \text{ L}} = 2.34 \ M$$

c. $5.05 \text{ L solution} \times \dfrac{6.01 \text{ mol HCl}}{1 \text{ L solution}} = 30.4 \text{ mol HCl}$

36. $M_1 \times V_1 = M_2 \times V_2$

a. $M_2 = \dfrac{(12.5 \text{ mL})(1.515 \ M)}{(12.5 + 25 \text{ mL})} = 0.505 \ (0.51) \ M$

b. $M_2 = \dfrac{(75.0 \text{ mL})(0.252 \ M)}{(225 \text{ mL})} = 0.0840 \ M$

c. $M_2 = \dfrac{(52.1 \text{ mL})(0.751 \ M)}{(52.1 + 250. \text{ mL})} = 0.130 \ M$

37. It would be convenient to first calculate the number of moles of NaOH present in the sample, because this information will be needed for each part of the answer. 36.2 mL = 0.0362 L

$$\text{mol NaOH} = 0.0362 \text{ L} \times \frac{0.259 \text{ mol NaOH}}{1 \text{ L}} = 9.38 \times 10^{-3} \text{ mol NaOH}$$

a. HCl + NaOH → NaCl + H₂0

$$9.38 \times 10^{-3} \text{ mol NaOH} \times \frac{1 \text{ mol HCl}}{1 \text{ mol NaOH}} = 9.38 \times 10^{-3} \text{ mol HCl}$$

$$9.38 \times 10^{-3} \text{ mol HCl} \times \frac{1 \text{ L}}{0.271 \text{ mol HCl}} = 0.0346 \text{ L} = 34.6 \text{ mL}$$

Cumulative Review: Chapters 13, 14, and 15

b. $H_2SO_4 + 2NaOH \rightarrow Na_2SO_4 + 2H_2O$

$$9.38 \times 10^{-3} \text{ mol NaOH} \times \frac{1 \text{ mol } H_2SO_4}{2 \text{ mol NaOH}} = 4.69 \times 10^{-3} \text{ mol } H_2SO_4$$

$$4.69 \times 10^{-3} \text{ mol } H_2SO_4 \times \frac{1 \text{ L}}{0.119 \text{ mol } H_2SO_4} = 0.0394 \text{ L} = 39.4 \text{ mL}$$

c. $H_3PO_4 + 3NaOH \rightarrow Na_3PO_4 + 3H_2O$

$$9.38 \times 10^{-3} \text{ mol } H_3PO_4 \times \frac{1 \text{ mol } H_3PO_4}{3 \text{ mol NaOH}} = 3.13 \times 10^{-3} \text{ mol } H_3PO_4$$

$$3.13 \times 10^{-3} \text{ mol } H_3PO_4 \times \frac{1 \text{ L}}{0.171 \text{ mol } H_3PO_4} = 0.0183 \text{ L} = 18.3 \text{ mL}$$

38. a. $125 \text{ mL solution} \times \dfrac{1.84 \text{ g solution}}{1 \text{ mL solution}} = 230. \text{ g solution}$

$230. \text{ g solution} \times \dfrac{98.3 \text{ g } H_2SO_4}{1 \text{ g solution}} = 226 \text{ g } H_2SO_4$

b. The concentrated solution contains 226 g of H_2SO_4 (molar mass 98.09 g) in 125 mL (0.125 L) of solution

$$226 \text{ g } H_2SO_4 \times \frac{1 \text{ mol } H_2SO_4}{98.09 \text{ g } H_2SO_4} = 2.304 \text{ mol } H_2SO_4$$

$$M = \frac{2.304 \text{ mol } H_2SO_4}{0.125 \text{ L solution}} = 18.4 \, M$$

c. $M_1 \times V_1 = M_2 \times V_2$

$$M_2 = \frac{(0.125 \text{ L})(18.4 \, M)}{3.01 \text{ L}} = 0.764 \, M$$

d. $\dfrac{0.764 \text{ mol}}{1 \text{ L}} \times \dfrac{2 \text{ equivalents}}{1 \text{ mol}} = 1.53 \, N$

e. $\text{mmol NaOH} = 45.3 \text{ mL} \times \dfrac{0.532 \text{ mmol NaOH}}{1 \text{ mL}} = 24.10 \text{ mmol}$

$H_2SO_4 + 2NaOH \rightarrow Na_2SO_4 + 2H_2O$

$\text{mmol } H_2SO_4 \text{ required} = 24.10 \text{ mmol NaOH} \times \dfrac{1 \text{ mmol } H_2SO_4}{2 \text{ mmol NaOH}} = 12.05 \text{ mmol } H_2SO_4$

$12.05 \text{ mmol } H_2SO_4 \times \dfrac{1 \text{ mL solution}}{0.764 \text{ mmol } H_2SO_4} = 15.8 \text{ mL of the sulfuric acid solution.}$

CHAPTER 16

Acids and Bases

1. Acids were recognized primarily from their sour taste. Bases were recognized from their bitter taste and slippery feel on skin.

2. $HCl(g) \xrightarrow{H_2O} H^+(aq) + Cl^-(aq)$

   $NaOH(s) \xrightarrow{H_2O} Na^+(aq) + OH^-(aq)$

3. The Arrhenius definitions of acid and base are restricted to aqueous solutions, in which a substance producing $H^+(aq)$ ions is considered an acid and a substance producing $OH^-(aq)$ ions is considered a base.

   The Brønsted-Lowry definitions of acid and base are less restrictive, and define an acid as a species capable of donating protons (regardless of the solvent), and a base as a species capable of receiving the protons being donated. In the equation $HA(aq) + H_2O(l) \rightarrow H_3O^+(aq) + A^-(aq)$, HA is a Brønsted-Lowry acid because it donates a proton to water; water is a Brønsted-Lowry base because it receives the proton from HA.

4. Conjugate acid–base pairs differ from each other by one proton (one hydrogen ion, H^+). For example, CH_3COOH (acetic acid), differs from its conjugate base, CH_3COO^- (acetate ion), by a single H^+ ion.

   $CH_3COOH(aq) \rightleftharpoons CH_3COO^-(aq) + H^+(aq)$

5. Water accepts a proton in going from H_2O to H_3O^+ which makes it a base in the Brønsted–Lowry model.

6. acids; bases

7. a. a conjugate pair: the two species differ by one proton

   b. not a conjugate pair

   $HClO, ClO^-$

   $HClO_2, ClO_2^-$

   c. not a conjugate pair

   $H_3PO_4, H_2PO_4^-$

   HPO_4^{2-}, PO_4^{3-}

   d. not a conjugate pair

   H_2CO_3, HCO_3^-

   HCO_3^-, CO_3^{2-}

Chapter 16: Acids and Bases

8.   a.   not a conjugate pair

           H_2SO_4, HSO_4^-

           HSO_4^-, SO_4^{2-}

    b.   a conjugate pair: the two species differ by one proton

    c.   not a conjugate pair

           $HClO_4$, ClO_4^-

           HCl, Cl^-

    d.   not a conjugate pair

           NH_4^+, NH_3

           NH_3, NH_2^-

9.   a.   HF (acid) + H_2O (base) ⇌ F^- (base) + H_3O^+ (acid)

    b. CN^- (base) + H_2O (acid) ⇌ HCN (acid) + OH^- (base)

    c. HCO_3^- (base) + H_2O (acid) ⇌ H_2CO_3 (acid) + OH^- (base)

10.   a. NH_3 (base) + H_2O (acid) ⇌ NH_4^+ (acid) + OH^- (base)

    b. PO_4^{3-} (base) + H_2O (acid) ⇌ HPO_4^{2-} (acid) + OH^- (base)

    c. $C_2H_3O_2^-$ (base) + H_2O (acid) ⇌ $HC_2H_3O_2$ (acid) + OH^- (base)

11.   The conjugate *acid* of the species indicated would have *one additional proton*:

    a. HPO_4^{2-}

    b. HIO_3

    c. HNO_3

    d. NH_3

12.   The conjugate *acid* of the species indicated would have *one additional proton*:

    a.   HClO

    b.   HCl

    c. $HClO_3$

    d. $HClO_4$

13.   The conjugate *bases* of the species indicated would have *one less proton*:

    a. HS^-

    b. S^{2-}

    c. NH_2^-

    d. HSO_3^-

Chapter 16: Acids and Bases

14. The conjugate *bases* of the species indicated would have *one less proton*:
    a. BrO$^-$
    b. NO$_2^-$
    c. SO$_3^{2-}$
    d. CH$_3$NH$_2$

15.
    a. HSO$_3^-$(aq) + H$_2$O(l) $\rightleftharpoons$ SO$_3^{2-}$(aq) + H$_3$O$^+$(aq)
    b. CO$_3^{2-}$(aq) + H$_2$O(l) $\rightleftharpoons$ HCO$_3^-$(aq) + OH$^-$(aq)
    c. H$_2$PO$_4^-$(aq) + H$_2$O(l) $\rightleftharpoons$ HPO$_4^{2-}$(aq) + H$_3$O$^+$(aq)
    d. C$_2$H$_3$O$_2^-$(aq) + H$_2$O(l) $\rightleftharpoons$ HC$_2$H$_3$O$_2$(aq) + OH$^-$(aq)

16.
    a. O^{2-}(aq) + H$_2$O(l) $\rightleftharpoons$ OH$^-$(aq) + OH$^-$(aq)
    b. NH$_3$(aq) + H$_2$O(l) $\rightleftharpoons$ NH$_4^+$(aq) + OH$^-$(aq)
    c. HSO$_4^-$(aq) + H$_2$O(l) $\rightleftharpoons$ SO$_4^{2-}$(aq) + H$_3$O$^+$(aq)
    d. HNO$_2$(aq) + H$_2$O(l) $\rightleftharpoons$ NO$_2^-$(aq) + H$_3$O$^+$(aq)

17. A strong acid is one for which the equilibrium in water lies far to the right. A strong acid is almost completely converted to its conjugate base when dissolved in water. A strong acid's anion (its conjugate base) must be very poor at attracting, or holding onto, protons. A regular arrow ($\rightarrow$) rather than a double arrow ($\rightleftharpoons$) is used when writing an equation for the dissociation of a strong acid to indicate this.

18. To say that an acid is *weak* in aqueous solution means that the acid does not easily transfer protons to water (and does not fully ionize). If an acid does not lose protons easily, then the acid's anion must be a strong attractor of protons (good at holding on to protons).

19. If water is a much stronger base than the anion of the acid, then protons will be attracted more strongly to water molecules than to the anions, and the acid will ionize well. If the anion of the acid is a much stronger base than water, then the anions will hold on to their protons (or will attract any protons present in the water), and the acid will ionize poorly.

20. A strong acid is one that loses its protons easily and fully ionizes in water; this means that the acid's conjugate base must be poor at attracting and holding on to protons, and is therefore a relatively weak base. A weak acid is one that resists loss of its protons and does not ionize well in water; this means that the acid's conjugate base attracts and holds onto protons tightly and is a relatively strong base.

21. The hydronium ion is H$_3$O$^+$.

    For a general acid HA: HA + H$_2$O $\rightarrow$ A$^-$ + H$_3$O$^+$

22. H$_2$SO$_4$ (sulfuric): H$_2$SO$_4$ + H$_2$O $\rightarrow$ HSO$_4^-$ + H$_3$O$^+$

    HCl (hydrochloric): HCl + H$_2$O $\rightarrow$ Cl$^-$ + H$_3$O$^+$

    HNO$_3$ (nitric): HNO$_3$ + H$_2$O $\rightarrow$ NO$_3^-$ + H$_3$O$^+$

Chapter 16: Acids and Bases

HClO$_4$ (perchloric): HClO$_4$ + H$_2$O → ClO$_4^-$ + H$_3$O$^+$

23. CH$_3$COOH + H$_2$O ⇌ CH$_3$COO$^-$ + H$_3$O$^+$

    CH$_3$CH$_2$COOH + H$_2$O ⇌ CH$_3$CH$_2$COO$^-$ + H$_3$O$^+$

24. An oxyacid is an acid containing a particular element which is bonded to one or more oxygen atoms. HNO$_3$, H$_2$SO$_4$, HClO$_4$ are oxyacids. HCl, HF, HBr are not oxyacids.

25. Acids that are *weak* have relatively strong conjugate bases.

    a. CN$^-$ is a relatively strong base; HCN is a weak acid.

    b. HS$^-$ is a relatively strong base; H$_2$S is a weak acid.

    c. BrO$_4^-$ is a very weak base; HBrO$_4$ is a strong acid.

    d. NO$_3^-$ is a very weak base; HNO$_3$ is a strong acid.

26. Salicylic acid is a monoprotic acid: only the hydrogen of the carboxyl group ionizes.

    [Structural equation: 2-hydroxybenzoic acid (COOH, OH on benzene ring) + H$_2$O ⇌ 2-hydroxybenzoate (COO$^-$, OH on benzene ring) + H$_3$O$^+$]

27. water as base: HF + H$_2$O ⇌ F$^-$ + H$_3$O$^+$

    water as acid: NH$_3$ + H$_2$O ⇌ NH$_4^+$ + OH$^-$

28. For example, HCO$_3^-$ can behave as an acid if it reacts with something that more strongly gains protons than does HCO$_3^-$ itself. For example, HCO$_3^-$ would behave as an acid when reacting with hydroxide ion (a much stronger base).

    HCO$_3^-$(aq) + OH$^-$(aq) → CO$_3^{2-}$(aq) + H$_2$O(l).

    On the other hand, HCO$_3^-$ would behave as a base when reacted with something that more readily loses protons than does HCO$_3^-$ itself. For example, HCO$_3^-$ would behave as a base when reacting with hydrochloric acid (a much stronger acid).

    HCO$_3^-$(aq) + HCl(aq) → H$_2$CO$_3$(aq) + Cl$^-$(aq)

    For H$_2$PO$_4^-$, similar equations can be written:

    H$_2$PO$_4^-$(aq) + OH$^-$(aq) → HPO$_4^{2-}$(aq) + H$_2$O(l)

    H$_2$PO$_4^-$(aq) + H$_3$O$^+$(aq) → H$_3$PO$_4$(aq) + H$_2$O(l)

29. The ion–product constant for water is the equilibrium constant for the reaction in which water autoionizes; the constant says that in water or an aqueous solution, there is an equilibrium between hydronium ion and hydroxide ion such that the product of their concentrations fulfills the value of the constant.

    H$_2$O + H$_2$O ⇌ H$_3$O$^+$ + OH$^-$ K_w = [H$_3$O$^+$][OH$^-$] = 1.0 × 10^{-14}

    H$_2$O ⇌ H$^+$ + OH$^-$ K_w = [H$^+$][OH$^-$] = 1.0 × 10^{-14}

Chapter 16: Acids and Bases

30. The hydrogen ion concentration and the hydroxide ion concentration of water are *not* independent: they are related by the equilibrium

$$H_2O(l) \rightleftharpoons H^+(aq) + OH^-(aq)$$

for which $K_w = [H^+][OH^-] = 1.0 \times 10^{-14}$ at 25°C.

If the concentration of one of these ions is increased by addition of a reagent producing H^+ or OH^-, then the concentration of the complementary ion will have to decrease so that the value of K_w will hold true. So if an acid is added to a solution, the concentration of hydroxide ion in the solution will decrease to a lower value. Similarly, if a base is added to a solution, then the concentration of hydrogen ion will have to decrease to a lower value.

31. $K_w = [H^+][OH^-] = 1.0 \times 10^{-14}$ at 25°C

   a. $[H^+] = \dfrac{1.0 \times 10^{-14}}{2.32 \times 10^{-4}\,M} = 4.3 \times 10^{-11}\,M$; solution is basic

   b. $[H^+] = \dfrac{1.0 \times 10^{-14}}{8.99 \times 10^{-10}\,M} = 1.1 \times 10^{-5}\,M$; solution is acidic

   c. $[H^+] = \dfrac{1.0 \times 10^{-14}}{4.34 \times 10^{-6}\,M} = 2.3 \times 10^{-9}\,M$; solution is basic

   d. $[H^+] = \dfrac{1.0 \times 10^{-14}}{6.22 \times 10^{-12}\,M} = 1.6 \times 10^{-3}\,M$; solution is acidic

32. $K_w = [H^+][OH^-] = 1.0 \times 10^{-14}$ at 25°C

   a. $[H^+] = \dfrac{1.0 \times 10^{-14}}{3.99 \times 10^{-5}\,M} = 2.5 \times 10^{-10}\,M$; solution is basic

   b. $[H^+] = \dfrac{1.0 \times 10^{-14}}{2.91 \times 10^{-9}\,M} = 3.4 \times 10^{-6}\,M$; solution is acidic

   c. $[H^+] = \dfrac{1.0 \times 10^{-14}}{7.23 \times 10^{-2}\,M} = 1.4 \times 10^{-13}\,M$; solution is basic

   d. $[H^+] = \dfrac{1.0 \times 10^{-14}}{9.11 \times 10^{-7}\,M} = 1.1 \times 10^{-8}\,M$; solution is basic

33. $K_w = [H^+][OH^-] = 1.0 \times 10^{-14}$ at 25°C

   a. $[OH^-] = \dfrac{1.0 \times 10^{-14}}{4.01 \times 10^{-4}\,M} = 2.5 \times 10^{-11}\,M$; solution is acidic

   b. $[OH^-] = \dfrac{1.0 \times 10^{-14}}{7.22 \times 10^{-6}\,M} = 1.4 \times 10^{-9}\,M$; solution is acidic

   c. $[OH^-] = \dfrac{1.0 \times 10^{-14}}{8.05 \times 10^{-7}\,M} = 1.2 \times 10^{-8}\,M$; solution is acidic

   d. $[OH^-] = \dfrac{1.0 \times 10^{-14}}{5.43 \times 10^{-9}\,M} = 1.8 \times 10^{-6}\,M$; solution is basic

Chapter 16: Acids and Bases

34. $K_w = [H^+][OH^-] = 1.0 \times 10^{-14}$ at 25°C

    a. $[OH^-] = \dfrac{1.0 \times 10^{-14}}{1.02 \times 10^{-7} M} = 9.8 \times 10^{-8} M$; solution is acidic

    b. $[OH^-] = \dfrac{1.0 \times 10^{-14}}{9.77 \times 10^{-8} M} = 1.02 \times 10^{-7} M \,(1.0 \times 10^{-7} M)$; solution is slightly basic

    c. $[OH^-] = \dfrac{1.0 \times 10^{-14}}{3.41 \times 10^{-3} M} = 2.9 \times 10^{-12} M$; solution is acidic

    d. $[OH^-] = \dfrac{1.0 \times 10^{-14}}{4.79 \times 10^{-11} M} = 2.1 \times 10^{-4} M$; solution is basic

35.
    a. $[H^+] = 1.2 \times 10^{-3} M$ is more acidic
    b. $[H^+] = 2.6 \times 10^{-6} M$ is more acidic
    c. $[H^+] = 0.000010 M$ is more acidic

36.
    a. $[OH^-] = 6.03 \times 10^{-4} M$ is more basic
    b. $[OH^-] = 4.21 \times 10^{-6} M$ is more basic
    c. $[OH^-] = 8.04 \times 10^{-4} M$ is more basic

37. Because the concentrations of $[H^+]$ and $[OH^-]$ in aqueous solutions tend to be expressed in scientific notation, and these numbers have negative exponents for their powers of ten, it tends to be clumsy to make comparisons between different concentrations of these ions (see questions 35 and 36 above). The pH scale converts such numbers into "ordinary" numbers between 0 and 14 that can be more easily compared. The pH of a solution is defined as the *negative* of the base 10 logarithm of the hydrogen ion concentration, $pH = -\log[H^+]$.

38. Answer depends on student choice.

39. As 2.33×10^{-6} has three significant figures, the pH should be expressed to the third decimal place. The figure *before* the decimal place in a pH is *not* one of the significant digits: the figure before the decimal place is related to the *power of ten* (exponent) of the concentration.

40. pH 1–2, deep red; pH 4, purple; pH 8, blue; pH 11, green

41. $pH = -\log[H^+]$

    a. $pH = -\log[4.02 \times 10^{-3} M] = 2.396$; solution is acidic
    b. $pH = -\log[8.99 \times 10^{-7} M] = 6.046$; solution is acidic
    c. $pH = -\log[2.39 \times 10^{-6} M] = 5.622$; solution is acidic
    d. $pH = -\log[1.89 \times 10^{-10} M] = 9.724$; solution is basic

42. $pH = -\log[H^+]$

    a. $pH = -\log[0.00100 M] = 3.000$; solution is acidic
    b. $pH = -\log[2.19 \times 10^{-4} M] = 3.660$; solution is acidic

c. pH = –log[9.18 × 10^{-11} M] = 10.037; solution is basic

d. pH = –log[4.71 × 10^{-7} M] = 6.327; solution is acidic

43. pOH = –log[OH$^-$]   pH = 14.00 – pOH

    a. pOH = –log[4.73 × 10^{-4} M] = 3.325

        pH = 14.00 – 3.325 = 10.675 = 10.68; solution is basic

    b. pOH = –log[5.99 × 10^{-1} M] = 0.223

        pH = 14.00 – 0.223 = 13.777 = 13.78; solution is basic

    c. pOH = –log[2.87 × 10^{-8} M] = 7.542

        pH = 14.00 – 7.542 = 6.458 = 6.46; solution is acidic

    d. pOH = –log[6.39 × 10^{-3} M] = 2.194

        pH = 14.00 – 2.194 = 11.806 = 11.81; solution is basic

44. pOH = –log[OH$^-$]   pH = 14.00 – pOH

    a. pOH = –log[8.63 × 10^{-3} M] = 2.064

        pH = 14.00 – 2.064 = 11.936 = 11.94; solution is basic

    b. pOH = –log[7.44 × 10^{-6} M] = 5.128

        pH = 14.00 – 5.128 = 8.872 = 8.87; solution is basic

    c. pOH = –log[9.35 × 10^{-9} M] = 8.029

        pH = 14.00 – 8.029 = 5.971 = 5.97; solution is acidic

    d. pOH = –log[1.21 × 10^{-11} M] = 10.917

        pH = 14.00 – 10.917 = 3.083 = 3.08; solution is acidic

45. pH = 14.00 – pOH

    a. pH = 14.00 – 4.32 = 9.68; solution is basic

    b. pH = 14.00 – 8.90 = 5.10; solution is acidic

    c. pH = 14.00 – 1.81 = 12.19; solution is basic

    d. pH = 14.00 – 13.1 = 0.9; solution is acidic

46. pOH = 14.00 – pH

    a. pOH = 14.00 – 9.78 = 4.22; solution is basic

    b. pOH = 14.00 – 4.01 = 9.99; solution is acidic

    c. pOH = 14.00 – 2.79 = 11.21; solution is acidic

    d. pOH = 14.00 – 11.21 = 2.79; solution is basic

47. a. pH = –log[4.76 × 10^{-8} M] = 7.322; solution is basic

$$[OH^-] = \frac{1.0 \times 10^{-14}}{4.76 \times 10^{-8}\ M} = 2.10 \times 10^{-7}\ M$$

Chapter 16: Acids and Bases

b. $pH = -\log[8.92 \times 10^{-3} \, M] = 2.050$; solution is acidic

$[OH^-] = \dfrac{1.0 \times 10^{-14}}{8.92 \times 10^{-3} \, M} = 1.1 \times 10^{-12} \, M$

c. $pH = -\log[7.00 \times 10^{-5} \, M] = 4.155$; solution is acidic

$[OH^-] = \dfrac{1.0 \times 10^{-14}}{7.00 \times 10^{-5} \, M} = 1.4 \times 10^{-10} \, M$

d. $pH = -\log[1.25 \times 10^{-12} \, M] = 11.903$; solution is basic

$[OH^-] = \dfrac{1.0 \times 10^{-14}}{1.25 \times 10^{-12} \, M} = 8.0 \times 10^{-3} \, M$

48. a. $pH = -\log[1.91 \times 10^{-2} \, M] = 1.719$; solution is acidic

$[OH^-] = \dfrac{1.0 \times 10^{-14}}{1.91 \times 10^{-2} \, M} = 5.2 \times 10^{-13} \, M$

b. $pH = -\log[4.83 \times 10^{-7} \, M] = 6.316$; solution is acidic

$[OH^-] = \dfrac{1.0 \times 10^{-14}}{4.83 \times 10^{-7} \, M} = 2.1 \times 10^{-8} \, M$

c. $pH = -\log[8.92 \times 10^{-11} \, M] = 10.050$; solution is basic

$[OH^-] = \dfrac{1.0 \times 10^{-14}}{8.92 \times 10^{-11} \, M} = 1.1 \times 10^{-4} \, M$

d. $pH = -\log[6.14 \times 10^{-5} \, M] = 4.212$; solution is acidic

$[OH^-] = \dfrac{1.0 \times 10^{-14}}{6.14 \times 10^{-5} \, M} = 1.6 \times 10^{-10} \, M$

49. $[H^+] = \{inv\}\{log\}[-pH]$ or 10^{-pH}

a. $[H^+] = \{inv\}\{log\}[-9.01] = 9.8 \times 10^{-10} \, M$

b. $[H^+] = \{inv\}\{log\}[-6.89] = 1.3 \times 10^{-7} \, M$

c. $[H^+] = \{inv\}\{log\}[-1.02] = 9.5 \times 10^{-2} \, M$

d. $[H^+] = \{inv\}\{log\}[-7.00] = 1.0 \times 10^{-7} \, M$

50. $[H^+] = \{inv\}\{log\}[-pH]$ or 10^{-pH}

a. $[H^+] = \{inv\}\{log\}[-1.04] = 9.1 \times 10^{-2} \, M$

b. $[H^+] = \{inv\}\{log\}[-13.1] = 8 \times 10^{-14} \, M$

c. $[H^+] = \{inv\}\{log\}[-5.99] = 1.0 \times 10^{-6} \, M$

d. $[H^+] = \{inv\}\{log\}[-8.62] = 2.4 \times 10^{-9} \, M$

51. $pH + pOH = 14.00$ $[H^+] = \{inv\}\{log\}[-pH]$ or 10^{-pH}

a. $pH = 14.00 - 4.95 = 9.05$ $[H^+] = \{inv\}\{log\}[-9.05] = 8.9 \times 10^{-10} \, M$

Chapter 16: Acids and Bases

    b.   pH = 14.00 – 7.00 = 7.00 $[H^+]$ = {inv}{log}[–7.00] = 1.0×10^{-7} M

    c.   pH = 14.00 – 12.94 = 1.06 $[H^+]$ = {inv}{log}[–1.06] = 8.7×10^{-2} M

    d.   pH = 14.00 – 1.02 = 12.98 $[H^+]$ = {inv}{log}[–12.98] = 1.0×10^{-13} M

52.   pH + pOH = 14.00 $[H^+]$ = {inv}{log}[–pH] or 10^{-pH}

    a.   pH = 14.00 – 4.99 = 9.01

           $[H^+]$ = {inv}{log}[–9.01] = 9.8×10^{-10} M

    b. $[H^+]$ = {inv}{log}[–7.74] = 1.8×10^{-8} M

    c.   pH = 14.00 – 10.74 = 3.26

           $[H^+]$ = {inv}{log}[–3.26] = 5.5×10^{-4} M

    d. $[H^+]$ = {inv}{log}[–2.25] = 5.6×10^{-3} M

53.   a.   pH = –log[4.78×10^{-2} M] = 1.321

    b.   pH = 14.00 – pOH = 14.00 – 4.56 = 9.44

    c.   pOH = –log[9.74×10^{-3} M] = 2.011   pH = 14.00 – 2.011 = 11.99

    d.   pH = –log[1.24×10^{-8} M] = 7.907

54.   a.   pH = –log[4.39×10^{-6} M] = 5.358

    b.   pH = 14.00 – pOH = 14.00 – 10.36 = 3.64

    c.   pOH = –log[9.37×10^{-9} M] = 8.028   pH = 14.00 – 8.028 = 5.97

    d.   pH = –log[3.31×10^{-1} M] = 0.480

55.   Effectively *no* molecules of HCl remain in solution. HCl is a strong acid. Its equilibrium with water lies far to the right. All the HCl molecules originally dissolved in the water will ionize.

56.   The solution contains water molecules, H_3O^+ ions (protons), and NO_3^- ions. Because HNO_3 is a strong acid, which is completely ionized in water, there are no HNO_3 molecules present.

57.   a.   HCl is a strong acid and completely ionized so

           $[H^+] = 1.04 \times 10^{-4}$ M

           pH = –log[1.04×10^{-4}] = 3.983.

    b. HNO_3 is a strong acid and completely ionized so

           $[H^+] = 0.00301$ M

           pH = –log[0.00301] = 2.521.

    c. $HClO_4$ is a strong acid and completely ionized so

           $[H^+] = 5.41 \times 10^{-4}$ M

           pH = –log[5.41×10^{-4}] = 3.267.

Chapter 16: Acids and Bases

      d. HNO_3 is a strong acid and completely ionized so

              $[H^+] = 6.42 \times 10^{-2}\ M$

              $pH = -\log[6.42 \times 10^{-2}] = 1.192$.

58.   a. HNO_3 is a strong acid and completely ionized so $[H^+] = 1.21 \times 10^{-3}\ M$ and pH = 2.917.

      b. $HClO_4$ is a strong acid and completely ionized so $[H^+] = 0.000199\ M$ and pH = 3.701.

      c.   HCl is a strong acid and completely ionized so $[H^+] = 5.01 \times 10^{-5}\ M$ and pH = 4.300.

      d.   HBr is a strong acid and completely ionized so $[H^+] = 0.00104\ M$ and pH = 2.983.

59.   A buffered solution is one that resists change in its pH even when a strong acid or base is added to it. A solution is buffered by the presence of the combination of a weak acid and its conjugate base.

60.   A buffered solution consists of a mixture of a weak acid and its conjugate base; one example of a buffered solution is a mixture of acetic acid (CH_3COOH) and sodium acetate ($NaCH_3COO$).

61.   The conjugate *base* component of the buffer mixture is capable of combining with any strong acid that might be added to the buffered solution. For the example of acetate ion ($C_2H_3O_2^-$) given in the solution to question 60, the equation is

        $HCl(aq) + C_2H_3O_2^-(aq) \rightarrow HC_2H_3O_2(aq) + Cl^-(aq)$.

62.   The weak acid component of a buffered solution is capable of reacting with added strong base. For example, using the buffered solution given as an example in Question 60, acetic acid would consume added sodium hydroxide as follows:

        $CH_3COOH(aq) + NaOH(aq) \rightarrow NaCH_3COO(aq) + H_2O(l)$.

Acetic acid *neutralizes* the added NaOH and prevents it from having much effect on the overall pH of the solution.

63.   a.   not a buffer: although HCl and Cl^- are conjugates, HCl is *not* a weak acid.

      b.   a buffer: CH_3COOH and CH_3COO^- are conjugates

      c.   a buffer: H_2S and HS^- are conjugates

      d.   not a buffer: S^{2-} (of Na_2S) is *not* the conjugate base of H_2S

64.   HCl: $H_3O^+ + C_2H_3O_2^- \rightarrow HC_2H_3O_2 + H_2O$

      NaOH: $OH^- + HC_2H_3O_2 \rightarrow C_2H_3O_2^- + H_2O$

65.   In whatever solvent is used, a Brønsted–Lowry acid will still be a proton donor. In liquid ammonia, HCl would still be an acid as shown in the following equation:

        $HCl + NH_3 \rightarrow Cl^- + NH_4^+$

in which the proton is transferred from HCl to NH_3. Similarly, OH^- would still be a base (proton acceptor) in liquid ammonia as indicated in the equation

        $OH^- + NH_3 \rightarrow H_2O + NH_2^-$

in which OH^- could receive a proton from ammonia.

Chapter 16: Acids and Bases

66. a. NaOH is completely ionized, so $[OH^-] = 0.10\ M$.

    $pOH = -\log[0.10] = 1.00$

    $pH = 14.00 - 1.00 = 13.00$

    b. KOH is completely ionized, so $[OH^-] = 2.0 \times 10^{-4}\ M$.

    $pOH = -\log[2.0 \times 10^{-4}] = 3.70$

    $pH = 14.00 - 3.70 = 10.30$

    c. CsOH is completely ionized, so $[OH^-] = 6.2 \times 10^{-3}\ M$.

    $pOH = -\log[6.2 \times 10^{-3}] = 2.21$

    $pH = 14.00 - 2.21 = 11.79$

    d. NaOH is completely ionized, so $[OH^-] = 0.0001\ M$.

    $pOH = -\log[0.0001] = 4.0$

    $pH = 14.00 - 4.0 = 10.0$

67. a, b, and d

68. b, c, and d

69. In order to consume added acid, a buffered solution must contain a species capable of strongly attracting protons. The conjugate base of a *weak* acid is capable of strongly attracting protons, but the conjugate base of a *strong* acid does *not* have a strong affinity for protons.

70. a. CH_3COO^- is a relatively strong base.

    b. F^- is a relatively strong base.

    c. HS^- is a relatively strong base.

    d. Cl^- is a very weak base (conjugate of a strong acid).

71. No. For any aqueous solution, the concentrations of $[H^+]$ and $[OH^-]$ are related by K_w. The product of the given concentrations would not equal the value of K_w.

72. Ordinarily in calculating the pH of strong acid solutions, the major contribution to the concentration of hydrogen ion present is from the dissolved strong acid; we ordinarily neglect the small amount of hydrogen ion present in such solutions due to the ionization of water. With $1.0 \times 10^{-7}\ M$ HCl solution, however, the amount of hydrogen ion present due to the ionization of *water* is *comparable* to that present due to the addition of *acid* (HCl) and must be considered in the calculation of pH.

73. hydroxide

74. accepts

75. proton

76. base

Chapter 16: Acids and Bases

77. strong

78. 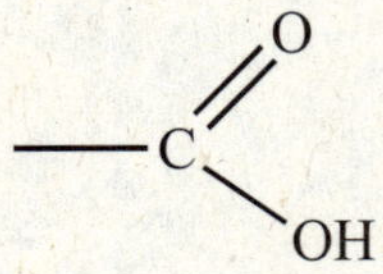 [structure: carboxylic acid group —C(=O)OH] $CH_3COOH + H_2O \rightleftharpoons C_2H_3O_2^- + H_3O^+$

79. autoionization

80. 1.0×10^{-14}

81. decimal places

82. lower

83. 0.20, 0.20

84. pH

85. buffered

86. weak acid

87. conjugate base

88. a. Equation 1:

   (acid1) + (base1) → (conjugate acid1) + (conjugate base1)

   Equation 2:

   (base2) + (acid2) → (conjugate acid2) + (conjugate base2)

   The acids are the proton donors and the bases are the proton acceptors. By looking at which species is positively charged or negatively charged in the products, it's possible to determine which reactant is the proton donor and which is the proton acceptor.

   b. An Arrhenius acid produces hydrogen ions. An Arrhenius base produces hydroxide ions. Therefore, acid1 is considered an Arrhenius acid. A Brønsted-Lowry acid is a proton donor and a Brønsted-Lowry base is a proton acceptor. Thus, acid1 and acid2 are both Brønsted-Lowry acids and base1 and base2 are both Brønsted-Lowry bases.

89. a. CH_3NH_2 (base), $CH_3NH_3^+$ (acid); H_2O (acid), OH^- (base)

   b. CH_3COOH (acid), CH_3COO^- (base); NH_3 (base), NH_4^+ (acid)

   c. HF (acid), F^- (base); NH_3 (base), NH_4^+ (acid)

90. The conjugate *acid* of the species indicated would have *one additional proton*:

   a. NH_4^+

   b. NH_3

   c. H_3O^+

   d. H_2O

91. The conjugate *bases* of the species indicated would have *one less proton*:

a. $H_2PO_4^-$

b. CO_3^{2-}

c. F^-

d. HSO_4^-

92. a. A buffer: HCN and NaCN are conjugates.

b. Not a buffer: PO_4^{3-} (of K_3PO_4) is not the conjugate base of H_3PO_4.

c. A buffer: HF and KF are conjugates.

d. A buffer: $HC_3H_5O_2$ and $NaC_3H_5O_2$ are conjugates.

93. Bases that are *weak* have relatively strong conjugate acids:

a. F^- is a relatively strong base; HF is a weak acid.

b. Cl^- is a very weak base; HCl is a strong acid.

c. HSO_4^- is a very weak base; H_2SO_4 is a strong acid.

d. NO_3^- is a very weak base; HNO_3 is a strong acid.

94. $K_w = [H^+][OH^-] = 1.0 \times 10^{-14}$ at 25°C

a. $[H^+] = \dfrac{1.0 \times 10^{-14}}{4.22 \times 10^{-3}\ M} = 2.4 \times 10^{-12}\ M$; solution is basic

b. $[H^+] = \dfrac{1.0 \times 10^{-14}}{1.01 \times 10^{-13}\ M} = 9.9 \times 10^{-2}\ M$; solution is acidic

c. $[H^+] = \dfrac{1.0 \times 10^{-14}}{3.05 \times 10^{-7}\ M} = 3.3 \times 10^{-8}\ M$; solution is basic

d. $[H^+] = \dfrac{1.0 \times 10^{-14}}{6.02 \times 10^{-6}\ M} = 1.7 \times 10^{-9}\ M$; solution is basic

95. $K_w = [H^+][OH^-] = 1.0 \times 10^{-14}$ at 25°C

a. $[OH^-] = \dfrac{1.0 \times 10^{-14}}{4.21 \times 10^{-7}\ M} = 2.4 \times 10^{-8}\ M$; solution is acidic

b. $[OH^-] = \dfrac{1.0 \times 10^{-14}}{0.00035\ M} = 2.9 \times 10^{-11}\ M$; solution is acidic

c. $[OH^-] = \dfrac{1.0 \times 10^{-14}}{0.00000010\ M} = 1.0 \times 10^{-7}\ M$; solution is neutral

d. $[OH^-] = \dfrac{1.0 \times 10^{-14}}{9.9 \times 10^{-6}\ M} = 1.0 \times 10^{-9}\ M$; solution is acidic

Chapter 16: Acids and Bases

96. pH + pOH = 14.00

   a. pH = 14.00 − 4.32 = 9.68; solution is basic
   b. pH = 14.00 − 8.90 = 5.10; solution is acidic
   c. pH = 14.00 − 1.81 = 12.19; solution is basic
   d. pH = 14.00 − 13.1 = 0.9; solution is acidic

97. pH = −log[H$^+$]   pOH = −log[OH$^−$]   pH + pOH = 14.00

   a. pH = −log[1.49 × 10^{-3} M] = 2.827; solution is acidic
   b. pOH = −log[6.54 × 10^{-4} M] = 3.184; pH = 14.00 − 3.184 = 10.816; solution is basic
   c. pH = −log[9.81 × 10^{-9} M] = 8.008; solution basic
   d. pOH = −log[7.45 × 10^{-10} M] = 9.128;   pH = 14.00 − 9.128 = 4.87; solution is acidic

98. pOH = −log[OH$^−$]   pH = 14.00 − pOH

   a. pOH = −log[1.4 × 10^{-6} M] = 5.85; pH = 14.00 − 5.85 = 8.15; solution is basic
   b. pOH = −log[9.35 × 10^{-9} M] = 8.029 = 8.03; pH = 14.00 − 8.029 = 5.97; solution is acidic
   c. pOH = −log[2.21 × 10^{-1} M] = 0.656 = 0.66; pH = 14.00 − 0.656 = 13.34; solution is basic
   d. pOH = −log[7.98 × 10^{-12} M] = 11.10; pH = 14.00 − 11.098 = 2.90; solution is acidic

99. pOH = 14.00 − pH

   a. pOH = 14.00 − 1.02 = 12.98; solution is acidic
   b. pOH = 14.00 − 13.4 = 0.6; solution is basic
   c. pOH = 14.00 − 9.03 = 4.97; solution is basic
   d. pOH = 14.00 − 7.20 = 6.80; solution is basic

100.
   a. [OH$^−$] = $\dfrac{1.0 \times 10^{-14}}{5.72 \times 10^{-4}\ M}$ = 1.75 × 10^{-11} M = 1.8 × 10^{-11} M

   pOH = −log[1.75 × 10^{-11} M] = 10.76

   pH = 14.00 − 10.76 = 3.24

   b. [H$^+$] = $\dfrac{1.0 \times 10^{-14}}{8.91 \times 10^{-5}\ M}$ = 1.12 × 10^{-10} M = 1.1 × 10^{-10} M

   pH = −log[1.12 × 10^{-10} M] = 9.95

   pOH = 14.00 − 9.95 = 4.05

   c. [OH$^−$] = $\dfrac{1.0 \times 10^{-14}}{2.87 \times 10^{-12}\ M}$ = 3.48 × 10^{-3} M = 3.5 × 10^{-3} M

   pOH = −log[3.48 × 10^{-3} M] = 2.46

   pH = 14.00 − 2.46 = 11.54

Chapter 16:   Acids and Bases

d. $[H^+] = \dfrac{1.0 \times 10^{-14}}{7.22 \times 10^{-8}\ M} = 1.39 \times 10^{-7}\ M = 1.4 \times 10^{-7}\ M$

   $pH = -\log[1.39 \times \times 10^{-7}\ M] = 6.86$

   $pOH = 14.00 - 6.86 = 7.14$

101. a. $[H^+] = \{inv\}\{log\}[-8.34] = 4.6 \times 10^{-9}\ M$

   b. $[H^+] = \{inv\}\{log\}[-5.90] = 1.3 \times 10^{-6}\ M$

   c. $[H^+] = \{inv\}\{log\}[-2.65] = 2.2 \times 10^{-3}\ M$

   d. $[H^+] = \{inv\}\{log\}[-12.6] = 2.5 \times 10^{-13}\ M$

102. $pH = 14.00 - pOH$ $[H^+] = \{inv\}\{log\}[-pH]$ or 10^{-pH}

   a. $[H^+] = \{inv\}\{log\}[-5.41] = 3.9 \times 10^{-6}\ M$

   b. $pH = 14.00 - 12.04 = 1.96$ $[H^+] = \{inv\}\{log\}[-1.96] = 1.1 \times 10^{-2}\ M$

   c. $[H^+] = \{inv\}\{log\}[-11.91] = 1.2 \times 10^{-12}\ M$

   d. $pH = 14.00 - 3.89 = 10.11$ $[H^+] = \{inv\}\{log\}[-10.11] = 7.8 \times 10^{-11}\ M$

103. a. $pH = 14.00 - 0.90 = 13.10$ $[H^+] = \{inv\}\{log\}[-13.10] = 7.9 \times 10^{-14}\ M$

   b. $[H^+] = \{inv\}\{log\}[-0.90] = 0.13\ M$

   c. $pH = 14.00 - 10.3 = 3.7$ $[H^+] = \{inv\}\{log\}[-3.7] = 2 \times 10^{-4}\ M$

   d. $[H^+] = \{inv\}\{log\}[-5.33] = 4.7 \times 10^{-6}\ M$

104. a. $HClO_4$ is a strong acid and completely ionized so $[H^+] = 1.4 \times 10^{-3}\ M$ and pH = 2.85.

   b. HCl is a strong acid and completely ionized so $[H^+] = 3.0 \times 10^{-5}\ M$ and pH = 4.52.

   c. HNO_3 is a strong acid and completely ionized so $[H^+] = 5.0 \times 10^{-2}\ M$ and pH = 1.30.

   d. HCl is a strong acid and completely ionized so $[H^+] = 0.0010\ M$ and pH = 3.00.

105. There are many examples. For a general weak acid (HA) and its conjugate base (A^-) the general equations illustrating the consumption of added acid and base would be:

   $HA(aq) + OH^-(aq) \rightarrow A^-(aq) + H_2O$

   $A^-(aq) + H_3O^+(aq) \rightarrow HA(aq) + H_2O$.

   For example, for a buffer consisting of equimolar HF/NaF:

   $HF(aq) + OH^-(aq) \rightarrow F^-(aq) + H_2O$

   $F^-(aq) + H_3O^+(aq) \rightarrow HF(aq) + H_2O$.

106. a and d; The conjugate base has one less proton (H^+) compared to its acid counterpart.

107.

	$[H^+]$	pH	pOH	$[OH^-]$
0.0070 M HNO_3	0.0070 M	2.15	11.85	$1.4 \times 10^{-12}\ M$
3.0 M KOH	$3.3 \times 10^{-15}\ M$	14.48	-0.48	3.0 M

Chapter 16: Acids and Bases

108. NaCl: neutral (Na^+ and Cl^- are very weak conjugates.)

RbOCl: basic (Rb^+ is a very weak conjugate. OCl^- is a weak base.)

KI: neutral (K^+ and I^- are very weak conjugates.)

$Ba(ClO_4)_2$: neutral (Ba^{2+} and ClO_4^- are very weak conjugates.)

NH_4NO_3: acidic (NH_4^+ is a weak acid. NO_3^- is a very weak conjugate.)

CHAPTER 17

Equilibrium

1. Answer depends on student choice of reaction.

2. Four C–H bonds in the CH_4 molecule and four Cl–Cl bonds in the four Cl_2 molecules must be broken. Four C–Cl bonds in CCl_4 and four H–Cl bonds in the four HCl molecules must form.

3. The collision model shows chemical reactions as taking place only when the reactant molecules *physically collide* with one another, with enough energy to break bonds in the reactant molecules. Not all collisions possess enough energy to break bonds in the reactant molecules. A minimum energy, the *activation energy* (E_a), is needed for a collision to result in reaction. If molecules do not possess this minimum energy when they collide, they bounce off one another without reacting. A simple reaction is illustrated in Figure 17.2.

4. The symbol E_a represents the *activation energy* of the reaction. The activation energy is the minimum energy two colliding molecules must possess in order for the collision to result in reaction. If molecules do not possess energies equal to or greater than E_a, a collision between these molecules will not result in a reaction.

5. A catalyst provides an alternative pathway, with a lower activation energy barrier, by which the reaction can take place.

6. Enzymes are biochemical catalysts that accelerate the complicated biochemical reactions in cells that would ordinarily be too slow to sustain life at normal body temperatures.

7. In an equilibrium system, two opposing processes are going on at the same time and speed. There is no net change in a system at equilibrium with time. Every time one process occurs, the opposite process occurs at the same time elsewhere in the system. A simple equilibrium might exist for the populations of two similar size towns connected by highway. Assuming there is no great attraction in one town compared to the other, we might assume the populations of the two towns would remain constant with time as individual people drive between them, but in such a way that new people are arriving in the first town from the second town as residents of the first town leave for the second town.

8. A state of equilibrium is attained when two opposing processes are exactly balanced so there is no further observable net change in the system.

9. We use a double-ended arrow (pointing both to the left and to the right) to indicate that the reaction is reversible and comes to equilibrium: ⇌

10. Chemical equilibrium occurs when two *opposing* chemical reactions reach the *same speed* in a closed system. When a state of chemical equilibrium has been reached, the concentrations of reactants and products present in the system remain *constant* with time, and the reaction appears to "stop." A chemical reaction that reaches a state of equilibrium is indicated by using a double

Chapter 17: Equilibrium

arrow ($\rightleftharpoons$). The points of the double arrow point in opposite directions, to indicate that two opposite processes are going on.

11. The word *dynamic* is used to describe physical and chemical states of equilibrium to emphasize that the system has not "stopped", but rather has reached the point where one process is exactly balanced by the opposing process. Once a chemical system has reached equilibrium the *net* concentration of product no longer increases because molecules of product already present react to form the original reactants.

12. a. The green line is H_2 because hydrogen is initially present in the greatest concentration. The blue line is N_2 because some nitrogen is initially present but not as much as the H_2 (a third of the amount). The pink line is NH_3 because at first no product is present but then N_2 and H_2 react to form NH_3.

    b. The concentrations of both N_2 and H_2 decrease at first because they react to form NH_3 (which then causes the concentration of NH_3 to go up). None of the concentrations become zero over time because eventually some of the NH_3 shifts back to form N_2 and H_2 again. Eventually the concentration of each species remains constant because the rate of the forward reaction equals the rate of the backward reaction (equilibrium is reached).

    c. Equilibrium is reached when the lines become straight (the concentration over time does not change). As stated in *b*, the rate of the forward reaction equals the rate of the backward reaction.

13. The equilibrium constant represents a ratio of the concentration of products present at the point of equilibrium to the concentration of reactants present, with the concentration of each species raised to the power of its coefficient in the balanced chemical equation for the reaction. For a general reaction

    $$aA + bB \rightleftharpoons cC + dD$$

    the equilibrium constant, K, has the algebraic form

    $$K = \frac{[C]^c[D]^d}{[A]^a[B]^b}$$

    The square brackets indicate the concentrations of the substances in moles per liter (molarity, M).

14. The equilibrium constant is a *ratio* of concentration of products to concentration of reactants, with all concentrations measured at equilibrium. Depending on the amount of reactant present at the beginning of an experiment, there may be different absolute amounts of reactants and products present at equilibrium, but the *ratio* will always be the same for a given reaction at a given temperature. For example, the ratios (4/2) and (6/3) are different absolutely in terms of the numbers involved, but each of these ratios has the *value* of 2.

15. a. $K = \dfrac{[C_2H_5Cl][HCl]}{[C_2H_6][Cl_2]}$

    b. $K = \dfrac{[NO]^4[H_2O]^6}{[NH_3]^4[O_2]^5}$

    c. $K = \dfrac{[PCl_3][Cl_2]}{[PCl_5]}$

Chapter 17: Equilibrium

16. a. $K = \dfrac{[NCl_3(g)]^2}{[N_2(g)][Cl_2(g)]^3}$

   b. $K = \dfrac{[HI(g)]^2}{[H_2(g)][I_2(g)]}$

   c. $K = \dfrac{[N_2H_4(g)]}{[N_2(g)][H_2(g)]^2}$

17. a. $K = \dfrac{[ClNO_2(g)][NO(g)]}{[NO_2(g)][ClNO(g)]}$

   b. $K = \dfrac{[BrF_5(g)]^2}{[Br_2(g)][F_2(g)]^5}$

   c. $K = \dfrac{[N_2(g)]^5[H_2O]^6}{[NH_3(g)]^4[NO(g)]^6}$

18. a. $K = \dfrac{[CH_3OH]}{[CO][H_2]^2}$

   b. $K = \dfrac{[NO]^2[O_2]}{[NO_2]^2}$

   c. $K = \dfrac{[PBr_3]^4}{[P_4][Br_2]^6}$

19. $PCl_5(g) \rightleftharpoons PCl_3(g) + Cl_2(g)$

   $K = \dfrac{[PCl_3][Cl_2]}{[PCl_5]} = \dfrac{[0.0302\ M][0.0491\ M]}{[0.0711\ M]} = 2.09 \times 10^{-2}$

20. $N_2(g) + 3H_2(g) \rightleftharpoons 2NH_3(g)$

   $K = \dfrac{[NH_3(g)]^2}{[N_2(g)][H_2(g)]^3} = \dfrac{[0.34\ M]^2}{[4.9 \times 10^{-4}\ M][2.1 \times 10^{-3}\ M]^3} = 2.5 \times 10^{10}$

21. $N_2(g) + O_2(g) \rightleftharpoons 2NO(g)$

   $K = \dfrac{[NO]^2}{[N_2][O_2]} = \dfrac{[4.7 \times 10^{-4}\ M]^2}{[0.041\ M][0.0078\ M]} = 6.9 \times 10^{-4}$

22. $2N_2O(g) + O_2(g) \rightleftharpoons 4NO(g)$

   $K = \dfrac{[NO]^4}{[N_2O]^2[O_2]} = \dfrac{[0.00341\ M]^4}{[0.0293\ M]^2[0.0325\ M]} = 4.85 \times 10^{-6}$

Chapter 17: Equilibrium

23. A *homogeneous* equilibrium system is one which all the substances present are in the same physical state. An example is

$$N_2(g) + O_2(g) \rightleftharpoons 2NO(g).$$

Heterogeneous equilibrium systems are those involving substances in more than one physical state (e.g., mixtures of liquids and gases, gases and solids, etc.). Examples are:

$$BaCO_3(s) \rightleftharpoons BaO(s) + CO_2(g)$$

$$NH_3(g) + HCl(g) \rightleftharpoons NH_4Cl(s).$$

24. Equilibrium constants represent ratios of the *concentrations* of products and reactants present at the point of equilibrium. The *concentration* of a pure solid or of a pure liquid is constant and is determined by the density of the solid or liquid. For example, suppose you had a liter of water. Within that liter of water are 55.5 mol of water (the number of moles of water that is contained in one liter of water *does not vary*).

25. a. $K = \dfrac{[PF_3]^4}{[F_2]^6}$

    b. $K = \dfrac{1}{[Xe][F_2]^2}$

    c. $K = \dfrac{[O_2]}{[Cl_2]^4}$

26. a. $K = [H_2O(g)][CO_2(g)]$

    b. $K = [CO_2(g)]$

    c. $K = \dfrac{1}{[O_2(g)]^3}$

27. a. $K = \dfrac{[H_2][CO]}{[H_2O]}$

    b. $K = [H_2O(g)]$

    c. $K = \dfrac{1}{[O_2]^3}$

28. a. $K = [N_2(g)][Br_2(g)]^3$

    b. $K = \dfrac{[H_2O(g)]}{[H_2(g)]}$

    c. $K = \dfrac{1}{[O_2(g)]^3}$

29. Le Châtelier's principle states that when a change is imposed on a system at equilibrium, the position of the equilibrium shifts in a direction that tends to reduce the effect of the change.

Chapter 17:   Equilibrium

30. When an additional amount of one of the reactants is added to an equilibrium system, the system shifts to the right and adjusts so as to consume some of the added reactant. This results in a net *increase* in the amount of product, compared to the equilibrium system before the additional reactant was added, and so the amount of $CO_2(g)$ in the system will be higher than if the additional $CO(g)$ had not been added. The numerical *value* of the equilibrium constant does *not* change when a reactant is added: the concentrations of all reactants and products adjust until the correct value of K is once again achieved.

31. When the volume of an equilibrium system involving gaseous substances is decreased suddenly, the pressure in the container increases. Reaction will occur, shifting in the direction that gives the smaller number of gas molecules, to reduce this increase in pressure.

32. If heat is applied to an endothermic reaction (i.e., the temperature is raised), the equilibrium is shifted to the right. More product will be present at equilibrium than if the temperature had not been increased. The value of K increases.

33.
   a. shifts left (system reacts to the left to get rid of excess H_2)
   b. shifts right (system reacts to the right to replace carbon monoxide as it is removed)
   c. no effect (carbon is in the *solid* state)

34.
   a. shifts right (system reacts to the right to get rid of excess fluorine)
   b. no change (P is in the *solid* state)
   c. shifts left (system reacts to get rid of excess PF_3)

35.
   a. no change (water is in the *liquid* state)
   b. Assuming the system is warm enough to convert the dry ice to the gaseous state, the equilibrium will shift to the left.
   c. shifts to right
   d. shifts to right

36.
   a. no change (B is solid)
   b. shift right (system reacts to replace removed C)
   c. shift left (system reacts by shifting in direction of fewer mol of gas)
   d. shift right (the reaction is endothermic as written)

37. As the reaction is exothermic, heat is effectively a product. An increase in temperature would basically fight *against* the forward reaction as it tries to release energy, and would disfavor the production of hydrogen chloride.

38. The answer is *d*. When hydrogen gas is added, equilibrium will shift away from the addition of reactant and toward the product side, producing more water vapor. The value of K does not change.

39. As the reaction is exothermic, heat is effectively a product of the reaction as written. Raising the temperature of the system (adding heat to the system) tends to disfavor the formation of products, and shifts the equilibrium to the left (toward reactants).

Chapter 17: Equilibrium

40. For an endothermic reaction, a decrease in temperature will shift the position of equilibrium to the left (toward the reactants).

41. The production of dextrose will be favored.

42. $CO(g) + 2H_2(g) \rightleftharpoons CH_3OH(l)$

    add additional $CO(g)$ or $H_2(g)$: the system will react in the forward direction to remove the excess
    decrease the volume of the system: the system will react in the direction of fewer moles of gas
    decrease the temperature: removal of heat favors products in an exothermic reaction

43. A large equilibrium constant means that the concentration of products is large compared to the concentration of remaining reactants. The position of the equilibrium lies far to the right. Reactions with numerically large equilibrium constants are greatly favored as a source of product. When we calculated the theoretical yield for a reaction in earlier chapters, we tacitly assumed that the reaction had a large equilibrium constant.

44. A small equilibrium constant implies that not much product forms before equilibrium is reached. The reaction would not be a good source of the products unless Le Châtelier's principle can be used to force the reaction to the right.

45. $K = \dfrac{[BrF_5]^2}{[Br_2][F_2]^5} = \dfrac{[1.01 \times 10^{-9}\, M]^2}{[2.41 \times 10^{-2}\, M][8.15 \times 10^{-2}\, M]^5} = 1.18 \times 10^{-11}$

46. $K = \dfrac{[SO_3][NO]}{[SO_2][NO_2]} = \dfrac{[4.99 \times 10^{-5}\, M][6.31 \times 10^{-7}\, M]}{[2.11 \times 10^{-2}\, M][1.73 \times 10^{-3}\, M]} = 8.63 \times 10^{-7}$

47. $K = \dfrac{[CO_2]^2}{[CO]^2[O_2]} = \dfrac{[1.1 \times 10^{-1}\, M]^2}{[2.7 \times 10^{-4}\, M]^2[1.9 \times 10^{-3}\, M]} = 8.7 \times 10^7$

48. $K = 5.21 \times 10^{-3} = \dfrac{[CO][H_2O]}{[CO_2][H_2]} = \dfrac{[4.73 \times 10^{-3}\, M][5.21 \times 10^{-3}\, M]}{[3.99 \times 10^{-2}\, M][H_2]}$

    $[H_2] = 0.119\, M$

49. $K = 2.1 \times 10^3 = \dfrac{[HF]^2}{[H_2][F_2]} = \dfrac{[HF]^2}{[0.0021\, M][0.0021\, M]}$

    $[HF]^2 = 9.26 \times 10^{-3}$
    $[HF] = 9.6 \times 10^{-2}\, M$

50. $K = 2.4 \times 10^{-3} = \dfrac{[H_2]^2[O_2]}{[H_2O]^2} = \dfrac{[1.9 \times 10^{-2}]^2[O_2]}{[1.1 \times 10^{-1}]^2}$

    $[O_2] = 8.0 \times 10^{-2}\, M$

51. a. The extent of conversion of O_2 to O_3 is very small.

    b. $K = \dfrac{[O_3]^2}{[O_2]^3} = 1.12 \times 10^{-54} = \dfrac{[O_3]^2}{[3.04 \times 10^{-2} \, M]^3}$

    $[O_3] = 5.61 \times 10^{-30} \, M$

52. $K = 8.1 \times 10^{-3} = \dfrac{[NO_2]^2}{[N_2O_4]} = \dfrac{[0.0021 \, M]^2}{[N_2O_4]}$

    $[N_2O_4] = 5.4 \times 10^{-4} \, M$

53. When a crystal of ionic solute M^+X^- is placed in water, initially the crystal simply dissolves producing $M^+(aq)$ ions and $X^-(aq)$ ions. As the concentration of the ions in solution increases, however, the likelihood of oppositely charged ions colliding and *re*forming the solid increases. Eventually, an equilibrium is reached when dissolving and reforming of the solid occur at the same speed. Past this point in time, there is no further net increase in the concentration of the dissolved ions.

54. solubility product, K_{sp}

55. The equilibrium represents the balancing of the dynamic processes of dissolving and of reformation of the solid. The amount of excess solid added in preparing a solution might affect the *speed* at which the point of equilibrium is reached, but it will not affect the net amount of solute present in solution at the point of equilibrium. A given amount of solvent can only "hold" a certain amount of solute.

56. Stirring or grinding the solute increases the speed with which the solute dissolves, but the ultimate *amount* of solute that dissolves is fixed by the equilibrium constant for the dissolving process, K_{sp}, which changes only with temperature. Therefore only the temperature will affect the solubility.

57. a. $AgIO_3(s) \rightleftharpoons Ag^+(aq) + IO_3^-(aq)$ $K_{sp} = [Ag^+(aq)][IO_3^-(aq)]$

    b. $Sn(OH)_2(s) \rightleftharpoons Sn^{2+}(aq) + 2OH^-(aq)$ $K_{sp} = [Sn^{2+}(aq)][OH^-(aq)]^2$

    c. $Zn_3(PO_4)_2(s) \rightleftharpoons 3Zn^{2+}(aq) + 2PO_4^{3-}(aq)$ $K_{sp} = [Zn^{2+}(aq)]^3[PO_4^{3-}(aq)]^2$

    d. $BaF_2(s) \rightleftharpoons Ba^{2+}(aq) + 2F^-(aq)$ $K_{sp} = [Ba^{2+}(aq)][F^-(aq)]^2$

58. a. $NiS(s) \rightleftharpoons Ni^{2+}(aq) + S^{2-}(aq)$ $K_{sp} = [Ni^{2+}(aq)][S^{2-}(aq)]$

    b. $CuCO_3(s) \rightleftharpoons Cu^{2+}(aq) + CO_3^{2-}(aq)$ $K_{sp} = [Cu^{2+}(aq)][CO_3^{2-}(aq)]$

    c. $BaCrO_4(s) \rightleftharpoons Ba^{2+}(aq) + CrO_4^{2-}(aq)$ $K_{sp} = [Ba^{2+}(aq)][CrO_4^{2-}(aq)]$

    d. $Ag_3PO_4(s) \rightleftharpoons 3Ag^+(aq) + PO_4^{3-}(aq)$ $K_{sp} = [Ag^+(aq)]^3[PO_4^{3-}(aq)]$

59. $Cu(OH)_2(s) \rightleftharpoons Cu^{2+}(aq) + 2OH^-(aq)$

    molar mass $Cu(OH)_2 = 97.57$ g

    Let x represent the solubility of $Cu(OH)_2$ in mol/L. Then $[Cu^{2+}] = x$ and $[OH^-] = 2x$ from the stoichiometry of the equation

    $K_{sp} = [Cu^{2+}][OH^-]^2 = 2.2 \times 10^{-20} = [x][2x]^2 = 4x^3$

Chapter 17: Equilibrium

x = the molar solubility of $Cu(OH)_2$ = 1.77×10^{-7} M (1.8×10^{-7} M)

$$1.77 \times 10^{-7} \frac{mol}{L} \times 97.57 \frac{g}{mol} = 1.7 \times 10^{-5} \text{ g/L}$$

60. $MgCO_3(s) \rightleftharpoons Mg^{2+}(aq) + CO_3^{2-}(aq)$

    Molar mass $MgCO_3$ = 84.32 g

    Let x represent the solubility of $MgCO_3$ in mol/L. Then $[CO_3^{2-}] = x$ and $[Mg^{2+}] = x$ from the stoichiometry of the equation.

    $K_{sp} = [Mg^{2+}][CO_3^{2-}] = 3.5 \times 10^{-8} = (x)(x) = x^2$

    then the molar solubility of $MgCO_3 = x = 1.87 \times 10^{-4}$ M (1.9×10^{-4} M)

    $$1.87 \times 10^{-4} \frac{mol}{L} \times \frac{84.32 \text{ g}}{1 \text{ mol}} = 0.016 \text{ g/L}$$

61. $NiS(s) \rightleftharpoons Ni^{2+}(aq) + S^{2-}(aq)$

    molar mass NiS = 90.77 g

    $$3.6 \times 10^{-4} \frac{g \, NiS}{L} \times \frac{1 \, mol \, NiS}{90.77 \, g \, NiS} = 3.97 \times 10^{-6} \, M$$

    $K_{sp} = [Ni^{2+}][S^{2-}] = (3.97 \times 10^{-6} \, M)(3.97 \times 10^{-6} \, M) = 1.6 \times 10^{-11}$

62. $Ni(OH)_2(s) \rightleftharpoons Ni^{2+}(aq) + 2OH^-(aq)$

    molar mass $Ni(OH)_2$ = 92.71 g

    let x represent the molar solubility of $Ni(OH)_2$: then $[Ni^{2+}] = x$ and $[OH^-] = 2x$.

    $K_{sp} = [Ni^{2+}][OH^-]^2 = 2.0 \times 10^{-15} = [x][2x]^2 = 4x^3$

    then the molar solubility of $Ni(OH)_2 = x = 7.9 \times 10^{-6} \, M$

    $$\text{gram solubility} = 7.98 \times 10^{-6} \frac{mol}{L} \times \frac{92.71 \text{ g}}{1 \text{ mol}} = 7.4 \times 10^{-4} \text{ g/L}$$

63. $CaCO_3(s) \rightleftharpoons Ca^{2+}(aq) + CO_3^{2-}(aq)$

    $K_{sp} = [Ca^{2+}][CO_3^{2-}] = 3.0 \times 10^{-9}$

    Let x represent the number of moles of $CaCO_3(s)$ that dissolve per liter, then $[Ca^{2+}] = x$ and $[CO_3^{2-}] = x$ also from the stoichiometry of the reaction;

    $K_{sp} = [x][x] = x^2 = 3.0 \times 10^{-9}$

    then $x = [CaCO_3] = 5.5 \times 10^{-5} \, M$ (5.5×10^{-3} g/L)

64. $CaSO_4(s) \rightleftharpoons Ca^{2+}(aq) + SO_4^{2-}(aq)$

    molar mass $CaSO_4$ = 136.15 g

    $$2.05 \frac{g}{L} \times \frac{1 \text{ mol}}{136.15 \text{ g}} = 1.506 \times 10^{-2} \, M$$

Chapter 17: Equilibrium

If $CaSO_4$ dissolves to the extent of 1.506×10^{-2} M, then $[Ca^{2+}]$ will be 1.506×10^{-2} M and $[SO_4^{2-}]$ will be 1.506×10^{-2} M also.

$K_{sp} = [Ca^{2+}][SO_4^{2-}] = [1.506 \times 10^{-2}\,M][1.506 \times 10^{-2}\,M] = 2.27 \times 10^{-4}$

65. $Fe(OH)_2(s) \rightleftharpoons Fe^{2+}(aq) + 2OH^-(aq)$

molar mass $Fe(OH)_2 = 89.87$ g

1.5×10^{-3} g/L $\times \dfrac{1\,mol\,Fe(OH)_2}{89.87\,g} = 1.67 \times 10^{-5}$ mol $Fe(OH)_2$/L

$K_{sp} = [Fe^{2+}][OH^-]^2$

If 1.67×10^{-5} M of $Fe(OH)_2$ dissolves then

$[Fe^{2+}] = 1.67 \times 10^{-5}$ M and $[OH^-] = 2 \times (1.67 \times 10^{-5}\,M) = 3.34 \times 10^{-5}$ M

$K_{sp} = (1.67 \times 10^{-5}\,M)(3.34 \times 10^{-5}\,M)^2 = 1.9 \times 10^{-14}$

66. $Cr(OH)_3(s) \rightleftharpoons Cr^{3+}(aq) + 3OH^-(aq)$

If $Cr(OH)_3$ dissolves to the extent of 8.21×10^{-5} M, then $[Cr^{3+}]$ will be 8.21×10^{-5} M and $[OH^-]$ will be $3(8.21 \times 10^{-5}\,M)$ in a saturated solution.

$K_{sp} = [Cr^{3+}][OH^-]^3 = [8.21 \times 10^{-5}\,M][2.46 \times 10^{-4}\,M]^3 = 1.23 \times 10^{-15}$

67. $MgF_2(s) \rightleftharpoons Mg^{2+}(aq) + 2F^-(aq)$

molar mass $MgF_2 = 62.31$ g

8.0×10^{-2} g MgF_2/L $\times \dfrac{1\,mol\,MgF_2}{62.31\,g\,MgF_2} = 1.28 \times 10^{-3}$ M $= 1.3 \times 10^{-3}$ M

$K_{sp} = [Mg^{2+}][F^-]^2$

If 1.28×10^{-3} M of MgF_2 dissolves, then $[Mg^{2+}] = 1.28 \times 10^{-3}$ M and $[F^-] = 2 \times 1.28 \times 10^{-3}$ M $= 2.56 \times 10^{-3}$ M.

$K_{sp} = (1.28 \times 10^{-3}\,M)(2.56 \times 10^{-3}\,M)^2 = 8.4 \times 10^{-9}$

68. $PbCl_2(s) \rightleftharpoons Pb^{2+}(aq) + 2Cl^-(aq)$

$K_{sp} = [Pb^{2+}][Cl^-]^2$

If $PbCl_2$ dissolves to the extent of 3.6×10^{-2} M, then $[Pb^{2+}] = 3.6 \times 10^{-2}$ M and $[Cl^-] = 2 \times (3.6 \times 10^{-2}) = 7.2 \times 10^{-2}$ M.

$K_{sp} = (3.6 \times 10^{-2}\,M)(7.2 \times 10^{-2}\,M)^2 = 1.9 \times 10^{-4}$

molar mass $PbCl_2 = 278.1$ g

$\dfrac{3.6 \times 10^{-2}\,mol}{1\,L} \times \dfrac{278.1\,g}{1\,mol} = 10.$ g/L

69. $Hg_2Cl_2(s) \rightleftharpoons Hg_2^{2+}(aq) + 2Cl^-(aq)$

$K_{sp} = [Hg_2^{2+}][Cl^-]^2$

Chapter 17: Equilibrium

Let x represent the number of moles of Hg_2Cl_2 that dissolve per liter; then $[Hg_2^{2+}] = x$ and $[Cl^-] = 2x$.

$K_{sp} = [x][2x]^2 = 4x^3 = 1.3 \times 10^{-18}$

$x^3 = 3.25 \times 10^{-19}$

$x = [Hg_2^{2+}] = 6.9 \times 10^{-7} \, M$

70. $Fe(OH)_3(s) \rightleftharpoons Fe^{3+}(aq) + 3OH^-(aq)$

    $K_{sp} = [Fe^{3+}][OH^-]^3 = 4 \times 10^{-38}$

    Let x represent the number of moles of $Fe(OH)_3$ that dissolve per liter; then $[Fe^{3+}] = x$.

    The amount of hydroxide ion that would be produced by the dissolving of the $Fe(OH)_3$ would then be $3x$, but pure water itself contains hydroxide ion at the concentration of $1.0 \times 10^{-7} \, M$ (see Chapter 17). The total concentration of hydroxide ion is then $[OH^-] = (3x + 1.0 \times 10^{-7})$. As x must be a very small number [because $Fe(OH)_3$ is not very soluble], we can save ourselves a lot of arithmetic if we use the approximation that

    $(3x + 1.0 \times 10^{-7} \, M) = 1.0 \times 10^{-7}$

    $K_{sp} = [x][1.0 \times 10^{-7}]^3 = 4 \times 10^{-38}$

    $x = 4 \times 10^{-17} \, M$

    molar mass $Fe(OH)_3 = 106.9$ g

    $\dfrac{4 \times 10^{-17} \text{ mol}}{1 \text{ L}} \times \dfrac{106.9 \text{ g}}{1 \text{ mol}} = 4 \times 10^{-15}$ g/L

71. Collision between molecules is *not* the only prerequisite for a reaction. The molecules must also possess sufficient energy to react with each other, and must have the proper spatial orientation for reaction.

72. An increase in temperature increases the fraction of molecules that possess sufficient energy for a collision to result in a reaction.

73. activation

74. catalyst

75. balancing

76. constant

77. To say a reaction is *reversible* means that, to one extent or another, the reaction may occur in either direction.

78. When we say that a chemical equilibrium is *dynamic*, we are recognizing the fact that even though the reaction has appeared macroscopically to have stopped, on a microscopic basis the forward and reverse reactions are still taking place, at the same speed.

79. equals

Chapter 17:   Equilibrium

80. heterogeneous

81. increase

82. position

83. pressure (and concentration)

84. The balanced equation is $H_2O(g) + CO(g) \rightleftharpoons H_2(g) + CO_2(g)$. Initially, 8 H_2O molecules are present and 6 CO molecules are present in the same 1.0-L container. The system reaches equilibrium by

$$H_2O(g) + CO(g) \rightleftharpoons H_2(g) + CO_2(g)$$

	H_2O	CO	H_2	CO_2
Initial	8	6	0	0
Change	$-x$	$-x$	$+x$	$+x$
Equilibrium	$8-x$	$6-x$	x	x

To determine the value of x, use the equilibrium expression

$$K = \frac{[H_2][CO_2]}{[H_2O][CO]} = \frac{(x)(x)}{(8-x)(6-x)} = 2.0$$

Use the quadratic equation to solve for x. $x = 24, 4$. The value of x cannot be 24 or else a negative equilibrium concentration for the reactants would result. Therefore, x must equal 4. The number of each type of molecule in the container at equilibrium is

$H_2O = 8 - x = 8 - 4 = 4$

$CO = 6 - x = 6 - 4 = 2$

$H_2 = x = 4$

$CO_2 = x = 4$

85. The solubility product (K_{sp}) for a sparingly soluble salt is the equilibrium constant for the dynamic equilibrium between solution and undissolved solute in a saturated solution of the salt. Consider the sparingly soluble salt AgCl.

$$AgCl(s) \rightleftharpoons Ag^+(aq) + Cl^-(aq) \qquad K_{sp} = [Ag^+(aq)][Cl^-(aq)]$$

86. An equilibrium reaction may come to many *positions* of equilibrium, but at each possible position of equilibrium, the numerical value of the equilibrium constant is fulfilled. If different amounts of reactant are taken in different experiments, the *absolute amounts* of reactant and product present at the point of equilibrium reached will differ from one experiment to another, but the *ratio* that defines the equilibrium constant will be the same.

87. alpha $\rightleftharpoons$ beta

$$K = \frac{[\text{beta}]}{[\text{alpha}]}$$

At equilibrium, [alpha] = 2 × [beta]

$$K = \frac{[\text{beta}]}{2 \times [\text{beta}]} = 0.5$$

Chapter 17: Equilibrium

88. $PCl_5(g) \rightleftharpoons PCl_3(g) + Cl_2(g)$

$$K = \frac{[PCl_3][Cl_2]}{[PCl_5]} = 4.5 \times 10^{-3}$$

The concentration of PCl_5 is to be twice the concentration of PCl_3: $[PCl_5] = 2 \times [PCl_3]$

$$K = \frac{[PCl_3][Cl_2]}{2 \times [PCl_3]} = 4.5 \times 10^{-3}$$

$$K = \frac{[Cl_2]}{2} = 4.5 \times 10^{-3} \quad \text{and} \quad [Cl_2] = 9.0 \times 10^{-3} \, M$$

89. density $CaCO_3(s) = 2.930$ g/cm^3

molar mass $CaCO_3 = 100.1$ g

$$\frac{2.930 \text{ g}}{1 \text{ cm}^3} \times \frac{1 \text{ mol}}{100.1 \text{ g}} \times \frac{1000 \text{ cm}^3}{1 \text{ L}} = 29.27 \, M \text{ for the solid}$$

density $CaO(s) = 3.30$ g/cm^3 (lime)

molar mass $CaO = 56.08$ g

$$\frac{3.30 \text{ g}}{1 \text{ cm}^3} \times \frac{1 \text{ mol}}{56.08 \text{ g}} \times \frac{1000 \text{ cm}^3}{1 \text{ L}} = 58.8 \, M \text{ for the solid}$$

90. As all of the metal carbonates indicated have the metal ion in the +2 oxidation state, we can illustrate the calculations for a general metal carbonate, MCO_3:

$MCO_3(s) \rightleftharpoons M^{2+}(aq) + CO_3^{2-}(aq)$ $\quad\quad K_{sp} = [M^{2+}(aq)][CO_3^{2-}(aq)]$

If we then let x represent the number of moles of MCO_3 that dissolve per liter, then $[M^{2+}(aq)] = x$ and $[CO_3^{2-}(aq)] = x$ also because the reaction is of 1:1 stoichiometry. Therefore,

$K_{sp} = [M^{2+}(aq)][CO_3^{2-}(aq)] = x^2$ for each salt. Solving for x gives the following results.

$[BaCO_3] = x = 7.1 \times 10^{-5} \, M$

$[CdCO_3] = x = 2.3 \times 10^{-6} \, M$

$[CaCO_3] = x = 5.3 \times 10^{-5} \, M$

$[CoCO_3] = x = 3.9 \times 10^{-7} \, M$

91. $Ca_3(PO_4)_2(s) \rightleftharpoons 3Ca^{2+}(aq) + 2PO_4^{3-}(aq)$

Let x represent the number of moles of $Ca_3(PO_4)_2$ that dissolve per liter; then $[Ca^{2+}] = 3x$ and $[PO_4^{3-}] = 2x$.

$K_{sp} = [Ca^{2+}]^3[PO_4^{3-}]^2 = 1.3 \times 10^{-32}$

$[3x]^3[2x]^2 = 1.3 \times 10^{-32}$

$108x^5 = 1.3 \times 10^{-32}$

$x^5 = 1.20 \times 10^{-34}$

$x = 1.64 \times 10^{-7} \, M$

$[Ca^{2+}] = 3x = 3 \times 1.64 \times 10^{-7} \, M = 4.9 \times 10^{-7} \, M$

92. Although a small solubility product generally implies a small solubility, comparisons of solubility based directly on K_{sp} values are only valid if the salts produce the same numbers of positive and negative ions per formula when they dissolve. For example, one can compare the solubilities of AgCl(s) and NiS(s) directly using K_{sp}, because each salt produces one positive and one negative ion per formula when dissolved. One could not directly compare AgCl(s) with a salt such as $Ca_3(PO_4)_2$, however.

93. A higher concentration means there are more molecules present, which results in a greater frequency of collision between molecules.

94. At higher temperatures, the average kinetic energy of the reactant molecules is larger. At higher temperatures, the probability that a collision between molecules will be energetic enough for reaction to take place is larger. On a molecular basis, a higher temperature means a given molecule will be moving faster.

95. When a liquid is confined in an otherwise empty closed container, the liquid begins to evaporate, producing molecules of vapor in the empty space of the container. As the amount of vapor increases, molecules in the vapor phase begin to condense and reenter the liquid state. Eventually the opposite processes of evaporation and condensation will be going on at the same speed: beyond this point, for every molecule that leaves the liquid state and evaporates, there is a molecule of vapor which leaves the vapor state and condenses. We know the state of equilibrium has been reached when there is no further change in the pressure of the vapor.

96. a. $K = \dfrac{[HBr]^2}{[H_2][Br_2]}$

    b. $K = \dfrac{[H_2S]^2}{[H_2]^2[S_2]}$

    c. $K = \dfrac{[HCN]^2}{[H_2][C_2N_2]}$

97. a. $K = \dfrac{[O_2]^3}{[O_3]^2}$

    b. $K = \dfrac{[CO_2][H_2O]^2}{[CH_4][O_2]^2}$

    c. $K = \dfrac{[C_2H_4Cl_2]}{[C_2H_4][Cl_2]}$

Chapter 17: Equilibrium

98. $N_2(g) + 3Cl_2(g) \rightleftharpoons 2NCl_3(g)$

$$K = \frac{[NCl_3(g)]^2}{[N_2(g)][Cl_2(g)]^3} = \frac{[1.9 \times 10^{-1} \, M]^2}{[1.4 \times 10^{-3} \, M][4.3 \times 10^{-4} \, M]^3} = 3.2 \times 10^{11}$$

99. $K = \dfrac{[PCl_3][Cl_2]}{[PCl_5]} = \dfrac{[0.325 \, M][3.9 \times 10^{-3} \, M]}{[1.1 \times 10^{-2} \, M]} = 0.12$

100. a. $K = \dfrac{1}{[O_2]^3}$

   b. $K = \dfrac{1}{[NH_3][HCl]}$

   c. $K = \dfrac{1}{[O_2]}$

101. a. $K = \dfrac{1}{[O_2]^5}$

   b. $K = \dfrac{[H_2O]}{[CO_2]}$

   c. $K = [N_2O][H_2O]^2$

102. The second snapshot is the first to represent an equilibrium mixture because after this point, the concentrations of the reactant and products remain constant. 6 molecules of A_2B reacted initially.

$$2A_2B(g) \rightleftharpoons 2A_2(g) + B_2(g)$$

	$2A_2B$	$2A_2$	B_2
Initial	?	0	0
Change	$-2x$	$+2x$	$+x$
Equilibrium	$? - 2x = 2$	$2x = 4$	$x = 2$

Therefore ? = 6.

103. $2NO(g) + O_2(g) \rightleftharpoons 2NO_2(g)$

   a. shifts to the right

   b. shifts to the right

   c. no effect (He is not involved in the reaction)

104. The reaction is *exo*thermic as written. An increase in temperature (addition of heat) will shift the reaction to the left (toward reactants).

105. $K = \dfrac{[CO_2][H_2]}{[CO][H_2O]} = \dfrac{[1.3 \, M][1.4 \, M]}{[0.71 \, M][0.66 \, M]} = 3.9$

Chapter 17:   Equilibrium

106. $K = \dfrac{[NH_3]^2}{[N_2][H_2]^3} = 1.3 \times 10^{-2} = \dfrac{[NH_3]^2}{[0.1M][0.1M]^3}$

$[NH_3]^2 = 1.3 \times 10^{-6}$

$[NH_3] = 1.1 \times 10^{-3}\ M$

107. $K = \dfrac{[NO]^2[Cl_2]}{[NOCl]^2} = 9.2 \times 10^{-6} = \dfrac{[1.5 \times 10^{-3}\ M]^2[Cl_2]}{[0.44\ M]^2}$

$[Cl_2] = 0.79\ M$

108. a. $Cu(OH)_2(s) \rightleftharpoons Cu^{2+}(aq) + 2OH^-(aq)$

$K_{sp} = [Cu^{2+}][OH^-]^2$

b. $Cr(OH)_3(s) \rightleftharpoons Cr^{3+}(aq) + 3OH^-(aq)$

$K_{sp} = [Cr^{3+}][OH^-]^3$

c. $Ba(OH)_2(s) \rightleftharpoons Ba^{2+}(aq) + 2OH^-(aq)$

$K_{sp} = [Ba^{2+}][OH^-]^2$

d. $Sn(OH)_2(s) \rightleftharpoons Sn^{2+}(aq) + 2OH^-(aq)$

$K_{sp} = [Sn^{2+}][OH^-]^2$

109. As the calculations for each of the silver halides would be similar, we can illustrate the calculations for a general silver halide, AgX:

$AgX(s) \rightleftharpoons Ag^+(aq) + X^-(aq) \qquad K_{sp} = [Ag^+(aq)][X^-(aq)]$

If we then let x represent the number of moles of AgX that dissolve per liter, then $[Ag^+(aq)] = x$ and $[X^-(aq)] = x$ also because the reaction is of 1:1 stoichiometry. Therefore, $[Ag^+(aq)][X^-(aq)] = x^2 = K_{sp}$ for each salt. Solving for x gives the following results.

$[AgCl] = x = 1.3 \times 10^{-5}\ M$

$[AgBr] = x = 7.1 \times 10^{-7}\ M$

$[AgI] = x = 9.1 \times 10^{-9}\ M$

110. molar mass AgCl = 143.4 g

$9.0 \times 10^{-4}\ \text{g AgCl/L} \times \dfrac{1\ \text{mol AgCl}}{143.4\ \text{g AgCl}} = 6.28 \times 10^{-6}\ \text{mol AgCl/L}$

$AgCl(s) \rightleftharpoons Ag^+(aq) + Cl^-(aq)$

$K_{sp} = [Ag^+][Cl^-] = (6.28 \times 10^{-6}\ M)(6.28 \times 10^{-6}\ M) = 3.9 \times 10^{-11}$

111. $HgS(s) \rightleftharpoons Hg^{2+}(aq) + S^{2-}(aq)$

$K_{sp} = [Hg^{2+}][S^{2-}] = 1.6 \times 10^{-54}$

Let x represent the number of moles of HgS that dissolve per liter; then $[Hg^{2+}] = x$ and $[S^{2-}] = x$.

Chapter 17: Equilibrium

$K_{sp} = [x][x] = x^2 = 1.6 \times 10^{-54}$

$x = 1.26 \times 10^{-27}\ M$

molar mass HgS = 232.7 g

$1.26 \times 10^{-27}\ M \times \dfrac{232.7\ g}{1\ mol} = 2.9 \times 10^{-25}\ g/L$

112. molar mass Ni(OH)$_2$ = 92.71 g

$\dfrac{0.14\ g\ Ni(OH)_2}{1\ L} \times \dfrac{1\ mol}{92.71\ g\ Ni(OH)_2} = 1.510 \times 10^{-3}\ M$

$Ni(OH)_2(s) \rightleftharpoons Ni^{2+}(aq) + 2OH^-(aq)$

$K_{sp} = [Ni^{2+}][OH^-]^2$

If $1.510 \times 10^{-3}\ M$ of Ni(OH)$_2$ dissolves, then $[Ni^{2+}] = 1.510 \times 10^{-3}\ M$ and $[OH^-] = 2 \times (1.510 \times 10^{-3}\ M) = 3.020 \times 10^{-3}\ M$.

$K_{sp} = (1.510 \times 10^{-3}\ M)(3.020 \times 10^{-3}\ M)^2 = 1.4 \times 10^{-8}$

113. The nitrogen–nitrogen triple bond in N$_2$ and the three hydrogen–hydrogen bonds (in the three H$_2$ molecules) must be broken. Six nitrogen–hydrogen bonds must form (in the two ammonia molecules).

114. The activation energy is the minimum energy two colliding molecules must possess in order for the collision to result in reaction. If molecules do not possess energies equal to or greater than E_a, a collision between these molecules will not result in a reaction.

115. Living cells contain biological catalysts called *enzymes*. Such enzymes speed up the complicated biochemical processes that must occur in cells (which would be too slow to sustain life without the enzyme's action).

116. Once a system has reached equilibrium the net concentration of product no longer increases because molecules of product already present react to form the original reactants. This is not to say that the *same* product molecules are necessarily always present.

117. $N_2(g) + 3H_2(g) \rightleftharpoons 2NH_3(g)$

$\dfrac{[NH_3]^2}{[N_2][H_2]^3} = \dfrac{[0.34\ M]^2}{[4.9 \times 10^{-4}\ M][2.1 \times 10^{-3}\ M]^3} = 2.5 \times 10^{10}$

118. $K = \dfrac{[CO]^2[O_2]}{[CO_2]^2} = \dfrac{[0.11\ M]^2[0.055\ M]}{[1.4\ M]^2} = 3.4 \times 10^{-4}$

119. A small equilibrium constant means that the concentration of products is small, compared to the concentration of reactants. The position of equilibrium lies far to the left. Reactions with very small equilibrium constants are generally not very useful as a source of the products unless the equilibrium position can be shifted by application of Le Châtelier's principle.

Chapter 17: Equilibrium

120. $3H_2(g) + N_2(g) \rightleftharpoons 2NH_3(g)$

$$K = \frac{[NH_3(g)]^2}{[N_2(g)][H_2(g)]^3} = \frac{\left[\dfrac{1.16 \text{ mol}}{3.50 \text{ L}}\right]^2}{\left[\dfrac{1.14 \text{ mol}}{3.50 \text{ L}}\right]\left[\dfrac{2.40 \text{ mol}}{3.50 \text{ L}}\right]^3} = 1.05$$

121. $A_2(g) + 2B(g) \rightleftharpoons 2AB(g)$

$$K = \frac{[AB(g)]^2}{[A_2(g)][B(g)]^2} = \frac{[5.3 \times 10^{-4} \text{ }M]^2}{[0.0090 \text{ }M][0.940 \text{ }M]^2} = 3.5 \times 10^{-5}$$

122. $3O_2(g) \rightleftharpoons 2O_3(g)$

$$K = \frac{[O_3(g)]^2}{[O_2(g)]^3}$$

$$1.8 \times 10^{-7} = \frac{[x]^2}{[0.062 \text{ }M]^3}$$

$$x = [O_3(g)] = \sqrt{4.3 \times 10^{-11} \text{ }M} = 6.5 \times 10^{-6} \text{ }M$$

123. $H_2(g) + I_2(g) \rightleftharpoons 2HI(g)$

$$K_p = \frac{P_{HI}^2}{P_{H_2} P_{I_2}}$$

$$45.9 = \frac{(4.94 \text{ atm})^2}{(0.628 \text{ atm})(x)}$$

$$x = P_{I_2} = 0.847 \text{ atm}$$

124. $H_2(g) + F_2(g) \rightleftharpoons 2HF(g)$

$$K = \frac{[HF(g)]^2}{[H_2(g)][F_2(g)]}$$

$$2.1 \times 10^{-3} = \frac{[x]^2}{[0.083 \text{ }M][0.083 \text{ }M]}$$

$$x = [HF(g)] = \sqrt{1.447 \times 10^{-5} \text{ }M} = 3.8 \times 10^{-3} \text{ }M$$

125. The correct answer is *a*. For an endothermic reaction, an increase in temperature will shift the position of equilibrium to the right (toward the products), which will increase *K*.

Chapter 17: Equilibrium

126.

	N_2	H_2	NH_3
Add N_2	Increase	Decrease	Increase
Remove H_2	Increase	Decrease	Decrease
Add NH_3	Increase	Increase	Increase
Add Ne	No change	No change	No change
Increase T	Increase	Increase	Decrease
Decrease V	Decrease	Decrease	Increase
Add catalyst	No change	No change	No change

CUMULATIVE REVIEW

Chapters 16 and 17

1. The Arrhenius and Brønsted-Lowry definitions both treat acids as a source of protons. The Arrhenius definition, however, is restricted to aqueous solutions, whereas the Brønsted-Lowry definition extends the idea of a "proton donor" to other situations. The two definitions differ in their treatment of bases. In the Arrhenius concept, the only real base is hydroxide ion because of the restriction to aqueous media. In the Brønsted-Lowry concept, a base is anything that receives a proton from an acid. Any Arrhenius acid will also be a Brønsted-Lowry acid. But since the Brønsted-Lowry definition applies in all media, there are Brønsted-Lowry acids which would not be acids in the Arrhenius sense.

2. A conjugate acid–base pair consists of two species related to each other by donation or acceptance of a single proton, H^+. An acid has one more H^+ than its conjugate base; a base has one less H^+ than its conjugate acid.

   Brønsted-Lowry acids:

   $HCl(aq) + H_2O(l) \rightarrow Cl^-(aq) + H_3O^+(aq)$

   $H_2SO_4(aq) + H_2O(l) \rightarrow HSO_4^-(aq) + H_3O^+(aq)$

   $H_3PO_4(aq) + H_2O(l) \rightleftharpoons H_2PO_4^-(aq) + H_3O^+(aq)$

   $NH_4^+(aq) + H_2O(l) \rightleftharpoons NH_3(aq) + H_3O^+(aq)$

   Brønsted-Lowry bases:

   $NH_3(aq) + H_2O(l) \rightleftharpoons NH_4^+(aq) + OH^-(aq)$

   $HCO_3^-(aq) + H_2O(l) \rightleftharpoons H_2CO_3(aq) + OH^-(aq)$

   $NH_2^-(aq) + H_2O(l) \rightarrow NH_3(aq) + OH^-(aq)$

   $H_2PO_4^-(aq) + H_2O(l) \rightleftharpoons H_3PO_4(aq) + OH^-(aq)$

3. When we say that acetic acid is a weak acid, we can take either of two points of view. Usually we say that acetic acid is a weak acid because it doesn't ionize very much when dissolved in water: we say that not very many acetic acid molecules dissociate. However, we can describe this situation from another point of view. We could say that the reason acetic acid doesn't dissociate much when we dissolve it is water is because the acetate ion (the conjugate base of acetic acid) is extremely effective at holding on to protons, and specifically is better at holding on to protons than water is in attracting them. So, acetic acid doesn't dissociate much because the acetate ion won't let the proton go!

   $HC_2H_3O_2 + H_2O \rightleftharpoons C_2H_3O_2^- + H_3O^+$

   Now what would happen if we had a source of free acetate ions (for example, sodium acetate) and placed them into water? Since acetate ion is better at attracting protons than is water, the acetate ions would pull protons out of water molecules, leaving hydroxide ions. That is,

   $C_2H_3O_2^- + H_2O \rightarrow HC_2H_3O_2 + OH^-$

Cumulative Review: Chapters 16 and 17

As an increase in hydroxide ion concentration would take place in the solution, the solution would be basic. Although acetic acid is a weak acid, the acetate ion is a base in aqueous solution.

4. The strength of an acid is a direct result of the position of the acid's ionization equilibrium. Strong acids are those whose ionization equilibrium positions lie far to the right, whereas weak acids are those whose equilibrium positions lie only slightly to the right. For example, HCl, HNO_3, and $HClO_4$ are all strong acids, which means they are completely ionized in aqueous solution (the position of equilibrium is very far to the right):

$$HCl(aq) + H_2O(l) \rightarrow Cl^-(aq) + H_3O^+(aq)$$

$$HNO_3(aq) + H_2O(l) \rightarrow NO_3^-(aq) + H_3O^+(aq)$$

$$HClO_4(aq) + H_2O(l) \rightarrow ClO_4^-(aq) + H_3O^+(aq)$$

As these are very strong acids, we know their anions (Cl^-, NO_3^-, ClO_4^-) must be very weak bases, and that solutions of the sodium salts of these anions would *not* be appreciably basic. As these acids have a strong tendency to lose protons, there is very little tendency for the anions (bases) to gain protons.

5. When we say that water is an amphoteric substance, we are just recognizing that water will behave as a Brønsted-Lowry base if a strong acid is added to it, or will behave as a Brønsted-Lowry acid if a strong base is added to it. For example, water behaves as a base when HCl is dissolved in it

$$HCl + H_2O \rightarrow Cl^- + H_3O^+.$$

However, water would behave as an acid if the strong base $NaNH_2$ were added to it

$$NH_2^- + H_2O \rightarrow NH_3 + OH^-.$$

We can also demonstrate the amphoterism of water by the fact that water undergoes autoionization, in which some water molecules behave as an acid and some behave as a base

$$H_2O + H_2O \rightleftharpoons H_3O^+ + OH^-.$$

Because this reaction is so important to our understanding the relative acidity and basicity of aqueous solution, the equilibrium constant for the autoionization of water is given a special symbol, K_w

$$K_w = [H_3O^+][OH^-].$$

This equilibrium constant has the value 1.0×10^{-14} at 25°C. Because of the fact that hydronium ions and hydroxide ions are produced in equal numbers when water molecules undergo autoionization, and with the value for the equilibrium constant given, we know that in pure water

$$[H_3O^+] = [OH^-] = 1.0 \times 10^{-7}\ M$$

If a solution has a higher concentration of H_3O^+ ion than OH^- ion, we say the solution is acidic. If a solution has a lower concentration of H_3O^+ ion than OH^- ion we say the solution is basic.

6. The pH of a solution is defined as the negative of the base 10 logarithm of the hydrogen ion concentration in the solution; that is

$$pH = -\log_{10}[H^+].$$

As in pure water, the amount of $H^+(aq)$ ion present is equal to the amount of $OH^-(aq)$ ion, we say that pure water is *neutral*. As $[H^+] = 1.0 \times 10^{-7}\ M$ in pure water, this means that the pH of pure water is $-\log[1.0 \times 10^{-7}\ M] = 7.00$. Solutions in which the hydrogen ion concentration is greater than 1.0×10^{-7} M (pH < 7.00) are *acidic*; solutions in which the hydrogen ion concentration is less than 1.0×10^{-7} M (pH > 7.00) are *basic*. The pH scale is logarithmic. When the pH changes by one unit, this corresponds to a change in the hydrogen ion concentration by a factor of *ten*.

In some instances, it may be more convenient to speak directly about the hydroxide ion concentration present in a solution, and so an analogous logarithmic expression is defined for the hydroxide ion concentration:

$$pOH = -\log_{10}[OH^-].$$

The concentrations of hydrogen ion and hydroxide ion in water (and in aqueous solutions) are *not* independent of one another, but rather are related by the dissociation equilibrium constant for water,

$$K_w = [H^+][OH^-] = 1.0 \times 10^{-14} \text{ at } 25°C.$$

From this constant it is obvious that pH + pOH = 14.00 for water (or an aqueous solution) at 25°C.

7. A buffered solution is one that resists a change in its pH even when a strong acid or base is added to it. Buffered solutions consist of approximately equal amounts of two components: a weak acid (or base) and its conjugate base (or acid). The weak acid component of the buffered solution is capable of reacting with added strong base. The conjugate base component of the buffered solution is able to react with added strong acid. By reacting with (and effectively neutralizing) the added strong acid or strong base, the buffer is able to maintain its pH at a relatively constant level. Consider the following three buffered solutions:

   0.10 M $HC_2H_3O_2$/0.10 M $NaC_2H_3O_2$

   0.50 M HF/0.50 M KF

   0.25 M NH_4Cl/0.25 M NH_3.

The *acidic* component of each of these buffered solutions can neutralize added OH^- ion as shown below:

$$HC_2H_3O_2 + OH^- \rightarrow C_2H_3O_2^- + H_2O$$

$$HF + OH^- \rightarrow F^- + H_2O$$

$$NH_4^+ + OH^- \rightarrow NH_3 + H_2O$$

Note that in each case, the added hydroxide ion has been neutralized and converted to a water molecule. The *basic* component of the buffered solutions can neutralize added H^+ ion as shown below:

$$C_2H_3O_2^- + H^+ \rightarrow HC_2H_3O_2$$

$$F^- + H^+ \rightarrow HF$$

$$NH_3 + H^+ \rightarrow NH_4^+$$

Note too that in each case the added hydrogen ion is converted to some other species.

Buffered solutions are crucial to the reactions in biological systems because many of these reactions are extremely pH dependent: a change in pH of only one or two units can make some reactions impossible or extremely slow. Many biological molecules have three-dimensional structures that are extremely dependent on the pH of their surroundings. For example, protein molecules can lose part of their necessary structure and shape if the pH of their environment changes (if a protein's structure is changed it may not work correctly). For example, if a small

Cumulative Review: Chapters 16 and 17

amount of vinegar is added to whole milk, the milk instantly curdles: the solid that forms is the protein portion of the milk, which becomes less soluble at lower pH values.

8. Chemists envision that a reaction can only take place between molecules if the molecules physically *collide* with each other. Furthermore, when molecules collide, the molecules must collide with enough force for the reaction to be successful (there must be enough energy to break bonds in the reactants), and the colliding molecules must be positioned with the correct relative orientation for the products (or intermediates) to form. Reactions tend to be faster if higher concentrations are used for the reaction; because, if there are more molecules present per unit volume there will be more collisions between molecules in a given time period. Reactions are faster at higher temperatures because at higher temperatures the reactant molecules have a higher average kinetic energy, and the number of molecules that will collide with sufficient force to break bonds increases.

9. A graph illustrating the activation energy barrier for a reaction is given as Figure 17.4 in the text. The activation energy for a reaction represents the minimum energy the reactant molecules must possess for a reaction to occur when the molecules collide. Although an increase in temperature does not change the activation energy for a reaction itself, at higher temperatures, the reactant molecules have a higher average kinetic energy, and a larger fraction of reaction molecules will possess enough energy for a collision to be effective. You will remember that in the kinetic-molecular theory, temperature was a direct measure of average kinetic energy. A catalyst speeds up a reaction by providing an alternate mechanism or pathway by which the reaction can occur, with such a mechanism or pathway having a lower activation energy than the original pathway.

10. Chemists define equilibrium as the exact balancing of two exactly opposing processes. When a chemical reaction is begun by combining pure reactants, the only process possible initially is

    reactants → products

    However, for many reactions, as the concentration of product molecules increases, it becomes more likely that product molecules will collide and react with each other

    products → reactants

    giving back molecules of the original reactants. At some point in the process the rates of the forward and reverse reactions become equal, and the system attains chemical equilibrium. To an outside observer, the system appears to have stopped reacting. On a microscopic basis, though, both the forward and reverse processes are still going on: every time additional molecules of the product form, however, somewhere else in the system molecules of product react to give back molecules of reactant.

    Once the point is reached that product molecules are reacting at the same speed at which they are forming, there is no further net change in concentration. A graph showing how the rates of the forward and reverse reactions change with time is given in the text as Figure 17.8. At the start of the reaction, the rate of the forward reaction is at its maximum, whereas the rate of the reverse reaction is zero. As the reaction proceeds, the rate of the forward reaction gradually decreases as the concentration of reactants decreases, whereas the rate of the reverse reaction increases as the concentration of products increases. Once the two rates have become equal, the reaction has reached a state of equilibrium.

11. The expression for the equilibrium constant for a reaction has as its numerator the concentrations of the products (raised to the powers of their stoichiometric coefficients in the balanced chemical equation for the reaction), and as its denominator the concentrations of the reactants (also raised to the powers of their stoichiometric coefficients). In general terms, for a reaction

$$aA + bB \rightleftharpoons cC + dD$$

the equilibrium constant expression has the form

$$K = \frac{[C]^c[D]^d}{[A]^a[B]^b}$$

in which square brackets [] indicate molar concentration. For example, here are three simple reactions and the expression for their equilibrium constants:

$N_2(g) + O_2(g) \rightleftharpoons 2NO(g)$ $K = [NO]^2/[N_2][O_2]$

$2SO_2(g) + O_2(g) \rightleftharpoons 2SO_3(g)$ $K = [SO_3]^2/[SO_2]^2[O_2]$

$N_2(g) + 3H_2(g) \rightleftharpoons 2NH_3(g)$ $K = [NH_3]^2/[N_2][H_2]^3$

12. The equilibrium constant for a reaction is a *ratio* of the concentration of products present at the point of equilibrium to the concentration of reactants still present. A *ratio* means that we have one number divided by another number (for example, the density of a substance is the ratio of a substance's mass to its volume). As the equilibrium constant is a ratio, there are an infinite number of sets of data that can give the same ratio: for example, the ratios 8/4, 6/3, 100/50 all have the same value, 2. The actual concentrations of products and reactants will differ from one experiment to another involving a particular chemical reaction, but the ratio of the amount of product to reactant at equilibrium should be the same for each experiment.

Consider this simple example: suppose we have a reaction for which $K = 4$, and we begin this reaction with 100 reactant molecules. At the point of equilibrium, there should be 80 molecules of product and 20 molecules of reactant remaining (80/20 = 4). Suppose we perform another experiment involving the same reaction, only this time we begin the experiment with 500 molecules of reactant. This time, at the point of equilibrium, there will be 400 molecules of product present and 100 molecules of reactant remaining (400/100 = 4). As we began the two experiments with different numbers of reactant molecules, it's not troubling that there are different absolute numbers of product and reactant molecules present at equilibrium; however, the ratio, K, is the same for both experiments. We say that these two experiments represent two different positions of equilibrium: an equilibrium position corresponds to a particular set of experimental equilibrium concentrations that fulfill the value of the equilibrium constant. Any experiment that is performed with a different amount of starting material will come to its own unique equilibrium position, but the equilibrium constant ratio, K, will be the same for a given reaction regardless of the starting amounts taken.

13. In a homogeneous equilibrium, all the reactants and products are in the same phase and have the same physical state (solid, liquid, or gas). For a heterogeneous equilibrium, however, one or more of the reactant or products exists in a phase or physical state different from the other substances. When we have a heterogeneous equilibrium, the concentrations of solids and pure liquids are left out of the expression for the equilibrium constant for the reaction: the concentration of a solid or pure liquid is constant.

$C(s) + O_2(g) \rightleftharpoons CO_2(g)$ heterogeneous (solid, gases)

$$K = [CO_2]/[O_2]$$

Cumulative Review: Chapters 16 and 17

$$2CO(g) + O_2(g) \rightleftharpoons 2CO_2(g) \quad \text{homogeneous (all gases)}$$

$$K = [CO_2]^2/[CO]^2[O_2]$$

14. Your paraphrase of Le Châtelier's principle should go something like this: "when you make any change to a system in equilibrium, this throws the system temporarily out of equilibrium, and the system responds by reacting in whichever direction will be able to reach a new position of equilibrium". There are various changes that can be made to a system in equilibrium. Following are examples:

    a. The concentration of one of the reactants is increased.

    Consider the reaction: $2SO_2(g) + O_2(g) \rightleftharpoons 2SO_3(g)$

    Suppose the reactants have already reacted and a position of equilibrium has been reached that fulfills the value of K for the reaction. At this point there will be present particular amounts of each reactant and of product. Suppose then one additional mole of O_2 is added to the system from outside. At the instant the additional O_2 is added, the system will not be in equilibrium: there will be too much O_2 present in the system to be compatible with the amounts of SO_2 and SO_3 present. The system will respond by reacting to get rid of some of the excess O_2 until the value of the ratio K is again fulfilled. If the system reacts to get rid of the excess of O_2, additional product SO_3 will form. The net result is more SO_3 produced than if the change had not been made.

    b. The concentration of one of the products is decreased by selectively removing it from the system.

    Consider the reaction: $CH_3COOH + CH_3OH \rightleftharpoons H_2O + CH_3COOCH_3$

    This reaction is typical of many reactions involving organic chemical substances, in which two organic molecules react to form a larger molecule, with a molecule of water split out during the combination. This type of reaction on its own tends to come to equilibrium with only part of the starting materials being converted to the desired organic product (which effectively would leave the experimenter with a mixture of materials). A technique that is used by organic chemists to increase the effective yield of the desired organic product is to *separate* the two products (if the products are separated, they cannot react to give back the reactants). One method used is to add a drying agent to the mixture: such a drying agent chemically or physically absorbs the water from the system, removing it from equilibrium. If the water is removed, the reverse reaction cannot take place, and the reaction proceeds to a greater extent in the forward direction than if the drying agent had not been added. In other situations, an experimenter may separate the products of the reaction by distillation (if the boiling points make this possible): again, if the products have been separated, then the reverse reaction will not be possible, and the forward reaction will occur to a greater extent.

    c. The reaction system is compressed to a smaller volume.

    Consider the example: $3H_2(g) + N_2(g) \rightleftharpoons 2NH_3(g)$

    For equilibria involving gases, when the volume of the reaction system is compressed suddenly, the pressure in the system increases. However, if the reacting system can relieve some of this increased pressure by reacting, it will do so. This will happen by the reaction occurring in whichever direction will give the smaller number of moles of gas (if the number of moles of gas is decreased in a particular volume, the pressure will decrease).

For the reaction above, there are two moles of the gas on the right side of the equation, but there is a total of four moles on the left side. If this system at equilibrium were to be suddenly compressed to a smaller volume, the reaction would proceed further to the right (in favor of more ammonia being produced).

d. The temperature is increased for an endothermic reaction.

Consider the reaction: $2NaHCO_3 + heat \rightleftharpoons Na_2CO_3 + H_2O + CO_2$

Although a change in temperature actually does change the *value* of the equilibrium constant, we can simplify reactions involving temperature changes by treating heat energy as if it were a chemical substance: for this endothermic reaction, heat is one of the reactants. As we saw in the example in part (a) of this question, increasing the concentration of one of the reactants for a system at equilibrium causes the reaction to proceed further to the right, forming additional product. Similarly for the endothermic reaction given above, increasing the temperature causes the reaction to proceed further in the direction of products than if no change had been made. It is as if there were too much "heat" to be compatible with the amount of substances present. The substances react to get rid of some of the energy.

e. The temperature is decreased for an exothermic process.

Consider the reaction: $PCl_3 + Cl_2 \rightleftharpoons PCl_5 + heat$

As discussed in part (d) above, although changing the temperature at which a reaction is performed does change the numerical value of K, we can simplify our discussion of this reaction by treating heat energy as if it were a chemical substance. Heat is a product of this reaction. If we are going to lower the temperature of this reaction system, the only way to accomplish this is to remove energy from the system. Lowering the temperature of the system is really working with this system in its attempt to release heat energy. So lowering the temperature should favor the production of more product than if no change were made.

15. When a slightly soluble salt is placed in water, ions begin to leave the crystals of salt and to enter the solvent to form a solution. As the concentration of ions in solution increases, eventually ions from the solution are attracted to and rejoin the crystals of undissolved salt. Eventually things get to the point that every time ions leave the crystals to enter the solution in one place in the system, somewhere else ions are leaving the solution to rejoin the solid. At the point where dissolving and "undissolving" are going on at the same rate, we arrive at a state of dynamic equilibrium. We write the equilibrium constant (the solubility product), K_{sp}, for the dissolving of a slightly soluble salt in the usual manner: because the concentration of the solid material is constant, we do not include it in the expression for K_{sp}:

$AgCl(s) \rightleftharpoons Ag^+(aq) + Cl^-(aq)$ $K_{sp} = [Ag^+][Cl^-]$

$PbCl_2(s) \rightleftharpoons Pb^{2+}(aq) + 2Cl^-(aq)$ $K_{sp} = [Pb^{2+}][Cl^-]^2$

$BaSO_4(s) \rightleftharpoons Ba^{2+}(aq) + SO_4^{2-}(aq)$ $K_{sp} = [Ba^{2+}][SO_4^{2-}]$

If the solubility product constant for a slightly soluble salt is known, the solubility of the salt in mol/L or g/L can be easily calculated. For example, for $BaSO_4$, $K_{sp} = 1.5 \times 10^{-9}$ at 25°C. Suppose x moles of $BaSO_4$ dissolve per liter; because the stoichiometric coefficients for the dissolving of $BaSO_4$ are all the same, this means that x moles of $Ba^{2+}(aq)$ and x moles of $SO_4^{2-}(aq)$ will be produced per liter when $BaSO_4$ dissolves.

Cumulative Review: Chapters 16 and 17

$$K_{sp} = [Ba^{2+}][SO_4^{2-}] = [x][x] = 1.5 \times 10^{-9}$$

$$x^2 = 1.5 \times 10^{-9} \text{ and therefore } x = 3.9 \times 10^{-5} \ M$$

So the molar solubility of $BaSO_4$ is $3.9 \times 10^{-5}\ M$; this could be converted to the number of grams of $BaSO_4$ that dissolve per liter using the molar mass of $BaSO_4$ (233.4 g):

$$(3.9 \times 10^{-5} \text{ mol/L})(233.4 \text{ g/mol}) = 9.10 \times 10^{-3} \text{ g/L}$$

16. Specific answer depends on student choice of examples. In general, for a weak acid, HA, and a weak base, B:

    $HA + H_2O \rightleftharpoons H_3O^+ + A^-$ $\qquad$ $B + H_2O \rightleftharpoons HB^+ + OH^-$

17. a. The conjugate base of each species will have one less proton:

    NO_3^-, HSO_4^-, ClO_4^-, NH_3, HCO_3^-

    b. The conjugate acid of each species will have one additional proton:

    HCl, H_2SO_4, NH_3, NH_4^+, HCO_3^-

18. a. $NH_3(aq)$(base) $+ H_2O(l)$(acid) $\rightleftharpoons NH_4^+(aq)$(acid) $+ OH^-(aq)$(base)

    b. $H_2SO_4(aq)$(acid) $+ H_2O(l)$(base) $\rightleftharpoons HSO_4^-(aq)$(base) $+ H_3O^+(aq)$(acid)

    c. $O^{2-}(s)$(base) $+ H_2O(l)$(acid) $\rightleftharpoons OH^-(aq)$(acid) $+ OH^-(aq)$(base)

    d. $NH_2^-(aq)$(base) $+ H_2O(l)$(acid) $\rightleftharpoons NH_3(aq)$(acid) $+ OH^-(aq)$(base)

    e. $H_2PO_4^-(aq)$(acid) $+ OH^-(aq)$(base) $\rightleftharpoons HPO_4^{2-}(aq)$(base) $+ H_2O(l)$(acid)

19. a. $[H^+] = \dfrac{1.0 \times 10^{-14}}{[OH^-]} = \dfrac{1.0 \times 10^{-14}}{[2.11 \times 10^{-4}]} = 4.7 \times 10^{-11}\ M$

    b. $pOH = -\log[OH^-] = -\log[7.34 \times 10^{-6}] = 5.134$

    $pH = 14.00 - pOH = 14.00 - 5.134 = 8.87$

    c. $pOH = -\log[OH^-] = -\log[9.81 \times 10^{-8}] = 7.008$

    d. $pOH = 14.00 - pH = 14.00 - 9.32 = 4.68$

    e. $pH = -\log[H^+] = -\log[5.87 \times 10^{-11}] = 10.231$

    f. $[H^+] = \{inv\}\{log\}[-5.83] = 1.5 \times 10^{-6}\ M$

20. a. HNO_3 is a strong acid, so $[H^+] = 0.00141\ M$

    $pH = -\log(0.00141) = 2.851$

    $pOH = 14.00 - 2.851 = 11.15$

    b. NaOH is a strong base, so $[OH^-] = 2.13 \times 10^{-3}\ M$

    $pOH = -\log(2.13 \times 10^{-3}) = 2.672$

    $pH = 14.00 - 2.672 = 11.33$

Cumulative Review: Chapters 16 and 17

c. HCl is a strong acid, so [H$^+$] = 0.00515 M

pH = –log(0.00515) = 2.288

pOH = 14.00 – 2.288 = 11.71

d. Ca(OH)$_2$ is a strong, but not very soluble base. Each formula unit of Ca(OH)$_2$ produces two formula units of OH$^-$ ion.

[OH$^-$] = 2 × 5.65 × 10^{-5} M = 1.13 × 10^{-4} M

pOH = –log(1.13 × 10^{-4}) = 3.947

pH = 14.00 – 3.947 = 10.05

21. a. $\dfrac{[N_2O]^2[O_2]}{[NO]^4}$

b. $\dfrac{[F_2]^6}{[PF_3]^4}$

c. $\dfrac{[CH_4][H_2O]}{[CO][H_2]^3}$

d. $\dfrac{[Br_2][F_2]^5}{[BrF_5]^2}$

e. $\dfrac{[H_2S][Cl_2]}{[HCl]^2}$

22. Br$_2$(g) + Cl$_2$(g) ⇌ 2BrCl(g)

$K = \dfrac{[BrCl(g)]^2}{[Br_2][Cl_2]} = \dfrac{[4.9 \times 10^{-4}]^2}{[7.2 \times 10^{-8}][4.3 \times 10^{-6}]} = 7.8 \times 10^5$

23. a. $K_{sp} = [Cu^{2+}][OH^-]^2$ b. $K_{sp} = [Co^{3+}]^2[S^{2-}]^3$

c. $K_{sp} = [Hg_2^{2+}][OH^-]^2$ d. $K_{sp} = [Ca^{2+}]^2[CO_3^{2-}]$

e. $K_{sp} = [Ag^+]^2[CrO_4^-]$ f. $K_{sp} = [Hg^{2+}][OH^-]^2$

24. MgCO$_3$ ⇌ Mg^{2+}(aq) + CO$_3^{2-}$(aq)   molar mass of MgCO$_3$ = 84.32 g

K_{sp} = [Mg^{2+}][CO$_2^{2-}$]

let x represent the number of moles of MgCO$_3$ that dissolve per liter. Then [Mg^{2+}] = x and [CO$_3^{2-}$] = x also

K_{sp} = [x][x] = x^2 = 6.82 × 10^{-6}

x = [MgCO$_3$] = 2.61 × 10^{-3} M

$\dfrac{2.61 \times 10^{-3} \text{ mol}}{1 \text{ L}} \times \dfrac{84.32 \text{ g}}{1 \text{ mol}} = 0.220$ g/L

CHAPTER 18

Oxidation–Reduction Reactions: Electrochemistry

1. Oxidation reduction reactions are involved in electrical batteries, corrosion and its prevention, combustion of fuels, many cleaning agents, and metabolic processes, to name a few.

2. Oxidation can be defined as the loss of electrons by an atom, molecule, or ion. Oxidation may also be defined as an increase in oxidation state for an element, but because elements can only increase their oxidation states by losing electrons, the two definitions are equivalent. The following equation shows the oxidation of copper metal to copper(II) ion

    $$Cu \rightarrow Cu^{2+} + 2e^-$$

    Reduction can be defined as the gaining of electrons by an atom, molecule, or ion. Reduction may also be defined as a decrease in oxidation state for an element, but naturally such a decrease takes place by the gaining of electrons (so the two definitions are equivalent). The following equation shows the reduction of sulfur atoms to sulfide ion.

    $$S + 2e^- \rightarrow S^{2-}$$

3. Each of these reactions involves one or more *free* elements on one side of the equation; on the other side of the equation, however, the element(s) is(are) *combined* in a compound. This is a clear sign that an oxidation–reduction process is taking place.

    a. chlorine is reduced; iodine is oxidized

    b. chlorine is reduced, lithium is oxidized

    c. sodium is oxidized; hydrogen is reduced

    d. bromine is oxidized; chlorine is reduced

4. Each of these reactions involves one or more *free* elements on one side of the equation; on the other side of the equation, however, the element(s) is(are) *combined* in a compound. This is a clear sign that an oxidation–reduction process is taking place.

    a. sodium is oxidized; nitrogen is reduced

    b. magnesium is oxidized; chlorine is reduced

    c. aluminum is oxidized; bromine is reduced

    d. magnesium is oxidized; copper is reduced

5. Each of these reactions involves one or more *free* elements on one side of the equation; on the other side of the equation, however, the element(s) is(are) *combined* in a compound. This is a clear sign that an oxidation–reduction process is taking place.

    a. calcium is oxidized; hydrogen is reduced

    b. hydrogen is oxidized; fluorine is reduced

Chapter 18: Oxidation–Reduction Reactions: Electrochemistry

    c.   iron is oxidized; oxygen is reduced

    d.   iron is oxidized; chlorine is reduced

6. Each of these reactions involves one or more *free* elements on one side of the equation; on the other side of the equation, however, the element(s) is(are) *combined* in a compound. This is a clear sign that an oxidation–reduction process is taking place.

    a.   magnesium is oxidized; bromine is reduced

    b.   sodium is oxidized; sulfur is reduced

    c.   hydrogen is oxidized; carbon is reduced

    d.   potassium is oxidized; nitrogen is reduced

7. The assignment of oxidation states is a bookkeeping method. *Charges* are assigned to the various atoms in a compound; this method allows us to keep track of electrons transferred between species in oxidation–reduction reactions.

8. A neutral molecule has an overall charge of zero.

9. For the very electronegative elements (such as oxygen), we assign an oxidation state equal to its charge when the element forms an anion. As oxygen forms O^{2-} ions, we assign oxygen an oxidation state of –2 even in compounds where no ions exist. The most common situation in which oxygen is *not* assigned the –2 oxidation state occurs with the *peroxide* ion, O_2^{2-}, as each oxygen atom is in the –1 oxidation state.

10. Fluorine is always assigned a negative oxidation state (–1) because all other elements are less electronegative. The other halogens are *usually* assigned an oxidation state of –1 in compounds. In interhalogen compounds such as ClF, fluorine is assigned oxidation state –1 (F is more electronegative than Cl). Chlorine, therefore, must be assigned a +1 oxidation state in this instance.

11. Oxidation states represent a bookkeeping method to assign electrons in a molecule or ion. As a neutral molecule has an overall charge of zero, the sum of the oxidation states in a neutral molecule must be zero. So in H_3PO_4, the sum of all the oxidation states must be zero.

12. The sum of all the oxidation states of the atoms in a polyatomic ion must equal the overall charge on the ion. The sum of all the oxidation states of all the atoms in PO_4^{3-} is –3.

13. The rules for assigning oxidation states are given in Section 18.2 of the text. The rule that applies for each element in the following answers is given in parentheses after the element and its oxidation state.

    a.   C, +4 (Rule 6); Br, –1 (Rule 5)

    b.   H, +1 (Rule 4); Cl, +7 (Rule 6); O, –2 (Rule 3)

    c.   K, +1 (Rule 2); P, +5 (Rule 6); O, –2 (Rule 3)

    d.   N, +1 (Rule 6); O, –2 (Rule 3)

Chapter 18: Oxidation–Reduction Reactions: Electrochemistry

14. The rules for assigning oxidation states are given in Section 18.2 of the text. The rule that applies for each element in the following answers is given in parentheses after the element and its oxidation state.

    a. N, +3 (Rule 6); Cl, –1 (Rule 5)
    b. S, +6 (Rule 6); F, –1 (Rule 5)
    c. P, +5 (Rule 6); Cl, –1 (Rule 5)
    d. Si, –4 (Rule 6); H, +1 (Rule 4)

15. The rules for assigning oxidation states are given in Section 18.2 of the text. The rule that applies for each element in the following answers is given in parentheses after the element and its oxidation state.

    a. 0 (Rule 1)
    b. +6 (Rule 6)
    c. +6 (Rule 6)
    d. –2 (Rule 5)

16. a. 0 (Rule 1)
    b. –3 (using Rule 4 for H)
    c. +4 (using Rule 3 for O)
    d. +5 (using Rule 3 for O and Rule 2 for Na)

17. a. +1 (using Rule 5 for F)
    b. 0 (Rule 1)
    c. –1 (Rule 5)
    d. +1 (using Rule 3 for O and Rule 4 for H)

18. a. +2 (using Rule 5 for Cl)
    b. +7 (using Rule 3 for O and Rule 2 for K)
    c. +4 (using Rule 3 for O)
    d. +3 (realizing that the acetate ion has 1– charge and apply Rule 6)

19. The rules for assigning oxidation states are given in Section 18.2 of the text. The rule that applies for each element in the following answers is given in parentheses after the element and its oxidation state.

    a. Cu, +2 (Rule 6); Cl, –1 (Rule 5)
    b. K, +1 (Rule 2); Cl, +5 (Rule 6); O, –2 (Rule 3)
    c. K, +1 (Rule 2); Cl, +7 (Rule 6); O, –2 (Rule 3)
    d. Na, +1 (Rule 2); C, +4 (Rule 6); O, –2 (Rule 3)

Chapter 18:   Oxidation–Reduction Reactions: Electrochemistry

20. The rules for assigning oxidation states are given in Section 18.2 of the text. The rule that applies for each element in the following answers is given in parentheses after the element and its oxidation state.

   a. Ca, +2 (Rule 6); O, –2 (Rule 3)
   b. Al, +3 (Rules 6 and 2); O, –2 (Rule 3)
   c. P, +3 (Rule 6); F, –1 (Rule 5)
   d. P, +5 (Rule 6); O, –2 (Rule 3)

21. The rules for assigning oxidation states are given in Section 18.2 of the text. The rule that applies for each element in the following answers is given in parentheses after the element and its oxidation state.

   a. C, +4 (Rule 7); O, –2 (Rule 3)
   b. N, +5 (Rule 7); O, –2 (Rule 3)
   c. P, +5 (Rule 7); O, –2 (Rule 3)
   d. S, +6 (Rule 7); O, –2 (Rule 3)

22. The rules for assigning oxidation states are given in Section 18.2 of the text. The rule that applies for each element in the following answers is given in parentheses after the element and its oxidation state.

   a. H, +1 (Rule 4); S, +6 (Rule 7); O, –2 (Rule 3)
   b. Mn, +7 (Rule 7); O, –2 (Rule 3)
   c. Cl, +5 (Rule 7); O, –2 (Rule 3)
   d. Br, +7 (Rule 7; O, –2 (Rule 3)

23. Consider the following simple oxidation reaction

   $Na \rightarrow Na^+ + e^-$

   Clearly the sodium atom on the left side of the equation is losing an electron in forming the sodium ion on the right side of the equation. The sodium atom on the left side of the equation is in the *zero* oxidation state because it represents a pure element. The sodium ion on the right side of the equation is in the +1 oxidation state (the same as the charge on the simple ion). Thus, by losing one electron, sodium has increased in oxidation state by one unit.

24. Electrons are negative; when an atom gains electrons, it gains one negative charge for each electron gained. For example, in the reduction reaction $Cl + e^- \rightarrow Cl^-$, the oxidation state of chlorine decreases from 0 to –1 as the electron is gained.

25. An oxidizing *agent* is a molecule, atom, or ion that *causes* the oxidation of some other species, while itself being reduced. A reducing *agent* is a molecule, atom, or ion that *causes* the reduction of some other species, while itself being oxidized.

26. An oxidizing agent decreases its oxidation state. A reducing agent increases its oxidation state.

27. Oxidizing agents gain the electrons they cause other species to lose. Reducing agents donate the electrons needed for the reduction of some other species.

Chapter 18: Oxidation–Reduction Reactions: Electrochemistry

28. An antioxidant is a substance that prevents oxidation of some molecule(s) in the body. It is not certain how all antioxidants work, but one example is in preventing oxygen molecules and other substances from stripping electrons from cell membranes, leaving them vulnerable to destruction by the immune system.

29.
   a. $Fe(s) + CuSO_4(aq) \rightarrow FeSO_4(aq) + Cu(s)$

   iron is being oxidized, copper is being reduced

   b. $Cl_2(g) + 2NaBr(aq) \rightarrow 2NaCl(aq) + Br_2(l)$

   bromine is being oxidized, chlorine is being reduced

   c. $3CuS(s) + 8HNO_3(aq) \rightarrow 3CuSO_4(aq) + 8NO(g) + 4H_2O(l)$

   sulfur is being oxidized, nitrogen is being reduced

   d. $2Zn(s) + O_2(g) \rightarrow 2ZnO(s)$

   zinc is being oxidized, oxygen is being reduced

30.
   a. $2Al(s) + 3S(s) \rightarrow Al_2S_3(s)$

   aluminum is being oxidized, sulfur is being reduced

   b. $CH_4(g) + 2O_2(g) \rightarrow CO_2(g) + 2H_2O(g)$

   carbon is being oxidized, oxygen is being reduced

   c. $2Fe_2O_3(s) + 3C(s) \rightarrow 3CO_2(g) + 4Fe(s, l)$

   carbon is being oxidized, iron is being reduced

   d. $K_2Cr_2O_7(aq) + 14HCl(aq) \rightarrow 2KCl(aq) + 2CrCl_3(s) + 7H_2O(l) + 3Cl_2(g)$

   chlorine is being oxidized, chromium is being reduced

31.
   a. $2Cu(s) + S(s) \rightarrow Cu_2S$

   copper is being oxidized; sulfur is being reduced

   b. $2Cu_2O(s) + O_2(g) \rightarrow 4CuO(s)$

   copper is being oxidized; oxygen (of O_2) is being reduced

   c. $4B(s) + 3O_2(g) \rightarrow 2B_2O_3(s)$

   boron is being oxidized; oxygen is being reduced

   d. $6Na(s) + N_2(g) \rightarrow 2Na_3N(s)$

   sodium is being oxidized; nitrogen is being reduced

32.
   a. $4KClO_3(s) + C_6H_{12}O_6(s) \rightarrow 4KCl(s) + 6H_2O(l) + 6CO_2(g)$

   carbon is being oxidized, chlorine is being reduced

   b. $2C_8H_{18}(l) + 25O_2(g) \rightarrow 16CO_2(g) + 18H_2O(l)$

   carbon is being oxidized, oxygen is being reduced

   c. $PCl_3(g) + Cl_2(g) \rightarrow PCl_5(g)$

   phosphorus is being oxidized, chlorine is being reduced

d. $Ca(s) + H_2(g) \rightarrow CaH_2(g)$

calcium is being oxidized, hydrogen is being reduced

33. Zinc is oxidized [0 in $Zn(s)$, +2 in $ZnCl_2(aq)$]; hydrogen is reduced [+1 in $HCl(aq)$, 0 in $H_2(g)$].

34. Iron is reduced [+3 in $Fe_2O_3(s)$, 0 in $Fe(l)$]; carbon is oxidized [+2 in $CO(g)$, +4 in $CO_2(g)$]. $Fe_2O_3(s)$ is the oxidizing agent; $CO(g)$ is the reducing agent.

35. Magnesium is oxidized [0 in $Mg(s)$, +2 in $Mg(OH)_2(s)$]; hydrogen is reduced [+1 in $H_2O(l)$, 0 in $H_2(g)$].

36.
   a. chlorine is being reduced, iodine is being oxidized; chlorine is the oxidizing agent, iodide ion is the reducing agent

   b. iron is being reduced, iodine is being oxidized; iron(III) is the oxidizing agent, iodide ion is the reducing agent

   c. copper is being reduced, iodine is being oxidized; copper(II) is the oxidizing agent, iodide ion is the reducing agent

37. Oxidation–reduction reactions must be balanced with respect to *mass* (the total number of each type of atom on each side of the balanced equation must be the same); with respect to *charge*, whatever number of electrons is lost in the oxidation process must be gained in the reduction process, with no "extra" electrons.

38. Oxidation–reduction reactions are often more complicated than "regular" reactions; frequently the coefficients necessary to balance the number of electrons transferred come out to be large numbers. We also have to make certain that we account for the electrons being transferred.

39. When an overall equation is split into separate partial equations representing the oxidation and the reduction processes, these partial equations are called *half–reactions*.

40. Under ordinary conditions it is impossible to have "free" electrons that are not part of some atom, ion, or molecule. For this reason, the total number of electrons lost by the species being oxidized must equal the total number of electrons gained by the species being reduced.

41.
   a. $Cu \rightarrow Cu^{2+}$

   Balance charge: $Cu \rightarrow Cu^{2+} + 2e^-$

   Balanced half–reaction: $Cu \rightarrow Cu^{2+} + 2e^-$

   b. $Fe^{3+} \rightarrow Fe^{2+}$

   Balance charge: $e^- + Fe^{3+} \rightarrow Fe^{2+}$

   Balanced half–reaction: $e^- + Fe^{3+} \rightarrow Fe^{2+}$

   c. $Br^- \rightarrow Br_2$

   Balance mass: $2Br^- \rightarrow Br_2$

   Balance charge: $2Br^- \rightarrow Br_2 + 2e^-$

   Balanced half–reaction: $2Br^- \rightarrow Br_2 + 2e^-$

Chapter 18: Oxidation–Reduction Reactions: Electrochemistry

    d. $Fe^{2+} \rightarrow Fe$

        Balance charge: $Fe^{2+} + 2e^- \rightarrow Fe$

        Balanced half–reaction: $Fe^{2+} + 2e^- \rightarrow Fe$

42.   a. $N_2(g) \rightarrow N_3^-(aq)$

        balance nitrogen: $\mathbf{3}N_2(g) \rightarrow \mathbf{2}N_3^-(aq)$

        balance charge: $3N_2(g) + \mathbf{2e^-} \rightarrow 2N_3^-(aq)$

        balanced half-reaction: $3N_2(g) + 2e^- \rightarrow 2N_3^-(aq)$

   b. $O_2^{2-}(aq) \rightarrow O_2(g)$

        balance charge: $O_2^{2-}(aq) \rightarrow O_2(g) + 2e^-$

        balanced half-reaction: $O_2^{2-}(aq) \rightarrow O_2(g) + 2e^-$

   c. $Zn(s) \rightarrow Zn^{2+}(aq)$

        balance charge: $Zn(s) \rightarrow Zn^{2+}(aq) + 2e^-$

        balanced half-reaction: $Zn(s) \rightarrow Zn^{2+}(aq) + 2e^-$

   d. $F_2(g) \rightarrow F^-(aq)$

        balance flourine: $F_2(g) \rightarrow \mathbf{2}F^-(aq)$

        balance charge: $F_2(g) + \mathbf{2e^-} \rightarrow 2F^-(aq)$

        balanced half-reaction: $F_2(g) + 2e^- \rightarrow 2F^-(aq)$

43.   a. $HClO(aq) \rightarrow Cl^-(aq)$

        Balance oxygen: $HClO \rightarrow Cl^- + H_2O$

        Balance hydrogen: $H^+ + HClO \rightarrow Cl^- + H_2O$

        Balance charge: $2e^- + H^+ + HClO \rightarrow Cl^- + H_2O$

        Balanced half–reaction: $2e^- + H^+(aq) + HClO(aq) \rightarrow Cl^-(aq) + H_2O(l)$

   b. $NO(aq) \rightarrow N_2O(g)$

        Balance nitrogen: $2NO \rightarrow N_2O$

        Balance oxygen: $2NO \rightarrow N_2O + H_2O$

        Balance hydrogen: $2H^+ + 2NO \rightarrow N_2O + H_2O$

        Balance charge: $2e^- + 2H^+ + 2NO \rightarrow N_2O + H_2O$

        Balanced half–reaction: $2e^- + 2H^+(aq) + 2NO(aq) \rightarrow N_2O(g) + H_2O(l)$

   c. $N_2O(aq) \rightarrow N_2(g)$

        Balance oxygen: $N_2O \rightarrow N_2 + H_2O$

        Balance hydrogen: $2H^+ + N_2O \rightarrow N_2 + H_2O$

        Balance charge: $2e^- + 2H^+ + N_2O \rightarrow N_2 + H_2O$

        Balanced half–reaction: $2e^- + 2H^+(aq) + N_2O(aq) \rightarrow N_2(g) + H_2O(l)$

Chapter 18: Oxidation–Reduction Reactions: Electrochemistry

d. $ClO_3^-(aq) \rightarrow HClO_2(l)$

Balance oxygen: $ClO_3^- \rightarrow HClO_2 + H_2O$

Balance hydrogen: $3H^+ + ClO_3^- \rightarrow HClO_2 + H_2O$

Balance charge: $2e^- + 3H^+ + ClO_3^- \rightarrow HClO_2 + H_2O$

Balanced half–reaction: $2e^- + 3H^+(aq) + ClO_3^-(aq) \rightarrow HClO_2(aq) + H_2O(l)$

44. a. $O_2(g) \rightarrow H_2O(l)$

balance oxygen: $O_2 \rightarrow 2H_2O$

balance hydrogen: $4H^+ + O_2 \rightarrow 2H_2O$

balance charge: $4e^- + 4H^+ + O_2 \rightarrow 2H_2O$

balanced half-reaction: $4e^- + 4H^+(aq) + O_2(g) \rightarrow 2H_2O(l)$

b. $SO_4^{2-}(aq) \rightarrow H_2SO_3(aq)$

balance oxygen: $SO_4^{2-} \rightarrow H_2SO_3 + H_2O$

balance hydrogen: $4H^+ + SO_4^{2-} \rightarrow H_2SO_3 + H_2O$

balance charge: $2e^- + 4H^+ + SO_4^{2-} \rightarrow H_2SO_3 + H_2O$

balanced half-reaction: $2e^- + 4H^+(aq) + SO_4^{2-}(aq) \rightarrow H_2SO_3(aq) + H_2O(l)$

c. $H_2O_2(aq) \rightarrow H_2O(l)$

balance oxygen : $H_2O_2 \rightarrow 2H_2O$

balance hydrogen : $2H^+ + H_2O_2 \rightarrow 2H_2O$

balance charge : $2e^- + 2H^+ + H_2O_2 \rightarrow 2H_2O$

balanced half-reaction: $2e^- + 2H^+(aq) + H_2O_2(aq) \rightarrow 2H_2O(l)$

d. $NO_2^-(aq) \rightarrow NO_3^-(aq)$

balance oxygen : $H_2O + NO_2^- \rightarrow NO_3^-$

balance hydrogen: $H_2O + NO_2^- \rightarrow NO_3^- + 2H^+$

balance charge: $H_2O + NO_2^- \rightarrow NO_3^- + 2H^+ + 2e^-$

balanced half-reaction: $H_2O(l) + NO_2^-(aq) \rightarrow NO_3^-(aq) + 2H^+(aq) + 2e^-$

45. a. $Mg(s) + Hg^{2+}(aq) \rightarrow Mg^{2+}(aq) + Hg_2^{2+}(aq)$

$Mg \rightarrow Mg^{2+}$

Balance charge: $Mg \rightarrow Mg^{2+} + \mathbf{2e^-}$

Balanced half–reaction: $Mg \rightarrow Mg^{2+} + 2e^-$

$Hg^{2+} \rightarrow Hg_2^{2+}$

Balance mercury: $\mathbf{2}Hg^{2+} \rightarrow Hg_2^{2+}$

Balance charge: $2Hg^{2+} + \mathbf{2e^-} \rightarrow Hg_2^{2+}$

Balanced half–reaction: $2Hg^{2+} + 2e^- \rightarrow Hg_2^{2+}$

Chapter 18: Oxidation–Reduction Reactions: Electrochemistry

As the number of electrons is the same in both half–reactions, the half–reactions can be directly combined for the overall equation.

$$Mg(s) + 2Hg^{2+}(aq) \rightarrow Mg^{2+}(aq) + Hg_2^{2+}(aq).$$

b. $NO_3^-(aq) + Br^-(aq) \rightarrow NO(g) + Br_2(l)$

$2NO_3^- \rightarrow NO$

Balance oxygen: $NO_3^- \rightarrow NO + \mathbf{2H_2O}$

Balance hydrogen: $\mathbf{4H^+} + NO_3^- \rightarrow NO + \mathbf{2H_2O}$

Balance charge: $4H^+ + NO_3^- + \mathbf{3e^-} \rightarrow NO + 2H_2O$

Balanced half–reaction: $4H^+ + NO_3^- + 3e^- \rightarrow NO + 2H_2O$

$Br^- \rightarrow Br_2$

Balance bromine: $\mathbf{2Br^-} \rightarrow Br_2$

Balance charge: $2Br^- \rightarrow Br_2 + \mathbf{2e^-}$

Balanced half–reaction: $2Br^- \rightarrow Br_2 + 2e^-$

Combine the half–reactions:

$2 \times (4H^+ + NO_3^- + 3e^- \rightarrow NO + 2H_2O)$

$3 \times (2Br^- \rightarrow Br_2 + 2e^-)$

$8H^+(aq) + 2NO_3^-(aq) + 6Br^-(aq) \rightarrow 3Br_2(l) + 2NO(g) + 4H_2O(l)$

c. $Ni(s) + NO_3^-(aq) \rightarrow Ni^{2+}(aq) + NO_2(g)$

$Ni \rightarrow Ni^{2+}$

Balance charge: $Ni \rightarrow Ni^{2+} + \mathbf{2e^-}$

Balanced half–reaction: $Ni \rightarrow Ni^{2+} + 2e^-$

$NO_3^- \rightarrow NO_2$

Balance oxygen: $NO_3^- \rightarrow NO_2 + \mathbf{H_2O}$

Balance hydrogen: $NO_3^- + \mathbf{2H^+} \rightarrow NO_2 + H_2O$

Balance charge: $NO_3^- + 2H^+ + \mathbf{e^-} \rightarrow NO_2 + H_2O$

Balanced half–reaction: $NO_3^- + 2H^+ + e^- \rightarrow NO_2 + H_2O$

Combine the half–reactions:

$2 \times (NO_3^- + 2H^+ + e^- \rightarrow NO_2 + H_2O)$

$Ni \rightarrow Ni^{2+} + 2e^-$

$2NO_3^-(aq) + 4H^+(aq) + Ni(s) \rightarrow 2NO_2(g) + 2H_2O(l) + Ni^{2+}(aq)$

d. $ClO_4^-(aq) + Cl^-(aq) \rightarrow ClO_3^-(aq) + Cl_2(g)$

$ClO_4^- \rightarrow ClO_3^-$

Balance oxygen: $ClO_4^- \rightarrow ClO_3^- + \mathbf{H_2O}$

Balance hydrogen: $\mathbf{2H^+} + ClO_4^- \rightarrow ClO_3^- + H_2O$

Chapter 18: Oxidation–Reduction Reactions: Electrochemistry

Balance charge: $2H^+ + ClO_4^- + 2e^- \rightarrow ClO_3^- + H_2O$

Balanced half–reaction: $2H^+ + ClO_4^- + 2e^- \rightarrow ClO_3^- + H_2O$

$Cl^- \rightarrow Cl_2$

Balance chlorine: $2Cl^- \rightarrow Cl_2$

Balance charge: $2Cl^- \rightarrow Cl_2 + 2e^-$

Balanced half–reaction: $2Cl^- \rightarrow Cl_2 + 2e^-$

Since the number of electrons transferred is the same in both half–reactions, the half–reactions can be combined directly to give the overall equation for the reaction:

$2H^+(aq) + ClO_4^-(aq) + 2Cl^-(aq) \rightarrow ClO_3^-(aq) + H_2O(l) + Cl_2(g)$

46. For simplicity, the physical states of the substances have been omitted until the final balanced equation is given.

   a. $Al(s) + H^+(aq) \rightarrow Al^{3+}(aq) + H_2(g)$

   $Al \rightarrow Al^{3+}$

   Balance charge: $Al \rightarrow Al^{3+} + 3e^-$

   $H^+ \rightarrow H_2$

   Balance hydrogen: $2H^+ \rightarrow H_2$

   Balance charge: $2e^- + 2H^+ \rightarrow H_2$

   Combine half–reactions:

   $3 \times (2e^- + 2H^+ \rightarrow H_2)$

   $2 \times (Al \rightarrow Al^{3+} + 3e^-)$

   $2Al(s) + 6H^+(aq) \rightarrow 2Al^{3+}(aq) + 3H_2(g)$

   b. $S^{2-}(aq) + NO_3^-(g) \rightarrow S(s) + NO(g)$

   $S^{2-} \rightarrow S$

   Balance charge: $S^{2-} \rightarrow S + 2e^-$

   $NO_3^- \rightarrow NO$

   Balance oxygen: $NO_3^- \rightarrow NO + 2H_2O$

   Balance hydrogen: $4H^+ + NO_3^- \rightarrow NO + 2H_2O$

   Balance charge: $3e^- + 4H^+ + NO_3^- \rightarrow NO + 2H_2O$

   Combine half–reactions:

   $3 \times (S^{2-} \rightarrow S + 2e^-)$

   $2 \times (3e^- + 4H^+ + NO_3^- \rightarrow NO + 2H_2O)$

   $8H^+ + 3S^{2-}(aq) + 2NO_3^-(g) \rightarrow 3S(s) + 2NO(g) + 4H_2O$

Chapter 18: Oxidation–Reduction Reactions: Electrochemistry

    c. $I_2(aq) + Cl_2(aq) \rightarrow IO_3^-(aq) + HCl(g)$

        $I_2 \rightarrow IO_3^-$

        Balance iodine: $I_2 \rightarrow \mathbf{2IO_3^-}$

        Balance oxygen: $\mathbf{6H_2O} + I_2 \rightarrow 2IO_3^-$

        Balance hydrogen: $6H_2O + I_2 \rightarrow 2IO_3^- + \mathbf{12H^+}$

        Balance charge: $6H_2O + I_2 \rightarrow 2IO_3^- + 12H^+ + \mathbf{10e^-}$

        $Cl_2 \rightarrow HCl$

        Balance chlorine: $Cl_2 \rightarrow \mathbf{2HCl}$

        Balance hydrogen: $\mathbf{2H^+} + Cl_2 \rightarrow 2HCl$

        Balance charge: $\mathbf{2e^-} + 2H^+ + Cl_2 \rightarrow 2HCl$

        Combine half–reactions:

        $5 \times (2e^- + 2H^+ + Cl_2 \rightarrow 2HCl)$

        $6H_2O + I_2 \rightarrow 2IO_3^- + 12H^+ + 10e^-$

        $6H_2O(l) + 2I_2(aq) + 5Cl_2(aq) \rightarrow 2IO_3^-(aq) + 10HCl(g) + 2H^+(aq)$

    d. $AsO_4^-(aq) + S^{2-}(aq) \rightarrow AsO_3^-(s) + S(s)$

        $AsO_4^- \rightarrow AsO_3^-$

        Balance oxygen: $AsO_4^- \rightarrow AsO_3^- + \mathbf{H_2O}$

        Balance hydrogen: $\mathbf{2H^+} + AsO_4^- \rightarrow AsO_3^- + H_2O$

        Balance charge: $\mathbf{2e^-} + 2H^+ + AsO_4^- \rightarrow AsO_3^- + H_2O$

        $S^{2-} \rightarrow S$

        Balance charge: $S^{2-} \rightarrow S + \mathbf{2e^-}$

        $2H^+(aq) + AsO_4^-(aq) + S^{2-}(aq) \rightarrow AsO_3^-(s) + S(s) + H_2O(l)$

47.   For simplicity, the physical states of the substances have been omitted until the final balanced equation is given.

    For the oxidation of iodide ion, I^-, in acidic solution, the half–reaction is always the *same*:

        $I^- \rightarrow I_2$

        Balance iodine: $\mathbf{2I^-} \rightarrow I_2$

        Balance charge: $2I^- \rightarrow I_2 + \mathbf{2e^-}$

        Balanced half–reaction: $2I^- \rightarrow I_2 + 2e^-$

    a. $IO_3^- \rightarrow I_2$

        Balance iodine: $\mathbf{2IO_3^-} \rightarrow I_2$

        Balance oxygen: $2IO_3^- \rightarrow I_2 + \mathbf{6H_2O}$

        Balance hydrogen: $2IO_3^- + \mathbf{12H^+} \rightarrow I_2 + 6H_2O$

        Balance charge: $2IO_3^- + 12H^+ + \mathbf{10e^-} \rightarrow I_2 + 6H_2O$

Balanced half–reaction: $2IO_3^- + 12H^+ + 10e^- \rightarrow I_2 + 6H_2O$

Combine the half–reactions:

$2IO_3^- + 12H^+ + 10e^- \rightarrow I_2 + 6H_2O$

$5 \times (2I^- \rightarrow I_2 + 2e^-)$

$2IO_3^-(aq) + 12H^+(aq) + 10I^-(aq) \rightarrow 6I_2(aq) + 6H_2O(l)$

$IO_3^-(aq) + 6H^+(aq) + 5I^-(aq) \rightarrow 3I_2(aq) + 3H_2O(l)$

b. $Cr_2O_7^{2-} \rightarrow Cr^{3+}$

Balance chromium: $Cr_2O_7^{2-} \rightarrow \mathbf{2}Cr^{3+}$

Balance oxygen: $Cr_2O_7^{2-} \rightarrow 2Cr^{3+} + \mathbf{7H_2O}$

Balance hydrogen: $Cr_2O_7^{2-} + \mathbf{14H^+} \rightarrow 2Cr^{3+} + 7H_2O$

Balance charge: $Cr_2O_7^{2-} + 14H^+ + \mathbf{6e^-} \rightarrow 2Cr^{3+} + 7H_2O$

Balanced half–reaction: $Cr_2O_7^{2-} + 14H^+ + 6e^- \rightarrow 2Cr^{3+} + 7H_2O$

Combine the half–reactions:

$3 \times (2I^- \rightarrow I_2 + 2e^-)$

$Cr_2O_7^{2-} + 14H^+ + 6e^- \rightarrow 2Cr^{3+} + 7H_2O$

$6I^-(aq) + Cr_2O_7^{2-}(aq) + 14H^+(aq) \rightarrow 3I_2(aq) + 2Cr^{3+}(aq) + 7H_2O(l)$

c. $Cu^{2+} \rightarrow CuI$

Balance iodine: $Cu^{2+} + I^- \rightarrow CuI$

Balance charge: $Cu^{2+} + I^- + e^- \rightarrow CuI$

Balanced half–reaction: $Cu^{2+} + I^- + e^- \rightarrow CuI$

Combine the half–reactions:

$2I^- \rightarrow I_2 + 2e^-$

$2 \times (Cu^{2+} + I^- + e^- \rightarrow CuI)$

$2Cu^{2+}(aq) + 4I^-(aq) \rightarrow 2CuI(s) + I_2(aq)$

48. $Cu(s) + 2HNO_3(aq) + 2H^+(aq) \rightarrow Cu^{2+}(aq) + 2NO_2(g) + 2H_2O(l)$

    $Mg(s) + 2HNO_3(aq) \rightarrow Mg(NO_3)_2(aq) + H_2(g)$

49. Answer depends on student choice of reaction.

50. A salt bridge typically consists of a *U*–shaped tube filled with an inert electrolyte (one involving ions that are not part of the oxidation–reduction reaction). A salt bridge is used to complete the electrical circuit in a cell. Any method that allows transfer of charge without allowing bulk mixing of the solutions may be used (another common method is to set up one half–cell in a porous cup, which is then placed in the beaker containing the second half–cell).

Chapter 18: Oxidation–Reduction Reactions: Electrochemistry

51. In a galvanic cell, electrons flow from the anode (where oxidation occurs) to the cathode (where reduction occurs).

52. Reduction takes place at the cathode and oxidation takes place at the anode.

53. A diagram of the cell is shown below:

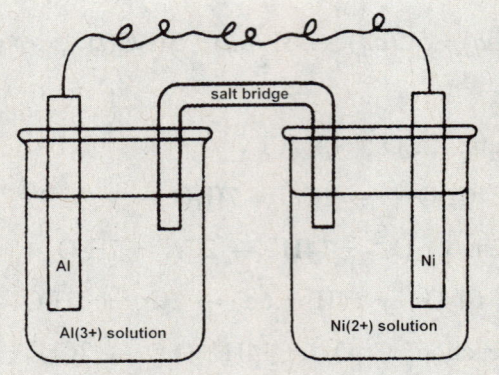

$Ni^{2+}(aq)$ ion is reduced; $Al(s)$ is oxidized.

The reaction at the anode is $Al(s) \rightarrow Al^{3+}(aq) + 3e^-$.

The reaction at the cathode is $Ni^{2+}(aq) + 2e^- \rightarrow Ni(s)$.

54. A diagram of the cell is shown below:

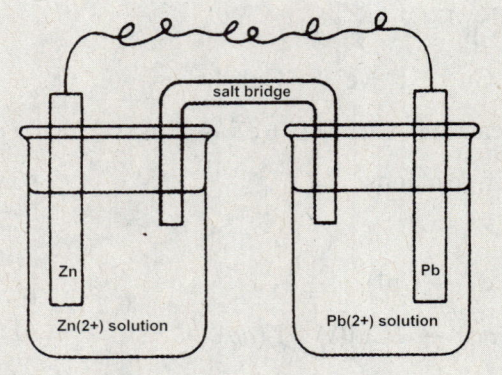

$Pb^{2+}(aq)$ ion is reduced; $Zn(s)$ is oxidized.

The reaction at the anode is $Zn(s) \rightarrow Zn^{2+}(aq) + 2e^-$.

The reaction at the cathode is $Pb^{2+}(aq) + 2e^- \rightarrow Pb(s)$

55. The overall reaction is

$$Pb(s) + PbO_2(s) + 2H_2SO_4(aq) \rightarrow 2PbSO_4(s) + 2H_2O(l)$$

in which $Pb^0(s)$ is oxidized to Pb^{2+} and $Pb^{IV}O_2$ is reduced to Pb^{2+}. This reaction can be reversed by *electrolysis* of the mixture of water and $PbSO_4(s)$ (passing electrical energy into the mixture from the outside).

56. $Cd + 2OH^- \rightarrow Cd(OH)_2 + 2e^-$ (oxidation)

    $NiO_2 + 2H_2O + 2e^- \rightarrow Ni(OH)_2 + 2OH^-$ (reduction)

57. Corrosion represents returning metals to the natural state (ore) and involves *oxidation* of the metal. Corrosion of a metal is undesirable because, as the metal is converted to its oxide, the bulk of the metal loses its strength, flexibility, and other metallic properties. If the metal were part of some constructed item, the item would slowly disintegrate.

58. Aluminum is a very reactive metal when freshly isolated in the pure state. However, on standing for even a relatively short period of time, aluminum metal forms a thin coating of Al_2O_3 on its surface from reaction with atmospheric oxygen. This coating of Al_2O_3 is much less reactive than the metal and serves to protect the surface of the metal from further attack.

59. Most steels contain additives such as chromium or nickel. These additives are able to form protective oxide coatings on the surface of the steel that tend to prevent further oxidation.

60. Chromium protects stainless steel by forming a thin coating of chromium oxide on the surface of the steel, which prevents oxidation of the iron in the steel.

61. In an electrolysis cell, an electric current from outside the cell is used to force an otherwise non-spontaneous oxidation–reduction reaction to occur. In a galvanic cell, a spontaneous oxidation–reduction reaction is used as a source of electrical current.

62. The main recharging reaction for the lead storage battery is

    $2PbSO_4(s) + 2H_2O(l) \rightarrow Pb(s) + PbO_2(s) + 2H_2SO_4(aq)$.

    A major side reaction is the electrolysis of water

    $2H_2O(l) \rightarrow 2H_2(g) + O_2(g)$.

    This results in production of an explosive mixture of hydrogen and oxygen, which accounts for many accidents in recharging of such batteries.

63. Aluminum is so reactive towards oxygen that there proved to be no convenient chemical reducing agent which could reduce aluminum ores to the metal. Widespread commercial preparation of aluminum was only possible when an electrolytic process was developed

64. The balanced equation is $2H_2O(l) \rightarrow 2H_2(g) + O_2(g)$. Oxygen is oxidized (going from -2 oxidation state in water to zero oxidation state in the free element). Hydrogen is reduced (going from $+1$ oxidation state in water to zero oxidation state in the free element). Heat is produced by burning the hydrogen gas produced by the electrolysis: since energy must be applied to water to electrolyze it, energy is released when hydrogen gas produced by the electrolysis and oxygen gas combine to form water in the fireplace.

65. electrons

66. loss; oxidation state

67. gain, oxidation number

68. electronegative

Chapter 18: Oxidation–Reduction Reactions: Electrochemistry

69. charge

70. An *oxidizing agent* is an atom, molecule, or ion that causes the oxidation of another species. During this process, the oxidizing agent is reduced.

71. oxidation, reduced

72. lose

73. equal

74. separate from

75. galvanic

76. oxidation

77. cathode

78. reducing; oxidizing

79. voltage or potential

80. electrolysis; In a galvanic cell, chemical energy is converted to electrical energy by means of an oxidation-reduction reaction. In electrolysis, electrical energy is used to produce a chemical change.

81. zinc

82. oxidation

83. aluminum oxide

84. a. $4Fe(s) + 3O_2(g) \rightarrow 2Fe_2O_3(s)$

    iron is oxidized; oxygen is reduced

    b. $2Al(s) + 3Cl_2(g) \rightarrow 2AlCl_3(s)$

    aluminum is oxidized; chlorine is reduced

    c. $6Mg(s) + P_4(s) \rightarrow 2Mg_3P_2(s)$

    magnesium is oxidized; phosphorus is reduced

85. a. zinc is oxidized; hydrogen (as H^+ in HCl) is reduced

    b. copper(I) is both oxidized to copper(II) and reduced to copper(0)

    c. iron (as Fe^{2+}) is oxidized to Fe^{3+}; chromium(VI) (in $Cr_2O_7^{2-}$) is reduced to Cr^{3+}

86. a. aluminum is oxidized; hydrogen is reduced

    b. hydrogen is reduced; iodine is oxidized

    c. copper is oxidized; hydrogen is reduced

Chapter 18: Oxidation–Reduction Reactions: Electrochemistry

87. a. $CH_2=CH_2(g) + Cl_2(g) \rightarrow ClCH_2–CH_2Cl(l)$

   carbon is oxidized (–2 to –1); chlorine is reduced (0 to –1)

   Cl_2 is the oxidizing agent; $CH_2=CH_2$ is the reducing agent

   b. $CH_2=CH_2(g) + Br_2(g) \rightarrow BrCH_2–CH_2Br(l)$

   carbon is oxidized (–2 to –1); bromine is reduced (0 to –1)

   Br_2 is the oxidizing agent; $CH_2=CH_2$ is the reducing agent

   c. $CH_2=CH_2(g) + HBr(g) \rightarrow CH_3–CH_2Br(l)$

   The process appears not to involve electron transfer: there are no changes in oxidation number.

   d. $CH_2=CH_2(g) + H_2(g) \rightarrow CH_3–CH_3$

   carbon is reduced (–3 to –3); hydrogen is oxidized (0 to +1)

   hydrogen is the reducing agent; $CH_2=CH_2$ is the oxidizing agent

88. a. $C_3H_8(g) + 5O_2(g) \rightarrow 3CO_2(g) + 4H_2O(g)$

   b. $CO(g) + 2H_2(g) \rightarrow CH_3OH(l)$

   c. $SnO_2(s) + 2C(s) \rightarrow Sn(s) + 2CO(g)$

   d. $C_2H_5OH(l) + 3O_2(g) \rightarrow 2CO_2(g) + 3H_2O(g)$

89. a. $2MnO_4^-(aq) + 6H^+(aq) + 5H_2O_2(aq) \rightarrow 2Mn^{2+}(aq) + 8H_2O(l) + 5O_2(g)$

   b. $6Cu^+(aq) + 6H^+(aq) + BrO_3^-(aq) \rightarrow 6Cu^{2+}(aq) + Br^-(aq) + 3H_2O(l)$

   c. $2HNO_2(aq) + 2H^+(aq) + 2I^-(aq) \rightarrow 2NO(g) + I_2(aq) + 2H_2O(l)$

90. Each of these reactions involves a *metallic* element in the form of the *free* element on one side of the equation; on the other side of the equation, the metallic element is *combined* in an ionic compound. If a metallic element goes from the free metal to the ionic form, the metal is oxidized (loses electrons).

   a. sodium is oxidized; oxygen is reduced

   b. iron is oxidized; hydrogen is reduced

   c. oxygen (O^{2-}) is oxidized; aluminum (Al^{3+}) is reduced (this reaction is the reverse of the type discussed above)

   d. magnesium is oxidized; nitrogen is reduced

91. Each of these reactions involves a *metallic* element in the form of the *free* element on one side of the equation; on the other side of the equation, the metallic element is *combined* in an ionic compound. If a metallic element goes from the free metal to the ionic form, the metal is oxidized (loses electrons).

   a. zinc is oxidized; nitrogen is reduced

   b. cobalt is oxidized; sulfur is reduced

Chapter 18: Oxidation–Reduction Reactions: Electrochemistry

    c.   potassium is oxidized; oxygen is reduced

    d.   silver is oxidized; oxygen is reduced

92. The rules for assigning oxidation states are given in Section 18.2 of the text. The rule that applies for each element in the following answers is given in parentheses after the element and its oxidation state.

    a.   H +1 (Rule 4); N –3 (Rule 6)

    b.   C +2 (Rule 6); O –2 (Rule 3)

    c.   C +4 (Rule 6); O –2 (Rule 3)

    d.   N +3 (Rule 6); F –1 (Rule 5)

93. The rules for assigning oxidation states are given in Section 18.2 of the text. The rule that applies for each element in the following answers is given in parentheses after the element and its oxidation state.

    a.   P +3 (Rule 6); Br –1 (Rule 5)

    b.   C –(8/3) (Rule 6); H +1 (Rule 4)

    c.   K +1 (Rule 2); Mn +7 (Rule 6); O –2 (Rule 3)

    d.   C 0 (Rule 6); H +1 (Rule 4); O –2 (Rule 3)

94. The rules for assigning oxidation states are given in Section 18.2 of the text. The rule that applies for each element is that given in parentheses after the element and its oxidation state.

    a.   Mn +4 (Rule 6); O –2 (Rule 3)

    b.   Ba +2 (Rule 2); Cr +6 (Rule 6); O –2 (Rule 3)

    c.   H +1 (Rule 4); S +4 (Rule 6); O –2 (Rule 3)

    d.   Ca +2 (Rule 2); P +5 (Rule 6); O –2 (Rule 3)

95. The rules for assigning oxidation states are given in Section 18.2 of the text. The rule that applies for each element is that given in parentheses after the element and its oxidation state.

    a.   Cr +3 (Rule 6); Cl –1 (Rule 2)

    b.   K +1 (Rule 2); Cr +6 (Rule 6); O –2 (Rule 3)

    c.   K +1 (Rule 2); Cr +6 (Rule 6); O –2 (Rule 3)

    d.   Cr +2 (Rule 6); C 0 (Rule 7); H +1 (Rule 4); O –2 (Rule 3)

          For chromous acetate, first the oxidation state of carbon in the acetate ion, $C_2H_3O_2^-$, is determined by Rule 7 (the sum of the oxidation numbers must equal the charge on the ion), then the oxidation state of Cr may be determined by Rule 6).

96. The rules for assigning oxidation states are given in Section 18.2 of the text. The rule that applies for each element is that given in parentheses after the element and its oxidation state.

    a.   Bi +3 (Rule 7); O –2 (Rule 3)

    b.   P +5 (Rule 7); O –2 (Rule 3)

Chapter 18: Oxidation–Reduction Reactions: Electrochemistry

    c.   N +3 (Rule 7); O –2 (Rule 3)

    d.   Hg +1 (Rule 7)

97.   a. $C(s) + O_2(g) \rightarrow CO_2(g)$

        carbon is oxidized (0 to +4); oxygen is reduced (0 to –2)

    b. $2CO(g) + O_2(g) \rightarrow 2CO_2(g)$

        carbon (of CO) is oxidized (+2 to +4); oxygen (of O_2) is reduced (0 to –2)

    c. $CH_4(g) + 2O_2(g) \rightarrow CO_2(g) + 2H_2O(g)$

        carbon (of CH_4) is oxidized (–4 to +4); oxygen (of O_2) is reduced (0 to –2)

    d. $C_2H_2(g) + 2H_2(g) \rightarrow C_2H_6(g)$

        hydrogen (of H_2) is oxidized (0 to +1); carbon (of C_2H_2) is reduced (–1 to –3)

98.   a. $2B_2O_3(s) + 6Cl_2(g) \rightarrow 4BCl_3(l) + 3O_2(g)$

        oxygen is oxidized (–2 to 0); chlorine is reduced (0 to –1)

    b. $GeH_4(g) + O_2(g) \rightarrow Ge(s) + 2H_2O(g)$

        germanium is oxidized (–4 to 0); oxygen is reduced (0 to –2)

    c. $C_2H_4(g) + Cl_2(g) \rightarrow C_2H_4Cl_2(l)$

        carbon is oxidized –2 to –1); chlorine is reduced (0 to –1)

    d. $O_2(g) + 2F_2(g) \rightarrow 2OF_2(g)$

        oxygen is oxidized (0 to +2); fluorine is reduced (0 to –1)

99.   a. $I^-(aq) \rightarrow I_2(s)$

        Balance iodine: $\mathbf{2}I^-(aq) \rightarrow I_2(s)$

        Balance charge: $2I^-(aq) \rightarrow I_2(s) + \mathbf{2e^-}$

        Balanced half–reaction: $2I^-(aq) \rightarrow I_2(s) + 2e^-$

    b. $O_2(g) \rightarrow O^{2-}(s)$

        Balance oxygen: $O_2(g) \rightarrow \mathbf{2}O^{2-}(s)$

        Balance charge: $O_2(g) + \mathbf{4e^-} \rightarrow 2O^{2-}(s)$

        Balanced half–reaction: $O_2(g) + 4e^- \rightarrow 2O^{2-}(s)$

    c. $P_4(s) \rightarrow P^{3-}(s)$

        Balance phosphorus: $P_4(s) \rightarrow \mathbf{4}P^{3-}(s)$

        Balance charge: $P_4(s) + \mathbf{12e^-} \rightarrow 4P^{3-}(s)$

        Balanced half–reaction: $P_4(s) + 12e^- \rightarrow 4P^{3-}(s)$

Chapter 18: Oxidation–Reduction Reactions: Electrochemistry

      d. $Cl_2(g) \rightarrow Cl^-(aq)$

          Balance chlorine: $Cl_2(g) \rightarrow \mathbf{2Cl^-}(aq)$

          Balance charge: $Cl_2(g) + \mathbf{2e^-} \rightarrow 2Cl^-(aq)$

          Balanced half–reaction: $Cl_2(g) + 2e^- \rightarrow 2Cl^-(aq)$

100.   a. $SiO_2(s) \rightarrow Si(s)$

          Balance oxygen: $SiO_2(s) \rightarrow Si(s) + \mathbf{2H_2O}(l)$

          Balance hydrogen: $SiO_2(s) + \mathbf{4H^+}(aq) \rightarrow Si(s) + 2H_2O(l)$

          Balance charge: $SiO_2(s) + 4H^+(aq) + \mathbf{4e^-} \rightarrow Si(s) + 2H_2O(l)$

          Balanced half–reaction: $SiO_2(s) + 4H^+(aq) + 4e^- \rightarrow Si(s) + 2H_2O(l)$

      b. $S(s) \rightarrow H_2S(g)$

          Balance hydrogen: $S(s) + \mathbf{2H^+}(aq) \rightarrow H_2S(g)$

          Balance charge: $S(s) + 2H^+(aq) + \mathbf{2e^-} \rightarrow H_2S(g)$

          Balanced half–reaction: $S(s) + 2H^+(aq) + 2e^- \rightarrow H_2S(g)$

      c. $NO_3^-(aq) \rightarrow HNO_2(aq)$

          Balance oxygen: $NO_3^-(aq) \rightarrow HNO_2(aq) + \mathbf{H_2O}(l)$

          Balance hydrogen: $NO_3^-(aq) + \mathbf{3H^+}(aq) \rightarrow HNO_2(aq) + H_2O(l)$

          Balance charge: $NO_3^-(aq) + 3H^+(aq) + \mathbf{2e^-} \rightarrow HNO_2(aq) + H_2O(l)$

          Balanced half–reaction: $NO_3^-(aq) + 3H^+(aq) + 2e^- \rightarrow HNO_2(aq) + H_2O(l)$

      d. $NO_3^-(aq) \rightarrow NO(g)$

          Balance oxygen: $NO_3^-(aq) \rightarrow NO(g) + \mathbf{2H_2O}(l)$

          Balance hydrogen: $NO_3^-(aq) + \mathbf{4H^+}(aq) \rightarrow NO(g) + 2H_2O(l)$

          Balance charge: $NO_3^-(aq) + 4H^+(aq) + \mathbf{3e^-} \rightarrow NO(g) + 2H_2O(l)$

          Balanced half–reaction: $NO_3^-(aq) + 4H^+(aq) + 3e^- \rightarrow NO(g) + 2H_2O(l)$

101.   For simplicity, the physical states of the substances have been omitted until the final balanced equation is given.

      a. $I^-(aq) + MnO_4^-(aq) \rightarrow I_2(aq) + Mn^{2+}(aq)$

          $I^- \rightarrow I_2$

          Balance iodine: $2I^- \rightarrow I_2$

          Balance charge: $2I^- \rightarrow I_2 + \mathbf{2e^-}$

          $MnO_4^- \rightarrow Mn^{2+}$

          Balance oxygen: $MnO_4^- \rightarrow Mn^{2+} + \mathbf{4H_2O}$

          Balance hydrogen: $\mathbf{8H^+} + MnO_4^- \rightarrow Mn^{2+} + 4H_2O$

          Balance charge: $8H^+ + MnO_4^- + \mathbf{5e^-} \rightarrow Mn^{2+} + 4H_2O$

Combine the half–reactions:

$5 \times (2I^- \rightarrow I_2 + 2e^-)$

$2 \times (8H^+ + MnO_4^- + 5e^- \rightarrow Mn^{2+} + 4H_2O)$

$16H^+(aq) + 2MnO_4^-(aq) + 10I^-(aq) \rightarrow 2Mn^{2+}(aq) + 8H_2O(l) + 5I_2(aq)$

b. $S_2O_8^{2-} + Cr^{3+} \rightarrow SO_4^{2-} + Cr_2O_7^{2-}$

$S_2O_8^{2-} \rightarrow SO_4^{2-}$

Balance sulfur: $S_2O_8^{2-} \rightarrow \mathbf{2}SO_4^{2-}$

Balance charge: $S_2O_8^{2-} + \mathbf{2e^-} \rightarrow 2SO_4^{2-}$

$Cr^{3+} \rightarrow Cr_2O_7^{2-}$

Balance chromium: $\mathbf{2}Cr^{3+} \rightarrow Cr_2O_7^{2-}$

Balance oxygen: $\mathbf{7H_2O} + 2Cr^{3+} \rightarrow Cr_2O_7^{2-}$

Balance hydrogen: $7H_2O + 2Cr^{3+} \rightarrow Cr_2O_7^{2-} + \mathbf{14H^+}$

Balance charge: $7H_2O + 2Cr^{3+} \rightarrow Cr_2O_7^{2-} + 14H^+ + \mathbf{6e^-}$

Combine the half–reactions:

$3 \times (S_2O_8^{2-} + 2e^- \rightarrow 2SO_4^{2-})$

$7H_2O + 2Cr^{3+} \rightarrow Cr_2O_7^{2-} + 14H^+ + 6e^-$

$7H_2O(l) + 2Cr^{3+}(aq) + 3S_2O_8^{2-}(aq) \rightarrow Cr_2O_7^{2-}(aq) + 14H^+(aq) + 6SO_4^{2-}(aq)$

c. $BiO_3^- + Mn^{2+} \rightarrow Bi^{3+} + MnO_4^-$

$BiO_3^- \rightarrow Bi^{3+}$

Balance oxygen: $BiO_3^- \rightarrow Bi^{3+} + \mathbf{3H_2O}$

Balance hydrogen: $\mathbf{6H^+} + BiO_3^- \rightarrow Bi^{3+} + 3H_2O$

Balance charge: $6H^+ + BiO_3^- + \mathbf{2e^-} \rightarrow Bi^{3+} + 3H_2O$

$Mn^{2+} \rightarrow MnO_4^-$

Balance oxygen: $\mathbf{4H_2O} + Mn^{2+} \rightarrow MnO_4^-$

Balance hydrogen: $4H_2O + Mn^{2+} \rightarrow MnO_4^- + \mathbf{8H^+}$

Balance charge: $4H_2O + Mn^{2+} \rightarrow MnO_4^- + 8H^+ + \mathbf{5e^-}$

Combine the half–reactions:

$5 \times (6H^+ + BiO_3^- + 2e^- \rightarrow Bi^{3+} + 3H_2O)$

$2 \times (4H_2O + Mn^{2+} \rightarrow MnO_4^- + 8H^+ + 5e^-)$

$2Mn^{2+}(aq) + 14H^+(aq) + 5BiO_3^-(aq) \rightarrow 2MnO_4^-(aq) + 5Bi^{3+}(aq) + 7H_2O(l)$

102. The correct answer is d. $Al^{3+}(aq)$ ion is reduced. $Mg(s)$ is oxidized. Reduction occurs at the cathode and oxidation occurs at the anode. The reaction at the cathode is $2Al^{3+}(aq) + 6e^- \rightarrow 2Al(s)$. The reaction at the anode is $3Mg(s) \rightarrow 3Mg^{2+}(aq) + 6e^-$.

Chapter 18: Oxidation–Reduction Reactions: Electrochemistry

103. A diagram of the cell is shown below:

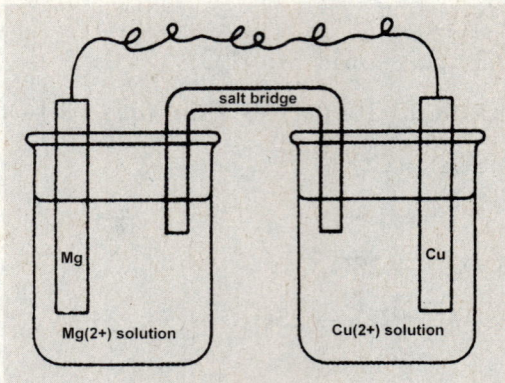

Cu^{2+}(aq) ion is reduced; Mg(s) is oxidized.

The reaction at the anode is Mg(s) → Mg^{2+}(aq) + 2e$^-$.

The reaction at the cathode is Cu^{2+}(aq) + 2e$^-$ → Cu(s).

104. Notice that both dyes include "C$_{16}$N$_2$H$_{10}$O$_2$". Since leucoindigo is Na$_2$C$_{16}$N$_2$H$_{10}$O$_2$, the "C$_{16}$N$_2$H$_{10}$O$_2$" portion has a 2– charge while indigo (C$_{16}$N$_2$H$_{10}$O$_2$) is neutral. Since the sum of the oxidation states equals the charge, the oxidation state of one or more of the elements must increase, thus the molecule must be oxidized.

105.

	Oxidation State
S in MgSO$_4$	+6
Pb in PbSO$_4$	+2
O in O$_2$	0
Ag in Ag	0
Cu in CuCl$_2$	+2

CHAPTER 19

Radioactivity and Nuclear Energy

1. The chemical properties of an atom are due almost exclusively to its electron structure and the nucleus has little or no effect.

2. The radius of a typical atomic nucleus is on the order of 10^{-13} cm, which is about one hundred thousand times smaller than the radius of an atom overall.

3. The atomic number represents the number of protons in an atom's nucleus.

4. The mass number represents the total number of protons and neutrons in a nucleus.

5. Many elements consist of atoms that are not identical. These atoms contain the same number of protons and electrons, so they are *chemically* the same, but they contain different numbers of neutrons, which modifies some *physical* properties.

6. The atomic number (Z) is written in such formulas as a left subscript, whereas the mass number (A) is written as a left superscript. That is, the general symbol for a nuclide is $^A_Z X$. As an example, consider the isotope of oxygen with 8 protons and 8 neutrons: its symbol would be $^{16}_{8} O$.

7. An alpha particle is essentially a helium atom nucleus, containing 2 protons and 2 neutrons: $^4_2 He$

8. When a nucleus produces an alpha particle, the atomic number of the parent nucleus decreases by two units.

9. A beta particle is essentially an electron, $^{\ 0}_{-1} e$. Emission of a beta particle increases the atomic number of the parent nucleus by one unit, but does not change the mass number of the parent nucleus.

10. Emission of a neutron, $^1_0 n$, does not change the atomic number of the parent nucleus, but causes the mass number of the parent nucleus to decrease by one unit.

11. Some radionuclei do not become stable on the emission of only a single particle. Some nuclei may decay through a *series* of particle emissions until they finally reach stability.

12. Gamma rays are high-energy photons of electromagnetic radiation. They are not normally considered to be particles. When a nucleus produces only gamma radiation, the atomic number and mass number remain unchanged.

13. A positron is a particle with the same mass as an electron, but with the opposite charge: a positron is *positively* charged. Its mass number is therefore zero, and its "atomic number" is +1. When an unstable nucleus produces a positron, the mass number of the original nucleus is unchanged, but the atomic number of the original nucleus decreases by one unit.

Chapter 19: Radioactivity and Nuclear Energy

14. Electron capture occurs when one of the inner orbital electrons is pulled into and becomes part of the nucleus.

15. The fact that the average atomic mass of sulfur is only slightly above 32 amu reflects the fact that the isotope of mass number 32 predominates.

Isotope	Number of neutrons
$^{32}_{16}S$	16 neutrons
$^{33}_{16}S$	17 neutrons
$^{34}_{16}S$	18 neutrons
$^{36}_{16}S$	20 neutrons

16. The fact that the average atomic mass of potassium is only slightly above 39 amu reflects the fact that the isotope of mass number 39 predominates.

Isotope	Number of neutrons
$^{39}_{19}K$	20 neutrons
$^{40}_{19}K$	21 neutrons
$^{41}_{19}K$	22 neutrons

17. $^{24}_{12}Mg$ (12 protons, 12 neutrons)

    $^{25}_{12}Mg$ (12 protons, 13 neutrons)

    $^{26}_{12}Mg$ (12 protons, 14 neutrons)

18. The approximate atomic molar mass could be calculated as follows:

    $0.79(24) + 0.10(25) + 0.11(26) = 24.3$.

    This is *only* an approximation because the mass numbers, rather than the actual isotopic masses, were used. The fact that the approximate mass calculated is slightly above 24 shows that the isotope of mass number 24 predominates.

19. a. $^{0}_{-1}e$ or $^{0}_{-1}\beta$

    b. $^{4}_{2}He$ or $^{4}_{2}\alpha$

    c. $^{1}_{0}n$

    d. $^{1}_{1}H$

20. The correct answer is *a*. In beta-particle production, $^{0}_{-1}e$ is produced (e.g. $^{234}_{90}Th \rightarrow {^{0}_{-1}e} + {^{234}_{91}Pa}$). The atomic number of the parent nuclide goes up, thus decreasing the neutron to proton ratio.

Chapter 19:   Radioactivity and Nuclear Energy

21. a. $^{4}_{2}He$

    b. $^{6}_{3}Li$

    c. $^{0}_{-1}e$

22. a. $^{192}_{83}Bi$

    b. $^{204}_{82}Pb$

    c. $^{206}_{84}Po$

23. a. $^{206}_{87}Fr$

    b. $^{0}_{-1}e$

    c. $^{1}_{0}n$

24. a. $^{0}_{-1}e$

    b. $^{0}_{-1}e$

    c. $^{210}_{83}Bi$

25. a. $^{14}_{6}C \rightarrow {}^{0}_{-1}e + {}^{14}_{7}N$

    b. $^{140}_{55}Cs \rightarrow {}^{0}_{-1}e + {}^{140}_{56}Ba$

    c. $^{234}_{90}Th \rightarrow {}^{0}_{-1}e + {}^{234}_{91}Pa$

26. a. $^{234}_{92}U \rightarrow {}^{4}_{2}He + {}^{230}_{90}Th$

    b. $^{222}_{86}Rn \rightarrow {}^{4}_{2}He + {}^{218}_{84}Po$

    c. $^{162}_{75}Re \rightarrow {}^{4}_{2}He + {}^{158}_{73}Ta$

27. a. $^{188}_{74}W \rightarrow {}^{0}_{-1}e + {}^{188}_{75}Re$

    b. $^{40}_{19}K \rightarrow {}^{0}_{-1}e + {}^{40}_{20}Ca$

    c. $^{198}_{79}Au \rightarrow {}^{0}_{-1}e + {}^{198}_{80}Hg$

28. a. $^{136}_{53}I \rightarrow {}^{0}_{-1}e + {}^{136}_{54}Xe$

    b. $^{133}_{51}Sb \rightarrow {}^{0}_{-1}e + {}^{133}_{52}Te$

    c. $^{117}_{49}In \rightarrow {}^{0}_{-1}e + {}^{117}_{50}Sn$

Chapter 19: Radioactivity and Nuclear Energy

29. A nuclear transformation represents the change of one element into another. Nuclear transformations are generally accomplished by bombardment of a target nucleus with some small, energetic particle that causes the desired transformation of the target nucleus.

30. In a nuclear bombardment process, a target nucleus is bombarded with high-energy particles (typically subatomic particles or small atoms) from a particle accelerator. This may result in the transmutation of the target nucleus into some other element. For example, nitrogen-14 may be transmuted into oxygen-17 by bombardment with alpha particles. There is often considerable repulsion between the target nucleus and the particles being used for bombardment (especially if the bombarding particle is positively charged like the target nucleus). Using accelerators to increases the kinetic energy of the bombarding particles can overcome this repulsion.

31. $^{9}_{4}Be + ^{4}_{2}He \rightarrow ^{12}_{6}C + ^{1}_{0}n$

32. $^{24}_{12}Mg + ^{2}_{1}H \rightarrow ^{22}_{11}Na + ^{4}_{2}He$

33. Geiger (Geiger-Müller) counters contain a probe that contains argon gas. The argon atoms themselves have no charge, but they can be ionized by high-energy particles from a radioactive decay process. Although a sample of normal uncharged argon gas does not conduct an electrical current, argon gas that has been ionized will briefly conduct an electrical current (until the argon ions and electrons recombine). If an electric field is applied to the argon gas probe, then a brief pulse of electricity will be passed through the argon every time an ionization event occurs (i.e., every time a high-energy particle strikes the argon gas probe). The Geiger counter detects each pulse of current, and these pulses are then counted and displayed on the meter of the device. A scintillation counter uses a substance like sodium iodide, which emits light when struck by a high-energy particle from a radioactive decay. A detector senses the flashes of light from the sodium iodide, and these flashes are then counted and displayed on the meter of the device.

34. The half-life of a nucleus is the time required for one-half of the original sample of nuclei to decay. A given isotope of an element always has the same half-life, although different isotopes of the same element may have greatly different half-lives. Nuclei of different elements typically have different half-lives.

35. When we say that one nucleus is "hotter" than another, we mean that the "hot" nucleus undergoes more decay events per time period. The "hotness" of radionuclei is most commonly indicated by their *half-lives* (the amount of time required for half a sample to undergo the decay process). A nucleus with a short half-life will undergo more decay events in a given time than a nucleus with a long half-life.

36. $^{226}_{88}Ra$ is the most stable (longest half-life); $^{224}_{88}Ra$ is the "hottest" (shortest half-life)

37. highest $^{24}Na > ^{131}I > ^{60}Co > ^{3}H > ^{14}C$ lowest

38. With a half-life of 2.8 hours, strontium-87 is the hottest; with a half-life of 45.1 days, iron-59 is the most stable to decay.

39. Half-life, 10 minutes; starting amount, 100 μg

time, min	0	10	20
mass, μg	100	50	25

    After 2 half-life periods, 25 μg of the N-13 isotope remains.

Chapter 19:   Radioactivity and Nuclear Energy

40. Half-life, 1.5 min; let *x* represent the starting amount of isotope

time, min	0	1.5	3.0	4.5	6.0
mass	x	$\frac{1}{2}x$	$\frac{1}{4}x$	$\frac{1}{8}x$	$\frac{1}{16}x$

    After six minutes (four half-lives), $\frac{1}{16}$ of the original Co-62 sample [$(\frac{1}{2})^4$] will remain.

41. Kr-81 is the most stable (longest half life); Kr-73 is the "hottest" (shortest half life). Since the half-lives of Kr-73 and Kr-74 are so short, after 24 hours there would essentially be no detectable amount of these isotopes remaining. Since 24 hours is very approximately two half life periods for Kr-76, approximately one-fourth of the original sample would remain. As the half life of Kr-81 is much larger than the 24-hour time period under consideration, essentially all of the sample would remain.

42. For an administered dose of 100 µg, 0.39 µg remains after 2 days. The fraction remaining is 0.39/100 = 0.0039; on a percentage basis, less than 0.4% of the original radioisotope remains.

43. We assume that a organism exchanges carbon-14 with the atmosphere while alive. When the organism dies, no further exchange of carbon-14 takes place between the organism and the atmosphere. The level of carbon 14 in an artifact containing the remains of such organisms is then measured and is compared with the amount of carbon-14 in the atmosphere. Since the half-life of carbon-14 is known, the age of the artifact can be calculated.

44. Carbon-14 is produced in the upper atmosphere by the bombardment of ordinary nitrogen with neutrons from space:

    $$^{14}_{7}\text{N} + ^{1}_{0}\text{n} \rightarrow ^{14}_{6}\text{C} + ^{1}_{1}\text{H}$$

45. It is assumed that the concentration of Carbon-14 in the atmosphere remains constant. When analyzing an artifact, we assume living cells in the artifact stopped exchanging Carbon-14 with the atmosphere when they died, and we compare the amount of Carbon-14 in the cells to the amount of Carbon-14 in the atmosphere.

46. We assume that the concentration of C-14 in the atmosphere is effectively constant. A living organism is constantly replenishing C-14 either through the processes of metabolism (sugars ingested in foods contain C-14), or photosynthesis (carbon dioxide contains C-14). When a plant dies, it can no longer replenish, and as the C-14 undergoes radioactive decay, its amount decreases with time.

47. Because iodine is used almost exclusively in the body by the thyroid gland, a patient can be given a dose of a short half-life radioisotope of iodine (in sodium iodide). The speed of the uptake of the radioiodine by the thyroid gland is measured by tracking the increase in the level of radioactivity at the thyroid gland. An effective picture of the thyroid gland can also be taken by mapping the level of radioactivity using a detector (thyroid scanner) that will highlight a tumor in the thyroid if one is present. If a malignant tumor in the thyroid gland is detected, then a larger dosage of a longer half-life isotope of iodine can be administered to destroy the tumor cells. As iodine is *only* used by the thyroid gland, the radioisotope used to destroy the tumor cells is not likely to cause damage to other cells in the body.

48. 1 day is about 13 half-lives for $^{18}_{9}\text{F}$. If we begin with 6.02×10^{23} atoms (1 mol), then after 13 half-lives, 7.4×10^{19} atoms of $^{18}_{9}\text{F}$ will remain.

Chapter 19: Radioactivity and Nuclear Energy

49. The forces that hold protons and neutrons together in the nucleus are *much greater* than the forces that bind atoms together in molecules.

50. fission, fusion, fusion, fission

51. The energies released by nuclear processes are on the order of 106 times more powerful than those associated with ordinary chemical reactions.

52. $^{1}_{0}n + ^{235}_{92}U \rightarrow ^{142}_{56}Ba + ^{91}_{36}Kr + 3^{1}_{0}n$ is one possibility.

53. A chain reaction is a nuclear decay which, when initiated by an external particle, generates sufficient additional particles of the same type to continue the process on its own. In the fission of Uranium-235, the chain reaction is initiated by bombarding Uranium-235 with neutrons to start the reaction. However, the decay of the first bit of Uranium-235 then produces additional neutrons which continue the decay of the remaining Uranium-235.

54. A critical mass of a fissionable material is the amount needed to provide a high enough internal neutron flux to sustain the chain reaction (enough neutrons are produced to cause the continuous fission of further material). A sample with less than a critical mass is still radioactive, but cannot sustain a chain reaction.

55. The *moderator* in a uranium fission reactor surrounds the fuel rods and slows down the neutrons produced by the uranium decay process so that they can be absorbed more easily by other uranium atoms. The *control rods* are constructed of substances that absorb neutrons, and can be inserted into the reactor core to control the power level of the reactor. The *containment* of a reactor refers to the building in which the reactor core is located. The building is designed to contain the radioactive core in the event of a nuclear accident. A *cooling liquid* (usually water) is circulated through the reactor to draw off the heat energy produced by the nuclear reaction, so that this heat energy can be converted to electrical energy in the power plant's turbines.

56. An actual nuclear explosion, of the type produced by a nuclear weapon, cannot occur in a nuclear reactor because the concentration of the fissionable materials is not sufficient to form a supercritical mass. However, since many reactors are cooled by water, which can decompose into hydrogen and oxygen gases, a *chemical* explosion is possible that could scatter the radioactive material used in the reactor.

57. If the system used to cool a reactor core fails, the reactor may reach temperatures high enough to melt the core itself. In a scenario referred to as the "China Syndrome," the molten reactor core could become so hot as to melt through the bottom of the reactor building and into the earth itself (eventually the molten material would reach cool ground water and resolidify, with possible release of radioactivity). If water is used to cool the reactor core, and the cooling system becomes blocked, it is possible for the heat from the reactor to cause a steam explosion (which would also release radioactivity), or to break down the coolant water into hydrogen gas (which could also explode).

58. breeder

59. Nuclear *fusion* is the process of combining two light nuclei into a larger nucleus, with an energy release larger than that provided by fission processes.

60. In one type of fusion reactor, two ^2_1H atoms are fused to produce ^4_2He. Because the hydrogen nuclei are positively charged, extremely high energies (temperatures of 40 million K) are needed to overcome the repulsion between the nuclei as they are shot into each other.

61. Fusion produces an enormous amount of energy per gram of fused material. Hydrogen is the fuel of choice for fusion reactors, because of its plentiful supply (in water) and the fact that the product nuclei from the fusion of hydrogen (helium isotopes) are far less dangerous than those produced by fission processes.

62. protons (hydrogen); helium

63. Although the energy transferred per event to a living creature is small, the quantity of energy is enough to break chemical bonds that may cause malfunctioning of cellular systems. In particular, many biochemical processes are chain-like in nature, and the production of a single odd ion in a cell by a radioactive event may have a cumulative effect. For example, ionization of a single bond in a sex cell may cause a drastic mutation in the creature resulting.

64. Somatic damage is directly to the organism itself, causing nearly immediate sickness or death to the organism. Genetic damage is to the genetic machinery of the organism, which will be manifested in future generations of offspring.

65. Alpha particles are stopped by the outermost layers of skin; beta particles penetrate only about 1 cm into the body; gamma rays are deeply penetrating.

66. Gamma rays penetrate long distances, but seldom cause ionization of biological molecules. Alpha particles, because they are much heavier although less penetrating, are very effective at ionizing biological molecules and leave a dense trail of damage in the organism. Isotopes that release alpha particles can be ingested or breathed into the body where the damage from the alpha particles will be more acute.

67. Nuclei of atoms that are chemically inert, or which are not ordinarily found in the body, tend to be excreted from the body quickly and do little damage. Other nuclei of atoms which form a part of the body's structure or normal metabolic processes are likely to be incorporated into the body. When a radioactive nuclide is ingested into the body, its capacity to cause damage also depends on how long it remains in the body. If the nuclide has been incorporated into the body, the danger is greatest.

68. Nuclear waste may remain radioactive for thousands of years, and much of it is chemically poisonous as well as radioactive. Most reactor waste is still in "temporary storage." Various suggestions have been made for a more permanent solution, such as casting the spent fuel into glass bricks to contain it, and then storing the bricks in corrosion-proof metal containers deep underground. No agreement on a permanent solution to the disposal of nuclear waste has yet been reached.

69. atomic number

70. radioactive

71. electron

72. mass

Chapter 19: Radioactivity and Nuclear Energy

73. alpha

74. neutron; proton

75. gamma (γ)

76. radioactive decay

77. higher

78. mass number

79. particle accelerators

80. transuranium

81. Geiger

82. half-life

83. $^{14}_{6}C$ (carbon-14)

84. radiotracers

85. fusion

86. chain

87. $^{235}_{92}U$

88. Every 35 years, the sample has a mass half of what it had. Over 140 years, the sample will be cut in half 4 times, so there must have been 48 g initially to have 3.0 g after 140 years.

89. The decay series, in order from the top right of the diagram, is: alpha, beta, beta, alpha, alpha, alpha, alpha, alpha, beta, beta, alpha, beta, beta, alpha. This decay is indicated in color in the figure.

90. a. $^{234}_{90}Th$; alpha-particle production

    b. $^{0}_{-1}e$; beta-particle production

91. a. cobalt is a component of Vitamin B-12

    b. bones consist partly of $Ca_3(PO_4)_2$

    c. red blood cells contain hemoglobin, an iron-protein compound

    d. mercury is absorbed by substances in the brain (this is part of the reason mercury is so hazardous in the environment)

92. 3.5×10^{-11} J/atom; 8.9×10^{10} J/g

Chapter 19: Radioactivity and Nuclear Energy

93. In order to sustain a nuclear chain reaction, the neutrons produced by the fission must be contained within the fissionable material, so that they can go on to cause other fissions. In order that the neutrons are contained, the fissionable material must be closely enough packed together that it is more likely for a neutron to encounter a fissionable nucleus than to be lost to the outside.

94. $^{90}_{40}Zr$, $^{91}_{40}Zr$, $^{92}_{40}Zr$, $^{94}_{40}Zr$, and $^{96}_{40}Zr$

95. $^{64}_{30}Zn$ (30 protons, 34 neutrons)

   $^{66}_{30}Zn$ (30 protons, 36 neutrons)

   $^{67}_{30}Zn$ (30 protons, 37 neutrons)

   $^{68}_{30}Zn$ (30 protons, 38 neutrons)

   $^{70}_{30}Zn$ (30 protons, 40 neutrons)

96. $^{27}_{13}Al$ (13 protons, 14 neutrons)

   $^{28}_{13}Al$ (13 protons, 15 neutrons)

   $^{29}_{13}Al$ (13 protons, 16 neutrons)

97. a. $^{4}_{2}He$

   b. $^{4}_{2}He$

   c. $^{1}_{0}n$

98. *Three* of the statements are true. Statements *a*, *b*, and *d* are true.

99. $^{14}_{7}N + ^{4}_{2}He \rightarrow ^{17}_{8}O + ^{1}_{1}H$

100. $^{9}_{4}Be + ^{4}_{2}He \rightarrow ^{12}_{6}C + ^{1}_{0}n$

101. Iodine-131 (^{131}I) is used in the diagnosis and treatment of thyroid cancer and other dysfunctions of the thyroid gland. The thyroid gland is the only place in the human body that uses and stores iodine. I-131 that is administered concentrates in the thyroid, and can be used to cause an image on a scanner or x-ray film, or in higher doses, to selectively kill cancer cells in the thyroid. ^{201}Tl concentrates in healthy muscle cells when administered, and can be used to detect and assess damage to heart muscles after a heart attack: the damaged muscles show a lower uptake of Tl-201 than normal muscles.

102. $^{238}_{92}U + ^{1}_{0}n \rightarrow ^{239}_{92}U$

   $^{239}_{92}U \rightarrow ^{239}_{93}Np + ^{0}_{-1}e$

   $^{239}_{93}Np \rightarrow ^{239}_{94}Pu + ^{0}_{-1}e$

Chapter 19: Radioactivity and Nuclear Energy

103.
   a. $^{232}_{90}\text{Th} \rightarrow {}^{4}_{2}\text{He} + {}^{228}_{88}\text{Ra}$

   b. $^{220}_{86}\text{Rn} \rightarrow {}^{4}_{2}\text{He} + {}^{216}_{84}\text{Po}$

   c. $^{216}_{84}\text{Po} \rightarrow {}^{4}_{2}\text{He} + {}^{212}_{82}\text{Pb}$

104. $^{4}_{2}\text{He}$; $^{0}_{-1}\text{e}$; $^{0}_{-1}\text{e}$; $^{0}_{-1}\text{e}$; $^{0}_{-1}\text{e}$

105. *b*, *d*, and *e* are correct.

106. Half-life, 80.9 years; let *x* represent the starting amount of isotope

time, years	0	80.9	162 (2×80.9)	243 (3×80.9)
mass decayed	*x*	(50%)*x*	(75%)*x*	(87.5%)*x*

After 243 years (three half-lives), 87.5% of the substance has decayed.

107. *a* and *b* are true. 2.5×10^9 is two half-lives away from 1.00×10^{10}.

CHAPTER 20

Organic Chemistry

1. Carbon has the unusual ability of bonding strongly to itself, forming long chains or rings of carbon atoms. As there are many different possible arrangements for a long chain of carbon atoms, there exists a great multitude of possible carbon compounds.

2. Carbon has only four valence electrons and can only make 4 bonds to other atoms.

3. A double bond represents the sharing of two pairs of electrons between two bonded atoms. The simplest example of an organic molecule with a double bond is ethene (ethylene), $CH_2=CH_2$.

4. A triple bond represents the sharing of three pairs of electrons between two bonded atoms. The sharing of three pairs imparts a linear geometry in the region of the triple bond. The simplest example of an organic molecule containing a triple bond is acetylene, H–C≡C–H.

5. When a carbon atom is bonded to four other atoms, the electron pairs of the carbon atom will be arranged with the tetrahedral configuration. This represents the electron pairs being as far away from each other as possible, separated by the tetrahedral angle of 109.5°.

6. $\ddot{O}=C=\ddot{O}$ C≡O

7. Molecules b and c contain multiple bonds and are therefore unsaturated.

8. Molecules a and c contain only carbon–carbon single bonds and are therefore saturated.

9. In butane or propane, the four electron pairs around the carbon atoms have a basically tetrahedral orientation, with 109.5° bond angles. This implies that the molecules cannot be linear.

10. 109.5°

11. The general formula for the alkanes is C_nH_{2n+2}. For an alkane with 20 carbon atoms, the number of hydrogen atoms would be 2(20) + 2 = 42, to give the alkane formula $C_{20}H_{42}$.

12. The general formula for the alkanes is C_nH_{2n+2}.
    a. 2(3) + 2 = 8
    b. 2(5) + 2 = 12
    c. 2(15) + 2 = 32
    d. 2(18) + 2 = 38

Chapter 20: Organic Chemistry

13. a. octane (eight carbons in a continuous chain)
    b. decane (ten carbons in a continuous chain)
    c. hexane (six carbons in a continuous chain)
    d. butane (four carbons in a continuous chain)

14. a. pentane $CH_3-CH_2-CH_2-CH_2-CH_3$
    b. undecane $CH_3-CH_2-CH_2-CH_2-CH_2-CH_2-CH_2-CH_2-CH_2-CH_2-CH_3$
    c. nonane $CH_3-CH_2-CH_2-CH_2-CH_2-CH_2-CH_2-CH_2-CH_3$
    d. heptane $CH_3-CH_2-CH_2-CH_2-CH_2-CH_2-CH_3$

15. Structural isomerism occurs when two molecules have the same atoms present, but those atoms are bonded differently. The molecules have the same formulas but different arrangements of the atoms. The alkane butane is the first alkane to have an isomer:

    Butane   2-methylpropane

16. A branched alkane contains one or more shorter carbon atom chains, attached to the side of the main (longest) carbon atom chain. The simplest branched alkane is 2-methylpropane.

17. The carbon skeletons for pentane, 2-methyl butane, and 2,2-dimethyl propane are shown:

    pentane   2-methylbutane   2,2-dimethylpropane

18. Carbon skeletons are shown.

Chapter 20: Organic Chemistry

19. propane (3), pentane (5), heptane (7), nonane (9)

20. The root name is derived from the number of carbon atoms in the *longest continuous chain* of carbon atoms.

21. An alkyl group represents a hydrocarbon *branch* occurring along the principal carbon atom chain of an organic molecule. An alkyl group has one fewer hydrogen atom than the corresponding alkane with the same carbon atom skeleton.

22. The position of a substituent is indicated by a number that corresponds to the carbon atom in the longest chain to which the substituent is attached.

23.
   a. *di-*
   b. *tetra-*
   c. *penta-*
   d. *tri-*

24. Multiple substituents are listed in alphabetical order, disregarding any prefix.

25.
   a. 4,5-dimethylnonane (look for the *longest* carbon chain)
   b. 3,3,4,4-tetramethylheptane
   c. 3,4,5,5-tetramethyloctane (look for the *longest* carbon chain)
   d. 2,3,4-trimethylhexane

26. Look for the *longest* continuous chain of carbon atoms.
   a. 3-ethylpentane
   b. 2,2-dimethylbutane
   c. 2,2-dimethylpropane
   d. 2,3,4-trimethylpentane

27.
   a. $CH_3-CH(CH_3)-CH(CH_3)-CH_2-CH_3$
   b. $CH_3-CH(CH_3)-CH_2-CH(CH_3)-CH_3$

Chapter 20: Organic Chemistry

c.
$$CH_3-\underset{\underset{CH_3}{|}}{\overset{\overset{CH_3}{|}}{C}}-CH_2-CH_2-CH_3$$

d.
$$CH_3-CH_2-\underset{\underset{CH_3}{|}}{\overset{\overset{CH_3}{|}}{C}}-CH_2-CH_3$$

28. a.
$$CH_3-\underset{\underset{CH_3}{|}}{\overset{\overset{CH_3}{|}}{C}}-CH_2-\underset{\underset{CH_3}{|}}{CH}-CH_2-CH_2-CH_2-CH_3$$

b.
$$CH_3-\underset{\underset{CH_3}{|}}{CH}-\overset{\overset{CH_3}{|}}{CH}-\underset{\underset{CH_3}{|}}{CH}-CH_2-CH_2-CH_2-CH_3$$

c.
$$CH_3-CH_2-\underset{\underset{CH_3}{|}}{\overset{\overset{CH_3}{|}}{C}}-\underset{\underset{CH_3}{|}}{CH}-CH_2-CH_2-CH_2-CH_3$$

d.
$$CH_3-\underset{\underset{CH_3}{|}}{CH}-CH_2-\underset{\underset{CH_3}{|}}{\overset{\overset{CH_3}{|}}{C}}-CH_2-CH_2-CH_2-CH_3$$

29. Petroleum is a thick, dark liquid composed largely of hydrocarbons containing from 5 to more than 25 carbon atoms. Natural gas consists mostly of methane, but also may contain significant amounts of ethane, propane, and butane. These substances were formed over the eons from the decay of living organisms.

30.

C atoms	Use
C_5–C_{12}	gasoline
C_{10}–C_{18}	kerosene, jet fuel
C_{15}–C_{25}	diesel fuel, heating oil, lubrication
C_{25}–	asphalt

31. In the pyrolytic cracking of petroleum, the more abundant kerosene fraction of petroleum is heated to about 700°C, which causes the large molecules of the kerosene fraction to break into the smaller molecules characteristic of the gasoline fraction.

32. Tetraethyl lead was added to gasolines to prevent "knocking" of high efficiency automobile engines. The use of tetraethyl lead is being phased out because of the danger to the environment posed by the lead in this substance.

33. Alkanes are relatively unreactive because the C–C and C–H bonds that characterize these substances are relatively strong and difficult to break.

34. The combustion of alkanes has been used as a source of heat, light, and mechanical energy.

$C_3H_8(g) + 5O_2(g) \rightarrow 3CO_2(g) + 4H_2O(g) + $ heat

35.
a. CH_3Cl
b. HCl
c. $CHCl_3$
d. Cl_2

36. When an alkane molecule is *dehydrogenated*, a double bond is introduced into the molecule, converting it to an alk*ene*. The simplest example is for the dehydrogenation of ethane, to produce ethene (ethylene): CH_3–$CH_3 \xrightarrow{\text{dehydrogenation}} CH_2{=}CH_2 + H_2$

37.
a. $2C_2H_6(g) + 7O_2(g) \rightarrow 4CO_2(g) + 6H_2O(g)$
b. $2C_4H_{10}(g) + 13O_2(g) \rightarrow 8CO_2(g) + 10H_2O(g)$
c. $C_3H_8(g) + 5O_2(g) \rightarrow 3CO_2(g) + 4H_2O(g)$

38.
a. $2C_6H_{14}(l) + 19O_2(g) \rightarrow 12CO_2(g) + 14H_2O(g)$
b. $CH_4(g) + Cl_2(g) \rightarrow CH_3Cl(l) + HCl(g)$
c. $CHCl_3(l) + Cl_2(g) \rightarrow CCl_4(l) + HCl(g)$

39. Alkenes are hydrocarbons that contain a carbon–carbon double bond.

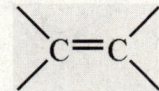

The general formula for alkenes is C_nH_{2n} where n is the number of carbon atoms present.

Chapter 20: Organic Chemistry

40. An alkyne is a hydrocarbon containing a carbon-carbon triple bond. The general formula is C_nH_{2n-2}.

41. To show that a hydrocarbon contains a double bond, the ending of the name of the corresponding alkane is changed to *-ene*. To show that a triple bond is present, the ending *-yne* is used.

42. The location of a double or triple bond in the longest chain of an alkene or alkyne is indicated by giving the *number* of the lowest number carbon atom involved in the double or triple bond.

43.
   a. $CH_2=CH–CH_3 + H_2 \xrightarrow{Pt} CH_3–CH_2–CH_3$

   b. $CH_2=CH–CH_3 + Br_2 \rightarrow CH_2Br–CHBr–CH_3$

   c. $2CH_2=CH–CH_3 + 9O_2 \rightarrow 6CO_2 + 6H_2O$

44. Hydrogenation converts unsaturated compounds to saturated (or less unsaturated) compounds. In the case of a liquid vegetable oil, this is likely to convert the oil to a solid.

    $C_2H_4(g) + H_2(g) \rightarrow C_2H_6(g)$

45.
   a. 2-pentene

   b. 2-pentene

   c. 2,5-dimethyl-3-hexene

   d. 3,4,5-trimethyl-1-heptene

46.
   a. 5,5-dichloro-3,4-dimethyl-1-pentene

   b. 4,5-dichloro-2-hexene (look for the *longest* chain)

   c. 2,2,5-trimethyl-3-heptene

   d. 5-methyl-1-hexyne

47. The most obvious choices would be the *normal* alkenes with seven carbon atoms:

$CH_2=CH–CH_2–CH_2–CH_2–CH_2–CH_3$	1-heptene
$CH_3–CH=CH–CH_2–CH_2–CH_2–CH_3$	2-heptene
$CH_3–CH_2–CH=CH–CH_2–CH_2–CH_3$	3-heptene

   Additional choices are shorter-chain alkenes with branches, such as:

   $$CH_2=C(CH_3)–CH_2–CH_2–CH_2–CH_3 \qquad CH_3–C(CH_3)=CH–CH_2–CH_2–CH_3$$

   2-methyl-1-hexene | 2-methyl-2-hexene

48. Shown are carbon skeletons:

C≡C—C—C—C—C   C—C≡C—C—C—C   C—C—C≡C—C—C

C≡C—C—C—C   C—C≡C—C—C   C≡C—C—C—C   C≡C—C—C
      | | | |
      C   C   C   C
                                                         |
                                                         C

(with the last structure having an extra C above: C≡C—C(C)(C)—C)

49. Aromatic hydrocarbons have in common the presence of the *benzene ring* (phenyl group).

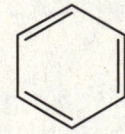

50. For benzene, a *set* of equivalent Lewis structures can be drawn, differing only in the *location* of the three double bonds in the ring. Experimentally, however, benzene does not demonstrate the chemical properties expected for molecules having *any* double bonds. We say that the "extra" electrons that would go into making the second bond of the three double bonds are delocalized around the entire benzene ring; this delocalization of the electrons explains benzene's unique properties.

51. The systematic method for naming monosubstituted benzenes uses the substituent name as a *prefix* for the word benzene. Examples are

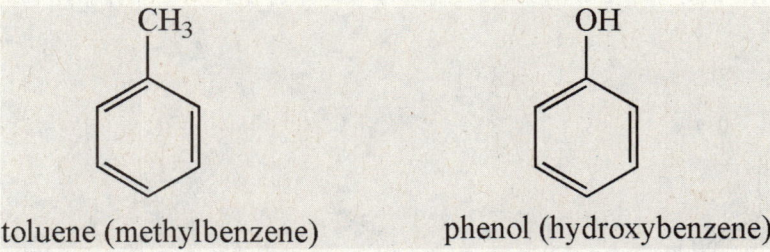

chlorobenzene   ethylbenzene

Two monosubstituted benzenes with their own special names are:

toluene (methylbenzene)   phenol (hydroxybenzene)

52. When named as a substituent, the benzene ring is called the *phenyl* group. Two examples are:

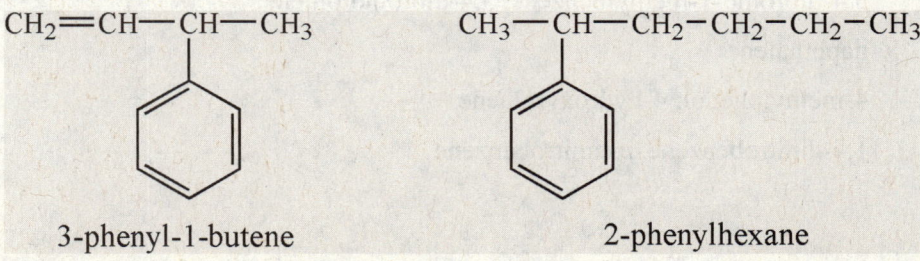

3-phenyl-1-butene   2-phenylhexane

Chapter 20: Organic Chemistry

53. For benzene rings with more than one substituent, *locator numbers* are used to indicate the position of the substituents around the ring. For this purpose, one of the carbon atoms holding a substituent is chosen to be carbon number-1, and the location of the other substituents is indicated relative to carbon number-1 (counting in the direction that leads to the smallest possible locator numbers).

54. *ortho–* refers to adjacent substituents (1,2–); *meta–* refers to two substituents with one unsubstituted carbon atom between them (1,3–); *para–* refers to two substituents with two unsubstituted carbon atoms between them (1,4–).

55. a. [naphthalene structure]

    b. [phenol with Br ortho structure]

    c. [benzene with CH=CH₂ and CH₃ meta structure]

    d. [benzene with Cl and NO₂ para structure]

    e. [benzene with two NO₂ groups meta structure]

56. a.   3,4-dibromo-1-methylbenzene, 3,4-dibromotoluene

    b.   naphthalene

    c.   4-methylphenol; 4-hydroxytoluene

    d.   1,4-dinitrobenzene, *p*-dinitrobenzene

57. Specific examples will depend on students' choice.
    a. R-NH$_2$
    b. R-OH
    c. R-COOH
    d. R-CHO
    e. R-C(=O)-OR
    f. R-C(=O)-R
    g. R-O-R

58. a. carboxylic (organic) acids
    b. aldehydes
    c. ketones
    d. alcohols

59. Alcohols are characterized by the presence of the hydroxyl group, –OH. To name an alcohol, the final -e is dropped from the name of the parent hydrocarbon, and the ending ol is added. A locator number may also be necessary to indicate the location of the hydroxyl group.

60. Primary alcohols have *one* hydrocarbon fragment (alkyl group) bonded to the carbon atom where the –OH group is attached. Secondary alcohols have *two* such alkyl groups attached, and tertiary alcohols contain *three* such alkyl groups. Examples are:

    ethanol (primary)

    $$CH_3-CH_2-OH$$

    2-propanol (secondary)

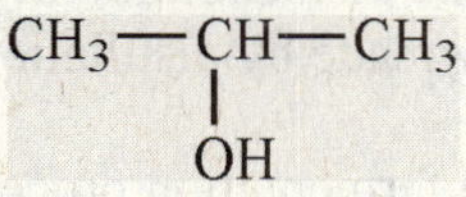

    $$CH_3-\underset{\underset{OH}{|}}{CH}-CH_3$$

    2-methyl-2-propanol (tertiary)

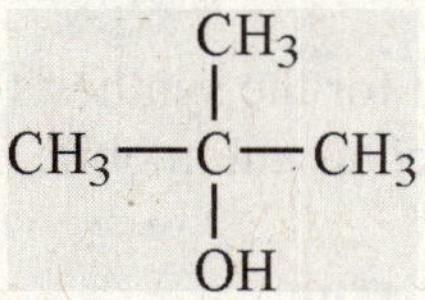

    $$CH_3-\underset{\underset{OH}{|}}{\overset{\overset{CH_3}{|}}{C}}-CH_3$$

61. a. 3-ethyl-3-pentanol; tertiary
    b. 2,4-dimethyl-3-pentanol; secondary
    c. 4,4-dimethyl-2-pentanol; secondary
    d. 2,2,4-trimethyl-1-pentanol; primary

Chapter 20: Organic Chemistry

62. Specific examples will depend on students' choice. Below are examples for alcohols with five carbon atoms.

   primary   (1-pentanol)

   $CH_3-CH_2-CH_2-CH_2-CH_2-OH$

   secondary   (2-pentanol)

   $CH_3-CH_2-CH_2-CH(OH)-CH_3$

   tertiary   (2-methyl-2-butanol)

   $CH_3-C(CH_3)(OH)-CH_2-CH_3$

63. Methanol is sometimes called "wood" alcohol, because it formerly was obtained by the heating of wood in the absence of air (this process was called destructive distillation of wood). Currently methanol is most commonly prepared by the catalyzed hydrogenation of carbon monoxide

   $CO(g) + 2H_2(g) \rightarrow CH_3OH(g)$

   Methanol is an important industrial chemical produced in large amounts every year. It is used as a starting material for the synthesis of acetic acid (CH_3COOH) and of many other important substances. Methanol is also used as a motor fuel for high-performance engines.

64. The reaction is

   $C_6H_{12}O_6 \xrightarrow{yeast} 2CH_3-CH_2-OH + 2CO_2$

   The yeast necessary for the fermentation process are killed if the concentration of ethanol is over 13%. More concentrated ethanol solutions are most commonly made by distillation.

65. Although much ethanol is produced each year by means of the fermentation process, ethanol is also produced synthetically by hydration of ethene (ethylene)

   $CH_2=CH_2 + H_2O \rightarrow CH_3-CH_2-OH$

   Ethanol is used in industry as a solvent and as a starting material for the synthesis of more complicated molecules. Mixtures of ethanol and gasoline are used as automobile motor fuels (gasohol).

66. methanol (CH_3OH) - starting material for synthesis of acetic acid and many plastics

   ethylene glycol (CH_2OH-CH_2OH) - automobile antifreeze

   isopropyl alcohol (2-propanol, $CH_3-CH(OH)-CH_3$) - rubbing alcohol

67. Aldehydes and ketones both contain the *carbonyl* functional group

   $>C=O$

68. Aldehydes and ketones both contain the carbonyl group C=O.

$$>\!C=O$$

Aldehydes and ketones differ in the *location* of the carbonyl function: aldehydes contain the carbonyl group at the end of a hydrocarbon chain (the carbon atom of the carbonyl group is bonded only to at most one other carbon atom); the carbonyl group of ketones represents one of the interior carbon atoms of a chain (the carbon atom of the carbonyl group is bonded to two other carbon atoms).

69. The simplest aldehyde, methanal (formaldehyde), is used as a tissue preservative. Several important aldehydes are used as artificial flavorings and aromas (e.g., benzaldehyde, cinnamaldehyde, vanillin). The most common ketone is propanone (acetone), which is used in great quantity as a solvent; acetone can be a product of metabolism in persons with certain diseases (e.g., diabetes). Butanone (methyl ethyl ketone) is another ketone that is used frequently as a solvent.

70. The specific answers depend on your choice of alcohols. Here are representative reactions involving general primary and secondary alcohols:

$$R\text{–}CH_2\text{–}OH \xrightarrow{\text{mild oxidation}} R\text{–}CHO$$

$$R\text{–}CHOH\text{–}R' \xrightarrow{\text{mild oxidation}} R\text{–}C(=O)\text{–}R'$$

71. The structures are:

$$CH_3\text{–}CH_2\text{–}C\!\!\begin{array}{c}\nearrow O \\ \searrow H\end{array} \qquad\qquad CH_3\text{–}\underset{\underset{O}{\|}}{C}\text{–}CH_3$$

$$\text{propanal} \qquad\qquad\qquad \text{propanone}$$

72. In addition to their systematic names (based on the hydrocarbon root, with the ending *–one*), ketones can also be named by naming the groups attached to either side of the carbonyl carbon as alkyl groups, followed by the word "ketone". Examples are:

$CH_3\text{–}C(=O)\text{–}CH_2CH_3$ methyl ethyl ketone (2-butanone, butanone)

$CH_3CH_2\text{–}C(=O)\text{–}CH_2CH_3$ diethyl ketone (3-pentanone)

73.   a.   2,2,4-trimethylpentanal

     b.   2,4-dimethyl-3-pentanone

     c.   4-chloro-3-methyl-2-butanone

     d.   ethanal (acetaldehyde)

74. The structures are:

     a. $CH_3\text{–}\underset{\underset{}{}}{\overset{\overset{O}{\|}}{C}}\text{–}CH_3$

Chapter 20: Organic Chemistry

b. 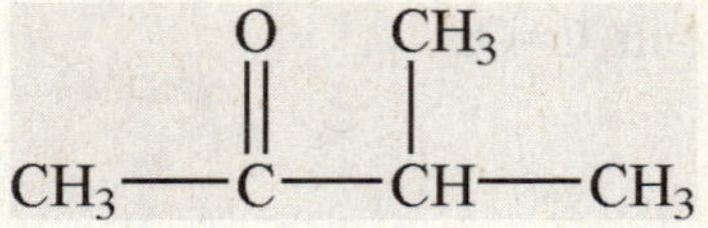 CH₃—C(=O)—CH(CH₃)—CH₃

c. CH₃—CH₂—CHO

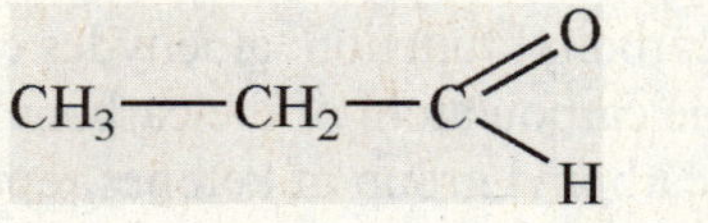

d. 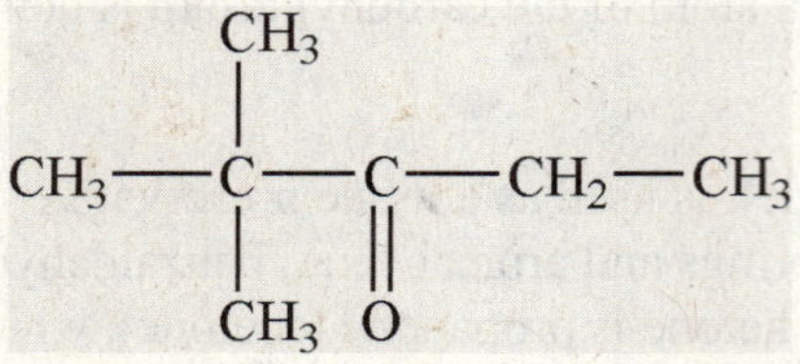 CH₃—C(CH₃)(CH₃)—C(=O)—CH₂—CH₃

75. Organic (carboxylic) acids contain the carboxyl group, –COOH.

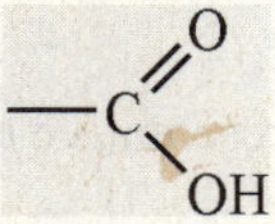

    —C(=O)—OH

    The general formula for organic acids is usually indicated in print as RCOOH, where R represents the hydrocarbon fragment.

76. Carboxylic acids are typically *weak* acids.

    $CH_3–CH_2–COOH(aq) \rightleftharpoons H^+(aq) + CH_3–CH_2–COO^-(aq)$

77. To name an organic acid, the final *-e* of the name of the parent hydrocarbon is dropped, and the ending *-oic acid* is added. For example, the organic acid CH₃CH₂COOH contains three carbon atoms and is considered as if it were derived from propan*e* - with the name: propan*oic acid*.

78. a. CH₃—CH₂—CH₂—CHO

    b. CH₃—CH₂—COOH

    c. 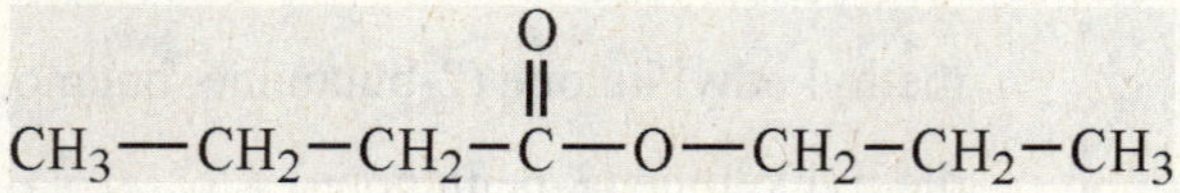 CH₃—CH₂—CH₂—C(=O)—O—CH₂—CH₂—CH₃

79. Esters are synthesized by the reaction of organic acids with alcohols,

    RCOOH + R´OH → RCOOR´ + H₂O.

    The general structural formula showing the linkage in esters is

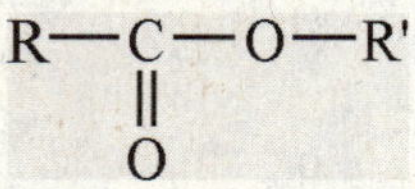

    R—C(=O)—O—R'

    Specific example and names depends on student choice of acid and alcohol

80. Acetylsalicylic acid is synthesized from salicylic acid (behaving as an alcohol through its –OH group) and acetic acid.

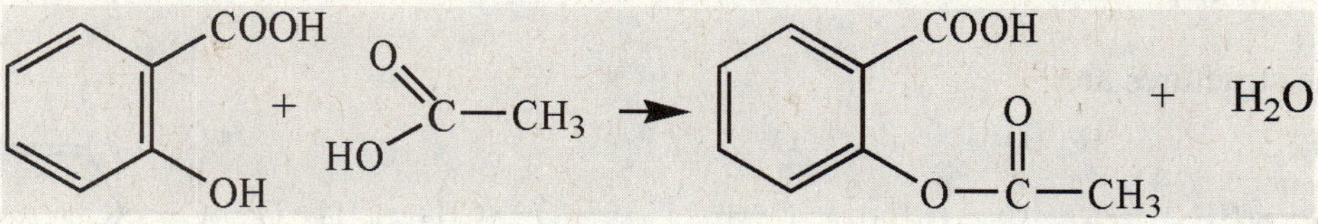

81. a. 2-methylbutanoic acid

b. 2-phenylethanoic acid

c. 3,3-dimethylpentanoic acid

d. 4-chloro-3,3-dimethylbutanoic acid

82. The structures are:

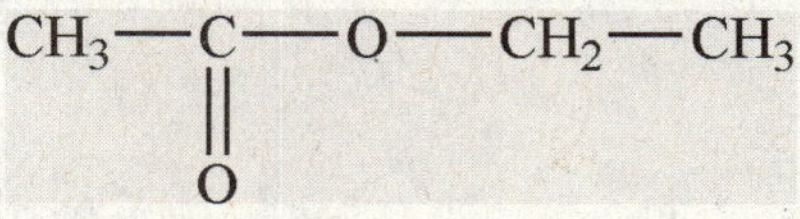

a. 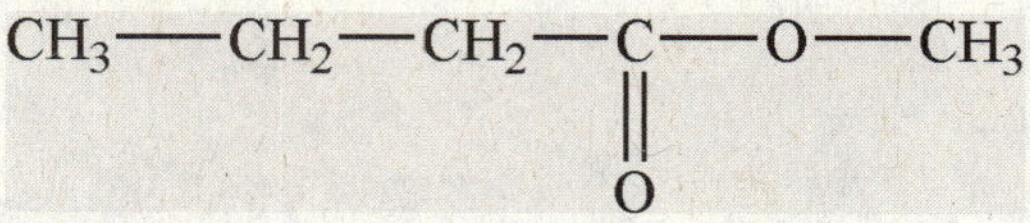 $CH_3-CH_2-CH_2-\underset{\underset{O}{\parallel}}{C}-O-CH_3$

b. $CH_3-\underset{\underset{O}{\parallel}}{C}-O-CH_2-CH_3$

c. (benzene ring with COOH and Cl ortho substituents)

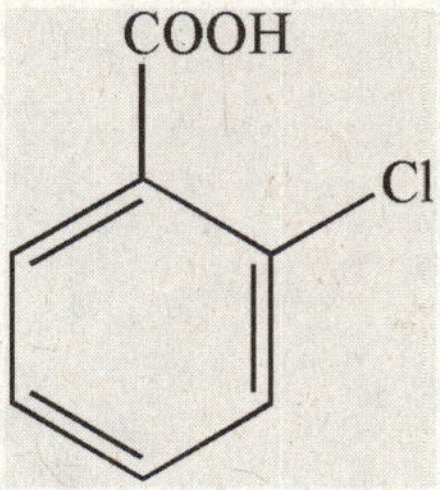

d. 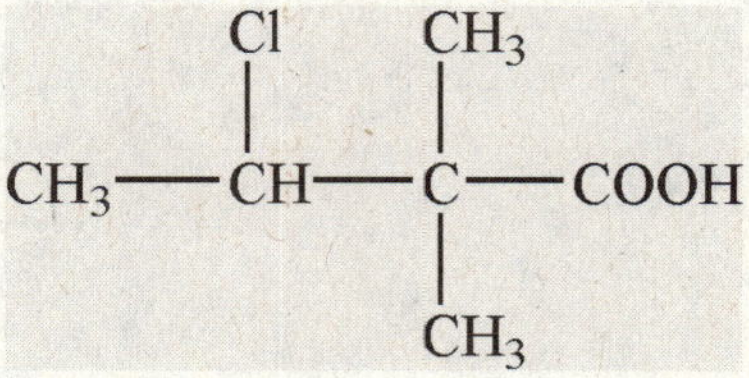 $CH_3-\underset{Cl}{\underset{|}{CH}}-\underset{\underset{CH_3}{|}}{\overset{\overset{CH_3}{|}}{C}}-COOH$

83. Polymers are large, usually chain-like molecules that are built up from smaller molecules. The smaller molecules may combine together and repeat in the chain of the polymer hundreds or thousands of time. The small molecules from which polymers are built are called *monomers*.

84. In addition polymerization, the monomer units simply add together to form the polymer, with no other products. Polyethylene and polytetrafluoroethylene (Teflon) are common examples.

85. In condensation polymerization, small molecules (such as H$_2$O) are formed and are split out as the monomer units combine to form the polymer chain. In addition polymerization, monomer units merely "add together" to form a longer chain. The polymer Nylon-66 is an example of a condensation polymer.

Chapter 20: Organic Chemistry

86. Kevlar is a *co*-polymer since two different types of monomers combine to generate the polymer chain.

87. Answer depends on students' choice. Common polymers are listed in Table 20.8.

88. The structures are:

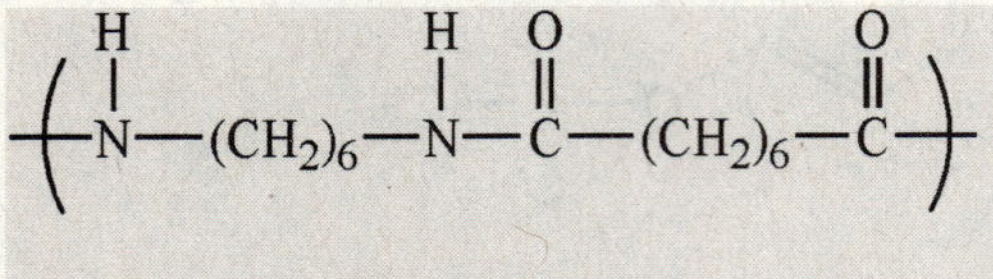

$$\left(-\underset{H}{N}-(CH_2)_6-\underset{H}{N}-\underset{O}{C}-(CH_2)_6-\underset{O}{C}-\right) \quad \left(-O-CH_2-CH_2-O-\underset{O}{C}-\underset{}{\bigcirc}-\underset{O}{C}-\right)$$

nylon 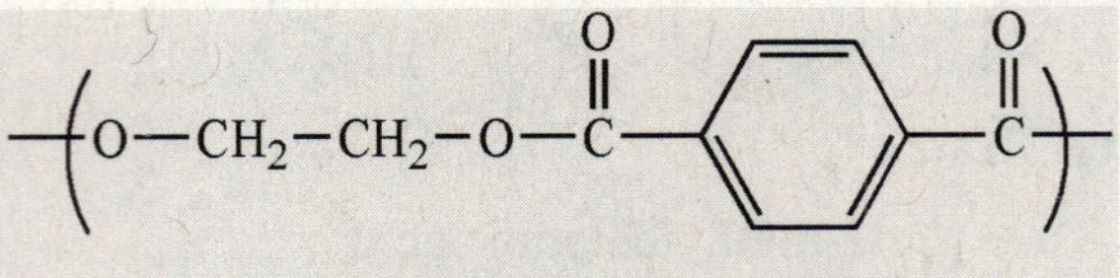   dacron

89. urea, ammonium cyanate

90. unsaturated

91. tetrahedral

92. straight-chain or normal

93. bonds

94. *-ane*

95. longest

96. number

97. pyrolytic cracking

98. anti-knocking

99. combustion

100. substitution

101. addition

102. hydrogenation

103. aromatic

104. functional

105. primary

106. carbon monoxide

107. fermentation

108. carbonyl

109. oxidation

110. The correct answer is *d*. Organic molecules must contain carbon. The formula for magnesium sulfate is MgSO$_4$.

111. esters, alcohol

112. a.   ethane, CH$_3$–CH$_3$

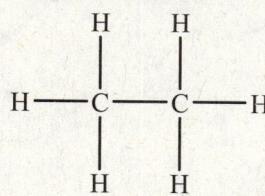

    b.   butane, CH$_3$–CH$_2$–CH$_2$–CH$_3$

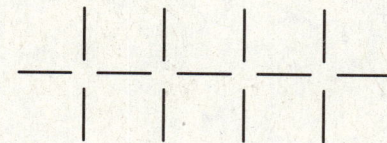

    c.   hexane, CH$_3$–CH$_2$–CH$_2$–CH$_2$–CH$_2$–CH$_3$

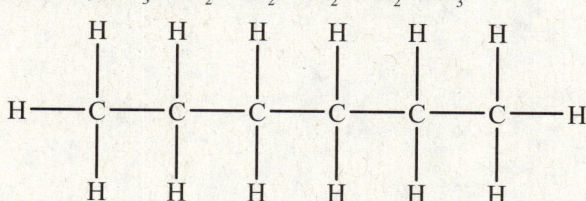

113. The general formula is C$_n$H$_{2n+2}$. Consider the following normal alkanes: each successive member differs from the previous by a –CH$_2$– unit:

CH$_3$–CH$_3$ CH$_3$–CH$_2$–CH$_3$ CH$_3$–CH$_2$–CH$_2$–CH$_3$ CH$_3$–CH$_2$–CH$_2$–CH$_2$–CH$_3$

114. A saturated hydrocarbon is one in which all carbon-carbon bonds are single bonds, with each carbon atom forming bonds to four other atoms. The saturated hydrocarbons are called alkanes.

115. 2; Esters and carboxylic acids must contain two oxygen atoms.

116. a.   2-butene

    b.   3-methyl-1-butene

    c.   1-butyne

    d.   3-chloro-1-butene

117. Several examples of molecules containing two or more fused benzene rings are shown in Table 20.4 of the text.

Chapter 20: Organic Chemistry

118.
a. 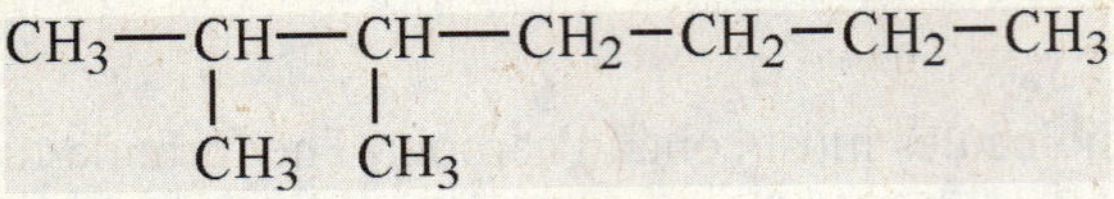 CH$_3$—CH—CH—CH$_2$—CH$_2$—CH$_2$—CH$_3$
           | |
          CH$_3$ CH$_3$

b. 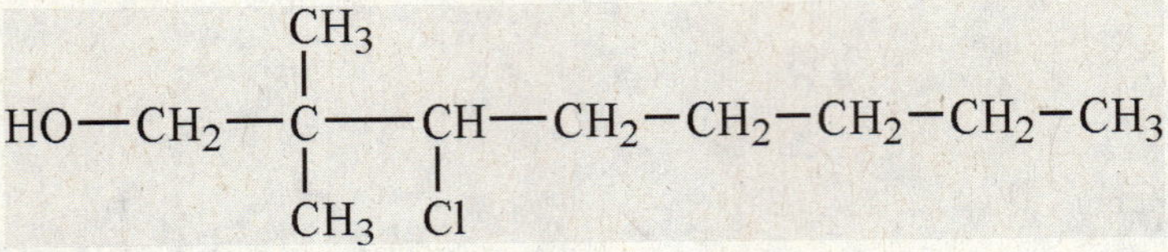 HO—CH$_2$—C—CH—CH$_2$—CH$_2$—CH$_2$—CH$_2$—CH$_3$
                | |
               CH$_3$ Cl
   (with CH$_3$ on central C)

c. CH$_2$=C—CH$_2$—CH$_2$—CH$_2$—CH$_3$
          |
         Cl

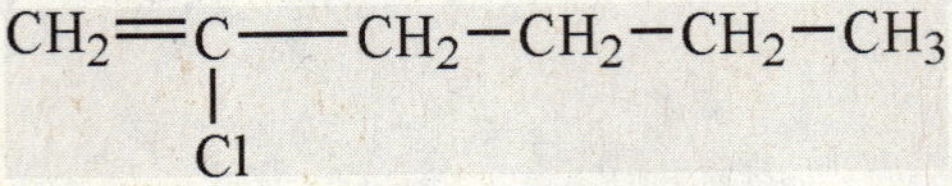

d. Cl—CH$_2$—CH=CH—CH$_2$—CH$_2$—CH$_3$

e. 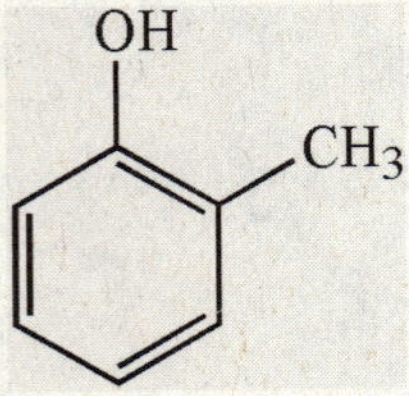 (ortho-methylphenol: benzene ring with OH and CH$_3$ in adjacent positions)

119.
a. CH$_3$–CH$_2$–Cl, CH$_2$Cl–CH$_2$–Cl, and various other chlorosubstituted ethanes.
b. CH$_3$–CH$_2$–CH$_2$–CH$_3$
c. 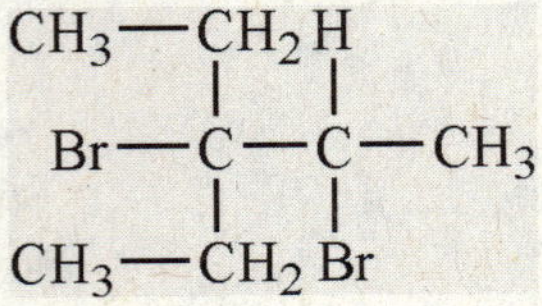
   CH$_3$—CH$_2$ H
        \   /
  Br—C—C—CH$_3$
        /   \
   CH$_3$—CH$_2$ Br

120. 1; Only *tert*-butyl alcohol is a tertiary alcohol.

121. 1,2,3-trihydroxypropane (1,2,3-propanetriol)

122. The correct answer is *e*. Amines contain –NH$_2$ groups.

123.
a. 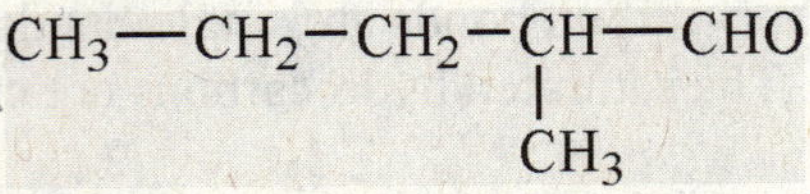 CH$_3$—CH$_2$—CH$_2$—CH—CHO
                         |
                        CH$_3$

b. 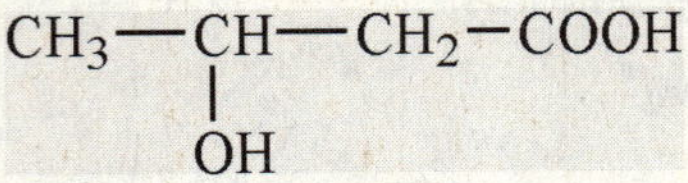 CH$_3$—CH—CH$_2$—COOH
           |
         OH

c. 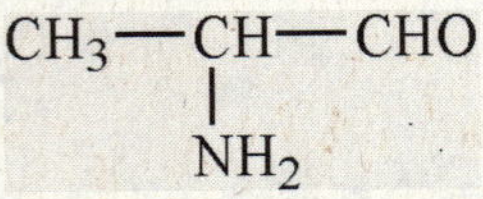 CH$_3$—CH—CHO
          |
        NH$_2$

d. 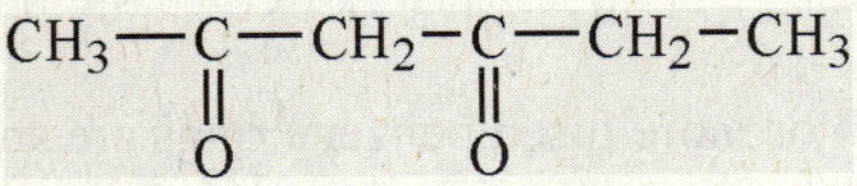 CH$_3$—C—CH$_2$—C—CH$_2$—CH$_3$
           ‖   ‖
           O   O

e. 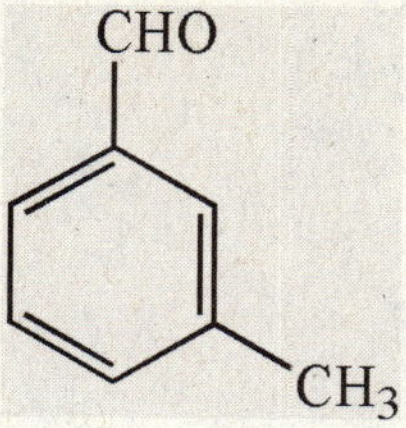

124. a. 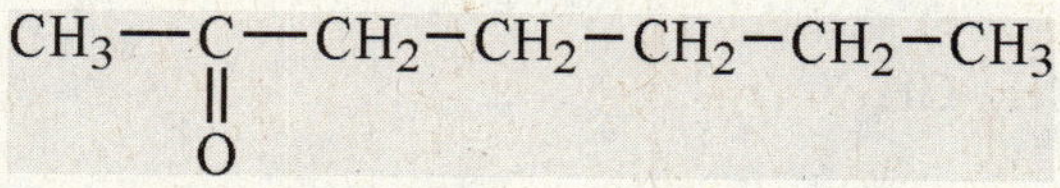

b. 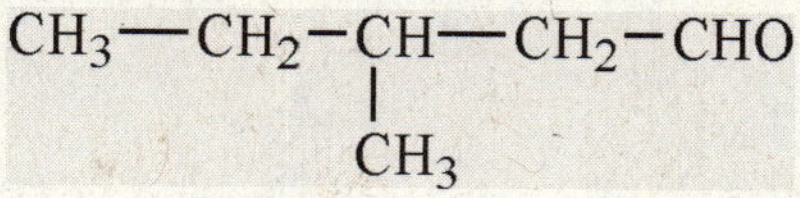

c. 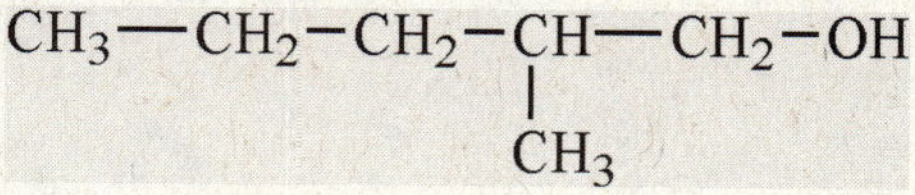

d. 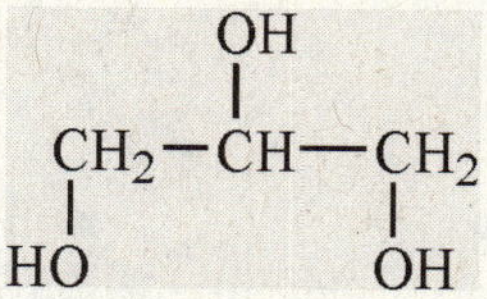

e. 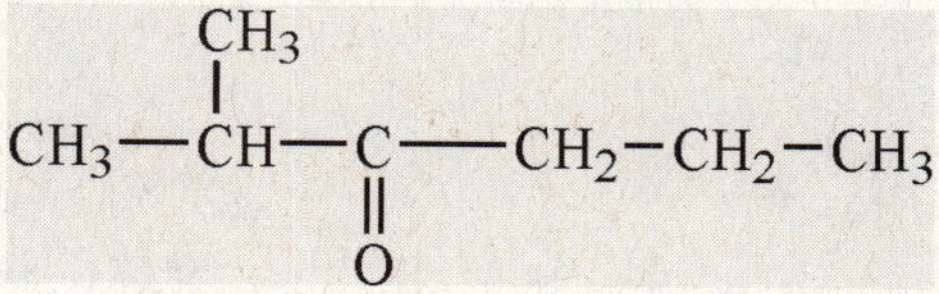

125. acetylsalicylic acid

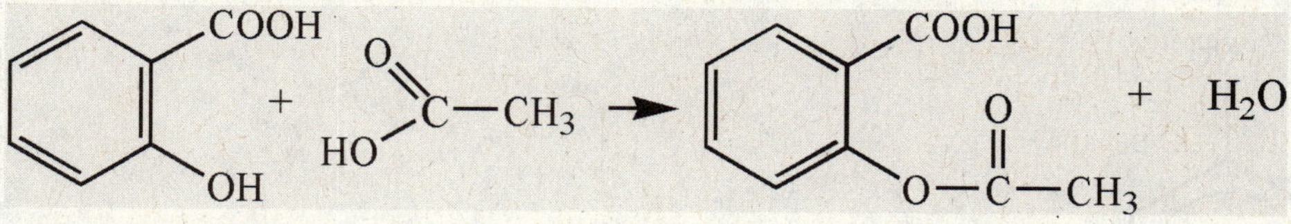

methyl salicylate

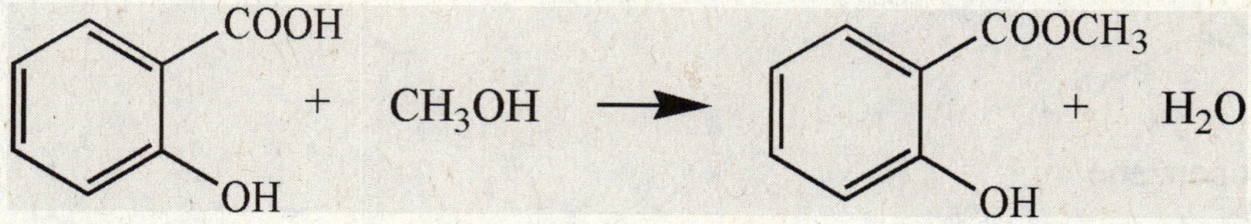

126. 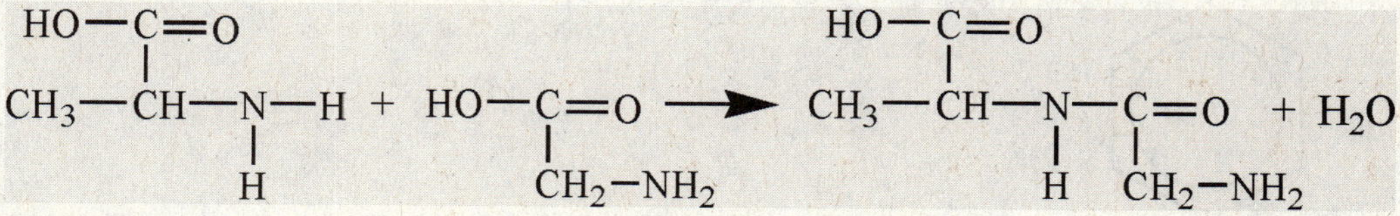

127.
Name	Structure
pentane	CH₃–CH₂–CH₂–CH₂–CH₃
octane	CH₃–CH₂–CH₂–CH₂–CH₂–CH₂–CH₂–CH₃
decane	CH₃–CH₂–CH₂–CH₂–CH₂–CH₂–CH₂–CH₂–CH₂–CH₃
pentane	CH₃–CH₂–CH₂–CH₂–CH₃
heptane	CH₃–CH₂–CH₂–CH₂–CH₂–CH₂–CH₃

128. a. 1,2-dichlorobenzene

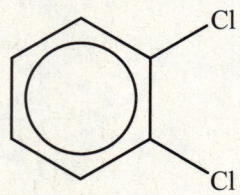

b. 1,3-dimethylbenzene

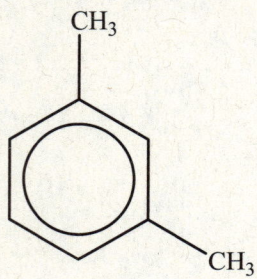

c. 3-nitrophenol

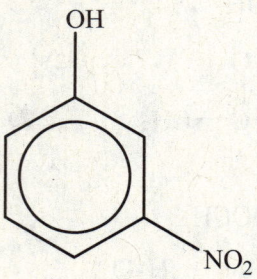

d. *p*-dibromobenzene

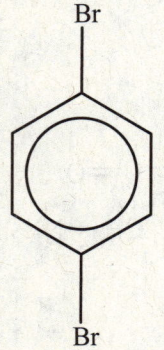

e. 4-nitrotoluene

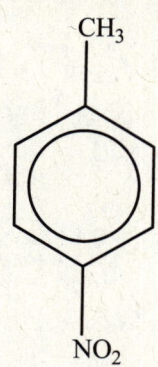

129. Some of the carbon skeletons are:

```
C—C—C—C—C—C        C—C—C—C—C        C—C—C—C—C—C
    |                  |   |              |
    C                  C   C              C

C—C—C—C—C—C—C                 C—C—C—C—C
                                 |   |
                                 C   C

        C                              C
        |                              |
C—C—C—C—C—C—C              C—C—C—C—C
        |                      |
        C                      C
```

130. a. 2,3-dimethylbutane

b. 3,3-diethylpentane

c. 2,3,3-trimethylhexane

d. 2,3,4,5,6-pentamethylheptane

131. a.
$$CH_3-\underset{\underset{CH_3}{|}}{\overset{\overset{CH_3}{|}}{C}}-CH_2-CH_2-CH_2-CH_3$$

b.
$$CH_3-CH-\underset{\underset{CH_3}{|}}{CH}-CH_2-CH_2-CH_3$$
(with CH₃ on first CH)

c.
$$CH_3-CH_2-\underset{\underset{CH_3}{|}}{\overset{\overset{CH_3}{|}}{C}}-CH_2-CH_2-CH_3$$

Chapter 20: Organic Chemistry

d. 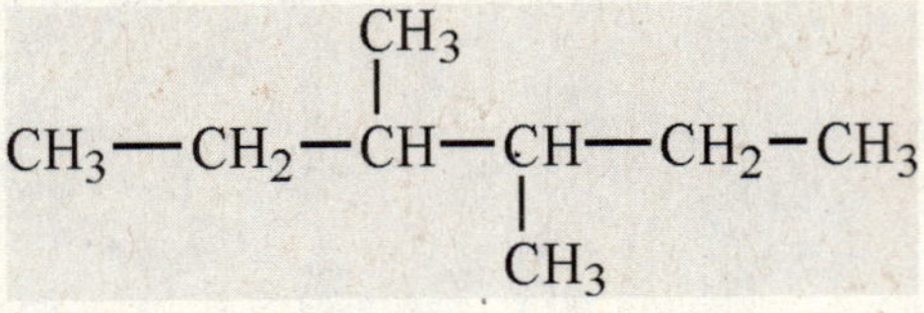
$$CH_3-CH_2-\underset{\underset{CH_3}{|}}{CH}-\underset{\underset{CH_3}{|}}{CH}-CH_2-CH_3$$

e. 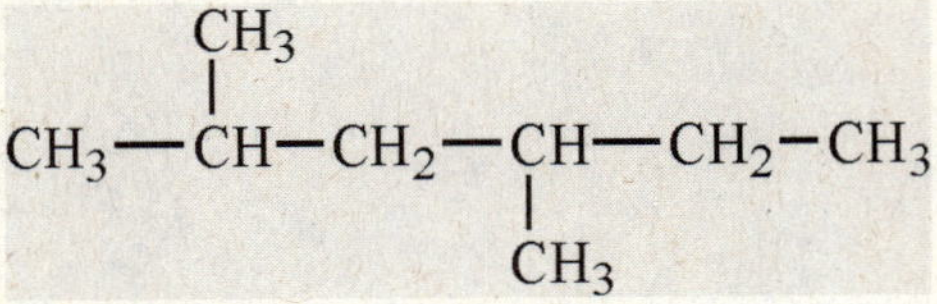
$$CH_3-\underset{\underset{CH_3}{|}}{CH}-CH_2-\underset{\underset{CH_3}{|}}{CH}-CH_2-CH_3$$

132. a. $CH_3Cl(g)$

 b. $H_2(g)$

 c. $HCl(g)$

133. Structures depend on student choices.

134.
$CH \equiv C-CH_2-CH_2-CH_2-CH_2-CH_2-CH_3$ 1-octyne

$CH_3-C \equiv C-CH_2-CH_2-CH_2-CH_2-CH_3$ 2-octyne

$CH_3-CH_2-C \equiv C-CH_2-CH_2-CH_2-CH_3$ 3-octyne

$CH_3-CH_2-CH_2-C \equiv C-CH_2-CH_2-CH_3$ 4-octyne

135. a. 1,2-dimethylbenzene (2-methyltoluene)

 b. 1,3,5-tribromobenzene

 c. 2-chloronitrobenzene

 d. naphthalene

136. a. carboxylic acid

 b. ketone

 c. ester

 d. alcohol (phenol)

137. a. 2-propanol (secondary)
$$CH_3-\underset{\underset{OH}{|}}{CH}-CH_3$$

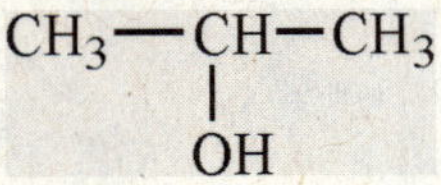

 b. 2-methyl-2-propanol (tertiary)
$$CH_3-\underset{\underset{OH}{\overset{\overset{CH_3}{|}}{|}}}{C}-CH_3$$

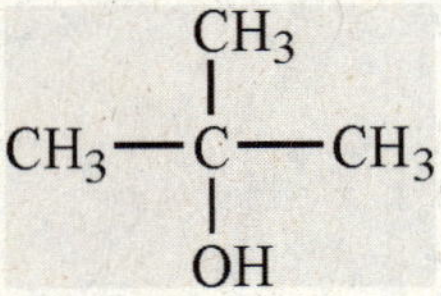

c. 4-isopropyl-2-heptanol (secondary)

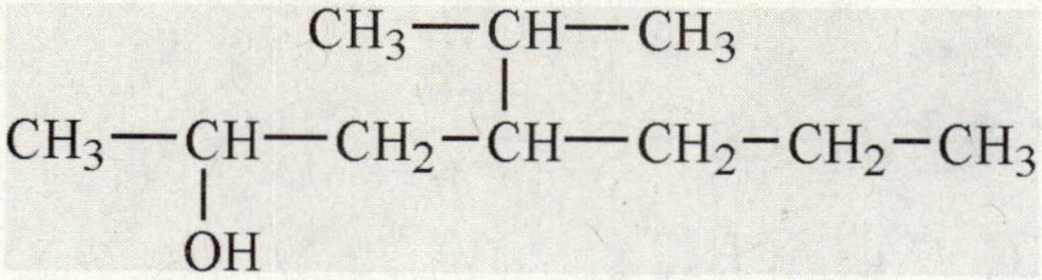

```
           CH₃—CH—CH₃
               |
CH₃—CH—CH₂—CH—CH₂—CH₂—CH₃
    |
    OH
```

d. 2,3-dichloro-1-pentanol (primary)

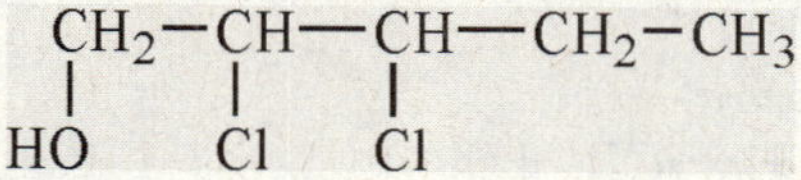

```
CH₂—CH—CH—CH₂—CH₃
 |   |   |
 HO  Cl  Cl
```

138. The correct answer is *e*. An ester has the general formula R-COO-R', and a carboxylic acid has the general formula R-COOH. A ketone has the general formula R-CO-R', and an aldehyde has the general formula R-COH.

139. Carboxylic acids are synthesized from the corresponding primary alcohol by strong oxidation with a reagent such as potassium permanganate

    $CH_3–CH_2–CH_2–OH \xrightarrow{strong\ oxidation} CH_3–CH_2–COOH$.

    The synthesis of carboxylic acids from alcohols is an oxidation/reduction reaction.

140. a.
```
CH₃—CH—CH₂—COOH
    |
    CH₃
```

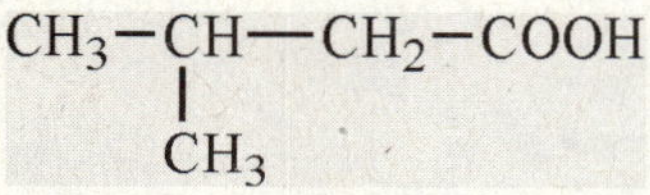

   b. (benzene ring with C(=O)—OH group and Cl substituent ortho)

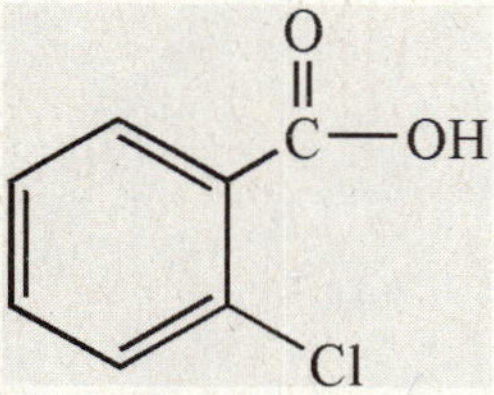

   c. $CH_3–CH_2–CH_2–CH_2–CH_2–COOH$

   d. $CH_3–COOH$

141. a. polyethylene

   $CH_2{=}CH_2$ \quad\quad *–(CH₂–CH₂)ₙ–*

   b. polyvinyl chloride

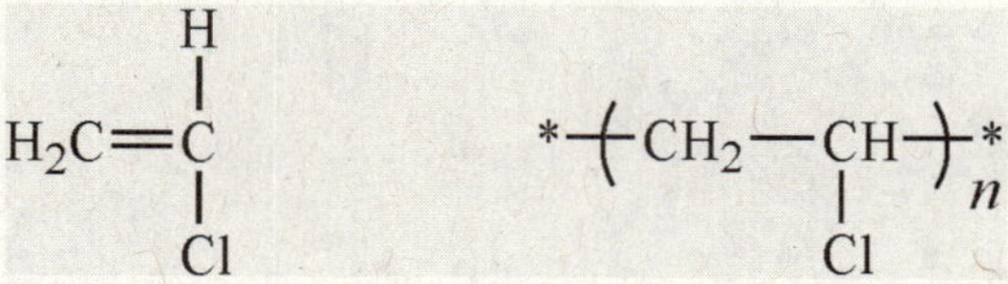

```
      H
      |
H₂C=C           *–(CH₂—CH)ₙ–*
      |                |
      Cl               Cl
```

   c. Teflon

   $F_2C{=}CF_2$ \quad\quad –(CF₂–CF₂)ₙ–

Chapter 20: Organic Chemistry

  d. polypropylene

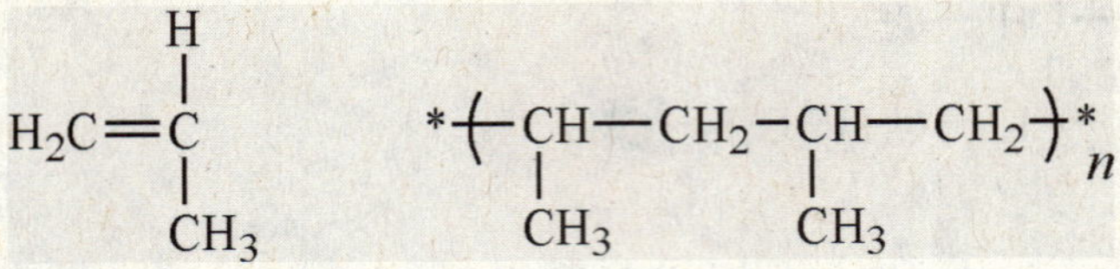

$$H_2C=C(H)(CH_3) \quad *{-(CH(CH_3)-CH_2-CH(CH_3)-CH_2)-}*_n$$

  e. polystyrene

$$H_2C=C(H)(C_6H_5) \quad *{-(CH(C_6H_5)-CH_2-CH(C_6H_5)-CH_2)-}*_n$$

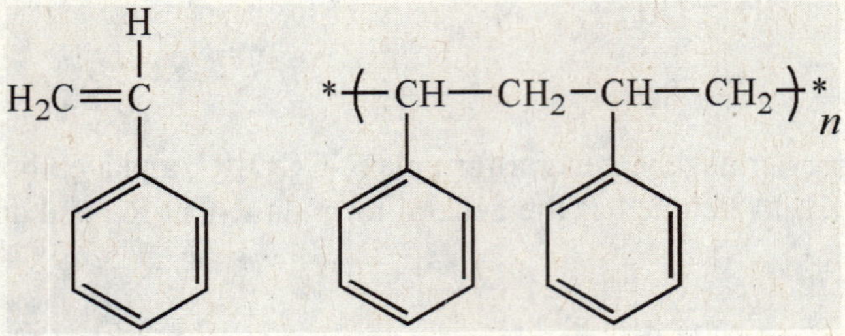

142. a. pentane

  b. 3-ethyl-2,5-dimethylhexane

  c. 4-ethyl-5-isopropyloctane

143. a. 2,2-dibromo-3-methylpentane

  b. 4-ethyl-3-iodo-2-methylhexane

  c. 4-fluoro-2-methylnonane

144. a. 2-methyl-1-butene

  b. 2,4-dimethyl-1,4-pentadiene

  c. 6-ethyl-2-methyl-4-octene

  d. 3-bromo-1-heptyne

  e. 7-chloro-2,5,5-trimethyl-3-heptyne

  f. 4-ethyl-3-methyl-1-octyne

145. a. heptanal

  b. 3-ethyl-5-methylhexanal

  c. 3-hexanone

  d. 4,5-dimethyl-3-hexanone

  e. 3,4-dimethylhexanoic acid

Chapter 20:   Organic Chemistry

146. 2-chloropropanoic acid

$$\begin{array}{c} \text{Cl} \\ | \\ \text{CH}_3-\text{CH}-\text{COOH} \end{array}$$

147. b, c, a; "Like dissolves like." Structure b is the least polar (and thus the least soluble with water which is polar). Structure a is the most polar (and thus most soluble with water) due to the carboxylic acid functional group on the end.

CHAPTER 21

Biochemistry

1. Biochemistry

2. Trace elements are those elements present in the body in only very small amounts, but which are essential to many biochemical processes in the body.

3. Proteins are large polymers of the α-amino acids, with molar masses in the range of 5000 to over 1 million g/mol, which constitute approximately 15% of our bodies by mass.

4. Fibrous proteins provide structural integrity and strength for many types of tissue and are the main components of muscle, hair, and cartilage. Globular proteins are the "worker" molecules of the body, performing such functions as transporting oxygen throughout the body, catalyzing many of the reactions in the body, fighting infections, and transporting electrons during the metabolism of nutrients.

5. The general structural formula for the alpha amino acids is

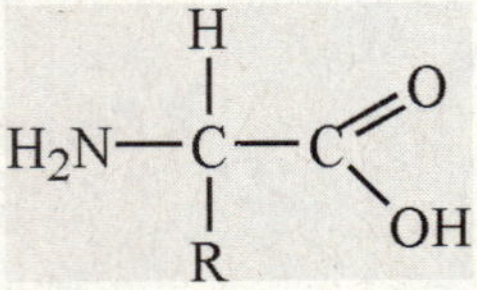

    $$H_2N-\underset{\underset{R}{|}}{\overset{\overset{H}{|}}{C}}-C\underset{OH}{\overset{O}{\diagup\!\!\!\diagdown}}$$

    All alpha amino acids contain the carboxyl group (–COOH) and the amino group (–NH$_2$) attached to the number-2 or "α-carbon" atom as indicated. In this general formula, R represents the remainder of the amino acid molecule (side-chain): it is this portion of the molecule that differentiates one amino acid from another.

6. The structures of the amino acids are given in Figure 21.2. A side chain is nonpolar if it is mostly hydrocarbon in nature (like alanine). Polar side chains may contain the hydroxyl group (–OH), the sulfhydryl group (–SH), or a second amino (–NH$_3$) or carboxyl (–COOH) group.

7. hydrophobic; hydrophilic

8. The amino acid will be hydrophilic if the R group is polar, and hydrophobic if the R group is nonpolar. Serine is a good example of an amino acid in which the R group is polar. Leucine is a good example of an amino acid with a nonpolar R group.

9. The two dipeptides are shown below

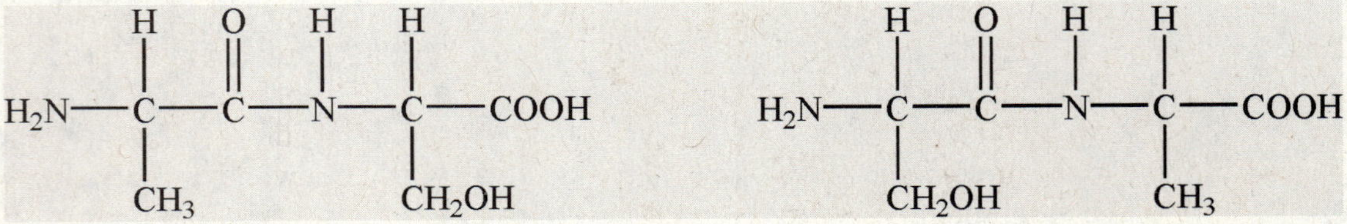

    $$H_2N-\underset{\underset{CH_3}{|}}{\overset{\overset{H}{|}}{C}}-\overset{\overset{O}{\|}}{C}-N-\underset{\underset{CH_2OH}{|}}{\overset{\overset{H}{|}}{C}}-COOH \qquad H_2N-\underset{\underset{CH_2OH}{|}}{\overset{\overset{H}{|}}{C}}-\overset{\overset{O}{\|}}{C}-N-\underset{\underset{CH_3}{|}}{\overset{\overset{H}{|}}{C}}-COOH$$

10. There are six tripeptides possible.

    cys-ala-phe   *ala-cys-phe*   *phe-ala-cys*

    cys-phe-ala   *ala-phe-cys*   *phe-cys-ala*

11. The "peptide linkage" is an amide linkage

    $$-\underset{\underset{O}{\|}}{C}-\underset{\overset{H}{|}}{N}-$$

    Specific example depends on student choice of amino acids.

12. The primary structure of a protein is the specific *sequence* of amino acids in the peptide chain. Adjacent amino acids are connected to each other by peptide (amide) linkages.

13. The secondary structure of a protein describes, in general, the arrangement in space of the protein's polypeptide chain. The most common secondary structures are the alpha-helix and the beta-pleated sheet.

14. Long, thin, resilient proteins, such as hair, typically contain elongated, elastic alpha-helical protein molecules. Other proteins, such as silk, which in bulk form sheets or plates, typically contain protein molecules having the beta pleated sheet secondary structure. Proteins that do not have a structural function in the body, such as hemoglobin, typically have a globular structure.

15. Collagen consists of three α-helical polypeptide chains twisted together to form a super helix.

16. Silk consists of a sheet structure where the individual chains of amino acids are lined up lengthwise next to each other to form the sheet.

17. The tertiary structure of a protein represents its specific, overall shape when it occurs in its natural environment and is influenced by that environment. To distinguish between the secondary and tertiary structures, consider this example: a given protein has an alpha helical secondary structure (the protein's own amino acid chain coils in a helix); in the body, however, this helix itself is folded and twisted by interactions with the protein's environment until it forms a tight sphere. The fact that the helical protein is folded into a tight sphere indicates the tertiary structure of the protein.

18. A disulfide linkage represents a S–S bond between two sulfur-containing amino acids in a peptide chain. It is the amino acid cysteine that forms such linkages. The presence of disulfide linkages produces bends and folds in the peptide chain and contributes greatly to the tertiary structure of a protein.

19. Denaturation of a protein represents the breaking down of the protein's tertiary structure. If the environment of a protein is changed from the normal environment of the protein, the specific folding and twisting of the protein's polypeptide chain will change to accommodate the new environment. If the protein's tertiary structure is changed, the protein most likely will no longer have whatever function in the body it ordinarily possesses. Cooking of an egg (adding heat to the environment of the protein) causes the protein albumin in the white of the egg to coagulate. Adding heavy metal ions (lead or mercury, for example) can disrupt the inter-chain linkages that contribute to the protein's tertiary structure. A permanent hair wave works by deliberately changing the hair protein's structure.

Chapter 21: Biochemistry

20. Oxygen is transported by the protein *hemoglobin*.

21. Proteins that catalyze biochemical processes are called *enzymes*.

22. ferritin

23. Antibodies

24. Amino acids contain both a weak-acid and a weak-base group, and thus they can neutralize both bases and acids, respectively.

25. Enzymes are typically 1 to 10 *million* times more efficient than inorganic catalysts.

26. When we say that an enzyme is selective for a particular substrate, we mean that the enzyme will catalyze the reactions of that molecule and that molecule only.

27. The action of an enzyme on its substrate takes place at a specific portion of the polypeptide chain called the *active site*.

28. The lock-and-key model for enzymes indicates that the structures of an enzyme and its substrate must be *complementary*, so that the substrate can approach and attach itself along the length of the enzyme at the enzyme's active sites. A given enzyme is intended to act upon a particular substrate: the substrate attaches itself to the enzyme, is acted upon, and then moves away from the enzyme. If a different molecule has a similar structure to the substrate, this other molecule may also be capable of attaching itself to the enzyme. But since this molecule is not the enzyme's proper substrate, the enzyme may not be able to act upon the molecule, and the molecule may remain attached to the enzyme preventing proper substrate molecules from approaching the enzyme (irreversible inhibition). If the enzyme cannot act upon its proper substrate, then the enzyme is said to be inhibited. Irreversible inhibition might be a desirable feature in an antibiotic, which would bind to the enzymes of a bacteria and prevent the bacteria from reproducing, thereby preventing or curing an infection.

29. Structures are shown for glucose (aldehyde) and fructose (ketone).

    ```
           CHO                    CH2OH
            |                      |
       H—C—OH                   C=O
            |                      |
      HO—C—H                  HO—C—H
            |                      |
       H—C—OH                  H—C—OH
            |                      |
       H—C—OH                  H—C—OH
            |                      |
          CH2OH                  CH2OH
         glucose                fructose
    ```

    Although we typically draw these monosaccharides stretched out as chains, in solution, they actually exist in a ring form (See Figure 21.12).

30. Simple sugars typically contain several hydroxyl groups as well as the carbonyl functional group.

31. These ring structures are shown in Figure 21.13. As indicated, the ring in glucose is not flat.

32. A pentose sugar is a carbohydrate containing five (*pent-*) carbon atoms in the chain. The pentose ribose is shown below.

```
        CHO
         |
    H — C — OH
         |
    H — C — OH
         |
    H — C — OH
         |
        CH₂OH
        ribose
```

33. A disaccharide consists of two monosaccharide units bonded together into a single molecule. Sucrose consists of a glucose molecule and a fructose molecule connected by an alpha-glycosidic linkage.

34. Starch is the form in which glucose is stored by plants for later use as cellular fuel. Cellulose is used by plants as their major structural component. Although starch and cellulose are both polymers of glucose, the linkage between adjacent glucose units differs in the two polysaccharides. Humans do not possess the enzyme needed to hydrolyze the linkage in cellulose.

35. Sucrose is a disaccharide formed from glucose and fructose by elimination of a water molecule to form a C–O–C linkage between the rings (called a glycosidic linkage). The structure of sucrose is shown in Figure 21.13.

36. ribose (aldopentose); arabinose (aldopentose); ribulose (ketopentose); glucose (aldohexose); mannose (aldohexose); galactose (aldohexose); fructose (ketohexose).

37. Deoxyribonucleic acid (DNA) is the molecule responsible for coding and storing genetic information in the cell and for subsequently transmitting the information needed to synthesize the proteins/enzymes the cell requires.

38. smaller

39. Deoxyribose is the pentose sugar found in DNA molecules, whereas the sugar ribose is found in RNA molecules.

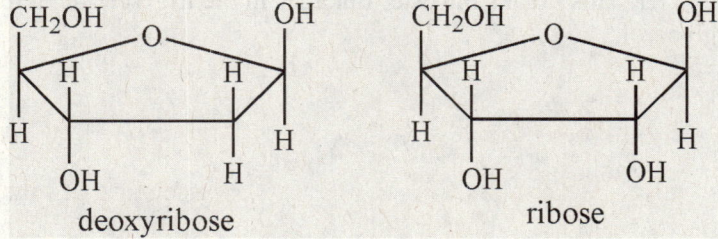

Chapter 21: Biochemistry

40. Uracil (RNA only); cytosine (DNA, RNA); thymine (DNA only); adenine (DNA, RNA); guanine (DNA, RNA)

41. In a strand of DNA, the phosphate group and the sugar molecule of adjacent nucleotides become bonded to each other. The chain-portion of the DNA molecule, therefore, consists of alternating phosphate groups and sugar molecules. The nitrogen bases are found protruding from the side of this phosphate-sugar chain, bonded to the sugar molecules. *Within* the chain of each strand, covalent bonding occurs. *Between* strands, hydrogen bonding occurs.

42. An overall DNA molecule consists of two chains of nucleotides, with the organic bases on the nucleotides arranged in complementary pairs (cytosine with guanine, and adenine with thymine). The structures and properties of the organic bases are such that these pairs fit together well and allow the two chains of nucleotides to form the double helix structure. When DNA replicates, it is assumed the double helix unwinds, and then new molecules of the organic bases come in and pair up with their respective partner on the separated nucleotide chains, thereby replicating the original structure. See Figure 21.20

43. A gene is a segment of the DNA molecule containing the code for synthesizing a particular protein.

44. Lipids are a group of substances defined in terms of their solubility characteristics: lipids are typically oily, greasy substances that are not very soluble in water.

45. The four types of lipids are fats (tristearin), phospholipids (lecithin), waxes (beeswax), and steroids (cholesterol).

46. A triglyceride typically consists of a glycerol backbone, to which three separate fatty acid molecules are attached by ester linkages.

$$\begin{array}{c} \text{O} \\ \| \\ \text{R—C—O—CH}_2 \quad\quad \text{O} \\ \quad\quad\quad\quad\quad | \quad\quad\quad \| \\ \quad\quad\quad\quad \text{CH—O—C—R}' \\ \text{O} \quad\quad | \\ \| \quad\quad\quad \\ \text{R''—C—O—CH}_2 \end{array}$$

47. Table 21.5 lists arachidic acid, butyric acid, caproic acid, lauric acid, and stearic acids as typical saturated fatty acids. Oleic acid, linoleic acid, and linolenic acid are given as examples of unsaturated fatty acids. Triglycerides from animals generally contain saturated fatty acids, whereas triglycerides from plant sources generally contain unsaturated fatty acids.

48. "Soaps" are the salts of long-chain organic acids ("fatty acids"), most commonly either the sodium or potassium salt. Soaps are prepared by treating a fat or oil (a triglyceride) with a strong base such as NaOH or KOH. This breaks the ester linkages in the triglyceride, releasing three fatty acid anions and glycerol.

Chapter 21: Biochemistry

$$\begin{array}{c}
CH_2-O-\overset{\overset{O}{\|}}{C}-R \\
CH-O-\overset{\overset{O}{\|}}{C}-R' \\
CH_2-O-\overset{\overset{O}{\|}}{C}-R''
\end{array}
\quad + 3NaOH \longrightarrow \quad
\begin{array}{c}
CH_2-OH \\
CH-OH \\
CH_2-OH
\end{array}
\quad + \quad
\begin{array}{c}
RCOONa \\
R'COONa \\
R''COONa
\end{array}$$

49. Fatty acid salts ("soaps") have both a long hydrocarbon chain (that tends to be nonpolar in nature), but also are ionic (at the carboxylate end). Fatty acids are able to dissolve both in nonpolar substances (oils and grease) and also in water. In water, a large number of fatty acid molecules combine to form a spherical grouping called a *micelle*, in which the polar (ionic) ends of the fatty acids face out into the water, and the nonpolar chains of the fatty acids are positioned into the interior of the micelle. Fatty acid salts work as soaps to remove grease and oil from clothing by taking the grease or oil molecules into the interior of the micelles, allowing the grease or oil to be dispersed in the water in which the fatty acid salts are themselves dispersed.

50. Soaps have both a nonpolar nature (due to the long chain of the fatty acid) and an ionic nature (due to the charge on the carboxyl group). In water, soap anions form aggregates called micelles, in which the water-repelling hydrocarbon chains are oriented towards the interior of the aggregate, with the ionic, water-attracting carboxyl groups oriented towards the outside. Most dirt has a greasy nature. A soap micelle is able to interact with a grease molecule, pulling the grease molecule into the hydrocarbon interior of the micelle. When the clothing is rinsed, the micelle containing the grease is washed away. See Figures 21.22 and 21.23.

51. The group of lipids called *steroids* all contain the same basic cluster of four rings.

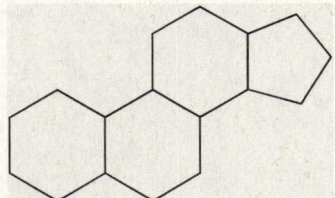

    This common structure is sometimes referred to as the steroid nucleus or kernel. Some important steroids are shown in Figure 21.25 in the text.

52. Cholesterol is the naturally occurring steroid from which the body synthesizes other needed steroids. As cholesterol is insoluble in water, it is thought that having too large a concentration of this substance in the bloodstream may lead to its deposition and build up on the walls of blood vessels, causing their eventual blockage.

53. testosterone: male sex hormone that controls development of male reproductive organs and secondary sex characteristics (deep voice, muscle structure, hair patterns)

    estradiol: female sex hormone that controls development of female reproductive organs and secondary sex characteristics

    progesterone: female hormone secreted during pregnancy that prevents further ovulation during and immediately after the pregnancy; birth-control pills are often synthetic progesterone derivatives

Chapter 21: Biochemistry

54. The bile acids are synthesized from cholesterol in the liver and are stored in the gall bladder. Bile acids such as cholic acid act as emulsifying agents for lipids and aid in their digestion.

55. v
56. i
57. t
58. m
59. x
60. u
61. q
62. f
63. k
64. g
65. y
66. r
67. c
68. p
69. n
70. o
71. e
72. b
73. s
74. d
75. h
76. a
77. deoxyribonucleic acid
78. The molar mass of DNA depends on the complexity of the species, but human DNA may have a molar mass as large as 2 billion g/mol. DNA is found in the nucleus of each cell.

79. ribose

80. ester

81. complementary paired

82. thymine, guanine

83. gene

84. transfer, messenger

85. DNA

86. cys-ala-phe; cys-phe-ala; phe-ala-cys; phe-cys-ala; ala-cys-phe; ala-phe-cys

87. triglycerides

88. unsaturated, saturated

89. saponification

90. ionic, nonpolar

91. micelles

92. cholesterol, adrenocorticoid hormones, sex hormones, and bile acids

93. cholesterol

94. The structures of steroids have a characteristic carbon ring structure of the type

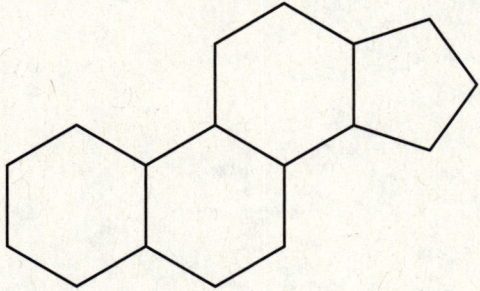

95. emulsifying

96. The polypeptide chain forms a coil or spiral. Such proteins are found in wool, hair, and tendons.

97. 24 (assuming no amino acid repeats)

98. tendons, bone (with mineral constituents), skin, cartilage, hair, fingernails.

99. hemoglobin

Chapter 21: Biochemistry

100. Collagen consists of three protein chains (each with α-helical structure) twisted together to form a superhelix. The result is a long, relatively narrow protein. Collagen functions as the raw material from which tendons are constructed.

101. An enzyme is inhibited if some other molecule, other than the enzyme's correct substrate, blocks the active sites of the enzyme. If the enzyme is inhibited irreversibly, the enzyme can no longer function and is said to have been inactivated.

102. pentoses (5 carbons); hexoses (6 carbons); trioses (3 carbons)

103. Although starch and cellulose are both polymers of glucose, the glucose rings in cellulose are connected in such a manner that the enzyme which ordinarily causes digestion of polysaccharides is not able to fit the shape of the substrate and is not able to act upon it.

104. In a strand of DNA, the phosphate group and the sugar molecule of adjacent nucleotides become bound to each other. The chain-portion of the DNA molecule, therefore, consists of alternating phosphate groups and sugar molecules. The nitrogen bases are found protruding from the side of this phosphate-sugar chain, bonded to the sugar molecules.

105. A wax is an ester of a fatty acid with a long chain monohydroxy alcohol. Waxes are solids that provide waterproof coatings on the fruits and leaves of plants, and on the skins and feathers of animals.

106. Phospholipids are esters of glycerol. Two fatty acids are bonded to the –OH groups of the glycerol backbone, with the third –OH group of glycerol bonded to a phosphate group. Having the two fatty acids, but also the polar phosphate group, makes the phospholipid lecithin a good emulsifying agent.